Lecture Notes in Artificial Intelligence   5991

Edited by R. Goebel, J. Siekmann, and W. Wahlster

Subseries of Lecture Notes in Computer Science

Ngoc Thanh Nguyen   Manh Thanh Le
Jerzy Świątek (Eds.)

Intelligent Information
and Database Systems

Second International Conference, ACIIDS
Hue City, Vietnam, March 24-26, 2010
Proceedings, Part II

 Springer

Series Editors

Randy Goebel, University of Alberta, Edmonton, Canada
Jörg Siekmann, University of Saarland, Saarbrücken, Germany
Wolfgang Wahlster, DFKI and University of Saarland, Saarbrücken, Germany

Volume Editors

Ngoc Thanh Nguyen
Wroclaw University of Technology, Institute of Informatics
Str. Wyb. Wyspianskiego 27, 50-370 Wroclaw, Poland
E-mail: Ngoc-Thanh.Nguyen@pwr.wroc.pl

Manh Thanh Le
Hue University, Str. Le Loi 3, Hue City, Vietnam
E-mail: lmthanh@hueuni.edu.vn

Jerzy Świątek
Wroclaw University of Technology
Faculty of Computer Science and Management
Str. Lukasiewicza 5, 50-370 Wroclaw, Poland
E-mail: Jerzy.Swiatek@pwr.wroc.pl

Library of Congress Control Number: Applied for

CR Subject Classification (1998): I.2, H.3, H.2.8, H.4, H.5, F.1, K.4

LNCS Sublibrary: SL 7 – Artificial Intelligence

ISSN   0302-9743
ISBN-10   3-642-12100-5 Springer Berlin Heidelberg New York
ISBN-13   978-3-642-12100-5 Springer Berlin Heidelberg New York

springer.com

© Springer-Verlag Berlin Heidelberg 2010

Typesetting: Camera-ready by author, data conversion by Scientific Publishing Services, Chennai, India
Printed on acid-free paper   06/3180

Preface

The 2010 Asian Conference on Intelligent Information and Database Systems (ACIIDS) was the second event of the series of international scientific conferences for research and applications in the field of intelligent information and database systems. The aim of ACIIDS 2010 was to provide an international forum for scientific research in the technologies and applications of intelligent information, database systems and their applications. ACIIDS 2010 was co-organized by Hue University (Vietnam) and Wroclaw University of Technology (Poland) and took place in Hue city (Vietnam) during March 24–26, 2010.

We received almost 330 papers from 35 countries. Each paper was peer reviewed by at least two members of the International Program Committee and International Reviewer Board. Only 96 best papers were selected for oral presentation and publication in the two volumes of the ACIIDS 2010 proceedings.

The papers included in the proceedings cover the following topics: artificial social systems, case studies and reports on deployments, collaborative learning, collaborative systems and applications, data warehousing and data mining, database management technologies, database models and query languages, database security and integrity, e-business, e-commerce, e-finance, e-learning systems, information modeling and requirements engineering, information retrieval systems, intelligent agents and multi-agent systems, intelligent information systems, intelligent internet systems, intelligent optimization techniques, object-relational DBMS, ontologies and information sharing, semi-structured and XML database systems, unified modeling language and unified processes, Web services and Semantic Web, computer networks and communication systems.

Accepted and presented papers highlight new trends and challenges of intelligent information and database systems. The presenters showed how new research could lead to new and innovative applications. We hope you will find these results useful and inspiring for your future research.

We would like to express our sincere thanks to the Honorary Chairs, Van Toan Nguyen (President of Hue University, Vietnam) and Paul P. Wang (Duke University, USA) for their support.

Our special thanks go to the Program Co-chairs, all Program and Reviewer Committee members and all the additional reviewers for their valuable efforts in the review process which helped us to guarantee the highest quality of selected papers for the conference. We cordially thank the organizers and Chairs of special sessions, which essentially contribute to the success of the conference.

We would like to thank our main sponsors, Hue University and Wroclaw University of Technology. Our special thanks are due also to Springer for publishing the proceedings, and other sponsors for their kind support.

We wish to thank the members of the Organizing Committee for their very substantial work, especially those who played essential roles: Huu Hanh Hoang, Radosław Katarzyniak (Organizing Chairs) and the members of the Local Organizing Committee for their excellent work.

Our special thanks go to the Foundation for Development of Wroclaw University of Technology for its efficiency in dealing with the registration and management issues.

We also would like to express our thanks to the Keynote Speakers (Leszek Rutkowski, A. Min Tjoa, Jerzy Świątek and Leon S L. Wang) for their interesting and informative talks of world-class standard. We cordially thank all the authors for their valuable contributions and the other participants of this conference. The conference would not have been possible without their support.

Thanks are also due to many experts who contributed to making the event a success.

Ngoc Thanh Nguyen
Manh Thanh Le
Jerzy Świątek

ACIIDS 2010 Conference Organization

Honorary Chairs

Van Toan Nguyen   President of Hue University, Vietnam
Paul P. Wang   Duke University, USA

General Chairs

Manh Thanh Le   Hue University, Vietnam
Świątek Jerzy   Wroclaw University of Technology, Poland

Program Chair

Ngoc Thanh Nguyen   Wroclaw University of Technology, Poland

Program Co-chairs

Shyi-Ming Chen   National Taiwan University of Science and
                    Technology, Taiwan
Kwaśnicka Halina   Wroclaw University of Technology, Poland
Huu Hanh Hoang   Hue University, Vietnam
Jason J. Jung   Yeungnam University, Korea
Edward Szczerbicki University of Newcastle, Australia

Local Organizing Co-chairs

Huu Hanh Hoang   Hue University, Vietnam
Radoslaw Katarzyniak Wroclaw University of Technology, Poland

Workshop Chairs

Radoslaw Katarzyniak Wroclaw University of Technology, Poland
Mau Han Nguyen   Hue University, Vietnam

Organizing Committee

Marcin Maleszka   Hoang Van Liem
Adrianna Kozierkiewicz-Hetmańska   Dao Thanh Hai
Anna Kozlowska   Duong Thi Hoang Oanh
Tran Dao Dong   Huynh Dinh Chien
Phan Duc Loc

Keynote Speakers

Leszek Rutkowski
Polish Academy of Sciences, Technical University of Czestochowa,
Poland

A. Min Tjoa
Vienna University of Technology,
Austria

Jerzy Świątek
Wroclaw University of Technology,
Poland

Leon S.L. Wang
National University of Kaohsiung,
Taiwan

Special Sessions

1. Multiple Model Approach to Machine Learning (MMAML 2010)

Oscar Cordón, European Centre for Soft Computing, Spain
Przemysław Kazienko, Wroclaw University of Technology, Poland
Bogdan Trawiński, Wroclaw University of Technology, Poland

2. Applications of Intelligent Systems (AIS 2010)
Shyi-Ming Chen, National Taiwan University of Science and Technology, Taipei,
    Taiwan

*3. Modeling and Optimization Techniques in Information Systems, Database
Systems and Industrial Systems (MOT 2010)*
Le Thi Hoai An, Paul Verlaine University – Metz, France
Pham Dinh Tao, INSA-Rouen, France

International Program Committee

Babica Costin	University of Craiova, Romania
Bielikova Maria	Slovak University of Technology, Slovakia
Bressan Stephane	National University of Singapore, Singapore
Bui The Duy	National University Hanoi, Vietnam
Cao Longbing	University of Technology Sydney, Australia
Cao Tru Hoang	Ho Chi Minh City University of Technology, Vietnam
Capkovic Frantisek	Slovak Academy of Sciences, Slovakia
Cheah Wooi Ping	Multimedia University, Malaysia

Chen Shyi-Ming	National Taiwan University of Science and Technology, Taiwan
Cuzzocrea Alfredo	University of Calabria, Italy
Dang Khanh Tran	Ho Chi Minh City University of Technology, Vietnam
Davidsson Paul	Blekinge Institute of Technology, Sweden
Forczmański Paweł	West Pomeranian University of Technology, Poland
Frejlichowski Dariusz	West Pomeranian University of Technology, Poland
Giorgini Paolo	University of Trento, Italy
Halina Kwasnicka	Wroclaw University of Technology, Poland
Helin Heikki	TeliaSonera, Finland
Ho Tu Bao	Japan Advanced Institute of Science and Technology, Japan
Hong Tzung-Pei	National University of Kaohsiung, Taiwan
Hoang Huu Hanh	Hue University, Vietnam
Hoang Trinh Hon	Ulsan University, Korea
Janiak Adam	Wroclaw University of Technology, Poland
Jezic Gordan	University of Zagreb, Croatia
Jung Jason J.	Yeungnam University, Korea
Kacprzyk Janusz	Polish Academy of Sciences, Poland
Karim S. Muhammad	Quaid-i-Azam University, Pakistan
Kim Cheonshik	Anyang University, Korea
Kim Chonggun	Yeungnam University, Korea
Krol Dariusz	Wroclaw University of Technology, Poland
Kuwabara Kazuhiro	Ritsumeikan University, Japan
Lau Raymond	Systems City University of Hong Kong, Hong Kong
Le Thi Hoai An	University Paul Verlaine - Metz, France
Lee Eun-Ser	Andong National University, Korea
Lee Huey-Ming	Chinese Culture University, Taiwan
Lee Zne-Jung	Huafan University, Taiwan
Lewis Rory	University of Colorado at Colorado Springs, USA
Lingras Pawan	Saint Mary's University, Canada
Luong Chi Mai	Institute of Information Technology, Vietnam
Matsuo Tokuro	Yamagata University, Japan
Narasimha Deepak Laxmi	University of Malaysia, Malaysia
Numao Masayuki	Osaka University, Japan
Nguyen Ngoc Thanh	Wroclaw University of Technology, Poland
Nguyen Thanh Binh	Hue University, Vietnam

Okraszewski Zenon   Wroclaw University of Technology, Poland
Ou Chung-Ming   Kainan University, Taiwan
Pan Jeng-Shyang   National Kaohsiung University of Applied Sciences,
                        Taiwan
Pandian Vasant   PCO Global
Paprzycki Marcin   Polish Academy of Sciences, Poland
Pedrycz Witold   Canada Research Chair, Canada
Prasad Bhanu   Florida A&M University, USA
Phan Cong Vinh   London South Bank University, UK
Selamat Ali   Universiti Teknologi Malaysia, Malaysia
Shen Victor R.L.   National Taipei University, Taiwan
Sobecki Janusz   Wroclaw University of Technology, Poland
Stinckwich Serge   UMI 209 UMMISCO • UPMC, IRD, MSI
Szczerbicki Edward   University of Newcastle, Australia
Takama Yasufumi   Tokyo Metropolitan University, Japan
Trawinski Bogdan   Wroclaw University of Technology, Poland
Truong Hong-Linh   Vienna University of Technology, Austria
Zhang Wen-Ran   Georgia Southern University, USA

Program Committees of Special Sessions

Special Session on Multiple Model Approach to Machine Learning (MMAML 2010)

Jesús Alcalá-Fdez   University of Granada, Spain
Oscar Castillo   Tijuana Institute of Technology, Mexico
Suphamit Chittayasothorn   King Mongkut's Institute of Technology Ladkrabang,
                            Thailand
Emilio Corchado   University of Burgos, Spain
Oscar Cordón   European Centre for Soft Computing, Spain
José Alfredo F. Costa   Federal University (UFRN), Brazil
Bogdan Gabryś   Bournemouth University, UK
Patrick Gallinari   Pierre et Marie Curie University, France
Lawrence O. Hall   University of South Florida, USA
Francisco Herrera   University of Granada, Spain
Tzung-Pei Hong   National University of Kaohsiung, Taiwan
Hisao Ishibuchi   Osaka Prefecture University, Japan
Yaochu Jin   Honda Research Institute Europe, Germany
Nikola Kasabov   Auckland University of Technology, New Zealand
Przemysław Kazienko   Wrocław University of Technology, Poland
Rudolf Kruse   Otto-von-Guericke University of Magdeburg, Germany
Mark Last   Ben-Gurion University of the Negev, Israel

Kun Chang Lee	Sungkyunkwan University, Korea
Kyoung Jun Lee	Kyung Hee University, Korea
Urszula Markowska-Kaczmar	Wrocław University of Technology, Poland
Kazumi Nakamatsu	University of Hyogo, Japan
Yew-Soon Ong	Nanyang Technological University, Singapore
Dymitr Ruta	British Telecom, UK
Robert Sabourin	University of Quebec, Canada
Ke Tang	University of Science and Technology of China, China
Bogdan Trawiński	Wrocław University of Technology, Poland
Pandian Vasant	University Technology Petronas, Malaysia
Shouyang Wang	Academy of Mathematics and Systems Science, China
Michał Wozniak	Wrocław University of Technology, Poland
Lean Yu	Academy of Mathematics and Systems Science, China
Zhongwei Zhang	University of Southern Queensland, Australia
Zhi-Hua Zhou	Nanjing University, China

Modeling and Optimization Techniques in Information Systems, Database Systems and Industrial Systems (MOT 2010)

Kondo Adjallah	Paul Verlaine University - Metz, France
Riad Aggoune	CRP-Tudor, Luxembourg
Aghezzaf El Houssaine	University of Gent, Belgium
Lydia Boudjeloud	Paul Verlaine University-Metz, France
Bouvry Pascal	University of Luxembourg, Luxembourg
Brzostowski Krzysztof	Wroclaw University of Technology, Poland
Brzykcy Grażyna,	Poznań University of Technology, Poland
Brieu Conan-Guez	Paul Verlaine University - Metz, France
Czarnowski Ireneusz	Gdynia Maritime University, Poland
Do Thanh Nghi	ENST Brest, France
Drapala Jarosław	Wroclaw University of Technology, Poland
Forczmański Pawel	West Pomeranian University of Technology, Poland
Fraś Dr. Mariusz	Wrocław University of Technology, Poland
Alain Gelly	Paul Verlaine University - Metz, France
Hao Jin-Kao	University of Algiers, France
Francois-Xavier Jollois	University of Paris V, France
Kubik Tomasz	Wrocław University of Technology, Poland
Le Thi Hoai An	Paul Verlaine University – Metz, France, Chair
Lisser Abdel	University of Paris 11, France
Marie Luong	University of Paris 13, France
Nalepa Grzegorz	AGH University of Science and Technology, Poland
Narasimha Deepak Laxmi	University of Malaya, Malaysia
Nguyen Vincent	The University of New South Wales, Australia
Orski Donat	Wrocław University of Technology, Poland
Pham Dinh Tao	INSA-Rouen, France, Co-chair
Jean-Marie	Research Director, INRIA-Metz, France

Rekuć Witold   Wrocław University of Technology, Poland
Ibrahima Sakho   Paul Verlaine University - Metz, France
Daniel Singer Paul   Verlaine University - Metz, France

Additional Reviewers

Chang Chung C.   Chinese Culture University, Taiwan
Chen Jr-Shian   Hungkuang University, Taiwan
Chen Rung-Ching   Chaoyang University of Technology, Taiwan
Chen Yen-Lin   National Taipei University of Technology, Taiwan
Chien Chih-Yao   National Taiwan University of Science and
                                Technology, Taiwan
Deutsch Alin   University of California, USA
Dobbie Gill   University of Auckland, New Zealand
Felea Victor   "Al.I.Cuza" University of Iasi, Romania
Garrigos Irene   University of Alicante, Spain
Gely Alain   Paul Verlaine University, France
Jeng Albert B.   Jinwen University of Science and Technology, Taiwan
Lee Li-Wei   National Taiwan University of Science and
                                Technology, Taiwan
Lee Ting-Kuei   National Taiwan University of Science and
                                Technology, Taiwan
Lehner Wolfgang   Technical University of Dresden, Germany
Leu Yungho   National Taiwan University of Science and
                                Technology, Taiwan
Lu Kun-Yung   National United University, Taiwan
Manthey Rainer   University of Bonn, Germany
Mazon Jose Norberto   University of Alicante, Spain
Morzy Tadeusz   Poznan University of Technology, Poland
Park Dong-Chul   Myong Ji University, South Korea
Proth Jean-Marie   INRIA-Metz, France
Shih An-Zen   Jin-Wen Science and Technology University, Taiwan
Suciu Dan   University of Washington, USA
Thomas Wojciech   Wrocław University of Technology, Poland
Vidyasankar K.   Memorial University of Newfoundland, Canada
Vossen Gottfried   University of Muenster, Germany
Wang Cheng-Yi   National Taiwan University of Science and
                                Technology, Taiwan
Wang Jia-wen   Nanhua University, Taiwan
Wang Yongli   North China Electric Power University, China
Woźniak Michał   Wrocław University of Technology, Poland
Wong Limsoon   National University of Singapore, Singapore

Table of Contents – Part II

Web-Based Systems for Data Management

Autonomous Systems

Collaborative Systems

Tools and Applications

Multiple Model Approach to Machine Learning

Modeling and Optimization Techniques in Information Systems, Database Systems and Industrial Systems

Table of Contents – Part I

Keynote Speeches

Intelligent Database Systems

Data Warehouses and Data Mining

Intelligent Information Retrieval

Technologies for Intelligent Information Systems

Applications of Intelligent Systems

Enhancing Accuracy of Recommender System through Adaptive Similarity Measures Based on Hybrid Features

Deepa Anand and Kamal K. Bharadwaj

School of Computer and System Sciences,
Jawaharlal Nehru University, New Delhi, India
deepanand209@gmail.com, kbharadwaj@gmail.co.in

Abstract. Collaborative Filtering techniques offer recommendations to users by leveraging on the preferences of like-minded users. They thus rely highly on similarity measures to determine proximity between users. However, most of the previously proposed similarity measures are heuristics based and are not guaranteed to work well under all data environments. We propose a method employing Genetic algorithm to learn user similarity based on comparison of individual hybrid user features. The user similarity is determined for each feature by learning a feature similarity function. The rating for each item is then predicted as an aggregate of estimates garnered from predictors based on each attribute. Our method differs from previous attempts at learning similarity, as the features considered for comparison take into account not only user preferences but also the item contents and user demographic data. The proposed method is shown to outperform existing filtering methods based on user-defined similarity measures.

Keywords: Recommender Systems, Collaborative Filtering, Learning Similarity, Hybrid User Features.

1 Introduction

The need to prune and filter large information spaces and to provide personalized services, has led to the emergence of recommender systems(RS). Collaborative Filtering(CF)[1][6] is a recommendation technique that emulates the process of word-of-mouth, where people glean opinions about objects not experienced by themselves, from like-minded friends and acquaintances. CF algorithms offer the advantage of cross-genre recommendations because of their non-dependence on content description. Breese et al.[5] categorized CF algorithms into memory-based, which make predictions based on the entire collection of previously rated items and model-based which depends on building a user model for making predictions. Memory-based algorithms are popular due to their ability to incorporate up-to-date preference information, when recommending items.

The performance of a memory-based CF depends on construction of a reliable set of neighbors who contribute towards prediction of measure of interest in an object for a user. The estimation of degree of concurrence between users is thus a crucial step in

N.T. Nguyen, M.T. Le, and J. Świątek (Eds.): ACIIDS 2010, Part II, LNAI 5991, pp. 1–10, 2010.

the filtering process. There have been several attempts [5][7][12][15] at capturing the elusive notion of user similarity based on a set of commonly preferred items. However, most of such measures are defined manually and their performance is often dataset dependent. Since the measures of similarity are used as weighs in determining the extent to which a user contributes to the prediction, the idea of learning such weights seems to be promising, since the weights learnt are adaptive, dataset dependent and should result in optimal/near optimal performance.

Machine learning techniques have been employed to learn similarities between entities in different domains such as content-based image retrieval[14] and case-based reasoning systems[16]. Cheung and Tian [9] propose to learn the optimal weights representing user similarities as well as the user bias, which removes the subjectivity from the ratings, in order to minimize a criterion function. Using gradient descent, the weights and user biases are updated and the process iterates till it converges. A bi-directional similarity metric computation technique[8], learns the similarity between users and items simultaneously by employing matrix factorization. A model-based technique[4] derives weights for all user-user/ item-item coefficients with the objective of minimizing a quadratic cost function. A related work [11] applies optimization techniques to learn significance weights for items for clustering users according to their preferences.

We propose a different approach to measuring the closeness between the tastes that two users share, by learning their similarity based on comparison of hybrid user attributes. The hybrid user features[2] allow us to construct a user profile by examining the item contents that the user has preferred thus supporting "collaboration via content"[13]. Such content-based features are then pooled with demographic user data to give a compact and hybrid set of features. The use of hybrid feature, instead of user ratings, offers twin advantage of a compact user profile representation and factoring in preference, content-description and user demographic information in the similarity estimation. Each attribute is treated as an independent predictor, thus allowing the feature to influence utility of an item for the active user. A distance-to-similarity function or a similarity table is learnt for each attribute depending on if the attribute is numeric or symbolic respectively. The predictions from each of the feature-level predictors are aggregated to give the final predictions. The experimental results support our ideas and demonstrate that the proposed methods are superior to the several predefined similarity measures.

The rest of the paper is organized as follows: Section 2 provides a overview into the different predefined similarity measures. A method for learning feature-level similarities is introduced in Section 3. Section 4 presents an experimental evaluation of the proposed scheme and compares them with several user-defined similarity measures. Finally, Section 5 presents conclusions and points out some directions for future research.

2   Similarity Measures

There have been several measures of similarity which gauge how closely the opinions of a user pair match. Such measures generally rely on computing the degree of agreement based on the set of items co-rated by the users. Pearson and Cosine similarity measures are two of the most popular among them. Cosine similarity measure is defined as follows;

$$sim\ (x,y) = \frac{\sum\limits_{i \in S_{xy}} r_{x,i} r_{y,i}}{\sqrt{\sum\limits_{i \in S_{xy}} r_{x,i}^2 \sum\limits_{i \in S_{xy}} r_{y,i}^2}}, \tag{1}$$

where S_{xy} is the set of items which users x and y have co-rated and $\bar{r}_x$ is the mean rating for user x. Whereas Pearson correlation coefficient is defined as;

$$sim(x,y) = \frac{\sum\limits_{i \in S_{xy}} (r_{x,i} - \bar{r}_x)(r_{y,i} - \bar{r}_y)}{\sqrt{\sum\limits_{i \in S_{xy}} (r_{x,i} - \bar{r}_x)^2 \sum\limits_{i \in S_{xy}} (r_{y,i} - \bar{r}_y)^2}}, \tag{2}$$

Jaccard is another similarity measure, which computes the similarity between users based on the number of items co-rated by them, regardless of the rating conferred on the item. Candillier et. al.[7], introduce several weighted similarity measures for user-based and item-based CF. The method proposed, uses jaccard similarity as a weighting scheme and combines it with other similarity measures such as Pearson correlation coefficient to emphasize similarity of users who share appreciation on several items with the active users.

3 Learning User Similarity

Similarity estimation is one of the important steps in the CF process. In this work, we propose to employ genetic algorithms to estimate this similarity between users, based on comparison of various hybrid features, for the movie recommendation domain. In the following subsections we briefly discuss the construction of a user profile based on hybrid features, the technique of learning the similarity for each feature and the proposed recommendation framework.

3.1 Hybrid User Features

To assess the similarity between users we intend to use hybrid user attributes as proposed in [2]. In the movie domain, each movie can belong to one or more than one genre. In particular MovieLens defines 18 movie genres. The inclination of a user towards a particular genre can be evaluated by examining the set of movies belonging to a particular genre, that a user has highly rated. The degree to which a user prefers a genre is captured through GIM(Genre Interestingness Measure)[2]. The GIM corresponding to the 18 genres are augmented with the available demographic attributes such as user age, occupation, gender and state to give a set of 22 hybrid features. Hence the user profile which initially consisted of the ratings of a large number of movies is squeezed to give a compact user profile, which involves content-based and demographic features as well.

3.2 Learning Similarity Function by Feature Comparison

Traditionally defined similarity measures such as Pearson and Cosine, treat all attributes (item ratings) similarly while trying to evaluate user closeness. This may not truly reflect the contribution of each attribute towards the similarity. For example, the attribute $GIM_{comendy}$, reveals the degree of interest of any user in genre "Comedy". It is possible that the predictions using this attribute alone gives best performance when all users who have shown interest in the genre are weighed equally, rather than weighing them by their degree of interest in the genre, i.e. the fact that the user "has liked" the genre is more important than "how much" he has liked it. Whereas for another genre, say romance, a small difference in the GIM values may imply a large reduction in the actual similarity between two users. Moreover traditional similarity measures cannot be easily extended when symbolic attributes such as occupation, gender, state etc are involved. When symbolic attributes are involved then two users are considered similar only if they have the same value for that attribute, thus leading to a coarse-grained approach to similarity computation. In real life, however, a person with occupation "teacher" and "student" may be quite similar in their tastes.

To overcome all these shortcomings we adopt a different approach by viewing each feature as a means to compute predictions for the user, learning an optimal similarity function for each attribute and aggregating the predictions so obtained for the final estimated prediction. The feature set consists of numeric attributes ($GIM_1,..,GM_{18}$, Age) and symbolic attributes(Occupation, State, Gender). A precise representation of the similarity function depends on the data type. We follow the similarity representation of tables for symbolic attributes(occupation,state) and vector for numeric attributes(GIMs, age) from [16]. For gender feature, two values are similar only if they have the same values, and hence no similarity vector/table needs to be learnt. Note that the definition for similarity table as well as distance based similarity function is slightly altered such that the similarity values lie in the range [-1,1]. This is to deter users whose values for the particular attribute are far apart in terms of preferred items, from contributing to each other's prediction.

3.3 Learning Similarity Function by Using Genetic Algorithms

Genetic algorithms base their operation on the Darwanian principle of "survival of the fittest" and utilize artificial evolution to get enhanced solutions with each iteration. The GA process starts with a population of candidate solutions known as chromosome or genotype. Each chromosome in the population has an associated fitness and these scores are used in a competition to determine which chromosomes are used to form new ones[10]. New individuals are created by using the genetics inspired operators of crossover and mutation. While crossover allows creation of two new individuals by allowing two parent chromosomes to exchange meaningful information, mutation is used to maintain the genetic diversity of the population by introducing a completely new member into the population. The process iterates over several generations till a convergence criterion is met. The task of learning an optimal similarity function for each of the attributes can be accomplished by means of Genetic Algorithms.

Chromosome Representation
Each individual of a similarity table I, used for representing similarity function for symbolic attributes, is represented by a matrix of floating numbers, in the range [-1,1], of size n x n where n is the number of distinct values taken by the attribute and I(a,b) represents the similarity between attribute values 'a' and 'b'. The similarity function for numeric attributes can be approximated by a similarity vector, which provides similarity values corresponding to a fixed number of distance values. The sampling points are chosen for each attribute by using "dynamic sampling" where an optimal distribution of sampling points is chosen from an interval depending on the number of difference values that fall in the interval, gauged from the training data.

Fitness Function
A fitness function quantifies the optimality of a chromosome and guides the process towards achieving its optimization goal by allowing fitter individuals to breed and thus hopefully improve the quality of individuals over the generations. The fitness of a similarity function represented as a table or a vector needs to be measured by the prediction accuracy offered by the selection and weighting of users according to the similarity function. To evaluate the fitness of chromosomes representing similarity functions, the training data for the user is divided into training and validation sets. The training data is utilized to build user neighborhood and prediction, whereas the validation set is used to learn the optimal similarity function by allowing the GA to search for the best similarity table/vector which leads to the least average prediction error for the validation set. The fitness of a similarity vector/table for an attribute A, is obtained by computing the average prediction error for the validation set, where the prediction is performed by constructing the neighborhood for the active user based solely on similarity of attribute A. The fitness function for an individual I based on attribute A is defined as;

$$fitness\ _A^I = \frac{1}{|V|} \sum_{i \in V} |\ r_{a,i} - pr_{a,i\ A}^I\ |,$$ (3)

where V is the set of all ratings in the validation set, $r_{a,i}$ is the actual rating for item i by the user a, and $pr_{a,i\ A}^I$ is the predicted score for the active user using similarity vector/table represented by the individual I based on attribute A.

Genetic operators
To maintain the genetic diversity of the population through generations, it desirable to generate new individuals from the ones in the current generation. Crossover and mutation are the two most common genetic transformations. Crossover works by letting a pair of chromosomes exchange meaningful information to create two offspring, while mutation involves a random manipulation of a single chromosome to create to a new individual. The discussion on genetic operators follows from [16].

In our framework, the crossover for similarity vectors is done either using simple crossover or arithmetic crossover, the probability of choosing among the two methods being equal. For similarity matrices, arithmetic crossover, row crossover and column crossover are performed with every method having equal probability of selection. Components of the vector/matrix are mutated by modifying their value randomly. The numbers of components thus modified are also random. Note that each time a new individual of type similarity vector is created, the constraint of values being non-increasing must hold.

3.4 Proposed Recommendation Framework

The proposed technique of learning similarity at the attribute level can be employed to obtain predictions for an active user at the feature level and aggregate them to arrive at the final predicted vote for the active user. The dataset is divided into training set, TR, validation set, V, and test set, T. The main steps of the proposed recommender system framework are given below:

Step1: Compute the GIM values of all users using formula (4) based on the training data set TR.

Step 2: Find the optimal similarity vector/table for each attribute

Step 3: Predict ratings based on feature level similarity function
The predicted rating for an item i for active user u is based on Resnick's prediction formula [15]. The final prediction for user i is obtained by aggregating predictions based on all attributes. Note that some attributes might not contribute to the predictions since the neighborhood set based on them might be empty.

4 Experimentation and Results

To demonstrate the effectiveness of proposed technique of learning user similarities employing GA, we conducted experiments on the popular MovieLens dataset. The experiments are conducted with the goal of establishing the superiority of the proposed similarity learning technique over predefined similarity measures.

4.1 Design of Experiments

The MovieLens dataset consists of 100,000 ratings provided by 943 users on 1682 movies. The ratings scale is in the range 1-5 with 1 - "bad" to 5 –"excellent". The ratings are discrete. Each user in the dataset has rated at least 20 movies. For our experiments we chose five subsets from the data, containing 100,200, 300,400, and 500 users called ML100, ML200, ML300, ML400 and ML500 respectively. This is to illustrate the effectiveness of the proposed scheme under varying number of participating users. Each of the datasets was randomly split into 60% training data, 20% validation data and 20% test data. The ratings of the items in the test set are treated as items unseen by the active user, while the ratings in the training set is used for neighborhood construction and for prediction of ratings. The ratings in the validation

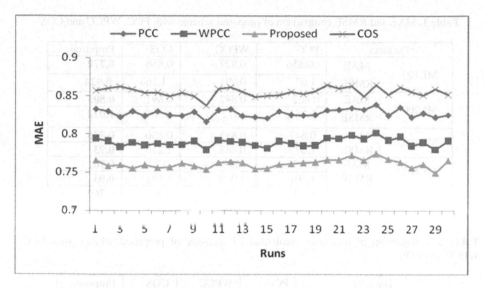

Fig. 1. MAE for ML400 over 30 runs

set are used to guide the GA learning process. For each dataset the experiment was run 30 times to eliminate the effect any bias in the data.

The effectiveness of the proposed scheme is compared with the Pearson correlation co-efficient(PCC)(Eq.2), cosine similarity(COS)(Eq. 1) and Weighted Pearson (WPCC) [7].

4.2 Performance Measurements

To compare the prediction accuracy we compare the various schemes via two metrics namely, Mean Absolute Error(MAE) and Root Mean Squared Error(RMSE). While MAE measures the average absolute deviation of the predicted rating from the actual ratings, RMSE uses a quadratic scoring rule to emphasize large errors. When preferences are binary i.e. when the task at hand is to guess if a user will or wont like an item, then classification metrics such as precision and recall are used to evaluate the performance of a recommendation algorithm. Precision estimates the proportion of useful recommendations among all items recommended to the user and Recall represents the fraction of useful items selected from among the number of actual useful items. In addition, F-Measure is an classification accuracy measure, which allows us to consider both recall and precision together by computing their harmonic mean.

4.3 Results

To demonstrate the ability of the proposed method to offer better prediction accuracy we compare the MAE and RMSE with that PCC, COS and WPCC. The results are as presented in Table 1. The MAE and RMSE are computed based on the average over 30 runs of the experiment over the different datasets. A lower value of MAE and RMSE corresponds to a better performance. As is clear from the results in Table 1 the

Table 1. MAE and RMSE comparison of proposed scheme with PCC, WPCC and COS

Datasets		PCC	WPCC	COS	Proposed
ML100	MAE	0.856	0.827	0.856	**0.770**
	RMSE	1.07	0.991	1.164	**0.924**
ML200	MAE	0.862	0.832	0.891	**0.803**
	RMSE	1.161	1.072	1.309	**1.022**
ML300	MAE	0.847	0.815	0.866	**0.776**
	RMSE	1.077	0.988	1.216	**0.93**
ML400	MAE	0.827	0.788	0.848	**0.761**
	RMSE	1.101	1.000	1.191	**0.93**
ML500	MAE	0.831	0.795	0.864	**0.760**
	RMSE	1.114	1.017	1.259	**0.939**

Table 2. Comparison of precision, recall and F1 measure of proposed scheme with PCC, WPCC and COS

Datasets		PCC	WPCC	COS	Proposed
ML100	Precision	63.2	64.6	65.3	**68.3**
	Recall	78.7	80.5	82.7	**86.1**
	F-Measure	67.6	68.8	69.8	**73.1**
ML200	Precision	62.7	**63.2**	62.6	62.7
	Recall	79.9	82.1	83.1	**84.7**
	F-Measure	67.7	68.9	68.4	**69.5**
ML300	Precision	62	**62.5**	62.1	**62.5**
	Recall	80.4	82.5	84.1	**86.1**
	F-Measure	67.6	68.7	68.5	**70**
ML400	Precision	61.9	62.6	62.4	**62.8**
	Recall	79.4	82	83.1	**85.3**
	F-Measure	66.9	68.3	68	**69.6**
ML500	Precision	61.2	61.6	61.6	**61.9**
	Recall	79.2	81.4	81.8	**85.6**
	F-Measure	66.7	67.7	67.1	**69.5**

proposed scheme considerably outperforms other user-defined similarity measures for all datasets with respect to both MAE and RMSE. This is due to the ability of the proposed technique to adapt the user similarity computation according to the dataset.

Table 2 presents the performance comparison(in percentage) based on the classification accuracy by comparing the precision, recall and F-Measure for each of the different techniques. A higher value of these measures imply better performance. The proposed scheme again outperforms the user-defined similarity measures in terms of precision, recall and F-Measure in almost all cases, with the only exception being for ML200 where WPCC has a higher precision. The MAE and F1 measure for the different runs of the experiment for ML400 are shown in Fig 1 and 2 respectively. A total of 30 runs were made for each dataset. For all the runs the proposed method performed better than any of the user-defined measures in terms of predictive as well as classification accuracy.

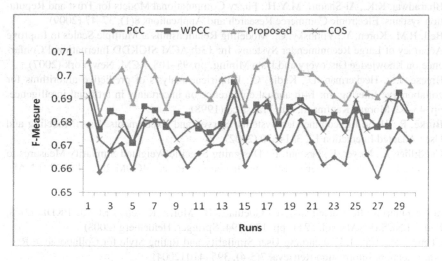

Fig. 2. F-Measure for ML400 over 30 runs

5 Conclusions

In our work we introduced a novel technique of evolving user similarity functions based on a set of hybrid user features. The method of evolving individual similarity function for each of the hybrid feature allows each feature to influence ratings prediction independently, thus allowing integration of predictions based on features whose type or range are vastly different. To evaluate our approach we tested it on the highly popular MovieLens dataset. The experiments establish the superiority of our method of learning user similarities over popular methods which are based on predefined similarity measures. Though the use of GA increases the time complexity of the proposed method, the learning process can be performed offline in a periodic manner, to adapt to the changes in data over a period of time.

In our future work we plan to integrate the current approach of learning of feature-wise similarity function with learning of the user-wise attribute weights for each of the hybrid attributes, thus quantifying the degree of importance of each feature for the active user. The current framework is specific to the movie domain and it would be interesting to explore the feasibility of extending the framework to other domains e.g. books, jokes etc. Another important direction for future work would be to incorporate the concepts of trust and reputation [3] to enhance recommendation accuracy.

References

1. Adomavicius, G., Tuzhilin, A.: Toward the Next Generation of Recommender Systems: A Survey of the State-of-the-Art and Possible Extensions. IEEE Transaction on Knowledge and Data Engineering 17(6), 734–749 (2005)
2. Al-Shamri, M.Y.H., Bharadwaj, K.K.: Fuzzy-Genetic Approach to Recommender System Based on a Novel Hybrid User Model. Expert Systems with Applications 35(3), 1386–1399 (2008)

3. Bharadwaj, K.K., Al-Shamri, M.Y.H.: Fuzzy Computational Models for Trust and Reputation systems. Electronic Commerce Research and Applications 8(1), 37–47 (2009)
4. Bell, R.M., Koren, Y., Volinsky, C.: Modeling Relationships at Multiple Scales to Improve Accuracy of Large Recommender Systems. In: 13th ACM SIGKDD International Conference on Knowledge Discovery and Data Mining, pp. 95–108. ACM, New York (2007)
5. Breese, J.S., Heckerman, D., Kadie, C.: Empirical analysis of predictive algorithms for collaborative filtering. In: 14th annual conference on uncertainty in artificial intelligence, pp. 43–52. Morgan Kaufmann, San Fransisco (1998)
6. Burke, R.: Hybrid Recommender Systems: Survey and Experiments. User Modeling and User-Adapted Interaction 12, 331–370 (2002)
7. Candillier, L., Meyer, F., Fessant, F.: Designing Specific Weighted Similarity Measures to Improve Collaborative Filtering Systems. In: Perner, P. (ed.) ICDM 2008. LNCS (LNAI), vol. 5077, pp. 242–255. Springer, Heidelberg (2008)
8. Cao, B., Sun, J., Wu, J., Yang, Q., Chen, Z.: Learning Bidirectional Similarity for Collaborative Filtering. In: Daelemans, W., Goethals, B., Morik, K. (eds.) ECML PKDD 2008, Part I. LNCS (LNAI), vol. 5211, pp. 178–194. Springer, Heidelberg (2008)
9. Cheung, K., Tian, L.F.: Learning User Similarity and Rating Style for Collaborative Recommendation. Information Retrieval 7(3-4), 395–410 (2004)
10. De Jong, K.A.: Learning with genetic algorithms: An overview. Machine Language 3(2-3), 121–138 (1988)
11. Jin, R., Chai, J.Y., Si, L.: An automatic weighting scheme for collaborative filtering. In: 27th annual international ACM SIGIR conference on research and development in information retrieval, pp. 337–344. ACM, New York (2004)
12. Ma, H., King, I., Lyu, M.R.: Effective missing data prediction for collaborative filtering. In: 30th annual international ACM SIGIR conference on research and development in information retrieval, pp. 39–46. ACM, New York (2007)
13. Pazzani, M.J.: A Framework for Collaborative, Content-Based and Demographic Filtering. Artificial Intelligence Review 13(5-6), 393–408 (1999)
14. Torres, R.S., Falcão, A.X., Zhang, B., Fan, W., Fox, E.A., Gonçalves, M.A., Calado, P.: A new framework to combine descriptors for content-based image retrieval. In: 14th ACM international conference on Information and knowledge management, pp. 335–336. ACM, New York (2005)
15. Resnick, P., Iacovou, N., Suchak, M., Bergstrom, P., Riedl, J.: Grouplens: An Open Architecture for Collaborative Filtering of Netnews. In: ACM CSCW 1994 Conference on Computer-Supported Cooperative Work, pp. 175–186. ACM, New York (1994)
16. Stahl, A., Gabel, T.: Using evolution programs to learn local similarity measures. In: Ashley, K.D., Bridge, D.G. (eds.) ICCBR 2003. LNCS, vol. 2689, pp. 537–551. Springer, Heidelberg (2003)

Exploring Wikipedia and Text Features for Named Entity Disambiguation

Hien T. Nguyen[1] and Tru H. Cao[2]

[1] Ton Duc Thang University, Vietnam
hien@tut.edu.vn
[2] Ho Chi Minh City University of Technology, Vietnam
tru@cse.hcmut.edu.vn

Abstract. Precisely identifying entities is essential for semantic annotation. This paper addresses the problem of named entity disambiguation that aims at mapping entity mentions in a text onto the right entities in Wikipedia. The aim of this paper is to explore and evaluate various combinations of features extracted from Wikipedia and texts for the disambiguation task, based on a statistical ranking model of candidate entities. Through experiments, we show which combinations of features are the best choices for disambiguation.

1 Introduction

This paper addresses the named entity disambiguation problem (NED) that is the task of determining whether two *entity mentions* refer to the same named entity in a world of discourse. For instance, it is to clarify if "J. Smith" and "John Smith" refer to the same person or different occurrences of "John Smith" refer to the same person. The aim of our method is to perform mapping mentions in a text onto the right entities in Wikipedia. For instance, for the text *"the computer scientist John McCarthy coined the term artificial intelligence in the late 1950's,"* the aim is to map "John McCarthy" in the text onto the entity John McCarthy, a computer scientist who is the inventor of LISP programming language, in Wikipedia.

The problem in reality is that one name in different occurrences may refer to different entities and one entity may have different names that may be written in different ways and with spelling errors. For example, the name "John McCarthy" in different occurrences may refer to different NEs such as a computer scientist from Stanford University, a linguist from University of Massachusetts Amherst, an Australian ambassador, a British journalist who was kidnapped by Iranian terrorists in Lebanon in April 1986, and so on. Such ambiguity makes identifying right entities in a text challenging and raises NED as a key research aspect not only in the Semantic Web, but also in Information Extraction, and areas of Natural Language Processing in general.

NED can be considered as an importantly special case of Word Sense Disambiguation (WSD). The aim of WSD is to identify which sense of a word is used in a given context when several possible senses of that word exist. In WSD, words to be disambiguated may either appear in a plain text or an existing knowledge base. Techniques

N.T. Nguyen, M.T. Le, and J. Świątek (Eds.): ACIIDS 2010, Part II, LNAI 5991, pp. 11–20, 2010.

for the latter use a dictionary, thesaurus, or an ontology as a sense inventory that defines possible senses of words. Having been emerging recently as the largest and widely-used encyclopedia in existence, Wikipedia is used as a knowledge source for not only WSD, but also Information Retrieval, Information Extraction, Ontology Building, Natural Language Processing, and so on ([9]).

Wikipedia is a free encyclopedia written by a collaborative effort of a large number of volunteer contributors. We describe here some of its resources of information for disambiguation. A basic entry in Wikipedia is a *page* (or *article*) that defines and describes a single entity or concept. It is uniquely identified by its title. In Wikipedia, every entity page is associated with one or more categories, each of which can have subcategories expressing meronymic or hyponymic relations. Each page may have several incoming links (henceforth *inlinks*), outgoing links (henceforth *outlinks*), and *redirect* pages. A redirect page typically contains only a reference to an entity or a concept page. Title of the redirect page is an alternative name of that entity or concept. For example, from redirect pages of the United States, we extract alternative names of the United States such as "US", "USA", "United States of America", etc. Other resources are disambiguation pages. They are created for ambiguous names, each of which denotes two or more entities in Wikipedia. Based on disambiguation pages one can detect all entities that have the same name in Wikipedia.

Mihalcea ([7]) showed that Wikipedia could be used as a sense inventory for sense disambiguation. In [8], Mihalcea and Csomai implemented and evaluated two different disambiguation algorithms that extracted terms in a document and linked them to Wikipedia articles using Wikipedia as a sense inventory. Then they reported the best performing algorithm was the one using a supervised learning model where Wikipedia articles, which had already been annotated, served as training data. This algorithm used the local context of three words to the left and right, with their parts-of-speech, as features for representing an ambiguous term. In 2007, we proposed an idea of exploiting identified entities to disambiguate remaining ones ([13]). Later on, in 2008, Medelyan et al. ([10]) proposed a method that bore a resemblance to our idea for disambiguating terms in a documents using Wikipedia. Milne and Witten ([11]) extended both works [8] and [10] by exploiting relatedness of a target term to its surrounding context, besides exploiting the feature as in the latter one. They reported the precision and recall of 75% approximately.

Bunescu and Paşca ([1]) and Cucerzan ([3]) proposed methods that mapped mentions in a text onto the right entities in Wikipedia by exploiting several of the disambiguation resources such as Wikipedia articles (entity pages), redirection pages, categories, and links in the articles. Bunescu and Paşca extracted words inside a 55-word window around a mention to form its feature vector. Based on the cosine similarity between feature vectors, they ranked candidate entities for a mapping and chose the one with the highest similarity score. Due to too low similarity scores with the cosine-based ranking in many cases, the authors employed the Support Vector Machine model (SVM) to learn a mapping from the context window to the specific categories of articles. They defined a threshold and assumed no mapping acceptable when the similarity scores fell below this threshold. The authors reported accuracies between 55.4% and 84.8% for sub-categories of the *people by occupation* category, depending on the model and the development/test data employed.

Cucerzan ([3]) proposed a method exploiting the same resources of information in Wikipedia for the disambiguation task as in [1]. This method simultaneously disambiguates all mentions in a document by maximizing the agreement among categories of candidate entities and maximizing the contextual similarity between contextual information in the document and context data stored for the candidate entities. The context data comprise appositives in the titles of articles and phrases that appear as anchor texts of links in the first paragraphs of the articles. The contextual information of a document contains all phrases occurring in the context data. Then the author reported accuracies of 88.3% and 91.4% on 5131 entities appearing in 350 Wikipedia articles and on 756 entities appearing in 20 news stories, respectively.

Overell and Rüger [12] proposed a method generating a co-occurrence model from *template*[1] (e.g., *infoboxes*[2]) that served as training data and then employed the SVM for place-name disambiguation. This method only works on co-occurrence place-names. It chooses a window size of ±10 location references regardless of other words that are not part of place-names. The authors reported that the best performing disambiguation achieved the precision and recall of 90.1% and 79.2%, respectively on 2,150 locations appearing in a dataset that included the two of Cucerzan's and 1,054 location entities in 121 documents of SemCor, a subset of Brown Corpus.

The common shortcoming of the works on NED reviewed above is that they do not exploit some significant pieces of contextual information and do not consider the used features in several settings to show how they effect the disambiguation performance. In this paper, we explore and evaluate various combinations of *Wikipedia features*[3] and *text features*[4] for the disambiguation task. Our contributions are as follows. First, we rigorously investigate various features and combine them in several ways to identify which combination gives the best disambiguation performance. Second, our analysis of text features does not depend on a given knowledge base, so it can be used for NED based on an arbitrary knowledge base. Third, we explore words around mentions that are coreferential with the mention to be disambiguated and then show that these features play an important role in disambiguation. In Section 2, we present our statistical ranking model with possible features to be explored. Experiments and results are presented in Section 3 and a conclusion is drawn in Section 4.

2 A Proposed Statistical Ranking Model

In this section, we present a statistical ranking model where we employ the Vector Space Model (VSM) to represent *ambiguous*[5] mentions in a text and entities in Wikipedia by their features. The VSM considers the set of features of entities as a 'bag of words'. Firstly we present extracted features and how we normalize them. Then we present how to weight words in the VSM and calculate the similarity between feature vectors of mentions and entities. Based on the calculated similarity, our disambiguation

---

[1] http://en.wikipedia.org/wiki/Help:Template
[2] http://en.wikipedia.org/wiki/Wikipedia:Infobox
[3] Features are extracted from Wikipedia
[4] Features are extracted from a text
[5] An *ambiguous* mention is a mention that is used to refer to two or more entities in Wikipedia. We call these entities *candidate entities* of that mention.

method ranks the candidate entities of each mention and chooses the best one. The quality of ranking depends on used features.

Text features
To construct the feature vector of a mention in a text, we extract all mentions co-occurring with it in the whole text, local words in a context window, and words in the context windows of those mentions that are co-referent with the mention to be disambiguated. Those features are presented below.

– *Entity mentions* (EM). After named entity recognition, mentions referring to named entities are detected. We extract these mentions in the whole text. After extracting the mentions, for the ones that are identical, we keep only one and remove the others. For instance, if "U.S" occurs twice in a text, we remove one.
– *Local words* (LW). All the words found inside a specified context window around the mention to be disambiguated. The window size is set to 55 words, not including special tokens such as $, #, ?, etc., which is the value that was observed to give optimum performance in the related task of cross-document coreference resolution ([6]). Then we remove those local words that are part of mentions occurring in the window context to avoid extracting duplicate features.
– *Coreferential words* (CW). All the words found inside the context windows around those mentions that are co-referent with the mention to be disambiguated in the text. For instance, if "John McCarthy" and "McCarthy" co-occur in the same text and are co-referent, we extract words not only around "John McCarthy" but also those around "McCarthy". The size of those context windows are also set to 55 words. Note that, when the context windows of mentions that are co-referent are overlapped, the words in the overlapped areas are extracted only once. We also remove those extracted words that are part of mentions occurring in the context windows to avoid extracting duplicate features.

Wikipedia features
For each entity in Wikipedia, serving as a candidate entity for an ambiguous mention in a text, we extract the following information to construct its feature vector.

– *Entity title* (ET). Each entity in Wikipedia has a title. For instance, "John McCarthy (computer scientist)" is the title of the page that describes Professor John McCarthy who is the inventor of LISP programming language. We extract "John McCarthy (computer scientist)" for the entity Professor John McCarthy.
– *Titles of redirect pages* (RT). Each entity in Wikipedia may have some redirect pages whose titles contain different names, i.e. aliases, of that entity. To illustrate, from the redirect pages of an entity John Williams in Wikipedia, we extract their titles: Williams, John Towner; John Towner Williams; Johnny Williams; Williams, John; John Williams (composer); etc.
– *Category labels* (CAT). Each entity in Wikipedia belongs to one or more categories. We extract labels of all its categories. For instance, from the categories of the entity John McCarthy (computer scientist) in Wikipedia, we extract the following category labels as follows: Turing Award laureates; Computer pioneers; Stanford University faculty; Lisp programming language; Artificial intelligence researchers; etc.

- *Outlink labels* (OL). In the page describing an entity in Wikipedia there are some links pointing to other Wikipedia entities. We extract labels (anchor texts) of those outlinks as features of that entity.
- *Inlink labels* (IL). For an entity in Wikipedia, there are some links from other Wikipedia entities pointing into it. We extract labels of those inlinks as its possible features.

Note that infoboxes of pages in Wikipedia are meaningful resources for disambiguation. However, these resources of information may be missed in many pages or information in many infoboxes is poor. Moreover, the information in infobox of each page can be distilled from the content of the page. Therefore, our disambiguation method does not extract information from infoboxes for disambiguation.

Normalization

After extracting features for a mention in a text or an entity, we put them into a 'bag of words'. Then we normalize the bag of words as follows. (i) Removing special characters in some tokens such as normalizing U.S to US, D.C (in "Washington, D.C" for instance) to DC, and so on; (ii) removing punctuation mark and special tokens such as commas, periods, question mark, \$, @, etc.; and (iii) removing stop words such as *a*, *an*, *the*, etc., and stemming words using Porter stemming algorithm. After normalizing the bag of words, we are already to convert it in to a token-based feature vector.

Term weighting

For a mention in a text, suppose there are N candidate entities for it in Wikipedia. We use the *tf-idf* weighting schema viewing each 'bag of words' as a document and using cosine similarity to calculate the similarity between the bag of words of the mention and the bag of words of each of the candidate entities respectively. Given two vector S_1 and S_2 for two bags of words, the similarity of the two bags of words is computed as:

$$Sim(S_1, S_2) = \sum_{common\ word\ t_j} w_{1j} \times w_{2j} \tag{1}$$

where t_j is a term present in both S_1 and S_2, w_{1j} is the weight of the term t_j in S_1 and w_{2j} is the weight of the term t_j in S_2.

The weight of a term t_j in vector S_i is given by:

$$w_{ij} = log(tf_j + 1) \times log(N/df_j) / \sqrt{s_{i1}^2 + s_{i2}^2 + ... + s_{iN}^2} \tag{2}$$

where tf_j is the frequency of the term t_j in vector S_i, N is the total number of candidate entities, df_j is the number of bags of words representing candidate entities in which the term t_j occurs, $s_{ij} = log(tf_j + 1) \times log(N/df_j)$.

Algorithm

For a mention m that we want to disambiguate, let C be the set of its candidate entities. We cast the named entity disambiguation problem as a ranking problem with the assumption that there is an appropriate scoring function to calculate semantic similarity between feature vectors of an entity $c \in C$ and the mention m. We build a ranking function that takes as input the feature vectors of the entities in C and the feature vector of the mention m, then based on the scoring function to return the entity $c \in C$ with the highest score. We use *Sim* function as given in Equation 1 as the scoring function.

16   H.T. Nguyen and T.H. Cao

What we have just described is implemented in Algorithm 1. *Sim* is used at Line 4 of the algorithm.

Algorithm 1. Statistical-Based Entity Ranking

1: let M be a set of ambiguous mentions
2: **for** each mention $m \in M$ **do**
3:    let C be a set of candidate entities of m
4:    $c^* \leftarrow \underset{c_i \in C}{arg\ max}\ Sim(\text{FeatureVector}(c_i), \text{FeatureVector}\ (m))$
5:    assign c^* to m
6: **end for**

---

3 Evaluation

We collect documents where there are occurrences of entities in Wikipedia that have the names "John McCarthy", "John Williams", "Georgia", or "Columbia". There are total 270 documents in the dataset. Table 1 presents information about mentions in the dataset that refer to entities having the names "John McCarthy", "John Williams", "Georgia", or "Columbia" in Wikipedia. The left column in the table shows the number of mentions in the dataset referring to the corresponding entity in the right column. For instance, as showed in the second row of the table, there are 23 mentions referring to the entity John "Hot Rod" Williams in Wikipedia.

Table 1. Statistics about ambiguous mentions in the dataset that refer to entities having the names "John McCarthy", "John Williams", "Georgia", or "Columbia" in Wikipedia

Entity (in Wikipedia)	# of mentions
John "Hot Rod" Williams	23
John Williams (actor)	2
John Williams (guitarist)	60
John Williams (composer)	154
John McCarthy (computer scientist)	30
John McCarthy (journalist)	79
John McCarthy (linguist)	16
John McCarthy (referee)	47
Georgia (country)	318
Georgia (U.S. state)	90
South Georgia and the South Sandwich Islands	59
British Columbia	34
Columbia Sportswear Company	65
Columbia University	13
Columbia, South Carolina	15
Space Shuttle Columbia	80
District of Columbia	1
Total	**1086**

In the Wikipedia data ([15]) that we use for experiments, there are 6 persons having the name "John McCarthy", and 41 persons having the name "John Williams", 17 entities belonging to different categories and having the name "Georgia", and 78 entities belonging to different categories and having the name "Columbia". The number of name that we use for testing to evaluate the impact of features on disambiguation performance is suitable and equal to that of names used in [14]. Table 2 presents information about four those names aforementioned in the dataset. The second column presents the total number of candidate entities of each name. The third column presents the total number of mentions whose referred entities have those names respectively. The last column presents the information about the total number of mentions disambiguated corresponding to each name in the first column.

Table 2. Statistics about total mentions and disambiguated mentions of each name

Name	# of candidates	# of total mentions	# disambiguated mentions
John McCarthy	6	172	170
John Williams	41	239	242
Georgia	17	468	453
Columbia	78	207	203
Total		**1086**	**1068**

Note that some mentions in the dataset that refer to entities listed in Table 1 are unambiguous. For instance, the "Columbia Sportswear" mention is unambiguous, because when sending it as a query to Wikipedia, we receive only one entity which is Columbia Sportswear Company and the right one. Since, the aim is to evaluate the impact of features on the disambiguation performance, for each case like that we use the corresponding ambiguous mention listed in the first column of Table 2 as a query. For instance, for the case "South Georgia", we send "Georgia" as a query to Wikipedia to retrieve entities having the name "Georgia". As a result, in experiments, our method deals with actually ambiguous mentions.

Prior to running our algorithm to disambiguate mentions in a text, we perform preprocessing tasks. In particular, we perform NE recognition and NE coreference resolution using natural language processing resources of Information Extraction engine based on GATE ([5]), a general architecture for developing natural language processing applications. The NE recognition applies pattern-matching rules written in JAPE's grammar of GATE to detect and tag boundaries of mentions occurring in the dataset and then categorize corresponding entities as Person, Location and Organization, etc. After detecting all mentions occurring in the text, we run NE co-reference resolution ([2]) module in the GATE system to resolve the different mentions of a NE into one group that uniquely represents the NE.

We consider entity mentions as the baseline features of text features. Then we combine the baseline with *local words* or both *local words* and *coreferential words*. Meanwhile, Wikipedia features are combined in four ways. First, *entity titles* (ET) is combined with *redirect page titles* (RT) as the baseline features, named *Wiki_baseline*. Then, Wiki_baseline is combined with *category labels* (CAT); CAT and *outlink labels* (OL); and then CAT, OL and *inlink labels* (IL), respectively. Table 3 presents the precision and recall measures calculated for the names "John McCarthy" and "John Williams". Table 4 presents the precision and recall measures calculated for the names "Georgia" and "Columbia". The experiment results on Table 3 and Table 4 show how we combine text features with each other, Wikipedia features with each other and text features with Wikipedia features. Table 5 presents the precision and recall measures calculated for all four names "John McCarthy", "John Williams", "Georgia", and "Columbia".

Table 3. Precision (P) and recall (R) for each name "John McCarthy" and "John Williams"

Name	Textual features	Wikipedia features	# of mentions correctly disambiguated	P (%)	R (%)
John McCarthy	EM	ET + RT	60	35.29	34.88
		ET + RT + CAT	60	35.29	34.88
		ET + RT + CAT + OL	156	91.76	90.70
		ET + RT + CAT + OL + IL	150	88.24	87.21
	EM + LW	ET + RT	7	42.35	41.86
		ET + RT + CAT	120	70.59	69.77
		ET + RT + CAT + OL	159	93.53	92.44
		ET + RT + CAT + OL + IL	159	93.53	92.44
	EM + LW + CW	ET + RT	108	63.53	62.79
		ET + RT + CAT	162	95.29	94.19
		ET + RT + CAT + OL	166	**97.65**	**96.51**
		ET + RT + CAT + OL + IL	159	**93.53**	**92.44**
John Williams	EM	ET + RT	3	01.24	01.26
		ET + RT + CAT	79	32.64	33.05
		ET + RT + CAT + OL	167	69.01	69.87
		ET + RT + CAT + OL + IL	199	82.23	83.26
	EM + LW	ET + RT	56	23.24	23.43
		ET + RT + CAT	157	64.88	65.69
		ET + RT + CAT + OL	204	84.30	85.36
		ET + RT + CAT + OL + IL	210	86.78	87.87
	EM + LW + CW	ET + RT	147	60.74	61.51
		ET + RT + CAT	208	85.95	87.03
		ET + RT + CAT + OL	229	**94.63**	**95.82**
		ET + RT + CAT + OL + IL	233	**96.28**	**97.49**

The figures in Table 3, Table 4 and Table 5 show that the combinations of baseline features with *local words* or both *local words* and *coreferential words* increasingly improve the performance. The figures in Table 3 and Table 4 show that the combination ET + RT + CAT + OL is the best, except for the name "John Williams", where IL gives a bit higher precision and recall measures. Nevertheless, the combination ET + RT + CAT + OL for the name "John Williams" also archives good performance, as the second best. While the improvement by IL in the case "John Williams" is little, using IL critically increases the disambiguation time. In fact, it runs in 48 times longer than that with the combination ET + RT + CAT + OL, to disambiguate a mention of John Williams. The figures in Table 5 show that the text features EM + LW + CW in combination with Wikipedia features ET + RT + CAT + OL give the best performance in terms of precision and recall.

Table 4. Precision (P) and recall (R) for each name "Georgia" and "Columbia"

Name	Textual features	Wikipedia features	# of mentions correctly disambiguated	P (%)	R (%)
Georgia	EM	ET + RT	95	20.97	20.30
		ET + RT + CAT	125	27.59	26.71
		ET + RT + CAT + OL	339	74.83	72.44
		ET + RT + CAT + OL + IL	284	62.69	60.68
	EM +LW	ET + RT	128	28.26	27.35
		ET + RT + CAT	144	31.79	30.77
		ET + RT + CAT + OL	356	78.59	76.07
		ET + RT + CAT + OL + IL	297	65.56	63.46
	EM +LW +CW	ET + RT	199	43.93	42.52
		ET + RT + CAT	183	40.4	39.1
		ET + RT + CAT + OL	393	**86.75**	**83.97**
		ET + RT + CAT + OL + IL	345	76.16	73.72
Columbia	EM	ET + RT	123	60.59	59.42
		ET + RT + CAT	101	49.75	48.79
		ET + RT + CAT + OL	147	72.41	71.01
		ET + RT + CAT + OL + IL	130	64.04	62.80
	EM +LW	ET + RT	143	70.44	69.08
		ET + RT + CAT	126	62.07	60.87
		ET + RT + CAT + OL	162	79.80	78.26
		ET + RT + CAT + OL + IL	152	74.88	73.43
	EM +LW +CW	ET + RT	178	87.68	85.99
		ET + RT + CAT	169	83.25	81.64
		ET + RT + CAT + OL	178	**87.68**	**85.99**
		ET + RT + CAT + OL + IL	171	84.24	82.61

Table 5. Precision (P) and recall (R) for four names "John McCarthy", "John Williams", "Georgia", and "Columbia"

Textual features	Wikipedia features	# of mentions correctly disambiguated	P (%)	R (%)
EM	ET + RT	281	26.31	25.87
	ET + RT + CAT	365	34.18	33.61
	ET + RT + CAT + OL	809	75.75	74.49
	ET + RT + CAT + OL + IL	763	71.44	70.26
EM +LW	ET + RT	399	37.36	36.74
	ET + RT + CAT	547	51.22	50.37
	ET + RT + CAT + OL	881	82.49	81.12
	ET + RT + CAT + OL + IL	818	76.59	75.32
EM +LW +CW	ET + RT	632	59.18	58.20
	ET + RT + CAT	722	67.60	66.48
	ET + RT + CAT + OL	966	**90.45**	**88.95**
	ET + RT + CAT + OL + IL	908	85.02	83.61

4 Conclusion

Based on our proposed statistical ranking model for named entity disambiguation, we have explored a range of features extracted from texts and Wikipedia, and vary combinations of those features to appraise which ones are good for the task. Our analysis of text features does not depend on a given knowledge base, so the results can be used for NED based on an arbitrary knowledge base. The experiment results show that the Wikipedia features ET, RT, CAT and OL in combination with the text features EM, LW and CW give the best performance. Our findings provide a guidance and direction for any statistical work on NED using Wikipedia and text features.

References

[1] Bunescu, R., Paşca, M.: Using encyclopedic knowledge for named entity disambiguation. In: Proc. of the 11th Conference of EACL, pp. 9–16 (2006)
[2] Bontcheva, K., et al.: Shallow methods for named entity coreference resolution. In: Proc. of TALN 2002 Workshop (2002)
[3] Cucerzan, S.: Large-scale named entity disambiguation based on Wikipedia data. In: Proc. of EMNLP-CoNLL Joint Conference (2007)
[4] Cohen, W., Ravikumar, P., Fienberg, S.: A comparison of string metrics for name-matching tasks. In: IJCAI-03 II-Web Workshop (2003)
[5] Cunningham, H., et al.: GATE: A framework and graphical development environment for robust NLP tools and applications. In: Proc. of ACL 2002 (2002)
[6] Gooi, C.H., Allan, J.: Cross-document coreference on a large-scale corpus. In: Proc. of HLT/NAACL 2004 (2004)
[7] Mihalcea, R.: Using Wikipedia for automatic word sense disambiguation. In: Proc. of HLT/NAACL 2007 (2007)
[8] Mihalcea, R., Csomai, A.: Wikify!: Linking documents to encyclopedic knowledge. In: Proc. of CIKM 2007, pp. 233–242 (2007)
[9] Medelyan, O., et al.: Mining meaning from Wikipedia. International Journal of Human-Computer Studies 67(9), 716–754 (2009)
[10] Medelyan, O., et al.: Topic indexing with Wikipedia. In: Proc. of WIKIAI 2008 (2008)
[11] Milne, D., Witten, I.H.: Learning to link with Wikipedia. In: Proc. of CIKM 2008, pp. 509–518 (2008)
[12] Overell, S., Rüger, S.: Using co-occurrence models for placename disambiguation. The IJGIS. Taylor and Francis, Abington (2008)
[13] Nguyen, H.T., Cao, T.H.: A Knowledge-based approach to named entity disambiguation in news articles. In: Orgun, M.A., Thornton, J. (eds.) AI 2007. LNCS (LNAI), vol. 4830, pp. 619–624. Springer, Heidelberg (2007)
[14] Chen, Y., Martin, J.: Towards robust unsupervised personal name disambiguation. In: Proc. of EMNLP-CoNLL Joint Conference (2007)
[15] Zesch, T., Gurevych, I., Mühlhäuser, M.: Analyzing and accessing Wikipedia as a lexical semantic resource. In: Rehm, G., Witt, A., Lemnitzer, L. (eds.) Data Structures for Linguistic Resources and Applications, pp. 197–205 (2007)

Telemedical System in Evaluation of Auditory Brainsteam Responses and Support of Diagnosis

Piotr Strzelczyk, Ireneusz Wochlik, Ryszard Tadeusiewicz, Andrzej Izworski, and Jarosław Bułka

AGH University of Science and Technology, Department of Automatics,
Al. Mickiewicza 30, 30-073 Krakow, Poland
pit@student.agh.edu.pl, wolo@agh.edu.pl, rtad@agh.edu.pl,
izwa@agh.edu.pl, bulek@agh.edu.pl

Abstract. The paper presents the use of telemedicine in intelligent supporting of otorhinolaryngologist in the diagnosis of auditory brainsteam responses. This test is easy to visualize but difficult to diagnose. The presented software system uses advanced methods of signal processing and an author's algorithm supporting the doctor by setting the characteristic points of the examination and reach the diagnosis. This paper describes the capabilities of the system and underline the benefits which are result of the nature of this application. It also highlights the benefits and opportunities introduced to this field of medicine by the described intelligent software system.

Keywords: Auditory brainsteam responses, telemedicine, intelligent software systems, diagnosis support systems.

1 Introduction

Nowadays, the progress of medicine is significant in the early detection of health defects. Recently audiology, the branch of medicine dedicated to the research field of hearing, experienced a rapid development. At the moment it is tend to reach a diagnose in disorder of sound reception as early as possible. Clearly visible is the desire to standardize the hearing testing and introducing them as obligatory for the whole country. Therefore, the hearing screening tests are going to be a norm by children in the preschool and school age. The detection of potential hearing failures by a patient at a such young age, has a significant impact on his psychological and intellectual development.

One of the most popular methods of hearing screening tests are the auditory brainsteam responses. Their results are not easy to interpret. For an unexperienced doctor it is not easy to reach a diagnose. A major diagnostic problem is a situation when the ABR (Auditory Brainsteam Responses) results are from a patient that has hearing

N.T. Nguyen, M.T. Le, and J. Świątek (Eds.): ACIIDS 2010, Part II, LNAI 5991, pp. 21–28, 2010.

disorder or the quality of the measurement is not good. The results are in this case very hard to diagnose. Audiologist, who is not sure of the final interpretation, does not reach a diagnosis without consultation with a person witch is an authority in this field.

In today's world of medicine, the problem of high-class specialist is visible, which are not much in the whole country. They are working in medical centers which are known as units specialized in the brunch of medicine. In the case of audiology such center in Poland is the International Center of Hearing and Speech in Kajetany near Warsaw. This means that in most cases, most consultations must be carried remotely.

Modern technology offer many opportunities for communication, but they are not sufficient for the needs of medicine. A physician, to remotely terminate its opinion, needs to see the results. In ABR the results are presented in graphs. They occupy a fairly large volume of memory. They lossy compression to a file of smaller size is not allowed, because there is a risk of losing important data. This loss may result in a completely incorrect diagnose. Sending a few results of a large capacity may take a long time. Another solution is to send data as a sequence of numbers obtained from the measurement equipment. This approach avoids the problem described about with the capacity of the files to be send. The disadvantage of this method is that the receiving side needs to have a tool to convert the numerical data to charts. In both approaches have significant discomfort. Doctors are forced to be involved in the process of initial data exchange, rather than take up the appropriate discussion about the results of the examination.

2 Auditory Brainsteam Responses

One of the most objective methods of hearing test are the auditory brainsteam responses. They are measuring the bioelectrical activity of the brain, forced by sound given into the tested ear. For this purpose the surface electrodes are placed in fixed points on the head. Then, using headphones a noise or short sound is made with a certain frequency. The end result is a chart that shows the stimulation of different hearing centers. At the moment it is used as a screening hearing test by newborns and infants. This method is also used in the differential diagnosis and in monitoring the function of the auditory nerve and brain stem during neurosurgical procedurs. The disadvantage of this test is the need that the patient is sleeping during it. This results from the fact that the brain registers various signals associated not only with hearing but for example with freedom thinking, work of muscles, work of eyes. This signals are disorders that prevent the execution of the test.

The test procedure consists of two main stages: the registration of a number of responses known as "intensity series" and analysis of recorded responses to determine the hearing threshold. The intensity series is created in the following way: in the first phase the ear is stimulated by a sound with high intensity for example 90 or 100 dB nHL, and in subsequences phases – responses for ever-lower levels of sound. The intensity of the stimulus is reduced in steps by usually by 20, 10 or 5 dB. For intensities near the hearing threshold the registration is reapeted two or three times. Then the results are listed on the screen in such a way that they are as to be closest to each other. Waveforms recorded for the same intensity are placed one after another.

A doctor or a trained technician can on this examination result mark characteristic waves, which are numbered in Roman. Peaks of every wave kind represents the place in anatomical structure which are responsible for their generation. Wave I is generated in the distal part of the auditory nerve, wave II in the proximal part, wave III mainly in the cochlear nucleus, wave IV mainly in the upper band olives, and wave V mainly in the nuclei sideband. Any pathology of the ear may change the morphology, time parameters and the amplitude of individual waves in the result. Only waves I, III and V are determined in clinical practice, which forms the basic for the diagnosis. The presence of wave V for an intensity means that the patient is hearing on this sound level. In this way, using the ABR test, the doctor sets the hearing threshold.

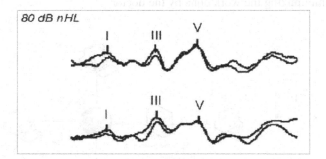

Fig. 1. An example measurement of ABR signal for the intensity of 80 dB. The upper chart is for the right ear and the other for the left ear. The measurement was made for every ear on this level two times.

3 Telemedical Application

In the last years internet technologies have been developing very fast. They become an internal part of success in business and projects undertaken by the people, institutions and companies. The main advantage of these solutions is the minimal requirements posted to the user.

Today, thanks to the web technologies, the only requirement is that the user have a web browser. To use the more advanced capabilities of these technologies, there is only the need to install the appropriate software which is free. It is often supplied as a plug-in to a concrete web browser.

Diversity and capabilities of these technologies were not yet found there use in many fields of medicine. At present there is in Poland no common online platform that allows the use of modern web technologies to support medical diagnosis of auditory brainsteam responses.

The platform presented here is an integrated telemedical system that supports the diagnosis of auditory brainsteam responses. This platform contains personal data of patients and connected with them hearing tests. The most important functions are: addition of examination, review of historical examinations with diagnosis, review of undiagnosed examinations, interactive presentation of the results and the ability to analysis them by many doctors at the same time, reaching a diagnosis. The reaching of diagnosis requires to analysis the intensity serics. This means that data are presented in interactive charts. They allow to mark characteristic points of the hearing test and adding description to them. The working wave detection algorithm in the background make it easier to reach a diagnose.

From the technical point of view the architecture of the application is three layer technology. In the view layer is embedded an interactive application used to present the results of ABR and analysis them. Characteristic for the system is: high level of security, easy and transparent operation, high interactive functionality for analyzing examinations, minimizing the work done by the doctor.

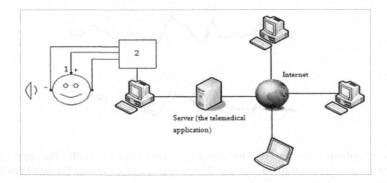

Fig. 2. The Flowchart of the telemedical application. Measurement electrodes (1), send the signal to the equipment (2) that filters out the ABR signal. The user can connect a computer to the measurement equipment and send the data to the telemedical application. Other doctors can via the internet watch, analysis and diagnose the hearing test results.

3.1 Interactive Presentation of ABR Results

The heart of the entire application is the page which presents the results of the tests. RIA (Rich Internet Application) were used to create this page which allows direct interaction with the user without refreshing the page. The communication between the page and the server is based on Web Service technology.

The interactive application is from the visual side divided in two parts: a chart and a option panel. The first part is used to present the data and allows the physician a comfortable and easy review of the hearing test. The second part provides options to perform a detailed analysis. Operations such as zooming or displaying selected measurements helps the doctor to reach a diagnosis.

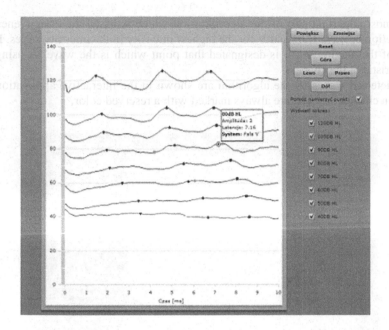

Fig. 3. The interactive application presenting the results of ABR. On the horizontal axis there is the latency of the signal and on the vertical axis the intensity level. On the right side of the chart there are buttons and checkbox that provides different options.

3.2 Preprocessing of ABR Signal

The ABR signal itself contains noise, which means that it is sometimes difficult to interpret by the physician. To improve the quality of the ABR signal there is used a few step preprocessing [1], which is performed before sending data to the interactive application that presents the results. The output from these operations is a smoother and more readable ABR signal.

The use of such a preprocessing algorithm cause that doctors can faster identify individual waves I, III and V. This element of the application is a next improvement, which make the diagnosis easier for a doctor.

The ABR signal converted by the preprocessing algorithm is the input data for the detection algorithm of wave I, III and V.

3.3 Detection Algorism of Wave I, III and V

Another improvement which is used in the application is the detection algorithm of wave I, III and V. He is mostly based on the stock indicator SMA (Simple Moving Average). This is one of the fundamental and first indicator used to determine the trend of shares. He is based on the arithmetic average of stock prices from a number of sessions. This indicator was slightly modified to fit the auditory brainsteam responses. He is basically used to assess the quality of the detected local maximums. Its value to decide if the found maximum can be a wave. This algorithm takes into account the delays in measurements for lower intensities.

The most important part of the algorithm, on which is based the effectiveness, is the detection of wave I, III and V for the highest intensity in the intensity series. From the set of the maximums, it is designated that point which is the wave V, using the characteristic of the chart.

The detected waves by the algorithm are shown in the interactive application described in chapter 3.1. The are always marked with a reserved color.

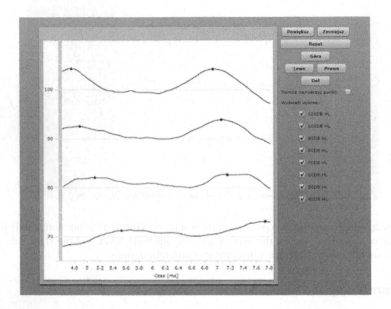

Fig. 4. The interactive application presenting the ABR result with the zoom effect on the chart component. The visible black points are waves detected by the system. In this case there are only visible wave III and V.

In the case of correct detection of the waves by the system the doctor accepts the proposal of the computer and proceed to the description of diagnosis. In this way the time used to analysis one examination is minimized.

A doctor is supported additional by the diagnosis with the highlighted information about the selected points. The exact value of the latency and amplitude of individual waves in the form of a table reduces errors that may result from the diagnosis based only on the analysis of the chart. These values also allows to specify the exact value of the intervals, which are one of the basic parameters important for the diagnosis.

The effectiveness of this algorithm is 87,4%. It has been tested on actual medical hearing tests, obtained from the International Institute Of Physiology And Pathology of Hearing in Kajetany near Warsaw. So a good result means that the waves marked by the system can often provide valuable suggestion to the physician and reduce the amount of work connected with marking of this waves.

4   The Benefits of the Application

The web based platform brings many opportunities in the field of otorhinolaryngology in the case of ABR. The platform was constructed in such a way that it is able to expand and extend it with new options. This possibility is very important because of the high dynamic of development of the ABR test and the possibility of need to extend the functionality of the application.

At present, diagnostic systems allows to mark only the wave V. This cause that every assessment of the measurement is an activity that is important for the health of the patient. The option of marking points on charts and adding comments to them, makes the application a place, where doctors can justify there decision. Such a examination cares not only value for the health condition of hearing, but is a source of information for doctors which needs additional training in this field.

Implementation of the preprocessing algorithm gives the user a better view in higher quality of the test results. The biggest advantage of it is the exposure of places suspected of occurrence of waves and the removal of small noise from the chart. This can have significant impact during the reaching of diagnosis, especially by an unexperienced otorhinolaryngologist.

The implemented wave detection algorithm of wave I, III and V is an innovative solution that combines mathematical and medical knowledge. The high performance of this algorithm, reaching more than 87%, can support the work done by the doctor.

The web platform allows to review hearing tests already diagnosed. This gives incredible possibilities. In first order it may be a source of training materials for young doctors. High functionality allows to leave a lot of information in each test, provides significantly more knowledge then the standard description of the diagnosis which is used at present. Another advantage is the possibilty of remote trainings for doctors. They need only access to the system resources to study and analysis the test on their own. This allows to save costs and reach in a short time a large group of people requiring training.

Another benefit of the historical examinations is the ability to verify the quality of the wave detection algorithm. On this basis it is possible to make statistics and find the cause of potential common errors. This will allow in future to propose amendments to it, making it even more effective. In case of confirmation of its effectiveness on a large number of samples, it may give a basis to create an algorithm to automatize the diagnosis.

The hearing test of auditory brainsteam responses are characterized by certain common attributes. After reviewing the appropriate number of measurements there is visible the occurrence of some regularities. At the moment there is no research on certain statistical regularity. Analysis of ABR results suggest that such dependencies exist. The absence of such information is mainly caused by the absence of a system collecting such data. This platform provides such a possibility. This may be a source in future to create another algorithm of wave detection and automatic diagnosis based on tools connected with probability and statistic. Statistical information can also make a major contribution to otorhinolaryngology introducing new theories related with diagnosis of ABR.

The most important feature of this system is the possibility of remote consultations. Such a solution can significantly minimize the mistakes made in clinics which have doctors with insufficient experience in difficult results of ABR.

The developed online platform will not replace a doctor in his work, but it provides tools that can improve the effectiveness of the diagnosis. It may also be a source, which opens up new perspectives in the field of otorhinolaryngology in the field of auditory brainsteam responses.

References

1. Izworski, A., Tadeusiewicz, R., Skarżyński, H., Kochanek, K., Bułka, J., Wochlik, I.: Automatic analysis and recognition of the auditory brainstem response signals. In: Proceedings of the 18th International Congress on Acoustics, Kyoto, Japan, vol. 2(5), II-1755–II-1758 (2004)
2. Izworski, A., Tadeusiewicz, R., Pasławski, A.: The utilization of context signals in the analysis of ABR potentials by application of neural networks. In: Lopez de Mantaras, R., Plaza, E. (eds.) ECML 2000. LNCS (LNAI), vol. 1810, pp. 195–202. Springer, Heidelberg (2000)
3. Wochlik, I., Bulka, J., Tadeusiewicz, R., Bania, P., Izworski, A.: Determination of diagnostic parameters in an automated system for ABR signal analysis. In: Proceedings of the international conference on Mathematics and Engineering Techniques in Medicine and Biological Sciences, USA, pp. 391–394. CSREA Press (2002)
4. Wochlik, I., Bulka, J., Izworski, A., Tadeusiewicz, R.: Features determination for an automated system for ABR signal analysis. In: Proceedings of the second IASTED international conference, pp. 68–71. ACTA Press (2002)
5. Wochlik, I., Bułka, J., Tadeusiewicz, R., Bania, P., Izworski, A.: Determination of diagnostic parameters in an automated system for ABR signal analysis. Bibliogr. 394, 391–394
6. Delgado, R.E., Ozdamar, O.: Automated Auditory Brainstem Response Interpretation. IEEE Engineering In Medicine And Biology (April/May 1994)

Service Discovery in the SOA System

Krzysztof Brzostowski, Witold Rekuć, Janusz Sobecki, and Leopold Szczurowski

Wroclaw University of Technology
{Krzysztof.Brzostowski,Witold.Rekuc,Janusz.Sobecki,
Leopold.Szczurowski}@pwr.wroc.pl

Abstract. In the paper we present an ontology based service discovery in the SOA system. First we show an application of the following models: business motivation model (BMM) and business process model as well as information for an identification of the business services in a selected area of the university activities has been described. Then we show an importance of the conceptual specification of the services for their subsequent retrieval has been emphasized. Then we present the representation of these services in WSMO and ontology based service discovery with WSM Toolkit.

Keywords: Educational institution, business motivation model, business service, SOA, service discovery, WSMO.

1 Introduction

SOA (Service Oriented Architecture) is such an approach to create systems which helps to make independent business solutions from technological constraints. We can say that SOA methodology enables different business sectors to make crucial decision supported by technology instead of making business decision which are determined or constrained by technology [6].

Systems designed accordant to SOA paradigm are composed of service which are executed on user's demand. Each service is an implementation of business functionality which can be used in different applications or, in general, in different (business) processes. We can further stress that each of services in SOA system can combine complex services out of atomic ones. Moreover, in modeled business process it is possible to specify which services can and how their elements interacting with each other.

Today we can say that paradigm of Service Oriented Knowledge Utilities (SOKU) requires development of methods for acquisition, processing and integration of knowledge. Because of the complexity of problems arising in SOKU systems, there is a great need for development of methods of representation of descriptive and procedural knowledge. Hence, ontologies are employed for representation of knowledge about users and services.

The main issue in the work is service modelling, then application of ontology for knowledge representation and processing in SOA/SOKU architecture [5] as well as application of ontology in the service discovery. The service retrieval is SOA/SOKU systems has ever increasing importance. The number of atomic and complex services

N.T. Nguyen, M.T. Le, and J. Świątek (Eds.): ACIIDS 2010, Part II, LNAI 5991, pp. 29–38, 2010.

increases rapidly so users may have a great difficulty in identifying the appropriate services, especially when there are no exact matches in service specification and we need ontology to specify them.

2 Discovering the Services in Business Process of Enrolment for the Courses in Education System

The university of technology that educates students at several departments and carries out the scientific researches is an educational system on the basis of which the service identification was made. „Its mission is to shape creative and critical personalities of the students and to chart development directions of science and technology. The university fulfils its mission through a high standard of teaching and scientific research ensuring its prominent place among the universities in Europe and the whole world." [11].

According to the SOA methodology, the service identification should be preceded by an analysis of the organization objectives. From these objectives the services are derived directly or indirectly (through processes). The Business Motivation Model (BMM) [2] was the notion tool used by the authors in the identification of services and objectives. It enabled the authors to create the organization objectives and to connect them with the mission, processes and business rules.

2.1 Process Areas in Subjective Educational System

As in the other organizations of scientific-didactic services sector there are many mutually associated courses of action operated at the Wroclaw's University of Technology's educational system. The most important courses of action are the programming and planning of education, recruitment of students, planning of didactic activities, realization of activities with students and a supervision and control of their education. The courses of action have been identified using diagram (map) of the hierarchy of business processes. This approach allowed us to use the hierarchy of purposes formulated according to BMM to find elements of information technology system compatible with the SOA paradigm. A separation of the elements led to an assignment of coherent business processes and further detailed (deductive-inductive) analyses. These analyses allowed us to disclose the SOA services together with elements of knowledge necessary for their execution.

The detailed analysis concerned two business processes proceeding in planning and realization of the didactic activities. The former one is a process operating the enrolment of students (the customers of the system) for the activities (courses) offered in the multi-semester contracts made with the University (plans of education at Faculties). The didactic programs of courses offered in these contracts have to meet the hierarchy of formal-substantial, qualitative and quantitative requirements. It results from the conditionings of a statutory (the rules written in the law of higher education, ministerial minima), trade (Universities' minima) and academic (rules of the University's senate) character. The other analysed process concerned the assignment of the supervisors to particular groups provided in semester's schedule. This is a relevant internal activity of an organization directly affecting the quality of the whole system's functioning. In a strategic way it decides of preservation of an ability to compete at the educational services market.

3   An Example Discovering Services in the Process of Enrolment for Courses

The mentioned above sub-processes in the educational system, at the Faculty of Computer Science and Management at Wroclaw's University of Technology are the subjects of the identification. The presented examples concern two relevant aspects of revealing the SOA services: identification of repetitive services (according to the *"Select a class from an enrolment array"* process) and conceptualizing of business elementary services and their compounds taking into account a compliance with the business rules, derived from BMM (on the basis of „*Select a set of courses"* process).

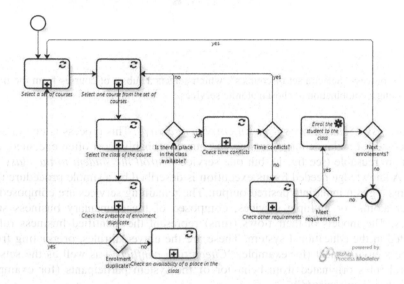

Fig. 1. Process "Select a class from an enrolment array"

In order to identify services a description of business processes, rules and information objects was necessary. The authors decided to use the widely known modeling standards: for business processes - BPMN [3] and for information objects – UML class diagram notation [10]. BizAgi Process Modeler (for BPMN) and Microsoft Office Visio 2007 (for class diagram UML) were the modeling tools: The service identification consisted in the process specification, establishing a relationship between activities from the process and the information objects defined by the class diagram. In the class diagram elaboration the authors tried to express the notions from the subject domain and relationships between them (ontologies). The specification of business rules was an important component of the process specification. Taking into account the goal of the Project, the decision was undertaken not to elaborate separated business rules description using Business Rule Management Tool. The rules were described right on the process diagram in the decision nodes and in the description of activities.

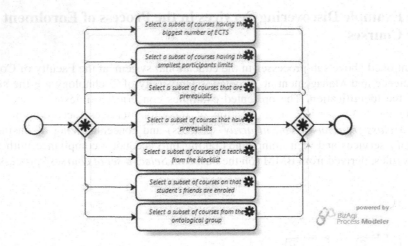

Fig. 2. The process "Select a set of courses", which selects a subset of courses from the input set by running a combination of chosen atomic services

The process "Select a class from an enrolment array". This process is one of several processes of a class choice for a given course, probably most often executed. All services are reusable (see fig. 1), but one service *"Enroll the student to the class"* is atomic. A knowledge needed for its execution is described by a simple procedure that transforms a given input into desired output. The remaining services are composed of the other atomic or multiple services, composed of the elementary business sub-processes. The models of their flows (runs) represent the identified business rules, formulated in the educational system. These are the universal rules, originating from the process environment (for example, *"Check time conflicts"*) as well as the sets of behavioral rules originated from behavior of the system participants (for example, *"Check other requirements"*).

The process "Select a set of courses". Many of the identified elementary processes are the combination of services from the set of atomic services. The sub-process „Select a set of courses" (see Fig. 2) is a good example. Each service represents a specific criterion of a different complexity. The output of the process is an intersection of the results generated by the subset of selected services. The compound service of such a structure is widely applied at the initial stage of the decision making process, based on the multi-criteria decision analysis, breaking a classical paradigm of the optimality of a choice to favor a soft modeling of the decision maker preferences.

The research carried out showed the usefulness of the approach based both on the BMM and on the usage of the process specification in the discovery of the services that were necessary for the business goal achievements. The analysis of the identified processes allowed us to disclose their common fragments and in this way to find the reusable services. The input/output and the input to output transformation specification in the categories of the UML class diagram allowed us to make a conceptual description of each particular service. This description, in connection with ontologies embedded in the class model, could be a good basis for a retrieval of the databases in a consistency with the service contents.

4 Services Representation in WSMO

In previous sections business process and set of services, utilized to fulfill it, are presented. The aim of this part of work is to specify services, used in considered business process, by use of semantic description. One of the method, which may be used, to convey the meaning of the formal terminologies applying to describe some real-world processes (or objects) is *ontology* [4].

How can we apply ontology in service descriptions? The first language which was used to semantically descriptions of the services is RDF (Resource Description Framework). Extension of the RDF language is OWL (Web Ontology Language) which is more expressive than RDF. In the work other method to define service with semantic information which is called WSMO (Web Service Modeling Ontology) is applied.

The concept of WSMO to describe services is continually developing. The ultimate set of feature will contain tools for discovery, selection, composition, mediation, execution and monitoring. Some of intended aims were reached but some of them such as composition had not be finished yet [4]. Despite the fact, that WSMO is a tool which is still being developed, it can be used to solve some problems connected with ontological service description and service discovery. The second task is considered in next section in details.

The main four top-levels elements of WSMO language are:

- *Ontologies* are used to provide the terminology utilized by each element of service to describe the relevant aspects of the certain domains;
- *Goals* state the intentions that should be solved by services and define objectives which have to be fulfilled in order to execute service;
- *Services* is used to achieve defined goal. Descriptions can be related to various aspects of a service and consist of functional, non-functional and the behavioral aspects of a service;
- *Mediators* resolve interoperability problems and describe elements to overcome incompatibility problems between different elements on data, process and protocol level.

In our point of view, the most interesting aspect of WSMO language are Goals. Goals is used to describe user's desires connected with requested functionality of the service. Because these functionalities allow us to specify properties of desired services, thus this tool is very useful in the discovery process. Considered process can be described as follows: in the service repository a set of accessible services are stored. Some of services, which are in repository, may be used to fulfill user's requirements. The problem is to find subset of accessible services, which are helpful to provide user's requirements. In order to solve this task, we need the tool to semantic description of service and another, to search repository for proper services.

To describe service by making the most of WSMO an ontology or set of ontologies are needed using to provide necessary common terminology for used concepts and relations between concepts.

The last element of WSMO language which is interested in our point of view is Service. It is useful tool to describe functional and nonfunctional properties of the service. This description is applied to represent service and can be utilized in process of discovery i.e. service's specification is matchmaking with user's requirements defined by Goals.

In the presented work, the problem of building the interface for service which is provided by Mediators, is not considered.

Now we are going to present main elements which constitute ontology by use of MOF notation (Meta Object Facility) [4]. The class of ontology has following elements: *nonFunctionalProperty, ooMediator, concept, relation, function, instance* and *axiom.*

It is worth stressing that all attributes are optional. In the listing shown below a fragment (header of *wsml* file and concept of *Student*) of proposed ontology for discussed Educational System is shown:

```
wsmlVariant _"http://www.wsmo.org/wsml/wsml-syntax/wsml-rule"
namespace { _"http://www.sti-innsbruck.at/ontologies#",
        dc _"http://purl.org/dc/elements/1.1/",
        xsd _"http://www.w3.org/2001/XMLSchema#",
        foaf _"http://xmlns.com/foaf/0.1/",
        wsml _"http://www.wsmo.org/wsml/wsml?syntax#" }
ontology EnrolmentFoTheCoursesOntology
concept Student subConceptOf Person
    nonFunctionalProperties
        dc#description hasValue "Student's Concept"
    endNonFunctionalProperties
            hasID ofType (1 1) _integer
            hasActivityStatus ofType (1 1) _integer
            hasRecordStatus ofType (1 1) _integer
            hasYearOfStudy ofType (1 1) _integer
            hasSemesterNumber ofType (1 1) _integer
            hasRegistrationStatus ofType (1 1) _boolean
            hasEnrolmentStatus ofType (1 1) _boolean
```

List. 1. Fragment of proposed ontology in WSMO

To describe functional properties of service in WSMO language the main functionality is Capability. Defining capabilities of the service is, in our point of view, crucial part of the process of designing semantic description of the services because it is used to discover services in repository.

Service definition in WSMO contains following elements:

- *precondition* – definition of input parameters;
- *postcondition* – definition of output parameters;
- *assumption* – definition of real world conditions required for the service to run correctly;
- *effect* – definition of the change in real world after service invocation.

As it can be seen, service definition in WSMO allows us to describe the *state of the World* before service's execution (*assumption*) and when it is delivered successfully (*effect*). *Preconditions* and *Postconditions* are used to specify required input and output data which are required by service. It is worth stressing that these elements of service's definition are crucial for discovery process applied in WSMO.

Taking into account structure of service in WSMO and presented in previous section Educational System service, following definition is proposed (see Fig. 2):

```
wsmlVariant _"http://www.wsmo.org/wsml/wsml-syntax/wsml-dl"
namespace { _"http://www.sti-innsbruck.at/services#",
                educl _"http://www.sti-innsbruck.at/ontologies#",
                discovery _"http://wiki.wsmx.org/index.php?title=DiscoveryOntology#"}
webService SelectASubsetOfCoursesHavingTheSmallestParticipantsLimitsService
importsOntology {educl#EnrolmentForTheCoursesOntology}
capability SelectASubsetOfCoursesHavingTheSmallestParticipantsLimitsServiceCapability
nonFunctionalProperties
                discovery#discoveryStrategy hasValue discovery#LightweightDLDiscovery
                discovery#discoveryStrategy hasValue discovery#NoPreFilter
endNonFunctionalProperties
sharedVariables        ?y1
precondition SelectASubsetOfCoursesHavingTheSmallestParticipantsLimitsServicePre
        definedBy
                ?x1 memberOf _"http://www.examples.org/ontologies/example#Student" and
                ?x2 memberOf _"http://www.examples.org/ontologies/example#Course" and
                ?x3 memberOf _"http://www.examples.org/ontologies/example#StudentEnrolmentVector".
postcondition SelectASubsetOfCoursesHavingTheSmallestParticipantsLimitsServicePost
        definedBy
                ?y1 memberOf _"http://www.examples.org/ontologies/example#StudentEnrolmentVector".
```

List. 2. Fragment of proposed service in WSMO

For illustrative purpose one of the proposed service (called "*Select A Subset Of Courses Having The Smallest Participants Limits*") for Educational System is shown in the framework of WSMO. Some elements of the considered service, such as precondition and postcondition, are not being seen in BizAgi diagrams (see for example Fig. 2) but they are specified by authors and are accessible under BizAgi program. Obviously, they are used in considered definition of the service in pre- and postcondition section.

The last part of top-levels elements which is interesting in point of view of discovery process is Goals. As it was written, they provide the means to specify requester objectives. It is worth stressing that WSMO allows to specify objectives that would potentially satisfy user's desires which is helpful in discovery process [4].

As it was stressed, at the beginning of this section, one of the main feature of the system such as system supported educational process is ability to find service aggregated in repository. To describe such services and goals, which are used to define service capabilities and user's requirements respectively, WSMO language may be used.

5 Service Discovery

Let us assume that in Educational System service repository a set of services are stored (in Fig. 1 and 2 some services for considered problem are shown). By discovery problem, we mean the problem of finding services that can comply with user's required. Motivation for service discovery process, in considered system, is the problem of executing all possible user's tasks which can be called in proposed system.

It is clear that it is not possible to specify all possible service which satisfies user's requirements exactly. Therefore, we need the tool which helping us to search repository and find services with different degree of relevance i.e. different types of logical relationship between semantic description of the services and user's goals. It helps us to propose such service for user in correct order.

One of the problem which has to be used in discovery task is the specification of user's requirements and, describe in previous section, service capabilities. Another issues, which has to be solved, is related to service matching algorithms with methods determining the degree of relevance of found services.

As it was stated earlier, WSMO supports the discovery process of services with semantic description. The general idea of services discovery is based on comparison of semantic description of the service's capabilities (*pre- and postconditions*) with description what the user wants to achieve (see Fig. 3).

As a result of using WSMO framework for service discovery, a set of candidate services which are met with user's demands are identified. It is worth stressing that services which are identified as useful, in user's point of view, providing with a relevance degree of semantic matchmaking. In WSMO there is a five matchmaking notions [7]:

– *Exact match* – holds if and only if all each possible ontology instance satisfies both service and user's goals and there is no ontology instance which satisfies only goal or service;
– *Plugin match* – hold if ontology instance is superset of relevant user's goal;
– *Subsumption match* – hold if ontology instance is subset of relevant user's goal;
– *Intersection match* – holds if one possible ontology instance which satisfies both service and goals;
– *No match* – there is any possible ontology instance which satisfies user's goals.

In List. 3 an example specification of user's goals ("*Select A Subset Of Courses Having The Smallest Participants Limits*") is presented. Taking into account presented definition of user's requirement and methods, which are used in WSMO, it is possible to discover and determine the degree of relevance of services from repository. As it was stated earlier to determine degree of relevance of service and user's goals five different notions are used. When discovery process is finished it is possible to consider which of accessible services are able to solve user problems defined as goals.

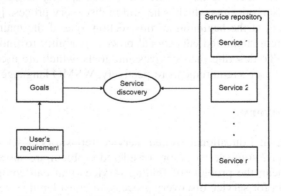

Fig. 3. General idea of service discovery in WSMO

```
wsmlVariant _"http://www.wsmo.org/wsml/wsml-syntax/wsml-dl"
namespace { _"http://www.sti-innsbruck.at/goals#",
                   educl _"http://www.sti-innsbruck.at/ontologies#",
                   discovery _"http://wiki.wsmx.org/index.php?title=DiscoveryOntology#"}
goal SelectASubsetOfCoursesHavingTheSmallestParticipantsLimitsGoal
importsOntology {educl#EnrolmentForTheCoursesOntology}
capability SelectASubsetOfCoursesHavingTheSmallestParticipantsLimitsGoalCapability
nonFunctionalProperties
          discovery#discoveryStrategy hasValue discovery#LightweightDLDiscovery
          discovery#discoveryStrategy hasValue discovery#NoPreFilter
endNonFunctionalProperties
sharedVariables       ?y1
precondition SelectASubsetOfCoursesHavingTheSmallestParticipantsLimitsGrupGoalPre
       definedBy
       ?x1 memberOf _"http://www.examples.org/ontologies/example#Student" and
       ?x2 memberOf _"http://www.examples.org/ontologies/example#Course" and
       ?x3 memberOf _ http://www.examples.org/ontologies/example#StudentEnrolmentVector".
postcondition SelectASubsetOfCoursesHavingTheSmallestParticipantsLimitsGoalPost
    definedBy
       ?y1 memberOf _"http://www.examples.org/ontologies/example#StudentEnrolmentVector".
```

List. 3. Fragment of proposed goals in WSMO

Solved problem of matching services from repository, which are met user's requirements, can be in different task connected with Educational Systems such service recommendation or service composition. These problems will be considered in future
works of the authors.

5 Summary

In this work we present the process of the selected educational services modeling
using BizAgi environment as well as application of WSMO and WSM toolkit for their
representation and the process of discovery based on the ontology. It was shown that
the selected tools are very useful for solving essential problems in the ever growing
SOA systems. The solution of the problem of automatic service discovery may be
used in the several different ways, first it may be used as a direct solution of the problem of retrieval of the relevant services, but also second, it may be used as an element
of the process of the service composition when we have to find and select all elementary services that may be composed in a larger aggregate.

Acknowledgement

The research presented in this paper has been partially supported by the European
Union within the European Regional Development Fund program no. POIG.01.03.01-
00-008/08.

References

[1] Barker, R., Longman, C.: Case Method SM-modelling of functions and processes. Wyd. WN-T, Warszawa (2001)

[2] Business Motivation Model (BMM) Specification, dtc/07-08-03, September 2007, Object Management Group (2007), http://www.omg.org/docs/dtc/07-08-03.pdf (29.03.2009)

[3] Business Process Modeling Notation, V1.1 OMG Available Specification OMG Document, http://www.bpmn.org/Documents/BPMN1-1Specification.pdf (25.03.2009)

[4] Fensel, D., et al.: Enabling Semantic Web Services. The Web Service Modeling Ontology. Springer, Berlin (2007)

[5] From Grids to Service-Oriented Knowledge Utilities. A critical infrastructure for business and the citizen in the knowledge society, European Commission brochure, Luxembourg: Office for Official Publications of the European Communities (2006)

[6] Hurtwitz, J., Bloor, R., Baroudi, C.: Service Oriented Architecture FOR DUMmIES, Indiana. Wiley, Chichester (2007)

[7] Keller, U., et al.: WSMO Discovery. Working draft 8.10. 2004

[8] Roy, B.: Multicriteria decission support, Wydawnictwo Naukowo-Techniczne, Warszawa (1990) (in Polish)

[9] Świtalski, Z.: Soft preferencje models and their application In the economy, Wydawnictwo Akademii Ekonomicznej w Poznaniu, Poznań (2002) (in Polish)

[10] UML 2.0 Superstructure Specification. OMG Adopted Specification ptc/03-08-02, http://www.omg.org/ (15.11.2004)

[11] Wroclaw University of Technology Mission, http://www.pwr.wroc.pl/24881.xml

An Information Theoretic Web Site Navigability Classification

Cheng-Tzu Wang[1], Chih-Chung Lo[2], Chia-Hsien Tseng[1], and Jong-Ming Chang[3]

[1] Department of Computer Science
National Taipei University of Education, Taipei, Taiwan, R. O. C.
[2] Department of Informatics
Fo-Gung University, Yilan, Taiwan, R. O. C.
[3] Yilan County Government, Yilan, Taiwan, R. O. C.
ctwang@tea.ntue.edu.tw, locc@mail.fgu.edu.tw,
g10963201@grad.ntue.edu.tw, postit@mail.e-land.gov.tw

Abstract. Usability is critical to the success of a web sit and good navigability enhances the usability. Hence the navigability is the most important issue in web sit design. Many navigability measures have been proposed with different aspects. Applying information theory, we propose a simple Markov model to represent the structure of a web site and use the users' log data to classify types of web pages in the model. Based on the web page classification, page navigability can be improved. The experimental results show that our model can provide effective measure and right classification.

Keywords: Navigability, web metric, classification, Markov model, information theory.

1 Introduction

Recently, with the growing usage of Internet, web sites play the important role of information providers to replace traditional ones. If people can not navigate through a web site effectively, they will leave quickly. Thus, effective web navigation is critical to the usability of a web site. Good usability assures the success of a web site. It is more important to make sure a web site is reaching its desired users and meets their expectations. The users of a web site are heterogeneous and geographically dispersed. To answer these questions, we may get knowledge from the content, usage, and structure of a web site [1]. Hence a web metric to understand, manage, and then improve web systems is critical to the quality and the success of web sites.

In the last decade, many web site metrics have been proposed to measure navigability with different ways. With user surveys, a web's navigability is evaluated by a set of criteria from a sample of the web site's users. There are many problems in this method, such as time consuming, geographical space, and representative of the sample [2]. Another is the way with usage analysis in which the usage data such as log data are investigated. Theoretically, the user behavior can be traced by examining the usage data. In fact, it is difficult to have complete trace of user behavior [3][4]. The

N.T. Nguyen, M.T. Le, and J. Świątek (Eds.): ACIIDS 2010, Part II, LNAI 5991, pp. 39–46, 2010.

structures of web sites are often used to measure the web navigability [5][6][7][8][9]. It is not enough to characterize the web navigability by examining the hyperlink structure of a web site statically [10].

In this paper, we propose an information theoretic model to measure a web site's navigability. In the proposed model, the static structure of a web site is represented as a first order of Markov model, and the dynamic user surfing behaviors are considered as the events or the emitted symbols in this model. Based on information theory, navigability of each page in a web site can be evaluated and classified into different categories in which structures of pages may be improved to enhance the navigability of pages.

The rest of this paper is organized as follows. In Sections 2, we summarize the related work for this study. We propose the Markov model for web navigability and discuss the web page classification based on the simulation on the proposed model in Section 3. Section 4 presents the experimental validation and analysis of the data set. In Section 5, conclusions and suggestions are presented for further studies.

2 Related Work

In this section, some related works are reviewed briefly. First, we give an overview of study in web site navigability. In this paper, we propose an information theory based measure, the related work in information theory is also introduced.

2.1 Web Site Navigability

Recently, many navigability measures based on the structures of web sites have been proposed. In these measures, a web sit is represented by a hypertext graph in which nodes represent web pages and directed edges represent hyperlinks. Botafogo et al. [5] proposed two measures Compactness C_p and stratum S_t to measure the navigability of a web site. In [5], the navigability of a web site was considered only based on the structure of the site. Yamada et al. [7] proposed three measures including aspects both of structure of a web site and its users' perspective. Zhang et al. [8] pointed out a web navigability is not proportional to the number of hyperlinks in the web site. They defined five structural measures of a web site's navigability. In [9], a visual tool based on structure of a web site has been proposed to measure the site. In another view, a user surfs from one web page to another can be considered as a dynamic model. This dynamic model can be described as a Markov source model. The stationary distribution can be used to infer the rank of web pages [10][11][12]. Dhawan et al. [13] proposed an approach for determining the reliability and maintainability of web-based systems.

Many empirical studies have been done to investigate the relation between a web site's structure and its usability. In these studies, two important results were found, the structure of a web site with moderate level of breadth and depth has better navigability and too few or too many hyperlinks cause lower navigability [14][15].

2.2 Information Sources

The first work in information theory that deals with symbols carrying information was done by Shannon [16]. Let S be a finite source emitting a sequence of symbols over the set of alphabet, $\{s_1, s_2, ..., s_n\}$. Assume that the successive symbols emitted from source S are independent and the probability of symbol s_i emitted is $p(s_i)$, where $0 \leq p(s_i) \leq 1$ and the summation of all $p(s_i)$ is 1. The information content of symbol s_i, $I(s_i)$, is the amount of information obtained if s_i occurs, and is defined as in (1). The entropy, $H(S)$, of the information source S is the average amount of information per symbol and is defined as Equ. (2).

$$I(s_i) = \log_2 \frac{1}{p(s_i)}.\tag{1}$$

$$H(S) = \sum_{i=1}^{n} p(s_i) \log_2 \frac{1}{p(s_i)}.\tag{2}$$

An mth-order *Markov source*, in which the probability of each symbol of the stream is dependent on the m preceding symbols, is distinguished from a *memoryless information source*, in which the probabilities of symbols in the stream are independent of previous symbols [17]. The information obtained, information content, if s_i occurs with m preceding symbols is defined as in (3). In (4), $H(S|s_{j1}, s_{j2}, ..., s_{jm})$ is the average amount information per symbol with the m preceding symbols, $s_{j1}, s_{j2}, ...,$ and s_{jm}. Thus the entropy of the m-th Markov source S is defined as Equ. (5).

$$I(s_i \mid s_{j_1}, s_{j_2}, ..., s_{j_m}) = \log_2 \frac{1}{p(s_i \mid s_{j_1}, s_{j_2}, ..., s_{j_m})}.\tag{3}$$

$$H(S \mid s_{j_1}, s_{j_2}, ..., s_{j_m}) = \sum_{S} p(s_i \mid s_{j_1}, s_{j_2}, ..., s_{j_m}) I(s_i \mid s_{j_1}, s_{j_2}, ..., s_{j_m}).\tag{4}$$

$$H(S) = \sum_{S^m} p(s_{j_1}, s_{j_2}, ..., s_{j_m}) H(S \mid s_{j_1}, s_{j_2}, ..., s_{j_m}).\tag{5}$$

3 Proposed Model for Navigability

A web site is comprised of many web pages in which the homepage is the first page. The structure of a web site is usually represented by a directed graph called web graph. As shown in Fig. 1, each node is a page in the web site, the node 1 represented by a rectangle is the homepage, and the directed edge from node x to y represents a hyperlink from page x to page y in a web site. Except the homepage node, each node in the web graph has at least 1 predecessor.

Applying information theory to web site navigability, user's navigation can be considered as a stream of symbols emitted by a first order Markov source to form a message. Suppose a web site has n pages labeled by 1 to n. For a user's navigation, this Markov source emits a sequence of symbols from a fixed finite source alphabet $S = \{1, 2, ..., n\}$, which represents the possible page labels. In the first order Markov model, the memory represents all possible predecessors and the event represents all nodes in a web site. The probability of event j occurs at memory i is denoted as $p(j|i)$. The user's behavior during navigation of a web site can be viewed as a sequence of events or symbols emitted by a Markov source corresponding to the web site.

3.1 Markov Model for Web Navigability

In order to represent the structure of a web site and the behavior of the web users, we use a tabular Markov model. First, we assume that user's navigation must begin at the homepage, and the user's surfing behavior only depends on the structure of hyperlinks in a web site. And we add an extra event, "exit", to represent the virtual terminating action. Table 1 shows a web structure corresponding to a web site as in Fig. 1. Each entry in Table 1 with non-zero conditional probability, $p(j|i)$, represents a hyperlink from page i to page j, and each entry with conditional probability 0 represents no hyperlink from page i to page j. In the Markov model as shown in Table 1, the entropy for each page can be computed by using Equ. (4).

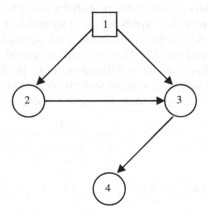

Fig. 1. An example of a web site

Table 1. The Markov model for Fig. 1

Memory	Next event				
	1	2	3	4	Exit
Empty	$p(1|e)$	0	0	0	0
1	0	$p(2|1)$	$p(3|1)$	0	$p(e|1)$
2	0	0	$p(3|2)$	0	$p(e|2)$
3	0	0	0	$p(4|3)$	$p(e|3)$
4	0	0	0	0	1

The amount of page's entropy represents average information included in the page. The higher the information means the hyperlinks in the page are navigated more average. The potential maximum entropy of a page s_i, $MaxH(S|s_i)$, happens when the page has equal conditional probability of each possible hyperlink, and can be calculated by the Equ. (6), where r is the number of possible hyperlinks in the page.

Then we define an entropy ratio of a page s_i, $R(s_i)$, by comparing the page's entropy with the page's potential maximum entropy, as shown in (7). The $R(s_i)$ can be used to represent the navigability of a page in a web site.

$$MaxH(S \mid s_i) = -\sum_{i=1}^{r} \frac{1}{r} \log_2 \frac{1}{r} = \log_2 r \ . \tag{6}$$

$$R(s_i) = \frac{H(S \mid s_i)}{MaxH(S \mid s_i)} \ . \tag{7}$$

3.2 Web Page Classification

In the previous section, an entropy ratio of a page s_i, $R(s_i)$, is defined to evaluate the page's navigability. In order to understand the relation between characteristics and entropy ratios of web pages, we classify web pages into 3 different types of pages in which type 1 represents hyperlinks in this type of pages are navigated fairly, type 2 represents hyperlinks in this type of pages are navigated unfairly but every one is navigated, and type3 represents at least one hyperlink in this type of pages is not navigated. The type 1 pages are well designed for their users. The type 3 pages can be improved at least by eliminating the hyperlinks that are not surfed. In type 2 pages, we may re-organize the structures of hyperlinks or eliminate the less used hyperlinks to enhance the navigability.

In order to classify different types of pages, we constructed an experiment in which 30000 times of page navigation were simulated randomly based on the different types of web site structures in [8]. After analyzing the results and tuning the threshold, we find the thresholds, 0.3 and 0.65, for classifying different types of pages have the best result. Pages with $R(s_i) \geq 0.65$ are classified to type 1, pages with $R(s_i) < 0.3$ are classified to type 3, and pages with $0.3 \leq R(s_i) < 0.65$ are classified to type 2.

As shown in Fig. 2, 8550 times of simulated page navigation belong to type 1, in which more than 96% of pages that with $R(s_i) \geq 0.65$. In Fig. 3, there are 11450 times of type 2 page navigation, in which about 96% of pages that with $0.3 \leq R(s_i) < 0.65$, and in Fig.4, there are 10000 times of type 3 simulated page navigation, in which about 87% of pages that with $R(s_i) < 0.3$.

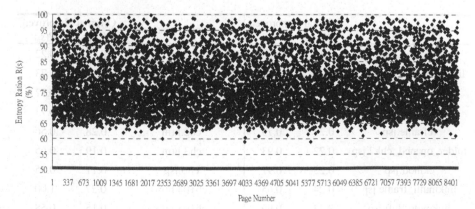

Fig. 2. Distribution of entropy ratio for simulated type 1 pages

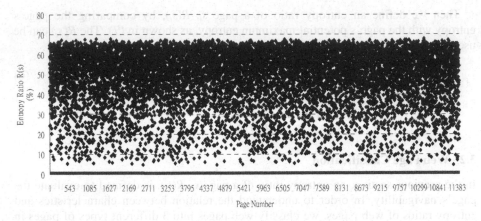

Fig. 3. Distribution of entropy ratio for simulated type 2 pages

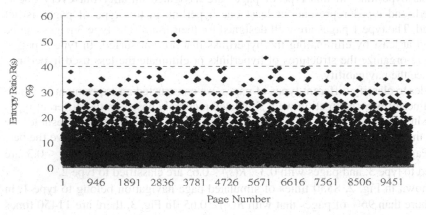

Fig. 4. Distribution of entropy ratio for simulated type 3 pages

Table 2. Original $R(s_i)$ vs. improved $R(s_i)$ for test web site

Page name	Original $R(s_i)$	Improved $R(s_i)$	Page name	Original $R(s_i)$	Improved $R(s_i)$
makepag2.jsp	0.35	0.60	linkmulti_item_new.jsp	0.13	0.22
linkmulti_add_item.jsp	0.36	0.78	file_selecter_frame.jsp	0.17	0.73
login_redir.jsp	0.15	0.63	file_selecter.jsp	0.02	0.77
login_chk.jsp	0.23	0.92	upload_media.jsp	0.28	0.74
reframe.jsp	0.27	0.73	ch2.htm	0.28	0.56
folder_opened_chk2.jsp	0.22	0.93	ch3.htm	0.19	0.65
ckmod.jsp	0.22	0.97	ch6.jsp	0.17	0.65
linkmulti_frame.jsp	0.24	0.65	ch104.htm	0.27	1.00
linkmulti_left.jsp	0.23	0.99	ch105.htm	0.15	0.58

4 Validation of Proposed Model

In this section, we present the empirical validation of our proposed model for navigability and classification. In this experiment, we applied the proposed model and thresholds to a university web site (it has been updated) [18] and collected users' log data of the web site for six months. This web site has 44 pages in which the maximum number of hyperlinks is 23, the minimum number of hyperlinks is 4, and the average number of hyperlinks is about 14. We used the analysis tool, Web Log Explorer, to trace users' navigation paths in which 35272 paths were found. Then those users' navigation paths were fed into our proposed model to evaluate the page navigability and to classify the types of pages.

The experimental result shows that 5 pages are classified to type 1, 9 pages are classified to type 2, and 30 pages are classified to type 3. By examining the 30 type 3 pages and the 5 type 1 pages, all the type 3 pages have at least one hyperlink not navigated and all the type1 pages are navigated fairly. This result shows both the accuracy of type3 and type 1 classification are 100%. In the 9 type 2 classified pages, twoe pages, "linkmulti_add_item.jsp" and "makepage2.jsp," have at least one hyperlink that is not navigated, and are classified incorrectly. The accuracy of type 2 classification is about 78% and the overall accuracy is above 95%.

After classifying pages, we can improve the structures of pages based on the types of pages. For example, there is a type 2 page, "linkmulti_add_item.jsp," with $R(s_i)$=0.36, in which 13 hyperlinks are not navigated. If we eliminate the 13 non-navigated hyperlinks, the $R(s_i)$ of this page is increasing to 0.78 and this page belongs to type 1. The comparison of original $R(s_i)$ and improved $R(s_i)$ of pages in the test web sit is shown in Table 2. The improved pages have much better navigability than that of original pages.

5 Conclusion and Future Work

In this paper, we propose an information based model to represent structures of web sites and to measure web page navigability. We also use the dynamic user behavior to classify types of pages. Based on the classification, navigability of web pages can be improved for users' expectation. Experiments are conducted to validate our model and results show the validity of the model.

In order to enhance the validity of ore model, more real web sites can be applied to our model in the future. The proposed model can be modified to reflect all the users' surfing behavior in real world. We also can extend the order of Markov source to investigate the relationship of user's behavior and to improve the accuracy of our model.

Acknowledgments. This work is partially supported by the National Science Council, Taiwan, R. O. C. under grant NSC 98-2221-E-431-002.

References

1. Lappas, G.: An Overview of Web Mining in Societal Benefit Areas. In: Proc. 9th IEEE International Conference on E-Commerce Technology, pp. 683–690 (2007)
2. McGovern, G.: Part 1: Navigation, Design, and Principles. Web Content Style Guide. Prentice Hall, Englewood Cliffs (2001)

3. Leven, M., Loizou, G.: Computing the Entropy of User Navigation in the Web. International J. Information Technology and Decision Making 2(3), 459–479 (2003)
4. Borges, J., Leven, M.: An Average Linear Time Algorithm foe Web Usage Mining. International J. Information Technology and Decision Making 3(2), 307–319 (2004)
5. Botafogo, R., Rivlin, E., Shneiderman, B.: Structural Analysis of Hypertexts: Identifying Hierarchies and Useful Metrics. ACM Trans. Information Systems 10(2), 142–180 (1992)
6. De Bra, P., Houben, G.J.: Hypertext Metrics Revisited: Navigational Metrics for Static and Adaptive Link Structures (2005),
   http://wwwis.win.tue.nl/~houben/pub/measures.ps.Z
7. Yamada, S., Hong, J., Sugita, S.: Development and Evaluation of Hypermedia for Museum Education: Validation of Metrics. ACM Trans. Computer-Human Interaction 2(4), 284–307 (1995)
8. Zhang, Y., Greenwood, S.: Website Complexity Metrics for Measuring Navigability. In: Proc. Fourth Int'l Conf. Quality Software, pp. 172–179 (2004)
9. Pascual-Cid, V.: An Information Visualisation System for the Understanding of Web Data. In: Proc. IEEE Symposium on Visual Analytics Science and Technology, pp. 183–184 (2008)
10. Zhou, Y., Leung, H.: MNav: A Markov Model-Based Web Site Navigability Measure. IEEE Trans. Software Engineering 33(12), 869–890 (2007)
11. Haveliwala, T.: Topic-Sensitive Pagerank: A Context-Sensitive Ranking Algorithm for Web Search. IEEE Trans. Knowledge and Data Eng. 15(4), 784–796 (2003)
12. Diligenti, M., Gori, M., Maggini, M.: A Unified Probabilistic Framework for Web Page Scoring Systems. IEEE Trans. Knowledge and Data Eng. 16(1), 4–16 (2004)
13. Dhawan, S., Kumar, R.: Analyzing Performance of Web-based Metrics for Evaluating Reliability and Maintainability of Hypermedia Applications. In: Proc. Third International Conference on Broadband Communications, Information Technology & Biomedical Applications, pp. 376–383 (2008)
14. Larson, K., Czerwinski, M.: Web Page Design: Implications of Memory, Structure and Scent for Information Retrieval. In: Proc. 16th ACM Conf. Human Factors in Computing Systems, pp. 18–23 (1998)
15. McDonald, S., Stevenson, R.: Effects of Text Structure and Prior Knowledge of the Learner on Navigation in Hypertext. Human Factors 40(1), 18–27 (1998)
16. Shannon, C.E.: A Mathematical Theory of Communication. Bell Systems Technology J. 27, 379–423, 623–656 (1948)
17. Abramson, N.: Information Theory and Coding. McGraw-Hill, New York (1963)
18. National Taipei University of Education, http://www.ntue.edu.tw

MACRO-SYS: An Interactive Macroeconomics Simulator for Advanced Learning*

César Andrés, Mercedes G. Merayo, and Yaofeng Zhang

Departamento Sistemas Informáticos y Computación
Universidad Complutense de Madrid, E28040 Madrid, Spain
{c.andres,mgmerayo,yaofeng}@fdi.ucm.es

Abstract. In this paper it is presented the features and behavior of the tutoring-training system MACRO-SYS. This system allows students to simulate experiments with complex macroeconomic environments. Users have to show their knowledge level by solving the proposed exercises. These exercises represent different economist behaviours. A big advantage of our system is that it allows students to be part of the simulation, interacting and modifying the behavior (parameters), both in the short and long-term, of a real scale economy. If MACRO-SYS detects that the simulated values are strongly deviating from the expected pattern, it will provide *hints* to the student so that she can change some parameters and bring the economy to the correct (according to the assignment) behavior. Finally, let us remark that, in contrast with most economic models, our system takes into account a huge amount of parameters in order to compute the current *state* of the economy.

1 Introduction

Macroeconomists try to forecast economic conditions to help consumers, factories and governments to make better decisions. In order to reach this goal, experts usually develop models that try to explain the relationship among different factors. A model in macroeconomics describes a logical, mathematical, and/or computational framework designed to manage the operation of an economy, and especially the dynamics of aggregate quantities, as for example the total amount of goods and produced services, total earned income, the level of employment, the level of productive resources, and the level of prices. Thus, models are used to clarify and illustrate basic theoretical principles in macroeconomics theory, to test, to compare and to quantify different theories, and to produce *what if scenarios* for generating economic forecasts.

Traditionally, the empirical study of new economic systems has been confronted with the difficulty of managing, with enough precision, environments containing a large number of relevant parameters. Thus, the definition of new

---

* Research partially supported by the Spanish MEC projects WEST/FAST (TIN2006-15578-C02-01) and TESIS (TIN2009-14312-C02-01), and UCM-BSCH programme to fund research groups (GR58/08 - group number 910606).

N.T. Nguyen, M.T. Le, and J. Świątek (Eds.): ACIIDS 2010, Part II, LNAI 5991, pp. 47–56, 2010.

models has been sometimes lessened because it is difficult to verify their validity. The main problem consists, on the one hand, in the impossibility of manipulating the real environment and, on the other hand, in the disability to generate realistic artificial environments. The students of Economics are usually confronted with exercises and assignments where they have to show their command, on the theoretical models which they have learned. A typical advanced exercise in microeconomics consists in calculating a general equilibrium for an economical, context having a producer, a consumer, and two goods. If we consider the macroeconomic side, students may expect to compute the GDP (Gross Domestic Product) according to some given data, or to discuss whether a given differential equation (with at most half a dozen parameters) adequately reflects reality. Obviously, the gap between these assignments, and the *real world* is huge. Nevertheless, the combination of *powerful computers* and *complex simulators* has favored both areas. It has had an important impact in the case of simulating the behavior of systems that cannot be manipulated, as the economic behavior of a society. Thus, the validity of a new theory can be contrasted by simulating the environment and by comparing the obtained and expected results. Afterwards, by using different test values, one can estimate the behavior of the real system in conditions that could never be applied to the real environment. Examples of such systems can be found in [7,4].

Simulators have been favorable in economic teaching in recent years. Usually an economic simulator incorporates a dynamic model that enables to experiment with strategies in a risk free environment and provides a useful extension to case study discussions. In fact, students may profit from the ability to manipulate environments that without these programs could not been studied. Besides, they can observe the relation among *low level* factors (that is, the value of the corresponding parameters) and high level factors (that is, the global behavior of the system). Thus, by using these simulators, students can have an overall vision of the studied system. In addition, they can even define new models and contrast their utility. A good candidate to be *simulated* is the economic behavior of a society. In fact, there are already good simulators and algorithms taking into account partial views of the economy (e.g. [1] at the microeconomic level). Thus, the use of computer simulation appears to be a promising method for learning macroeconomics. In addition to help the students to execute an advanced training relating to their specific area, the use of computers give the students more flexibility in learning, allowing it occurs in different places but also at different times. *E-Learning* technologies provide us all these benefits.

E-Learning technologies have evolved since computers were first used in education. There are several accepted definitions of E-Learning. From our point of view, a complete definition of E-Learning is "any form of distance learning that utilizes a net for delivery, interaction or facilitation. This net does not be only Internet. The learning takes an important place as part of a class. We refer as online courses as classes that are both synchronously (at the same time) and asynchronously (at different times), or some combination of the two."

Regarding, we present a new advanced tutoring system, called MACRO-SYS (acronym for MACROeconomic SYStem), for helping in the asynchronously E-Learning [8]. The students, whenever and wherever they were, can use the tool. In a traditional class, students can present questions or comments to the teacher, interact in verbal discussions, or collaborate in small groups. Similar forms of interaction and collaboration exist in online courses, with a few substantial differences. The first major difference is that written communication is much more prevalent in online courses. Regarding to this aspect, MACRO-SYS is proposed as a way for letting teachers and students to interact, to receive and to propose exercises, to review statistical information, etc. The next major difference is that the technologies can be asynchronous. This means that teachers and students will make their comments as little as a few minutes, and as much as a few weeks, apart from one another. Entirely new skills need to be developed by both teachers and students to make this asynchronous collaboration effective. Everyone in the discussion needs to follow new rules of etiquette, and each interactive tool is slightly different. If done well, asynchronous class discussions can be as effective as traditional class discussions. MACRO-SYS makes it possible recording all logs in the database and allowing the student to send the teacher a question or comment about the obtained result from a simulation.

MACRO-SYS allows users to interact with computer simulation models, that is, users are not limited to observe the behavior of the economy. Thus, adequate interfaces are provided where the user can take part in the evolution of the simulation. In addition, several *views* of the economic behavior are provided, by means of graphical representations, according to the different parameters to be considered (e.g. prices, employment levels, stocks, etc).

Let us remark two main characteristics of our system. First, our design methodology can be described as a *symbiosis* between the implementation of the simulator and the learning of advanced macroeconomic concepts. Initially, we identified the set of basic operators (at the human level) producing the economic process (i.e. purchasing, selling, work, demand of goods, etc). Next, we identified the main results giving raise to the correct interrelation among these concepts (i.e. business cycles and their control, state regulations by means of taxes and unemployment benefits, monopolies and oligopolies, etc). Afterwards, we looked for economic explanations to the basic forces producing the high level behavior of the system (i.e. if there is an excess of supply then the entrepreneur will decrease the prices). We generated preliminary models and studied their validity, so that we could incrementally improve the models until they reached the desired behavior.

It is worth to point out that the economic models that we consider are much more complex than most macroeconomics models in use. So, our system is able to explain how the economy of a society works in a more adequate way than models found in the literature, which usually are focused on one particular characteristic.

The rest of the paper is organized as follows. In Section 2 the objectives of MACRO-SYS are presented. In Section 3 we give some important details concerning the implementation of the system. We describe the main features of MACRO-SYS and the algorithm underlying its behavior. In addition, we introduce the main technologies applied in the design and the development of the tool. Finally, in Section 4 we present our conclusions.

2   Objectives of MACRO-SYS

The main aim of MACRO-SYS consists in providing students with an *easy-to-use* tool where they can practice the knowledge previously gained in the classroom. So, students will be able to check whether they have fully assimilated the concepts they were supposed to. The main objective pursued with the design and development of MACRO-SYS was to help students having another source of learning (i.e. a teacher, or even a good book).

In order to ensure a personalized treatment, students login into the system by giving their ID-number and password. This allows the system to recover the data from previous sessions. This mechanism tries to avoid *attacks* to previous sessions of students. For example, an attacker could ask the system for either previous or partially solved assignments and provide wrong answers. Then, when the *real* student logs in, she will find out that his economies are in a different (for worst) point.

Exercises and assignments are classified according to the difficulty level. MACRO-SYS offers four different categories of exercises. All of them are solved by filling successive forms. The exercises may not have a unique correct answer and, depending on the previous answers, the next form will be generated. The first category proposes a collection of exercises that allow the students become accustomed to use the tool. In the second and third categories, the student gets a running economy and is asked to modify some of the parameters under his control to change, in a given way, the behavior of the economy. Finally, in the last category, students are asked not only to modify the values of the parameters but to change the proposed model to vary the behavior of an economy.

3   Design and Development of MACRO-SYS

Next, we introduce the technological components used in the development of the system. Moreover, we explain the main features of our tool, and give a description of the logical aspects of it.

3.1   Technologies

First, we describe the software platform used to implement the structure of MACRO-SYS. The object-oriented paradigm has been used to design the application [5], considering features such as encapsulation, modularity, polymorphism, and inheritance. Among the wide spectrum of object-oriented languages available for the development of complex systems we have chosen a combination of Java and PHP5. The most important reason for selecting Java was that its implementation uses a virtual machine that is intended to run code unchanged on

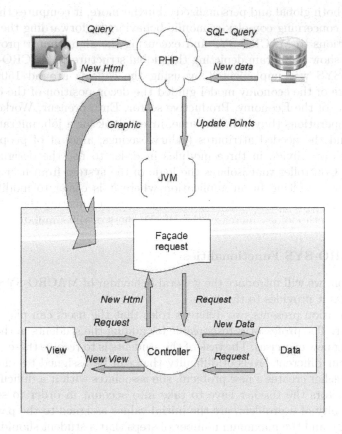

Fig. 1. Global structure of MACRO-SYS

many different platforms. Regarding to PHP, it is a general-purpose scripting language that is especially suited for web development. PHP can be deployed on most web servers, many operating systems and platforms, and can be used with many relational database management systems. The relational database management system MySQL was selected due to the fact that it works on many different system platforms. Thus, all these technologies provides MACRO-SYS with a high level of portability. In addition, we only use free software, that can be used and modified without restriction, and which can be copied and redistributed.

The architecture of MACRO-SYS is split into different components. First, the client-side component provides a Web interface for users. This component consists of HTML pages with embedded Java Applets, that are dynamically generated by the server in response to queries of users. The server-side component selects the data needed to respond to each user and composes the HTML pages before sending them to the client. In addition, it records the answers that the user propose in the different steps of the exercises, provides simple statistics of the system use (both global and personalized). Furthermore, it computes the results of questions concerning complex economic behaviors by forwarding the suitable initial conditions to MACRO-SYS and executing the algorithm it provides. In Figure 1 we show a diagram depicting the global structure of MACRO-SYS.

MACRO-SYS was implemented by using the code is around 15000 lines. The structure of the economy model guided the decomposition of the design in classes (State of the Economy, Productive sector, Entrepreneur, Worked, etc.), the needed operations (buy, sell, consume, invest, look for a job, migrate of sector, etc.), and the needed attributes (salary, savings, amount of people, etc.). These classes are divide in three modules in order to use the design pattern Model-View-Controller that isolates the logic of the system from user interface considerations, resulting in an application where it is easier to modify either the visual appearance or the underlying rules without affecting the other. This development methodology increases the effectiveness of the simulator [2].

3.2   MACRO-SYS Functionalities

In this section, we will introduce the general behaviour of MACRO-SYS and the functionalities it provides to the users.

The application presents two different roles that the users can play: student and professor. The professor is responsible for adding the students in the system and including new exercises. The task of the students is to resolve these exercises. There are four different levels of difficulty that can be assigned to an exercise. When the teacher creates a new problem, she associates with it a difficulty level. The main factors the teacher have to take into account in order to select the level of difficulty of a problem, are the initial values assigned to the parameters (see Figure 2), and the maximum number of steps that a student should perform in order to achieve the objectives. The *basic level* level introduces the student in the use of the system. The second level corresponds to exercises that include *differential equations*. The third level includes *feedback* and allows to change the values of some parameters. Finally, the fourth level allows the student complete control of all the elements in the exercise.

In the same way, when a new student is added to the system, the teacher assigns him a level of knowledge that ranges from 1 to 4. The assigned level avoids the student to access to exercises that he is not ready to resolve. When the student is able to resolve successfully more than 70% of the exercises of her level of knowledge, she is promoted by the system to the next one, according to adaptive E-Learning methodology [6]. Thanks to efficient management of the exercises and their solutions by the system, all students can interact simultaneously with it, that is, the application is multi-user which allows concurrent access by multiple users. A visual representation of the obtained results up to the actual step in the development of the exercise is provided to the student, by means of graphics (see Figure 3).

The system collects information from all sessions and store it in the database. This information corresponds to the number of problems that have been resolved, the number of attempts before obtaining a successful result, the time that the

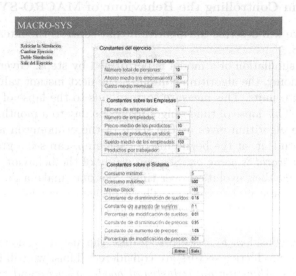

Fig. 2. Parameters in MACRO-SYS

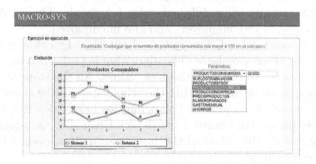

Fig. 3. MACRO-SYS Graphics

students take to resolve them, the date of resolution, etc. All these parameters are used by the system in order to elaborate statistics that can be examined by both students and teachers.

In addition, the system includes a *training mode*. At this stage the students can select the parameters he wants to handle and the maximum number of steps for each simulation. Finally, the students have the chance of performing a dual simulation with different initial values for the same set of parameters. The graphics module draws out the differences between the parameters along the evolution of the problem.

3.3   Algorithm Controlling the Behaviour of MACRO-SYS

In this section we will describe the algorithm that controls the internal behavior of MACRO-SYS.

The running simulation of a model is performed by steps. Given the current state of the economy, the algorithm computes the next instant values. The execution is split into units. This measure corresponds to the lapse of time of each simulation step. This lapse of time ranges from one day to a month. An iterated execution of the algorithm gives us the state of the economy in any arbitrary future time. A student, at the beginning of each step, can see a graphical summary of *how her world is going on*. She can study of the behavior of more than 10 parameters and their evolution over time. She may make a decision to take part in the evolution of the economy. These decisions will be included in the system by means of changes in the current values of certain parameters. Then, these new values will be considered in the next step of the simulation.

MACRO-SYS considers four sectors, unlike the usual economical point of view where three productive sectors are considered. Thus, we will consider four productive sectors: *Alimentation, industry of goods, services*, and *raw materials*. We denote the first three productive sectors by *sectors of goods*. The latter sector must be understood, in a broad sense, as those materials that are needed by companies to perform the productive process. For instance, it includes industries as energy, mining, heavy industry, construction, etc.

MACRO-SYS presents three different roles for people. These roles are: employees (the entrepreneurs), employers and the unemployed.

Next, we will make a brief description about the algorithm governing the economic behavior of our simulator. The algorithm is performed for each step of the simulation in order to compute the new values. All the steps are influenced by previous ones, as all of them modify the whole state of the economy.

The algorithm includes 5 phases. Following, we briefly describe the purpose of each of these phases. In Figure 4 we depict them. Let us remark that the algorithm may modify the values of the main parameters, even if the user did not do it in previous steps.

The first phase of the algorithm is the *manufacturing of products*. The process of manufacturing increases the amount of resources that can be sold. It depends on the number of employees assigned to the factories. The second phase is the *payment phase* where the employers' rents are estimated. The third phase corresponds to *compute the costs of the factory*, taking into account the salaries of the employees, the invested in stock, the incomes, etc. The most important parameters considered in this phase are the savings and the expenses of the employees, and the consumed products of the factories. The fourth phase *Updates the data company*, that is, the saves of the employers and the new rents of the employees. During the last phase all the parameters are updated according to the previous computations. In this phase MACRO-SYS provides adjustments about the inflation percentage, ratio of salaries, ratio of prices, etc.

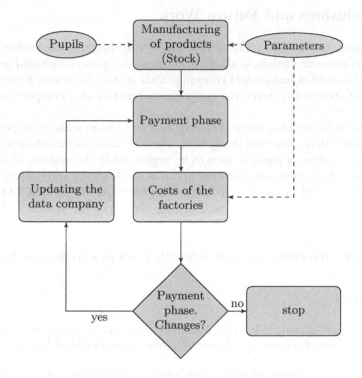

Fig. 4. Scheme of MACRO-SYS algorithm

3.4   Interface Design

Another important issue of E-Learning related to MACRO-SYS is how is the interaction between students and computers [3]. Often the students complain about computer-based training or express a preference for classroom-based instruction. Usually, their complaints are mainly related to the interface and the way they must interact with the tool, that in many cases are confused. According to E-Learning paradigm MACRO-SYS tries to improve this issue. The success of training in MACRO-SYS is largely dependent on the own motivation of the student and his attitude. We knew that a poorly design of the interface prompts the students feel lost, confused, or frustrated, and it will become a barrier to effective learning.

Nowadays, the worst faults behind bad interfaces are the designers themselves. Many designers are not able to put themselves in the place of the students when they try to create training systems easily to use. When the designers create the interface, sometimes, they assume that it is clear what each on-screen button does, and how the content is organized. Fortunately, most common errors are easily observed and repaired. The design of MACRO-SYS has been refined several times until the actual interface has been reached.

4   Conclusions and Future Work

We have presented MACRO-SYS, an intelligent tutoring/training system to deal with macroeconomic models. It allows to simulate complex economical behaviors using basic low level economical concepts. This system is indeed a very useful tool to help students to learn the main characteristics of a complex economic system.

As future work we have some proposals extracted from some tests performed to the students that have used the system. The first one s to introduce a *learning management system*. It enables users to be registered in the system, to be monitored, to be tested and so on. The system can assess learning and control content delivery. The second proposal is to extend the functionality of MACRO-SYS in order to add new learning contents. The new version of the system will store a wide variety of media types. These include course texts, graphics, sounds, videos, tests and so on. In addition, the students proposed to add a work flow to ensure that they have completed a module before they can pass to the next level.

References

1. Cheng, J.Q., Wellman, M.P.: The WALRAS algorithm: A convergent distributed implementation of general equilibrium outcomes. Computational Economics 12, 1–24 (1998)
2. Dumblekar, V.: Management simulations: Tests of effectiveness. An interdisciplinary Journal of Theory, Practice and Research (2004)
3. Entwistle, N., Peterson, E.: Conceptions of learning and knowledge in higher education: Relationships with study behaviour and influences of learning environments. International Journal of Educational Research 41(6), 407–428 (2004)
4. Frolova, J., Korobitsin, V.: Simulation of gender artificial society: Multi-agent models of subject-object interactions. In: Sloot, P.M.A., Tan, C.J.K., Dongarra, J., Hoekstra, A.G. (eds.) ICCS-ComputSci 2002. LNCS, vol. 2329, pp. 226–235. Springer, Heidelberg (2002)
5. Laurillard, D.: Modelling benefits-oriented costs for technology enhanced learning. Higher Education 54, 21–39 (2007)
6. Shute, V.J., Towle, B.: Adaptive e-learning. Educational Psychologist 38(2), 105–114 (2003)
7. Suppi, R., Munt, P., Luque, E.: Using PDES to simulate individual-oriented models in ecology: A case study. In: Sloot, P.M.A., Tan, C.J.K., Dongarra, J., Hoekstra, A.G. (eds.) ICCS-ComputSci 2002. LNCS, vol. 2329, pp. 107–116. Springer, Heidelberg (2002)
8. Whitehouse, K.: Web-enabled simulations: Exploring the learning process. Educause Quarterly 28(3), 20–29 (2005)

Managing Web Services in SOKU Systems

Agnieszka Prusiewicz

Institute of Computer Science, Wrocław University of Technology, Poland
agnieszka.prusiewicz@pwr.wroc.pl

Abstract. The management of the Web services refers to two levels: physical and business one. The first level involves the resources allocation necessary to service execute, and the second one service designing and modelling, then deploying, executing and monitoring. In this work the second level is discussed. In particular the problem of the services functionalities management is taken up. An original proposal of the system for managing the services execution in SOKU (Service Oriented Knowledge Utilities) architecture is presented. The aim of the proposed system is improving of the Web services system efficiency by the services discovery and adaptation according to the users' needs.

1 Introduction

Web services are currently seen as a solution for integration of the heterogeneous resources and making heterogeneous systems interoperable. They are self-contained, self-describing and modular applications that can be published and invoked across the Web [13]. According to their role they are grouped into three categories: business information, business integration and business process externalization. In case of business information Web services are used to share information with users through informational portals. Any transactional e-commerce systems such as auction e-marketplaces, reservation systems and credit checking represent the business integration. And finally in the business process externalization Web services are used to integrate and combine processes from different companies into a global chain of products sale and distribution [12].

Web services use Internet as a global platform where organisations and individual participants communicate with each other to carry out their commercial activities and achieve their business goals. The indubitable advantage of Web services is the ability of creation the applications using the reusable software components that are distributed among the Web. In this way the Web services provide a flexible solution to the problem of application integration [8].

Web services implement Service Oriented Architecture (SOA). SOA is an approach to build IT systems based on the concept of a service. "The primary goal of Service-Oriented Architecture (SOA) is to align the business world with the world of information technology (IT) in a way that makes both more effective." [11]. One of the basic assumptions of the SOA systems is to provide the services on the fly and on user request. In order to meet the users' requirements it is necessary to apply the mechanisms of users' habits retrieval.

N.T. Nguyen, M.T. Le, and J. Świątek (Eds.): ACIIDS 2010, Part II, LNAI 5991, pp. 57–64, 2010.

The enhancement of SOA concepts with the knowledge retrieval and usage is the SOKU approach. The aim of such extension is the automation of the service composition processes including users' preferences in order to supply personalised services.

The users' preferences are obtained on the basis of the users' behaviour analysis. In this way both the knowledge about the users' preferences and demographic characteristics are the input for the recommendations of the dedicated services increasing users' satisfaction. On the other hand the knowledge about the services requests is used to manage the services repositories and service composition processes.

In this work the problem of the services execution management is discussed. We introduce the framework of the proposal of the system for managing services execution. This paper is organised as follows. In Section 2 the problem of the service execution management is discussed and related work is given. Then in Section 3 the framework of the services execution management system is presented. In particular the functionality of the system and its two main modules are introduced. Finally in Section 4 some conclusions are drawn.

2 The Services Execution Management

Considering the services execution management it is necessary to take into account two levels of management: physical level and business level. The first one refers to the resources allocation necessary to service execution. The second one includes: service designing and modelling then deploying, executing and monitoring. In this work the second aspect of the service execution management is discussed.

The integral part of service execution management is the task of service composition. A considerable amount of research has been invested in developing the methods of Web services composition [1]. Different techniques have been proposed for the composition of Web services including manual, semi-automatic and automatic methods. A brief survey of automated composition methods is given in [4]. The service composition is defined as a process of discovering, collecting and executing atomic services that build together the required functionality.

Generally two kinds of service composition techniques are developed: workflow composition or AI planning. In the first case the composite service composition is equivalent to a workflow, where each component represents functionality of a single service including control and data flow among the services. In the second one the process of service composition is identified with AI planning and deduction theory proving. This includes automatic generation of service execution plan with assumption that each service is specified by its preconditions and effects in the planning context [5]. The services should be dynamically discovered, integrated and executed to fulfil users' demands [9]. The problem of searching the services that fulfil users requirements is described by the researchers as the service matchmaking process that is mainly based on discovering an appropriate service provider [2,7,9]. Recently a lot of attention has been paid to developing the methods for services matching. For example, in [6] Matching Engine for flexible matching of user requests and providers services is proposed. In [3] the logic-based method for service selection including hybrid matching filters and the generic matching algorithm are presented. If there is no possibility to find the service that match user's requirements then dedicated service must be composed.

When dealing with knowledge about users' preferences the personalized systems of services recommendation are concerned. In such systems the management of service execution requires applying the methods of users behaviour monitoring in order to adapt the system's parameters to functional and non-functional users' requirements that are changing over the time. The expected result of such monitoring is data that is necessary to discover the knowledge about the services requests and system loading. On the basis of the discovered knowledge the decisions about services proposal and execution are made.

The main aspects of services execution management are:

- the management of the services functionalities,
- the management of the functional services parameters i.e. the management of the time and places of the services execution, the resources allocation, the management of the services security etc.,
  the management of the services access.

3   The Services Execution Management System

The processes carried out in Web Services system are represented by the system consisting of the four layers:

- the business processes layer,
- the service requests layer,
- the service composition layer,
- the atomic service execution layer.

The services execution management system proposed in this work is set in the business process layer and the service request layer (Fig.1).

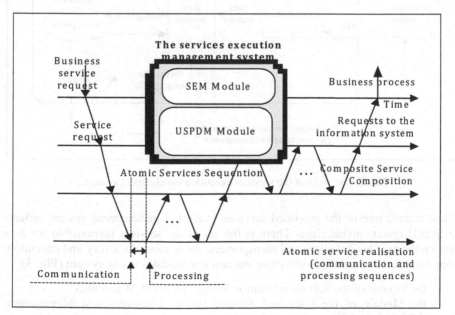

Fig. 1. The location of the services execution management system

The aim of the services execution management is to improve the Web services system efficiency by the services discovery and adaptation according to the users' needs. The users' needs are discovered and stored in our service execution management system as the user and service profiles. We define the user profile as aggregated characteristics describing the way of the services usage by one user including demographic user data and user interests. Whereas the service profile is aggregated characteristic describing the way the group of the users uses services. Both profiles are represented by the ontology. The user profiles are brought into play in the processes of the personalised Web services composition, while service profiles - in the processes of the service repository management.

Generally, the tasks of the services execution management system are as follows:

– collection of data from Web services system,
– data analysis and knowledge discovery in order to obtain two knowledge structures: user profiles and service profiles.
– generating the advices about the personalised services recommendation and management of the services repository.

The advices are sent as feedback to the Web services system (Fig.2).

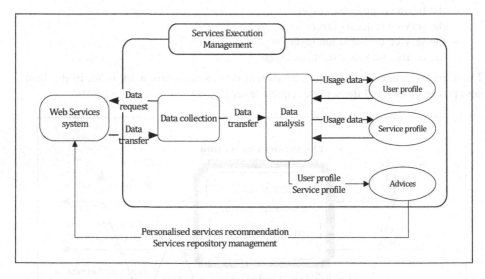

Fig. 2. The tasks of the services execution management system

The architecture of the proposed services execution management system reflects the SOKU system architecture. There is the set of the services responsible for data transfer, the profiles discovery and management, the advices discovery and execution. From the functional point of view there are two key modules in our system (Fig. 3):

– the Module of the Service Execution Management (SEM module),
– the Module of the User and Service Profile Discovery and Management (USPDM module).

The crucial role of the USPDM module is to discover the user and service profiles that are the basis for making decisions in the SEM module.

3.1 The SEM Module

The general aim of the SEM Module is the management of access to the services and management of the functional and non-functional services parameters. The information stored in the user and service profiles provides knowledge about who, when and in which way uses services. This knowledge is required to manage the services access rights and the services offer. We can adapt functional and non-functional services parameters (i.e. the security service level, the time of service accessibility, the time of service completion etc.) regarding the services popularity.

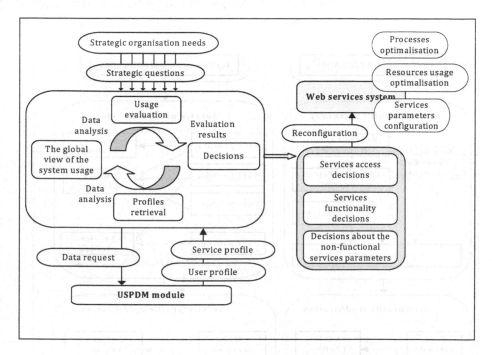

Fig. 3. The service management system module with its functionality

The knowledge represented by the service profile determines the global view of the service usage by the group of users. Constant agregation of the knowledge from all the service profiles provides the global view of the system usage. In this way we can monitor the system loading by the services, adaptively determine the services parameters, manage the service repository by means of adding new atomic services or granulating existing ones, achieving hereby the processes and resources usage optimalisation.

3.2 USPDM Module

The goal of USPDM module is knowlege about the users behaviour discovery (Fig. 4). This knowledge is represented by two structures: user profile and service profile.

Therefore we distinguish two kinds of functionalities of USPDM module: connected with the user profile and service profile. The functionality connected with the user profile includes: 1) user profile discovery that consists of automatic user profile determination on the basis of the system usage data and the user data such as demographic data and interests. In order to discover the user profile it is necessary to monitor the user behaviour such as the requested and executed services, their functional and non-functional parameters, the order of the services requests and then data analysis to discover pattern of the user behaviour; 2) user profile modification according to the discovered changes of the user behaviour.

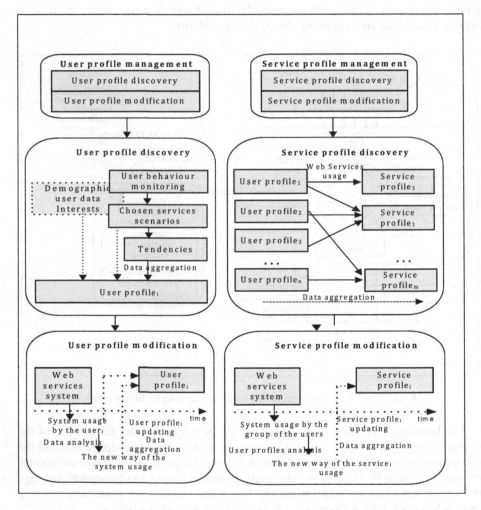

Fig. 4. The functionality of the module of User and Service Profile Discovery and Management

With the service profile the following functionality is connected: 1) service profile discovery that consists of the automatic services usage data aggregation. In order to discover the service profile user profiles are analysed. In this way the knowledge about usage of the chosen services by the users is discovered. This knowledge is the base for grouping the users according to their requirements and needs. As a consequence the proper services recommendation can be applied; 2) service profile modification according to the discovered changes of the services usage. In this case some semantic distance measures must be applied to determine the differences between the existing service profiles and the new discovered tendencies.

4 Conclusions

In this paper the problem of the services execution management was discussed. We concentrated on the SOKU system in which the key point is that the services are knowledge-assisted in the processes of the services delivery. The services are equipped with the semantic descriptions and the knowledge is applied for the services personalisation. Therefore it was assumed that knowledge discovered on the basis of the user behaviour analysis is applied to the services execution management and personalisation. Then the system for managing the services execution was proposed. The aim of this system was to improve the Web services system efficiency by the services discovery and adaptation according to the users' needs. The functionalities of two key system modules were described. Proposed solution requires further development, proposed modules functionality should be designed in details and then implemented.

References

1. Charif, Y., Sabouret, N.: An Overview of Semantic Web Services Composition Approaches. Electronic Notes in Theoretical Computer Science 146, 33–41 (2006)
2. Kawamura, J.A., De Blasio, T., Hasegawa, M.: Paolucci, and K. Sycara, A Preliminary Report of a Public Experiment of a Semantic Service Matchmaker combined with a UDDI Business Registry. In: 1st International Conference on Service Oriented Computing, Italy (2003)
3. Klusch, M., Fries, B., Sycara, K.: OWLS-MX: A hybrid Semantic Web service matchmaker for OWL-S services. Journal of Web Semantics: Science, Services and Agents on the World Wide Web 7, 121–133 (2009)
4. Ko, J.M., Kim, C.O., Kwon, I.-H.: Quality-of-service oriented of web service composition algorithm and planning architecture. The Journal of Systems and Software 81, 2079–2090 (2008)
5. Rao, J., Su, X.: A survey of automated web service composition methods. In: Cardoso, J., Sheth, A.P. (eds.) SWSWPC 2004. LNCS, vol. 3387, pp. 43–54. Springer, Heidelberg (2005)
6. Sycara, K., Paolucci, M., Ankolekar, A., Srinivasan, N.: Automated Discovery, Interaction and Composition of Semantic Web services. Journal of Web Semantics 1(1), 27–46 (2003)
7. Sycara, K., Widoff, S.: LARKS: dynamic matchmaking among heterogeneus software agents in cyberspace. Journal of Autonomous Agents and Multi-Agent Systems 5, 173–203 (2002)
8. Wang, H., Huang, J.Z., Qu, Y., Xie, J.: Web services: problems and future directions. Journal of Web Semantics: Science, Services and Agents on the World Wide Web 1, 309–320 (2004)

9. Younas, M., Chao, K.-M., Laing, C.: Composition of mismatched web services in distributed service oriented design activities. Advanced Engineering Informatics 19, 143–153 (2005)

10. Service Execution Management,
    http://www.casewise.com/Solutions/SOA/sem/

11. Service Oriented Business,
    https://www.ibm.com/developerworks/mydeveloperworks/blogs/woolf/entry/service_oriented_business

12. Web Application Development User Guide – Web services overview,
    http://help.eclipse.org/help32/index.jsp?topic=/org.eclipse.jst.ws.doc.user/concepts/cws.html

13. Web Services Overview,
    http://publib.boulder.ibm.com/infocenter/rtnlhelp/v6r0m0/index.jsp?topic=/com.ibm.etools.webservice.doc/concepts/cws.html

A Computational Analysis of Cognitive Effort

Luca Longo and Stephen Barrett

Department of Computer Science and Statistics - Trinity College Dublin
{llongo,stephen.barrett}@cs.tcd.ie

Abstract. Cognitive effort is a concept of unquestionable utility in understanding human behaviour. However, cognitive effort has been defined in several ways in literature and in everyday life, suffering from a partial understanding. It is common to say "Pay more attention in studying that subject" or "How much effort did you spend in resolving that task?", but what does it really mean? This contribution tries to clarify the concept of cognitive effort, by introducing its main influencing factors and by presenting a formalism which provides us with a tool for precise discussion. The formalism is implementable as a computational concept and can therefore be embedded in an artificial agent and tested experimentally. Its applicability in the domain of AI is raised and the formalism provides a step towards a proper understanding and definition of human cognitive effort.

Keywords: Cognitive Effort, Artificial Intelligence, Virtual Agents.

1 Introduction

Attention plays a central role in the behaviour of human beings. The concept of attention has been inconclusively studied in the history of psychology since the early years of the nineteen century. A plethora of definitions has been proposed and a large number of studies have been carried out in different directions. In 1908 Titchener [1] asserted that: "the doctrine of attention is the nerve of the whole psychological system, and that as men judge of it, so shall they be judged before the general tribunal of psychology". Behaviourists and Gestalt theories shared the conviction that the operations which relate output, such as response or percept, to input, such as stimulus or field, conform to a simple set of rules, such as isomorphism or conditioning [6]. By the end of the 1950s, the situation radically changed and the new concept of attention was a central topic in an emergent cognitive psychology that ascribed more spontaneity and autonomy to the organism implying some degree of local unpredictability than the previous classical doctrines. Post-behaviouristic psychology used the label of attention to denote some of the internal mechanisms that determine the significance of stimuli and therefore make it impossible to predict behaviour by stimulus consideration alone. In everyday language, attention is the act or faculty of attending, by directing the mind to an object or thought. Psychologists refer to attention as a state of consciousness characterised by such concentration.

N.T. Nguyen, M.T. Le, and J. Świątek (Eds.): ACIIDS 2010, Part II, LNAI 5991, pp. 65–74, 2010.
© Springer-Verlag Berlin Heidelberg 2010

Let's now consider the example of Luca, a young schoolboy. Luca does not like school that much, as most of his coetaneous, and for this reason he enjoys a pleasant state of drowsiness most of the time. When the teacher calls attention to him, Luca does not merely fail to pay attention but he has less attention to pay. This facts suggests that the drowsy schoolboy merely suffers from, or perhaps enjoys, a general low level of attention paying a small amount of cognitive effort. This example illustrates the dynamic construct of the concept of effort that changes within individuals in response to individual and environmental factors. This thesis is sustained by motivation theories [2] [3] and contrasts with recent empirical studies that have tended to treat effort as a static concept [4]. Berlyne suggested, in 1960, that the intensity of attention is related to the level of arousal that can be measured with electrophysiological techniques, and that is largely controlled by the properties of the stimuli to which the organism is exposed [5]. He was mainly concerned with involuntary attention. In voluntary attention, the subject attends to stimuli because they are relevant to a task that he has chosen to perform and not because of their arousing quality. This suggests that the intensive aspect of attention corresponds to effort rather than to mere wakefulness. Theories of information processing consider cognitive effort as a hypothetical construct, regarded as a limited capacity resources that affects the speed of information processing [6]. In the work of Norman and Bobrow [7], if a task is resource-limited, then the performance will improve if more cognitive effort is allocated to the task. Although cognitive effort may be a hypothetical construct, it is a subjective state that people have introspective access to [8]. In the literature there are several attempts towards the measurement of cognitive efforts. It is a multi-faceted phenomenon: it can be related to physiological states of stress and effort, to subjective experiences of stress, mental effort, time pressure, and to objective measures of performance levels. These various aspects of cognitive effort have led to distinct means for assessing workload including physiological criteria such as heart rate, skin temperature, pupils dilation, blood pressure, respiration, performance criteria such as quantity and quality of performance by using primary task and secondary task measures, and subjective criteria such as rating of level of effort, self-report measures [9]. Despite an extensive literature, there appears to be no attempt to formalise the concept of cognitive effort as a computational concept therefore our goal is to begin the development of a formalism suitable for computations. Our research question here is:

How can we formalise cognitive effort as a computational concept?

We propose to develop a formalism, suitable for ongoing refinement, that captures the core aspects of a more complete theory. The subjective nature of the concept is noted in section 2 where a literature review underlines the main factors that influence cognitive effort. The methodology adopted towards the formalisation of cognitive effort is presented in section 3. In section 4 we present our heuristic formalism built on these factors. We consider possible fields of applications in 5 and a synthesis of open issues and future works in section 6.

2 Related Work and Review of Cognitive Effort

Cognitive effort is a subjective phenomenon. One of the classic dilemmas of psychology concerns the division of attention among concurrent streams of mental activity. Humans often perform several activities in parallel. They suppress or queue stimuli on their behaviour organisation, underling an internal bottleneck characteristic on processing stimuli which can only operate on one stimulus or one response at a time [6]. Attention theories propose that the central neural system is limited, so humans are unable to think, remember, perceive or decide more than one thing at a time. Capacity theory provides a contrary view, assuming the existence of structural bottlenecks that supposes a limited humans' capacity allocable among concurrent activities [12]. The concept of *short-memory* is introduced, that refers to the capacity of holding a small amount of information in mind in an active, readily available state for a short period of time. The more one acquires experience the less cognitive effort he consumes to resolve the same task. If the amount of cognitive effort that individuals allocate to a task decreases, as they become more skilled, the rate of change in cognitive effort should depend on the rate of *skill acquisition* [13]. *Long-term memory* is the store of experience and results of skill acquisition. Arousal is an important factor in regulating attention because it is crucial for motivating certain behaviours [6]. Arousal is a physiological and psychological concept which refers to the state of being awake. Motivation, perceived difficulty, subjective experience, psychological stress are all example of factors that play a role in directing attention towards a certain task. For instance, anxiety or boredom may have impact on performing certain activities. High ability individuals have larger pool of cognitive resources than low ability individuals who need to make larger resource adjustments to achieve the same outcome. Self-regulation theories [14] [15] suggest that individuals with different levels of cognitive ability may react to changes in task difficulty in different ways. Low ability people with a high degree of perceived difficulty require

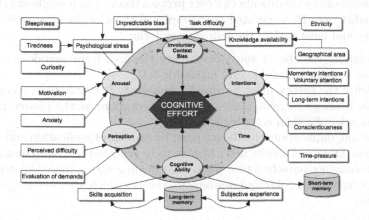

Fig. 1. Cognitive Effort influencing factors

more cognitive effort in performing a task [16]. The *conscientiousness* may moderate the level of attention on a task: highly conscientious individuals choose to work harder and persevere longer than individuals with lower level of conscientiousness [17]. We propose a summary of the main influencing factors in figure 1.

3   A Proposal Methodology

One of the main difficulties in discussing cognitive effort is that the phenomenon has a subjective nature. Studies in psychology and neuro-sciences have attempted to provide a detailed definition of the concept but they demonstrated restricted scope over limited aspects of the concept. The present study, on the contrary, differs from the previous ones because we adopt an approach based on synthesis, that seeks to develop a more comprehensive basis for computing cognitive effort. For any definition of cognitive effort we can formulate a test of the formalism to understand whether it is suitable to the accepted definition. This approach is powerful because it may be seen as a refining stage towards a formalism that satisfies most people's views of cognitive effort, what it is and how it works. The formalism is conceived from experience, intuitive expectations about cognitive effort from a subjective point of view, and conclusions to be found in the psychological, philosophical and sociological literature. The methodology's goal is to merge together different observations, intuitions and definitions to build a simple formalism that is supposed to model the way cognitive effort behaves. We make no affirmations about the validity of the formalism, even if based on literature's studies, but we claim that the results of the application of the formalism to a consideration of cognitive effort is the same as if we had been considering the problem using real cognitive effort.

To deal with the formalisation of cognitive effort we propose to adopt the Popper methodology, presented in *The Logic of Scientific Discovery* [10] where he asserted that: a scientific theory should be based on a "falsification" approach in which no number of experiments can ever prove a theory but a single experiment can contradict one. He suggested that empirical theories are characterised by falsifiability and must satisfy the following criteria:

- demarcation: the theory must demarcate the area from pseudo-science, it must be testable, refutable and falsifiable [11].
- simplicity: the theory must be simple. Simplicity is better than complexity because it allows extreme tests and experimentations on the theory, making it more scientific than complex theories.
- replication/duplication: the theory must be capable of replication and/or duplication. Obtained results must be able to be repeated and as a consequence we can convince ourselves that we are not dealing with a mere isolated coincidence but with regular and reproducible events which are inter-subjectively testable.

Scientific theories are not static and they change perpetually: the formalism presented here can be continually refined.

4   An Example Heuristic Formalism

Few studies have tried to measure cognitive effort and they can be classified in three groups [18]: subjective or self-report measurement, performance and physiological measures. Self-report measures have always attracted researchers: no one is able to provide a more accurate judgement with respect to experienced mental load than the person concerned. However, self-report measures suffer from different rating scales and personal judgements. Performance measures are task-dependent and primary and secondary task approaches have been widely used so far, producing good results [18]. Unfortunately, these techniques require laboratory tools to measure, for instance, the reaction-time useful to assess the amount of cognitive effort required for completing a task. Physiological measures represent the most accurate way of assessing mental workload, often unobtrusively, but they need appropriate equipment to measures physiological behaviours such as blood temperature, pupils dilation. The formalisation of cognitive effort as a computational concept needs to rely less on these classical measurements and it needs to focus on more general concepts easy to model. Indeed, we need a tool to monitor users' behaviour while performing a task, and we assume that this tool can be build up as a piece of software. We have analysed the factors involved in assessing cognitive effort and we propose a possible formalisation of each of them towards a general model.

Cognitive Ability: Some people obviously and consistently understand new concepts more quickly, solve new problem faster, see relationships and are more knowledgeable about a wider range of topics than others. Modern psychological theory views cognitive ability as a multidimensional concept and several studies, today known as IQ tests, tried to measure this trait [19]. Carroll suggested [20] that there is a tendency for people who perform well in a specific range of activities, to perform well in all others as well. Recent work [21] suggests that some aspects of people's cognitive ability peak around the age of 22 and begin a slow decline starting around age 27. However, as it is noted, there is a great deal of variance among people and most cognitive functions are at a highly effective level into their final years, even when living a long life. Some type of mental flexibility decreases relatively early in adulthood, but that how much knowledge one has, and the effectiveness of integrating it with one's abilities, may increase throughout all of adulthood if there are no pathological diseases. These considerations represent pieces of evidence that allow us to model cognitive ability with a growing function, i.e. a curve that starts at low levels and increases quickly to a growing rate threshold from which it still increases but moderately. The flexible sigmoid function proposed by Yin et al. in [22] is suitable for our purposes.

$$CA : [1..G_{th}]^3 \in \aleph^3 \rightarrow [0..1] \in \Re$$

$$CA(G_{th}, G_r, t) = CA_{max} \left(1 + \frac{G_{th} - t}{G_{th} - G_r}\right) \left(\frac{t}{G_{th}}\right)^{\frac{G_{th}}{G_{th} - G_r}}$$

where CA is the level of cognitive ability, G_{th} is the growing threshold, which we may set to an average of mortality of 85 years. G_r is the growing rate, which we may set to 22 years, i.e. the point where the curve reaches the maximum growing weight and from that, increases moderately. t is the age in years of a person and CA_{max} is the maximum level of cognitive ability an individual can reach, in this case equal to 1. The properties G_{th} and G_r are flexible because they can be chosen by considering environmental factors such as the degree of mortality or level of education of a country. For instance, if we consider a person 40 years old, with a growing rate of 22 years and a growing threshold of 85 years, by applying the formula above, we obtain 0.62 of cognitive ability.

Arousal: The concept of arousal was sometimes treated in literature as a unitary dimension, as if a subject's arousal state could be completely specified by a single measurement such as the size of his pupil [6]. However this is a oversimplification and arousal is a multidimensional concept that may vary in different situations and, above all, there are several kinds of individual-subjective factors to consider. This is a relevant problem in studying subjective cognitive effort, but the main goal of our contribution is to present the main factors that influence it. For this reason, we propose a simple *subjective arousal taxonomy* where different type of arousal, such as curiosity, motivation, anxiety, psychological stress, are organised in a multi-level tree. In other words this represents a map of an individual subjective status before performing a certain task. A *subjective arousal taxonomy* is a 3-tuple $< A, W, R >$ composed by a set of pieces of arousal factors A organised in a tree where their unidirectional relationship in the tree is defined in R by using the weights in W. Each node has at most one parent, except the root node which has no parent. Each internal node a_i has a fixed influence strength w_i towards his only parent. Just leaf nodes (node without children), with cardinality $Card_{ln}$, have a value in $[0..1] \in \Re$ which indicates the degree of arousal (eg. 0 is not motivated at all, 1 is highly motivated) while each internal node's and the root node's values are inferred by the relationship with their children along with the related strength. The root of the tree is the final level of arousal that influences the degree of cognitive effort.

$$A : \{a_1, a_2, ..., a_n \mid a_i : [0..1] \in \Re, \ 1 \le i \le n\}$$
$$W : \{w_1, w_2, ..., w_n \mid w_i : [0..1] \in \Re, \ 1 \le i \le n\}$$
$$R : \{\forall \ a_i \ \exists! \ r_i \mid \ r_i : \ A \times A \ \to W, \ r_i(a_i, \ a_p) = w_i\}$$

where a_p is a_i's parent. The degree of each arousal a_i for leaf nodes is an explicit input value while the degree of each arousal a_i for internal nodes is the weighted sum of its c children's values.

$$a_i^{int} = (\textstyle\sum_{z=0}^{c}(a_z \ \cdot \ w_z)) \le 1$$

Finally the root's weight w_{root} is 1 and, as it has no parent, its relation $r_{root} = \emptyset$. The root node inferred by applying the previous steps is:

$$A_{root} : A^{Card_{ln}} \times W^n \times R^n \to [0..1] \in \Re$$

Intentions: Subject's intentions have an important role in determining the level of cognitive effort while performing a task. This individual subjective concept may be splitted into short-term and long-term intentions. We propose to model these concepts with real values computing their average. We refer to short-term intentions or momentary intentions with I_{st} and to long-term intentions with I_{lt} which are subjective judgements in the range $[0..1]$ (0: no intentions at all, 1: highly intentioned). Intentional shades can be modelled: a person can be highly intentioned to get a degree (long-term) but in short-time does not like examinations.

$$I : [0..1]^2 \in \Re^2 \to [0..1] \in \Re, \; I_{ST} : [0..1] \in \Re, \; I_{LT}[0..1] \in \Re$$

$$I(I_{ST}, I_{LT}) = \frac{I_{ST} + I_{LT}}{2}$$

Involuntary Context Bias: Several factors can influence cognitive effort as pseudo-static and unpredictable biases. The latter refer to biases which are almost static and depend on environmental aspects. For instance, there is a large difference across ethnic groups and geographic areas in the available knowledge: people living in Africa have a reduced access on knowledge compared to people living in occidental countries so they may find a question more or less difficult. Similarly, another pseudo-static bias is the task difficulty. Even though it is hard to exactly estimate the complexity of different tasks, it is not expensive to claim that reading a newspaper demands less cognitive effort than resolving a math equation. Cognitive effort may be eventually influenced by unpredictable context biases. For instance, in a working context, phone ringing, questions from colleagues, e-mail delivering all represent involuntary context biases. We propose to use a ranking system to build up a task-difficulty dictionary and real values to model contextual available knowledge and unpredictable bias.

$$CB : [0..1] \in \Re^3 \to [0..1] \in \Re^3, \quad C_{know}, T_{diff}, U_{bias} : [0..1] \in \Re$$

$$CB(C_{know}, T_{diff}, U_{bias}) = \frac{1}{3} \cdot C_{know} + \frac{1}{3} \cdot T_{diff} + \frac{1}{3} \cdot U_{bias}$$

where CB is the total context bias, C_{know} is the contextual knowledge availability, T_{diff} is the task difficulty and U_{bias} is the unpredictable bias. This formula is flexible because provides a way to model particular situations. We may formalise a situation where a person in Central Africa can not use the Internet, so low level of knowledge availability, performing an hard physics task in a noisy library.

Perception: The same task may be perceived differently by two subjects. Perceived difficulty is higher when individuals are presented to new tasks: they may not know what the optimal amount of effort is, given a particular difficulty level. We propose to model this concept as a value $PD : [0..1] \in \Re$ where values near 0 indicate tasks perceived to be very easy and values tending to 1 represent tasks perceived highly complex.

Time: Time is a crucial factor that must be considered in modelling cognitive effort. Time-pressure is sometimes imposed by explicit instruction to hurry and sometimes by demand characteristics of the task. In the former case a real value is sufficient to model the concept, while in the latter we may easily add a task-related time-pressure value to the task-difficulty dictionary previously proposed. Formally: $T_{press} : [0..1] \in \Re$. Furthermore, time is essential because performing a task is not a single-instant action, rather is an action over time. This fact suggests that the described factors that influence the level of cognitive effort need to be considered within an interval of time. Several temporal formal methods and theories are available in literature and studying the temporal-related aspect of cognitive effort requires a separate contribution. We remind this investigation to future work and in this first attempt we propose a simple cognitive effort formalism that propagates all the proposed factors at the same level over time. We assume the existence of a function $F_{time} : \aleph \to [0..1] \in \Re$ that models the trajectory of focused attention for each task over time: it returns the level of attention at a given time. Finally, we propose to model cognitive effort as a discrete function:

$$CE : ([0..1]^6 \in \Re^6) \times (\Re) \times (f : \aleph \to [0..1] \in \Re) \to \Re$$

$$CA' = CA(G_{th}, G_r, t), A' = A_{root}, I' = I(I_{st}, I_{lt}),$$

$$CB' = CB(C_{know}, T_{diff}, U_{bias}), PD' = PD, t'_p = t_{press}, t' = t$$

$$CE(CA', A', I', CB', PD', t'_p, t') =$$

$$\sum_{i=0}^{t'} F_{time}(i) \cdot \frac{[CA' + A' + I' + CB' + PD' + t'_p]}{6}$$

where CA is *cognitive ability*, A_{root} is *arousals*, I is *intentions*, CB is *contextual bias*, t_{press} is *time pressure*, f_{time} is the function for *attention over time* and t is the effective *time* spent to complete a task. Therefore, the final level of cognitive effort elicited on a task is a function of time and of the individual subjective status along with environmental properties.

5 Possible Applications

The formalism proposed in this paper may be applied in several disciplines such as computer science, psychology, neuro-science, economy. Here we present some example of its application. We assume all the cognitive effort influencing factors are available and can be gathered by using unobtrusive appropriate tools, monitors/loggers or derived from works and studies in literature. In education contexts, we may use the formalism to monitor the learning rate of students based on the hypothesis that students should show less cognitive effort in performing similar tasks due to their skill acquisition level. The more they acquire knowledge, the less cognitive effort they should spend in similar activities. *Recommender systems* may benefit from using the cognitive effort formalism, such as the online encyclopaedia Wikipedia, that foresees interactions among users and web-pages. The more users spend cognitive effort in contributing towards the improvement of an article, the more their contribution may be considered qualitative. Therefore, we may hypothesise that the trustworthiness of a Wikipedia article may depend on the quality of contributions. Similarly, in *social search*, assuming the existence of a logger that captures Internet users' behaviour while surfing web-pages, cognitive effort may be adopted to predict users' interests on a particular web document. Here the hypothesis is that the more users show a positive degree of cognitive effort on a web-page, the more that web-page may be considered interesting. Yet, if most of the users show similar level of cognitive effort on a web page, that means similar behaviours, we may infer a level of trustworthiness to it, which can be either positive or negative. Extending this concept to the WWW, a *social search engine* may be conceived. In *online communities*, such as blogs, forums, social networks, people interact with each other leaving feedback. Here the amount of cognitive effort may help to classify most active and trustworthy users: finance forum may benefit from our formalism. Measuring cognitive effort may be helpful for *clinical purposes* as well. If we can assess a degree of cognitive effort spent on a certain task, and we are able to do this repeatedly over time, we may predict people's addiction to that task. This is based on the hypothesis that addicted people show persistence of cognitive effort, that means the same behaviour on a task over time. A *clinical addiction predictor* may be adopted to have a first insight into the degree of addictiveness of online game player or betters. Similarly, psychologists may use the formalism as a preliminary tool to study individuals' addictiveness. In *neuro-science*, the application of the formalism may avoid the use of fMRI scanners for patients who show low addicted behaviour.

6 Open Issues and Future Works

This contribution is an introduction of a formalism for cognitive effort which is useful in clarifying and motivating discussion of the concept and is extensible to take into consideration further studies in the area. In addition the formalism is implementable, it offers the basis for the first implementation of cognitive effort in an intelligent artificial agent. Being based on simple mathematics, it provides the ideal tool for artificial agents in making reasoned decisions. The main aim of this contribution is to increase the understanding of cognitive effort and to provide a tool of great importance as an indicator of work which could be done. Despite these considerations and the intrinsic complexity of the phenomenon, cognitive effort is also to a large extent automatic, unconscious so further studies need to be carried out. Subjective cognitive effort, in our opinion, is a non-monotonic concept, further influencing factors may be added to the formalism attacking or supporting previous ones, other may be grouped, other ones deleted. A defeasible reasoning logic may describe the relations among factors. The phenomenon may be modelled by a more appropriate algebra or first-order logic. Several decisions contributed in the formalisation of cognitive effort: some of these imposed a general structure of cognitive effort which may be not always valid. This contribution is the first attempt in formalising cognitive effort as a computational concept, so it does not aim to be the final implementation but, instead, a first basic clarification tool that need to be faced and refined over time. It is beneficial to social science because it allows the precise discussion of the concept of cognitive effort. In the new Distributed Artificial Intelligence field it allows robustness and sensible behaviour in unpredictable and patchy environments and it allows agents to reason sensibly about other agents, either human or artificial.

References

1. Titchener, E.B.: Lectures on the Elementary Psychology of Feeling and Attention, p. 173. Macmillian, New York
2. Locke, E.A.: Toward a Theory of Task Motivation and Incentives. Organizational Behavior and Human Decision Processes 3(2) (1968)
3. Vroom, V.H.: Work and Motivation. Wiley, New York (1964)
4. Fried, Y., Slowik, L.H.: Enriching goal-setting theory with time: An Integrated Approach. Academy Management Review 29(3), 404–422
5. Berlyne, D.E.: Conflict, Arousal and Curiosity. McGraw-Hill, New York (1960)
6. Kahneman, D.: Attention and Effort. Prentice Hall, NJ (1973)
7. Norman, D.A., Bobrow, D.B.: On Data-limited and Resource-limited Processes. Cognitive Psychology 7, 44–64 (1975)
8. Humphreys, M.S., Revelle, W.: Personality, Motivation and Performance: A Theory of the Relationship Between Individual Differences and Information Processing. Psychological Review 91(2) (1984)
9. Schvaneveldt, R.W., Reid, G.B., Gomez, R.L., Rice, S.: Modeling mental workload. Cognitive Technology, 3, 9–31 (March 1998)
10. Popper, K.R.: The Logic of Scientific Discovery. Hutchinson (1967)
11. Popper, K.R.: Conjectures and Refutations. Routledge, New York (1969)
12. Moray, N., O'Brien, T.: Signal-detection theory applied to selective listening. Journal of the Acoustical Society of America 42, 765–772
13. Deadrick, D.L., Bennet, N., Russel, C.J.: Using HLM to Examine Dynamic, Performance Criteria Over Time. Journal Management 23(6)
14. Carver, C.S., Scheiner, M.F.: On the Self-regulation of Behavior. Cambridge University Press, Cambridge
15. Karoly, P.: Mechanisms of Self-regulation: A Systems View. Annual Review of Psychology 44, 23–52
16. Yeo, G., Neal, A.: Subjective Cognitive Effort: a Model of States, Traits, and Time. Journal of Applied Psychology 93(3), 617–631 (2008)
17. Barrick, M.R., Mount, M.K., Strauss, J.P.: Conscientiousness and Performance of Sales Representatives: Test of the Mediating Effects of Goal Setting. Journal of Applied Psychology 78(5), 715–722
18. O' Donnell, R.D., Eggemeier, F.T.: Workload Assessment Methodology. Cognitive Processes & Performance II, 1–49 (1986)
19. Dickens, T.W.: Cognitive Ability. In: Durlauf, S. (ed.) The New Palgrave Dictionary of Economics (forthcoming)
20. Carrroll, J.B.: Human Cognitive Abilities: A Survey of Factor-analytic Studies. Cambridge University Press, UK (1993)
21. Salthouse, T.: When does age-related cognitive decline begin? Journal of Neurobiology of Aging 30(4), 507–514 (2009)
22. Yin, X., Goudriaan, J., Lantinga, E.A., Vos, J., Spiertz, H.J.: A Flexible Sigmoid Function of Determinate Growth. Annals of Botany 91 (2003)

An Algorithm for Computing Optimal Coalition Structures in Non-linear Logistics Domains

Chattrakul Sombattheera

Faculty of Informatics
Mahasarakham University
Mahasarakham
Thailand 44150
chattrakul.s@msu.ac.th

Abstract. We study computing optimal coalition structures in non-linear logistics domains where coalition values are not known a priori and computing them is NP-Hard problem. The common goal of the agents is to minimize the system's cost. Agents perform two steps: *i*) deliberate appropriate coalitions, and *ii*) exchange computed coalitions and generate coalition structures. We apply the concept of best coalition introduced in [1], to work in the non-linear logistics domain. We provide an algorithm, explain via examples to show it works. Lastly, we show the empirical results of our algorithm in terms of elapsed time and number of coalition structures generated.

1 Introduction

This research studies optimal coalition structures in non-linear domains. Here, we use an example of logistics providers distributing goods from manufacturing sites to end customers. We consider the operation of small independent logistics providers, whose individual resource is merely a truck, in a small but complex economy system. Note that these trucks are independent economy units which have to make their own living [2]. Hence it is important that the tasks will be allocated to fairly to agents. The state of the art in anytime algorithm [3] for the problem is not applicable here because the algorithm works in practice only for a small number of agents, i.e., it has to scan all the coalition values as input. While the number of coalitions involved is much larger, scanning all coalition values are impractical.[1] We follow the same principle proposed in [1], that is to compute only a relatively small number of coalitions whose values are more relevant to the process of computing optimal coalition structure values. Agents may choose any appropriate heuristics to generate coalitions. To examine how economically agents can address this problem in such a setting, we propose a set of distributed algorithms to tackle the task.

---

[1] For example, choosing a coalition of 10 agents from 100 agents requires at least $^{100}C_{10} = 1.731E + 13 \approx 17.3$ trillions bytes.

N.T. Nguyen, M.T. Le, and J. Świątek (Eds.): ACIIDS 2010, Part II, LNAI 5991, pp. 75–84, 2010.
© Springer-Verlag Berlin Heidelberg 2010

The structure of the paper is as follow. We explain the setting of the problem. We discuss in details about how the algorithm works and support this an example. Lastly, we show the experiment results and related work in the literature.

2 Distributed Algorithm for Distributing Goods

2.1 Setting

There are two levels of coalitions in our setting: i) task-plan in order to find the best solutions for each task at each point in time, and ii) task-agent in order to assign to each best task-plan a coalition of agents such that the total cost is minimal.

Task-Plan Solution: We define a graph $\mathcal{G} = (\mathcal{V}, \mathcal{E})$ where $\mathcal{V}$ is a set of vertices and $\mathcal{E}$ is a set of edges connecting each pair of vertices with a certain distance. The graph represents an economic map of a road network, i.e., a vertex represents a location in the road network, a vertex represents the cost in term of the shortest distance achieved optimally between each pair of locations. We define a task as a tuple $T = \langle \mathcal{S}, \{\mathcal{D}\} \rangle$, where $\mathcal{S} \in \mathcal{V}$ is the source and $\{\mathcal{D}\} \subseteq \mathcal{V}$ is a set of destinations associated with $\mathcal{S}$. We refer to the number of $\mathcal{D}$s in T as the size of T and denote it by $|T|$. We assume the smallest $|T| = 1$. The set of all tasks is denoted by $\mathcal{T}$. We define $\mathfrak{L}$ a set of n logistics provider (LP) agents. Each of these agents, $\mathcal{L} \in \mathfrak{L}$, is a truck with the same capacity load and travel distance. Agents are to cooperatively deliver the goods as per requests. We assume the number of agents is at least equal to the number of tasks and the maximal number of agents is not greater than $\sum_{T \in \mathcal{T}} |T|$. A task T can be partitioned into $1 \leq p \leq |T|$ parts, each of which to be executed by an agent. The agents assigned to a task is then a coalition. We shall refer to a set of partitions of the same p parts as *partition* P_p. We shall refer to each instance of P_p as *partition instance* and denote it by $P_{p.i}$, where i is the lexicographical order index of the partitions of the same p size. We denote by $|P_p|$ the number of all instances $P_{p.i}$ in P_p[2]. We shall denote the j-th part of $P_{p.i}$ by $P_{p.i.j}$, starting from left to right. Hence $P_{p.i.j}$ is the index indicating the number of destinations in the j-th part of $P_{p.i}$ and we denote such a number by $|P_{p.i.j}|$. For example, a task of 5 destinations can have 5 partitions, which can be broken down to 7 partition instanes altogether. Partition P_1, P_4 and P_5 has just one instance, while P_2 and P_3 which have two instances each, i.e. $P_{2.1}, P_{2.2}, P_{3.1}$ and $P_{3.2}$ respectively. Given $P_{2.1}$, for example, we partition the whole task into 2 parts: one part of 4 destinations and one part of 1 destination. In other words, we may assign a coalition of 2 agents for this task: one agent to execute $P_{2.1.1}$ which has 4 destinations, and the other one to execute $P_{2.1.2}$ which has 1 destination.

In each $P_{p.i.j}$, a collection of $|P_{p.i.j}|$ destinations can be choosen to build a route on which any agent can travel and distribute the goods. Let $D_{p.i.j}$ be a set

---

[2] The exact number of P_p and integer partition are well explained in Kreher, D., Stinson, D.: Combinatorial Algorithms Generation, Enumeration and Search. CRC Press, FA, USA (1999).

of destinations yet to be choosen into $P_{p.i.j}$. Hence there are $^{|D_{p.i.j}|}C_{|P_{p.i.j}|}$ alternative combinations. For each of these, we can have $|P_{p.i.j}|!$ alternative routes, which we denote as $R_{p.i.j}$. Given $D_{p.i.j}$, we can compute the number of alternative routes, $|R_{p.i.j}|$, as follows: $\frac{|D_{p.i.j}|!}{(|D_{p.i.j}|-|P_{p.i.j}|)! \cdot |P_{p.i.j}|!} \cdot |P_{p.i.j}|!$. Hence, for each P_p there are $\prod_{i=1}^{|P_p|} \prod_{j=1}^{p} (^{|D_{p.i.j}|}C_{|P_{p.i.j}|})$ alternatives. We shall refer to each of these alternative as a *plan*, $\mathcal{P}_p$. Given an optimization technology and a computation time t, we denote $R_{p.i.j}^t$ as the optimized route with mininal cost (distance). We denote this optimal cost by $C_{p.i.j}^t$. Hence, the execution cost of a coalition of agents for a $\mathcal{P}_p$ is $CE_{\mathcal{P}_p}^t = \sum_{j=1}^{p} C_{p.i.j}^t$. For each partition of task T, we are interested in acheiving the best plan $\mathcal{P}_{T,p}^*$, whose cost $CE_{\mathcal{P}_p}^* = arg_{min}C_{\mathcal{P}_p}^t$ is minimal. Agents are to cooperatively compute each $T.\mathcal{P}_p$. Rahwan et.al. [4] propose an algorithm for computing coalition values among cooperative agents. However, that algorithm seems inapplicable here because it computes the values of all coalitions, whose number is too large in our setting.

Task-Agent Coalition: The next step is to assign appropriate coalitions of agents to tasks to achieve the overall minimal cost for the system. We have defined a set of logistic providers $\mathfrak{L}$. There is a location function $Loc : \mathfrak{L} \rightarrow \mathcal{V}$ which associates an agent $\mathcal{L} \in \mathfrak{L}$ to a vertex $\mathfrak{n} \in \mathcal{V}$, where the agent is located. Let $RT \subseteq \mathfrak{T}$ be the set of tasks yet to be assigned with agents. Let $RL \subseteq \mathfrak{L}$ be the set of agents yet to be assigned tasks. In order to execute a task T, a coalition $S \subseteq \mathfrak{L}$ of $1 \leq |S| \leq (|RL| - (|RT| - 1))$ agents have to travel the source of the task and collect goods before distributing them. For each agent $\mathcal{L}$, this will incur the *access cost*, $CA_{T,\mathcal{L}}$, to the agent. Hence the access cost for S to T is $CA_{T,S} = \sum_{\mathcal{L} \in \mathfrak{L}} CA_{T,\mathcal{L}}$. This cost is consistent and is independent of time. Let $\mathfrak{L}_{\mathcal{P}_{T,p}^*} \subseteq \mathfrak{L}$ be the set of available agents for assigning to $\mathcal{P}_{T,p}^*$, there are $^{|\mathfrak{L}_{\mathcal{P}_{T,p}^*}|}C_{\mathcal{P}_{T,p}^*}$ ways of assigning a coalition of agents to $\mathcal{P}_{T,p}^*$. We shall refer to each pair of assigning $S \subseteq \mathfrak{L}_{\mathcal{P}_{T,p}^*}$ to $\mathcal{P}_{T,p}^*$ as a *task-agent* coalition and denote it by $S_{\mathcal{P}_{T,p}^*}$. Hence the total cost of executing plan $\mathcal{P}_{T,p}^*$ by S is $C_{\mathcal{P}_{T,p}^*,S} = CA_{T,S} + C_{\mathcal{P}_p}^*$.

Task Coalition Structure: Here, we are not only interested in finding the cheapest assignment to each T, but we are also interested in finding the cheapest assignment to $\mathfrak{T}$. We define a *task coalition structure*, $TCS = \{S_{\mathcal{P}_{T,p}^*} \mid$ where $\bigcup_T = \mathfrak{T}$, and $T_i \cap T_j = \emptyset$, and $\bigcup_S = \mathfrak{L}$, and $S_i \cap S_j = \emptyset\}$, is a set of task-agent assignments where all tasks are assigned with a a unique coalition of agents and all agents are assigned to tasks. We define the total assignment cost, $V(TCS) = \sum C_{\mathcal{P}_{T,p}^*,S}$ the sum of all assignments in a task coalition structure. We are interested in finding a task coalition structure such that the sum of total cost is minimal, i.e.,

$$TCS^* = arg_{min}V(TCS)$$

Best Task-Agent Assignment: In [1], we have defined the *best* candidate coalition, based on the ratio between the coalition value and its cardinality, in order to place the next coalition into CS. Here we apply the same principle. In order to find the most appropriate $S^*_{\mathcal{P}^*_{T,p}}$ as the next assignment to one of the remaining tasks, we define the *reduction contribution* as $\bar{A} = \frac{C_{\mathcal{P}^*_{T,p},S}}{|S|}$. Hence the *best task-agent* assignment is $S^*_{\mathcal{P}^*_{T,p}} = arg_{min}\bar{A}$, whose reduction contribution is the lowest among all the possible assignments.

Time Allocation Strategy for Overall Deliberation: We define a time allocation strategy as a tuple $\mathcal{T}_{ST} = \langle \mathcal{T}_{PA}, \mathcal{T}_{TA} \rangle$, where $\mathcal{T}_{PA}$ is the percentage of remaining time to be allocated to the deliberation to find $\mathcal{P}^*_{T,p}$ for each T, $\mathcal{T}_{TA}$ is the percentage of remaining time to be allocated to the deliberation to find $S^*_{\mathcal{P}^*_{T,p}}$, and $\mathcal{T}_{PA} + \mathcal{T}_{TA} \leq 100$. Agents may choose to split the remaining time equally, i.e., $\mathcal{T}_{ST} = \langle 50, 50 \rangle$, which will give agent just one round of each deliberation. Alternatively, agents may choose to spare some time for the exact computing, and repeat the process of both deliberations for the remaining time, e.g. $\langle 50, 40 \rangle$ which would recursively leave spare time for deliberations in later rounds as 10%.

Time Allocation for Task-Plan Deliberation: There are two issues for time allocation in the task-plan deliberation. i) Time allocation for each task has to be distributed to each task efficiently and has to take into account the tradeoff between the time spent on the computation and the quality of the result. ii) The distribution of each task has to meet the time constraints required by customers and all agents have to play their parts in distributing goods. We invent a hueristic strategy which will allocate time to each task based on the potential computation workload on each task, i.e. the sum of the size of the search space in each partition of the task. We compute for the weight of each partition and allocate the time to optimize the partition based on its proportional weight to the total weight of the task. Firstly, we compute the product of each part j in each partition instance j, $\prod |P_{p.i.j}|$. The weight for each partition is then defined by $W_p = \frac{\sum \prod |P_{p.i.j}|}{|i|}$, where $|i|$ is the number of instances in P_p. The actual time allocation for the each P_p in task T is $A(W_p) = \frac{W_p}{\sum_{k=1}^{|T|} W_k}$. Let $W_T = \sum_{k=1}^{|T|} W_k$ be the total weights of all P_p T. Let $W_{\mathfrak{T}} = \sum_{T \in \mathfrak{T}} W_T$ be the total weights of all $T \in \mathfrak{T}$. The proportional time allocation out of the available time for task-plan deliberation is merely $A(W_T) = \frac{W_T}{W_{\mathfrak{T}}}$.

Heuritic for Choosing Shrinking and Altering Point: Instead of doing thorough search by altering and shrinking once the algorithm cannot extend anymore, we introduce a heuristic to find the alter and shrink point which is more appropriate to our setting. This heuristic repeatedly bi-partitions TSC into sections. We define a bi-partitioned of a number $I \in \mathbb{I}^+, I > 1$ a tuple $\langle HH, LH \rangle$, where $HH = LH = \frac{I}{2}$ if I is and even number, or $HH = \lceil \frac{I}{2} \rceil$, $LH = \lfloor \frac{I}{2} \rfloor$ otherwise. At each time of the bi-partitioning, we refer to the partition as the level, l. We refer to each part of the partition as the section, s. The shrinking and altering of TCS will take place at the position indicated by this bi-partition. We compute the weight for each bi-partition as the following $W_{I_l} = \frac{\prod I_{l,j}}{I_l}$. The most appropriate bi-partition is $I_l^* = arg_{max}W_{I_l}$. The present TCS will be shrunk from task $|j|, |j-1|, \ldots, 1$. Each time TCS is shrunk, the alter takes place at $T_{|j|}$ in order to assign the next best S to the task, replacing the previous one.

2.2 Main Algorithm

In the beginning, the algorithm requires the time allocation strategy $\langle T_{ST} \rangle$ and the time available for deliberations, which we shall refer to as the remaining time, T_R. The function $AllocPlanTime(T_{ST}, T_R)$ will return the time allocated to the task plan deliberation process by $planDelibTime = T_R/100 * T_{PA}$. Agents spend at least $planDelibTime$ to solve the optimization problem in all tasks by calling the function . Once the solution deliberation process in this round is done, agents spend $assignDelibTime = T_R/100 * T_{TA}$ to compute the most efficient assignment task-agent for the time being. The algorithm then computes for T_R by deducting the time spent on both deliberation in that round from T_R. If there is remaining time, the algorithm goes into the loop and repeat the deliberation processes again until $T_R \leq 0$. The details processes are shown in algorithm 1.

Algorithm 1. Main Algorithm: Construct task-agent structures by repeatedly deliberate for optimal task-cardinality and task-agent assignment

Require: T_{ST}
Require: T_R
1: $elapsedTime \leftarrow 0$
2: $startTime \leftarrow presentTime$
3: $planDelibTime \leftarrow AllocPlanTime(T_{ST}, T_R)$
4: $DelibTaskPlan(\mathfrak{T}, planDelibTime)$
5: $assignDelibTime \leftarrow AllocAssignTime(T_{ST}, T_R)$
6: $DelibTaskAssign(Tasks, assignDelibTime)$
7: $elapsedTime \leftarrow presetTime - startTime$
8: $T_R \leftarrow T_R - elapsedTime$
9: **while** (**do**$T_R > 0$)
10:    $planDelibTime \leftarrow AllocateTime(T_{ST}, T_R)$
11:    $DelibTaskPlan(\mathfrak{T}, planDelibTime)$
12:    $assignDelibTime \leftarrow AllocateTime(T_{ST}, T_R)$
13:    $DelibTaskAssign(\mathfrak{T}, assignDelibTime)$
14:    $elapsedTime \leftarrow presetTime - startTime$
15:    $T_R \leftarrow T_R - elapsedTime$
16: **end while**

2.3 Algorithm to Deliberate Task-Plan

Here, agents have to split the available time to compute each $\mathcal{P}_{T,p}^*$. The detailed process of allocating time to task-plan is described in alogorithm 2. It takes the $ramainTime$ as an input. The algorithm initializes two arrays and one variable to store computed weights for Ps, Ts, and $\mathfrak{T}$. The algorithm then finds the weight of each P_p, total weight for each T, and total weight for $\mathfrak{T}$. In each of these processes, we do not show details on how to compute the weights because it is quite straghtforward from what we have described in section "time allocation for task-plan solution". The function $computeWeight()$ serves this purpose as a blackbox that returns just the weight of the respective P. Both the total weight of each T and $\mathfrak{T}$ will also be accumulated at the same time. Once all the weights are computed, the algorithm then goes to each T, finds the exact time for optimizating this T. The allocated time for this T will then be allocated to each P, which will be optimized for the the best route by the function $optimizePartn()$. This function is the call to the underpinning technology that takes T, P, and the available time as inputs, and return $\mathcal{P}$ for the given P. The detailed processes are shown in algorithm 2.

Algorithm 2. Optimize for the best possible plans for each task

Require: $remainTime$
Require: $maxTaskSize$
Require: $\mathfrak{T}$
 1: init $planWeightArray[|\mathfrak{T}|][maxTaskSize]$
 2: init $taskWeightArray[|\mathfrak{T}|]$
 3: $TotalWeight \leftarrow 0$
 4: **for** each $T \in \mathfrak{T}$ **do**
 5:     **for** each $P \in T$ **do**
 6:         $planWeightArray[T][P] \leftarrow computeWeight(\mathfrak{T}, T, P)$
 7:         $taskWeightArray[T] \leftarrow taskWeightArray[T] + planWeightArray[T][P]$
 8:         $TotalWeight \leftarrow TotalWeight + planWeightArray[T][P]$
 9:     **end for**
10: **end for**
11: **for** each $T \in \mathfrak{T}$ **do**
12:     $taskAllocTime \leftarrow getTaskAllocTime(remainTime, TotalWeight,$
13:     $taskWeightArray[T])$
14:     **for** each $P \in T$ **do**
15:         $partnAllocTime \leftarrow getPartnAllocTime(taskAllocTime, T, planWeightArray)$
16:         $\mathcal{P} \leftarrow optimizePartn(T, P, partnAllocTime)$
17:     **end for**
18: **end for**

2.4 Algorithm to Deliberate Task-Agent

This algorithm repeatedly searches for the next $S^*_{\mathcal{P}^*_{T,p}}$ to be placed in TCS, while meeting with other requirements such as all agents must be assigned to tasks. In princicle, it is similar to algorithm [1] that it searches for the next best task-agent assignment. However, there are two main differences from algorithm [1]: $i)$ the alter and shrink points, and $ii)$ the process of choosing best assignment is more complex than choosing the best candidate coalition. We define an array TCS of size $|\mathfrak{T}|$ to store the assignment of each task. Firstly, it creates the first TCS by calling $chooseBestAssignment()$ to receive the next $S^*_{\mathcal{P}^*_{T,p}}$. This new assignment will be placed in TCS by calling $assignTask()$, which will locate the next available element for the new assignment. This process repeats until all the tasks are assigned with coalitions of agents, i.e. $unassigned(TCS) = false$. We treat the function $unassigned()$ as a blackbox which can simply scan the all elements of TCS to locate the first empty element and returns true. Once no empty element is found, it returns false. The algorithm then calculates for the remaining time before it can improve TCS. While the remaining time is greater than zero, the algorithm calls function $chooseTaskToBeImproved()$. The function will locate the task $taskToBeImprove$ which will be the shrinking point and alter the present agent coalition with the next best one. The algorithm goes into another loop to assign agent coalitions to the remaining unassigned task, starting from $taskToBeImprove$.

Algorithm 3. Search for OTCS by assigning recursively the best task agent into the existing structure.

Require: $remainTime$
1: $initTCS[|\mathfrak{T}|]$
2: $elapsedTime \leftarrow 0$
3: $startTime \leftarrow presentTime$
4: **while** $unassigned(TCS) = true$ **do**
5:     $assignment \leftarrow chooseBestAssignment(Tasks)$
6:     $assignTask(TCS, assignment)$
7: **end while**
8: $elapsedTime \leftarrow presentTime - startTime$
9: $remainTime \leftarrow remainTime - elapsedTime$
10: **while** $remainTime > 0$ **do**
11:     $taskToBeImprove \leftarrow chooseTaskToBeImproved()$
12:     $unassignTask(taskToBeImproved)$
13:     **while** $unassigned(Tasks) = true$ **do**
14:         $assignment \leftarrow chooseBestAssignment(Tasks)$
15:         $assignTask(Tasks, assignment)$
16:     **end while**
17:     $elapsedTime \leftarrow presentTime - startTime$
18:     $remainTime \leftarrow remainTime - elapsedTime$
19: **end while**

2.5   Algorithm to Choose the Best Assignment

For each T, all agents will be ranked by their access costs in descending order, instead of scanning and keeping all the coalitions as in [1]. The next available coalition with minimal access cost of P_p can be found by collecting p available agents $\mathcal{L} \in \mathfrak{L}$ from the ranking. The algorithm begins by initiating $totalRatio$ as the benchmark for the best assignment. It then goes through, from large to small $|T|$, all of the remaining tasks. In each task, it goes through all the Ps whose size is not greater than $(|RL| - (|RT| - 1))$. For each P, the algorithm locates the next best coalition, S, by calling function $NextAvail()$, which will start looking for the first $|S|$ agents ranked by access cost to T. Note that we start scanning from the largest possible $|S|$ down to 1. The access cost $CA_{T,S}$ of S is then aggregated by calling the function $AccessCost(T, S)$. The total cost $,C_{\mathcal{P}^*_{T,p},S}$, for S executing T is then computed and followed by the reduction contribution $\bar{A}$ of this assignment $\langle T, S \rangle$. If $\bar{A}$ is smaller than $\bar{A}^*$, the value of $\bar{A}$ is kept as the new benchmark as well as $\langle T, S \rangle$ is kept is the best assignment. After going through all the remaining tasks and valid partitions, the best assignment is returned.

Algorithm 4. Choose best assignment

```
1:  Ā* ← MAX_DOUBLE
2:  for eachT ∈ RT do
3:      for each P ∈ T and |P| ≤ (|RL| − (|RT| − 1) do
4:          S ← NextAvail(T, |P|)
5:          CA_{T,S} ← AccessCost(T, S)
6:          C_{P*_{T,p},S} ← CA_{T,S} + CE*_{P_p}
7:          Ā ← C_{P*_{T,p},S}/|S|
8:          if Ā < Ā* then
9:              Ā* ← Ā
10:             assignment ← ⟨T, S⟩
11:         end if
12:     end for
13: end for
14: return assignment
```

3   Experiments

There are two dimensions in the data generation with respect to elapsed time; the quality of the solutions and the number of solutions generated. Since we are interested in the cost, we refer to the cost reduction as the quality of the solution. We assign to both dimensions the distribution patterns presented above. Hence we have NLRP-NLRP for an environment where the solutions are produced rarely in the early stage of the computation but are inc rapidly in later stage, and the costs are reduced rapidly in the early stage but are hardly reduced in later stage of the execution. In addition to this, we also take into account the coalition value distribution patterns used in [1]. However, we only limit the patterns to just two types, namely, CUP and CAP. Hence, we have 8 data distribution patterns altogether for this work. We allow 3,600,000,000 milliseconds (1 hour) for the agents to compute TCSs. The total cost at the begining of each data distribution pattern is around 100000. The time frame will be allocated to both deliberations according to the time allocation strategies. We used strategies already specified earlier in this work, i.e. $\langle 50, 40 \rangle, \langle 40, 40 \rangle, \langle 40, 30 \rangle, \langle 30, 30 \rangle, \langle 30, 25 \rangle, \langle 25, 25 \rangle$, which would recursively leave spare time for deliberations in later rounds as $10\%, 20\%, 30\%, 40\%, 50\%$, respectively. The results from our experiments are shown below. Note that we only choose to focus on the last part of the results because they are more important.

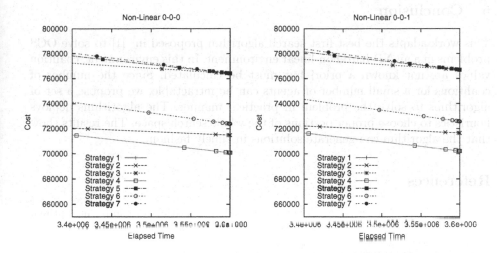

Fig. 1. Empirical Results NLRP-NLRP-CUP(0-0-0) and NLRP-NLRP-CAP(0-0-1). The graphs show solutions acheived from the seven time allocation strategies as per elapsed time.

4 Related Work

It is common in the literatures on multi-agent systems that agents optimize their tasks locally and deliberately exchange their tasks to increase the performance of the systems [2,5]. In [6], each agent in each possible coalition knows the set of the tasks they are required to do (agents just need to optimize the coalition values and the optimal coalition structures can be computed based on them.) Here, each agent seeks for a set of coalitions that are efficient for a given task. The optimal coalition structures will be searched from this larger set of coalitons. In [2], each agent is assigned a small set of tasks and the number of agents in the system is quite small. Our agents need to find a set of the most appropriate tasks (forming task-oriented coalitions before searching for optimal coalition structures) and the number of agents involved is larger. In [3], the algorithms split the whole search space into smaller portions based on the *configuration* of the coalition structure. By applying branch and bound, the algorithm can reach optimality rapidly. However, it can be mislead by the *upperbound* of portions and waste time for no improvement in the solutions. [1] proposes a best-first search algorithm to overcome the "mislead" problem. Their algorithm use the *agent contribution ratio* to decide which is the next *best* coalition to be chosen from available candidates and be placed in the coalition structure. Their extensive experiments show that the algorithm works much more robust than that of [3] in wider range of data distributions.

5 Conclusion

This work adapts the best first search algorithm proposed in [1] to solve OCS problems in a an NP-hard non-linear environment. In this environment, coalition values are not known a priori but must be calculated. Since the number of coalitions for a small number of agents can be intractable, we propose a set of algorithms to solve the problem in practical manner. The algorithms deploys heuristics to choose proper portions of the whole search space. The results show that the algorithm can generate solutions in timely fashion.

References

1. Sombattheera, C., Ghose, A.: A best-first anytime algorithm for computing optimal coalition structures. In: Proceedings of the 7th International Joint Conference on Autonomous Agents and Multiagent Systems (AAMAS 2008), pp. 1425–1428. ACM Press, New York (2008)
2. Shehory, O., Kraus, S.: Feasible formation of coalitions among autonomous agents in non-super-additive environments. Computational Intelligence 15(3), 218–251 (1999)
3. Rahwan, T., Ramchurn, S., Dang, V., Giovannucci, A., Jennings, N.: Anytime optimal coalition structure generation. In: Proceedings of the 22nd National Conference on Artificial Intelligence (AAAI 2007), Vancouver, Canada, pp. 1184–1190. AAAI Press, Menlo Park (2007)
4. Rahwan, T., Jennings, N.: Distributing coalitional value calculations among cooperating agents. In: Proceedings of the 25th National Conference on Artificial Intelligence (AAAI 2005), Pittsburgh, USA, pp. 152–157. AAAI Press, Menlo Park (2005)
5. Kraus, S., Shehory, O., Taase, G.: The advantages of compromising in coalition formation with incomplete information. In: Proceedings of the 3rd International Joint Conference on Autonomous Agent and Multi Agent Systems (AAMAS 2004), Washington DC, USA, pp. 588–595. IEEE Computer Society, Los Alamitos (2005)
6. Sandholm, T., Lesser, V.: Coalition formation among bounded rational agents. In: Proceedings of the 14th International Joint Conference on Artificial Intelligence (IJCAI 1995), Montreal, Canada, pp. 662–669. Kaufman Morgan, San Francisco (1995)

Moral Hazard Resolved by Common-Knowledge in S5n Logic

Takashi Matsuhisa*

Department of Natural Sciences, Ibaraki National College of Technology
Nakane 866, Hitachinaka-shi, Ibaraki 312-8508, Japan
mathisa@ge.ibaraki-ct.ac.jp

Facit indignatio versum
D. L. Luvenalis

Facit indignatio scientia
T. M.

Abstract. This article investigates the role of common-knowledge in the principal-agent model under asymmetric information. We treat the problem: How the common-knowledge condition will be able to settle a moral hazard problem in the principal-agents model under asymmetric information. We shall propose a solution program for the moral hazard in the principal-agents model under asymmetric information by common-knowledge. Let us assume that the agents have the knowledge structure induced from a partition relation associated with the multimodal logic **S5n**. In particular we consider the situation that the agents commonly know all decision values of the other agents. Under certain assumptions we shall show the moral hazard can be resolved in the principal-agents model when all the expected marginal costs are common-knowledge among the principal and agents.

Keywords: Agreeing to Disagree, Common-Knowledge, Moral hazard, Principal-agents model under uncertainty, **S5n**-logic.

AMS 2000 Mathematics Subject Classification: Primary 91A35, Secondary 03B45.

Journal of Economic Literature Classification: C62, C78.

1   Introduction

This article considers the relationship between common-knowledge and agreement in multi-agent system. How we capture the fact that the agents agree

* Partially supported by the 2009 Special Grant-in-Aid for Education and Science of the MEXT (Ministry of Education, Culture, Sports and Science and Technology) in Japan: 'Regeneration of National College of Technology as the Big Name: Building up Educational System of National College of Technology to Bringing up Excellent Students Playing an Active Role in the International Stage as Businesspersons.'

on an event or they get consensus on it? We treat the problem from Fuzzy set theoretical flavour. The purposes are first to introduce common-knowledge structure on multi-agent system, and by which we show that all agents can agree on an event, and second to apply the result to solving the moral hazard in a principal-agents model under asymmetric information. Let us consider that there are agents more than two and the agents have the structure of the Kripke semantics for the multi-modal logic **S5n**.

Assume that all agents have a common probability measure. By i's decision value of an event under agent i's private information, we mean the conditional probability value of the event under agents' private information. We say that all agent can agree on the event if all the the membership values are equal.

Aumann [1] considered the situation that the agents have common-knowledge of the membership values; that is, simultaneously everyone knows the membership values, and everyone knows that 'everyone knows the values' and everyone knows that "everyone knows that 'everyone knows the values'" and so on. He showed the famous agreement theorem for a partition information structure equivalent to the multi-modal logic **S5n**, and Samet [6] and Matsuhisa & Kamiyama [4] extend the theorem to models corresponding to weaker logics (e.g. **S4n** etc.)

Theorem 1 (Aumann [1]). *The agents can agree on an event if all membership values of the event under private information are common-knowledge among them.*

We shift our attention to the principal-agents model as follows: an owner (principal) of a firm hires managers (agents), and the owner cannot observe how much effort the managers put into their jobs. In this setting, the problem known as the moral hazard can arise: There is no optimal contract generating the same effort choices for the owner and the managers. We apply Theorem 1 to solve the problem. The aim is to establish that

Theorem 2. *The owner and the managers can reach consensus on their expected marginal costs for their jobs if their expected marginal costs are common-knowledge.*

This article is organised as follows. In Section 2 we describe the moral hazard in our principal-agents model. Sections 3 and 4 introduce the notion of common-knowledge associated with partitional (reflexive, transitive and symmetric) information structure and the notion of decision function. Section 5 gives the formal statement of Theorem 1. In Section 6 we introduce the formal description of a principal-agents model under asymmetric information. We will propose the program to solve the moral hazard in the model: First the formal statement of Theorem 2 is given, and secondly what further assumptions are investigated under which Theorem 2 is true. In the final section we conclude with remarks.

2   Moral Hazard

Let us consider the sinario: There are a president and n faculty members in a National College of Technology. The president wishes increasing the income of the college, and he/she frequently encourages faculty members to get any research grants awarded from outside organizations. Because the 30 percent of the amount of each research grant awarded from outside organizations goes for overhead costs to the college. Actually the research activity of almost faculty members is very low, and the amount of research grants is poor in comparing with the other Colleges. The president considers this situation comes from the lower stage of the incentives of the faculty members for their research activities. The president proposes to the faculty member the plan to keep their research activities up: The all overhead costs will be refunded to all faculty members for improving their incentive to research activities. The amount of the refund costs to each member shall be in proportion to not only his/her efforts to working scientific research but also his/her efforts to educational contribution to the college, where the former contribution shall be evaluated by the number of publishing academic papers and the latter contribution shall be evaluated by the grades of questionnaires for his/her lessons and by his/her school duties excluding his/her research activities.

Formally we intrduce the *principal-agents model* as follows: There are the principal P and n agents $\{1, 2, \cdots, k, \cdots, n\}$ $(n \geq 1)$ in a firm. The principal makes a profit by selling the productions made by the agents. He/she makes a contact with each agent k that the total amount of all profits is refunded to each agent k in proportion to the agent's contribution to the firm. The intended interpretation is as follows: the form is a College of Technology and a agent is a faculty member of the college.

Let e_k denote the measuring managerial effort for k's productive activities. The set of possible efforts for k is denoted by E_k with $E_k \subseteq \mathbb{R}$. Let $I_k(\cdot)$ be a real valued continuously differentiable function on E_k. It is interpreted as the profit by selling the productions made by the agent k with the cost $c(e_k)$. Here we assume $I'_k(\cdot) \geq 0$ and the cost function $c(\cdot)$ is a real valued continuously differentiable function on $E = \cup_{k=1}^n E_k$. Let I_P be the total amount of all the profits:

$$I_P = \sum_{k=1}^n I_k(c_k).$$

The principal P cannot observe these efforts e_k, and shall view it as a random variable on a probability space (Ω, μ). The optimal plan for the principal then solves the following problem:

$$\text{Max}_{e=(e_1, e_2, \cdots, e_k, \cdots, e_n)} \{\text{Exp}[I_P(e)] - \sum_{k=1}^n I_k(e_k)\}.$$

Let $W_k(e_k)$ be the total amount of the refund to agent k:

$$W_k(e_k) = r_k I_P(e),$$

with $\sum_{k=1}^n r_k = 1, 0 \leq r_k \leq 1$, where r_k denotes the proportional rate representing k's contribution to the firm. The optimal plan for each agent also solves the problem: For every $k = 1, 2, \cdots, n$,

$$\text{Max}_{e_k} \{\text{Exp}[W_k(e_k)] - c(e_k)\}$$

$$\text{subject to } \sum_{k=1}^{n} r_k = 1, 0 \le r_k \le 1.$$

We assume that r_k is independent of e_k, and the necessity conditions for critical points are as follows: For each agent $k = 1, 2, \cdots, n$, we obtain

$$\frac{\partial}{\partial e_k} \text{Exp}[I_k(e_k)] - c'(e_k) = 0$$

$$r_k \frac{\partial}{\partial e_k} \text{Exp}[I_k(e_k)] - c'(e_k) = 0$$

in contraction. This is called the **moral hazard** in the principal-agents model.

In the above mentioned scenario, the contradictory situation is interpreted as that the president thinks the faculty members not to do their best to produce scientific publications corresponding to the refund they obtained from the college, even when each member is in full activity for his/her scientific researches.

3   Common-Knowledge

Let N be a set of finitely many agents and k denote an agent. The specification is that $N = \{P, 1, 2, \cdots, k, \cdots, n\}$ consists of the principal P and the agens $\{1, 2, \cdots, k, \cdots, n\}$ in a firm. A state-space Ω is a non-empty set, whose members are called *states*. An *event* is a subset of the state-space. If Ω is a state-space, we denote by 2^{Ω} the field of all subsets of it. An event E is said to occur in a state ω if $\omega \in E$.

3.1   Information and Knowledge

By *partition information structure* we mean $\langle \Omega, (\Pi_i)_{i \in N} \rangle$ in which $\Pi_i : \Omega \to 2^{\Omega}$ satisfies the three postulates: For each $i \in N$ and for any $\omega \in \Omega$,

Ref $\omega \in \Pi_i(\omega)$;
Trn $\xi \in \Pi_i(\omega)$ implies $\Pi_i(\xi) \subseteq \Pi_i(\omega)$.
Sym   If $\xi \in \Pi_i(\omega)$ then $\omega \in \Pi_i(\xi)$.

This structure is equivalent to a Kripke semantics for the multi-modal logic **S5n**. The set $\Pi_i(\omega)$ will be interpreted as the set of all the states of nature that i knows to be possible at ω, or as the set of the states that i cannot distinguish from ω. We call $\Pi_i(\omega)$ i's *information set* at ω.

We will give the formal model of knowledge as follows (C.f.; Fagin et al [2].)

Definition 1. The **S5**-*knowledge structure* is a tuple $\langle \Omega, (\Pi_i)_{i \in N}, (K_i)_{i \in N} \rangle$ that consists of a partition information structure $\langle \Omega, (\Pi_i)_{i \in N} \rangle$ and a class of i's *knowledge operator* $K_i : 2^{\Omega} \to 2^{\Omega}$ defined by

$$K_i E = \{\omega \mid \Pi_i(\omega) \subseteq E\ \}$$

The event $K_i E$ will be interpreted as the set of states of nature for which i knows E to be possible.

We record the properties of i's knowledge operator: For every E, F of 2^Ω,

N $K_i \Omega = \Omega$;
K $K_i(E \cap F) = K_i E \cap K_i F$;
T $K_i E \subseteq E$
4 $K_i E \subseteq K_i(K_i E)$.
5 $\Omega \setminus K_i E \subseteq K_i(\Omega \setminus K_i E)$.

According to these properties we can say the structure $\langle \Omega, (K_i)_{i \in N} \rangle$ is a model for the multi-modal logic **S5n**.

3.2   Common-Knowledge and Communal Information

The *mutual knowledge operator* $K_E : 2^\Omega \to 2^\Omega$ is the intersection of all individual knowledge operators:

$$K_E F = \cap_{i \in N} K_i F,$$

which interpretation is that everyone knows E.

Definition 2. The *common-knowledge operator* $K_C : 2^\Omega \to 2^\Omega$ is defined by

$$K_C F = \cap_{n \in \mathbb{N}} (K_E)^n F.$$

The intended interpretations are as follows: $K_C E$ is the event that 'everyone knows E' and "everyone knows that 'everyone knows E'," and "'everybody knows that "everyone knows that 'everyone knows E'," .'" An event E is *common-knowledge* at $\omega \in \Omega$ if $\omega \in K_C E$.

Let $M : 2^\Omega \to 2^\Omega$ be the dual of the common-knowledge operator K_C:

$$ME := \Omega \setminus K_C(\Omega \setminus E).$$

By the *communal* information function we mean the function $M : \Omega \to 2^\Omega$ defined by $M(\omega) = M(\{\omega\})$. It can be plainly observed that the communal information function has the following properties:

Proposition 1. *Notations are the same as above. Then*

(i) $\omega \in K_C E$ *if and only if* $M(\omega) \subseteq E$
(ii) *For every* $i \in N$, $M(\omega)$ *can be decomposed into the disjoint union of the components* $\Pi_i(\xi)$ *for* $\xi \in M(\omega)$: *i.e.,* $M(\omega) = \sqcup_{\xi \in M(\omega)} \Pi_i(\xi)$.

Proof. Is easily given. (See, Fagin et al [2].)   □

4   Decision Function and Consensus

Let Z be a set of decisions, which set is common for all agents. By a *system decision function* we mean a mapping f of $2^\Omega \times 2^\Omega$ into the set of decisions Z. We refer the following property of the function f: Let X be an event.

Disjoint Union Consistency (DUC): For every pair of disjoint events S and T, if $f(X;S) = f(X;T) = d$ then $f(X;S \cup T) = d$;

By i's *decision function* associated with f under agent i's private information we mean the function d_i from $2^\Omega \times \Omega$ into Z defined by $d_i(X;\omega) = f(X;\Pi_i(\omega))$, and we call $d_i(X;\omega)$ the i's *decision value* of X associated with f under agent i's private information at ω.

Definition 3. We say that all agent can *agree on* an event X (or *consensus* on X can be guaranteed among all agents) if $d_i(X;\omega) = d_j(X;\omega)$ for any agent $i,j \in N$ and in all $\omega \in \Omega$.

Remark 1. If f is intended to be a posterior probability, we assume given a probability measure μ on a state-space Ω which is common for all agents; precisely, for some event X of Ω, $f(X;\cdot)$ is given by $f(X;\cdot) = \mu(X|\cdot)$. Then the i's decision value of X is the conditional probability value $d_i(X;\omega) = \mu(X|\Pi_i\omega))$. Consensus on X guaranteed among all agents can be interpreted as that the fuzzy sets (X, d_i) and (X, d_j) are equal for any $i,j \in N$.

5   Agreeing to Disagree Theorem

We can now state explicitly Theorem 1 as below: Let D be the event of the i's decision values of an event X for all agents at ω, which is defined by

$$D = \cap_{i \in N}\{\xi \in \Omega \mid d_i(X;\xi) = d_i(X;\omega)\}.$$

Theorem 3. *Assume that the agents have the* **S5n**-*knowledge structure and the system decision function f with satisfying the condition (DUC). If all the agent commonly know an event X then they cannot agree on disagree on it: Formally, if $\omega \in K_C D$ then $d_i(X;\omega) = d_j(X;\omega)$ for any agents $i,j \in N$ and in all $\omega \in \Omega$.*

Proof. By Proposition 1 it is plainly observed that

$$M(\omega) = \sqcup_{\xi \in M(\omega)}\Pi(\xi) \subseteq D \subseteq \{\xi \in \Omega \mid d_i(X;\xi) = d_i(X;\omega)\}.$$

On noting that $d_i(X;\xi) = d_i(X;\omega)$ for any $\xi \in M(\omega)$, it can be observed by (DUC) that $f(X;M(\omega)) = d_i(X;\omega)$ for every $i \in N$, and thus $d_i(X;\xi) = d_j(X;\omega)$ for any $i,j \in N$.   □

6 Moral Hazard Revisited

This section investigates the moral hazard problem from the common-knowledge view point. Let us reconsider the principal-agents model and let notations and assumptions be the same in Section 2. We show the evidence of Theorem 2 under additional assumptions **A1-2** below. This will give a possible solution of our moral hazard problem.

A1 The principal P has a partition information $\{\Pi_P(\omega) \mid \omega \in \Omega\}$ of Ω, and each agent k has also his/her a partition information $\{\Pi_k(\omega) \mid \omega \in \Omega\}$:

A2 For each $\omega, \xi \in \Omega$ there exists the decision function $f : 2^\Omega \times 2^\Omega \to \mathbb{R}$ satisfying the Disjoint Union Consistency together with

(a) $f(\{\xi\}; \Pi_P(\omega)) = \frac{\partial}{\partial e_0(\xi)} \mathrm{Exp}[I_P(e) | \Pi_P(\omega)]$;

(b) $f(\{\xi\}; \Pi_k(\omega)) = \frac{\partial}{\partial e_k(\xi)} \mathrm{Exp}[W_k(e) | \Pi_k(\omega)]$

We have now set up the *principal-agents model under asymmetric information*.

The optimal plans for principal P and agent k are then to solve

PE $\mathrm{Max}_{e=(e_1, e_2, \cdots, e_k, \cdots, e_n)} \{\mathrm{Exp}[I_P(e) | \Pi_P(\omega)] - \sum_{k=1}^n I_k(e_k)\}$;

AE $\mathrm{Max}_{e_k} \{\mathrm{Exp}[W_k(e_k) | \Pi_k(\omega)] - c(e_k)\}$ subject to $\sum_{k=1}^h r_k = 1, 0 \le r_k \le 1$.

From the necessity condition for critical points together with **A2** it can been seen that the principal's marginal expected costs for agent k is given by

$$c_P'(e_k(\xi)) = f(\xi; \Pi_P(\omega)),$$

and agent k's expected marginal costs is also given by

$$c_k'(e_k(\xi)) = f(\xi; \Pi_P(\omega)).$$

To establish this solution program we have to solve the problem: Construct the information structure together with decision function such that the above conditions **A1** and **A2** are true.

Under these circumstances, a resolution of the moral hazard given by Theorem 2 will be restate as follows: We denote

$$[c'(e(\xi))] = \cap_{i \in N} \{\zeta \in \Omega | f(\xi; \Pi_i(\zeta)) = f(\xi; \Pi_i(\omega))\}.$$

interpreted as the event of all he expected marginal costs.

Theorem 4. *Under the conditions* **A1** *and* **A2**, *we obtain that for each* $\xi \in \Omega$, *if* $\omega \in K_C([c'(e(\xi))])$ *then* $c_P'(e_k(\xi)) = c_k'(e_k(\xi))$ *for any* $k = 1, 2, \cdots, n$.

This interpreted as that the principal and the agents commonly know their expected marginal costs then they can reach consensus on them, and so there is no moral hazard.

Proof. Follows immediately from Theorem 3 in the case that $X = [c'(e(\xi))]$. $\square$

Remark 2. To establish Theorem 4 we have to solve the problem: Construct the information structure $(\Pi_i)_{i \in N}$ together with decision function f such that the above conditions **A1** and **A2** are true.

7 Concluding Remarks

It is well to end our remarks on additional problems for making further progresses:

1. If the proportional rate r_k representing k's contribution to the college depends only on his/her effort for research activities in the principal-agents model, what solution can we have for the moral hazard problem?
2. Can we construct a communication system for the principal agents model in the line of Parik &Krasucki [5]? Where each agent and the principal communicate privately each other about their expected marginal cost as messages: The principal sends his expected marginal cost as messages. The agent as the recipient of the message revises his/her information structure and recalculates the expected marginal cost under the revised information structure, and he/she sends the revised expected marginal cost to the principal. The principal as the recipient of the message revises his/her information structure and recalculates the expected marginal cost under the revised information structure, and he/she sends the revised expected marginal cost to the agent, and so on. In the circumstance does the limiting expected marginal costs actually coincide ? Matsuhisa (2008) introduces a fuzzy communication system and extends Theorem 3 in the communication model. By using this model Theorem 4 can be extended in the communication framework, and the detail will be reported in near future.

References

1. Aumann, R.J.: Agreeing to disagree. Annals of Statistics 4, 1236–1239 (1976)
2. Fagin, R., Halpern, J.Y., Moses, Y., Vardi, M.Y.: Reasoning about Knowledge. MIT Press, Cambridge (1995)
3. Matsuhisa, T.: Fuzzy communication reaching consensus under acyclic condition. In: Ho, T.-B., Zhou, Z.-H. (eds.) PRICAI 2008. LNCS (LNAI), vol. 5351, pp. 760–767. Springer, Heidelberg (2008)
4. Matsuhisa, T., Kamiyama, K.: Lattice structure of knowledge and agreeing to disagree. Journal of Mathematical Economics 27, 389–410 (1997)
5. Parikh, R., Krasucki, P.: Communication, consensus, and knowledge. Journal of Economic Theory 52, 178–189 (1990)
6. Samet, D.: Ignoring ignorance and agreeing to disagree. Journal of Economic Theory 52, 190–207 (1990)

An Algorithmic Approach to Social Knowledge Processing and Reasoning Based on Graph Representation – A Case Study

Zbigniew Tarapata, Mariusz Chmielewski, and Rafał Kasprzyk

Military University of Technology, Cybernetics Faculty,
Gen. S. Kaliskiego Str. 2, 00-908 Warsaw, Poland
zbigniew.tarapata@wat.edu.pl,
mariusz.chmielewski@wat.edu.pl,
rafal.kasprzyk@wat.edu.pl

Abstract. The paper concludes our ideas on new concepts of indirect association analysis to extract useful information for terrorist threat indication. The method introduces an original approach to knowledge representation as a semantic model, which is further processed by the inference algorithms and structure graph analysis towards a complex network. Described models consist of experience gathered from intelligence experts and several open Internet knowledge systems such as the Global Terrorism Database [15]. We have managed to extract core information from several ontologies and fuse them into one domain model aimed at providing basis for indirect associations identification method.

1 Introduction

Studies on this subject received government and military attention, which required use of sophisticated methods for detection of criminal networks and individuals associated with them. Conclusions gathered through the multi-analysis allow to effectively monitor the activities of criminal organizations and help to identify the phases of actions being prepared or in progress. Opposing irregular groups require the ability to analyze large data sets, which can be unreliable or incomplete. Researchers have shown that the analysis of activities of terrorist organizations is possible, however, requires the use of interdisciplinary knowledge [14]. In recent years rapid development of information technology, have highlighted the need to introduce methods, algorithms and systems providing precision tools for analysts to support the concept of Semantic Web [10]. This idea is associated with, what is being called, the next stage of development of the Internet grouped around the description languages and ontologies with a strong theoretical basis for inference mechanisms [7].

Semantic net gives one of the most important advantages for terrorist activities data representation – scalability and flexibility of knowledge representation. Presented method of semantic network analysis and association acquiring, aims at: providing a tool for operating on large information resources, eliminating the unreliable and unwanted information within the semantic network (essential requirement due to algorithm complexity), selecting significant nodes and relations between them (for the

N.T. Nguyen, M.T. Le, and J. Świątek (Eds.): ACIIDS 2010, Part II, LNAI 5991, pp. 93–104, 2010.

analysis) [2], searching the indirect relations between the nodes in the semantic network (building the new knowledge in the system).

The presented idea uses Social Network analysis in anti-terrorism applications and indicates the usefulness of Social Network analysis as a basis for quantitative methods for situation awareness and decision-making in law-enforcement applications. Figuring out nested connections across a known set of individuals or organizations is direct Social Network application. Since not all people who have had contact with a terrorist are criminals themselves, there is a need for techniques, which are able to efficiently filter those who have frequent contacts with known or suspected individuals, or with any member of a known or suspected group of terrorists. Such people become more or less suspects themselves, thereby potentially spreading the suspicion to even more individuals. One issue, thus becomes how one can automatically estimate, which individuals among a very large community, who have been "transitively" in contact with each other, need to be investigated further and who do not. We explore such methods of Social Networks analysis using Semantic Networks and Complex Networks as a model of a terrorist organization [13]. Moreover, we define the multicriteria weighted graph similarity decision problem (MWGSP). Presenting idea is a significant improvement of social network analysis and we show how to use it in anti-terrorism application. Figuring out indirect associations and in the end nested connection, across a known set of individuals or an organization is one example of our network centric analysis. Another problem we have been able to solve is the question how one can automatically estimate, which people among a very large community, who have been transitively in contact with each other, need to be investigated further and who do not.

2   Semantic Model and Ontology Application: A Case Study

Emerging new technologies introduced new standards for data representation, which extends relational and object models. Semantic description can improve the way information is presented. Search mechanisms are able to provide clustered data based on semantics and provide even more complex operations such as merging information from all relevant documents, removing redundancy, and summarizing where appropriate. Semantic data representation is based on a graph model and is based on RDF (Resource Definition Framework) concept of triples representing resources and data describing them. Triples identify subject-predicate-object expressions that is subject denotes the resource, predicate denotes traits or aspects of the resource and expresses a relationship between the subject and the object.

Supporting these features is the ability to reason, which can be explained as inferencing from the existing knowledge to create new facts. In case of semantic graphs inference mechanisms are based on Description Logics (DL) or specialized reasoner using developed graph-based algorithms such as path finding, vertices measures evaluation (sect.4.1), graph similarity (sect.4.2), etc. Considering the structure of knowledge representation, similar to association memory, building new knowledge can be described as introducing new nodes or links between nodes to existing graphs. The use of semantic metadata is also crucial for integrating information from heterogeneous sources, whether within one organisation or across organisations.

Ontology is used as a tool for describing and representing selected knowledge branches that is medicine, finances, battlefield etc. Definitions inside ontology must be defined precisely and uniquely from the logic point of view. For model representation we choose Description Logic and an appropriate dialect for expressing our ontology. For selecting currently available semantic languages we have based our model on OWL DL, which is the direct offspring of the description logic of $\mathcal{SHOIN}$ $(\mathcal{ALCR_+HOIN})$ [7]. We can present our model twofold: one is the graph representation of the ontology and the DL axioms provided by the OWL DL statements of the designed knowledge base. As an example we propose some definitions of identified classes reflecting modeling requirements of the terrorist domain presented in our approach (Fig.1).

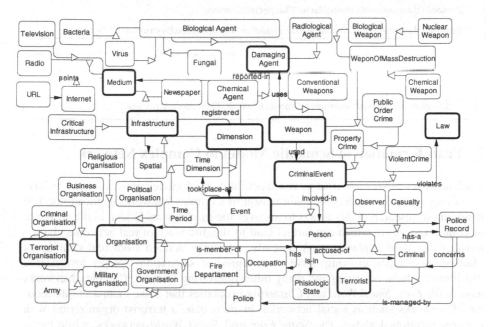

Fig. 1. Semantic graph model presenting core elements using ontology relations

Purpose and leading idea of the model is to track and infer elementary events registered in all available monitored mediums. Using reasoning facilities of DL and additional SRWL rules combined with Jess environment [8], we have been able to provide a knowledge base capable of representing registered events and people involved in them. For exploiting DL semantics we propose the primitive classes such as (Event, Person, Organization, Medium, etc) further constructing the defined classes using expressions consisting of value and cardinality restrictions. Proposed sentences reflect identified sample structural relations among main entities in the developed model [6]. As an example we define a criminal organization using this statement:

"An organization whose members took part in an event, in which registered law-breaking has taken place is said to be a Suspected Criminal Organisation."

Sentences further can be transformed into DL class expressions defining concepts in the TBox [7].

> Organisation⊓∃has-member.(Person⊓∃ *conviced-of.* CriminalEvent)≡
> SuspectedCriminalOrganisation

A similar sentence applies to the Terrorist, further capturing similar relations crucial to the social relations analysis:

> Person⊓∃is-member-of.CriminalOrganisation≡ Criminal
>
> Person⊓∃ *conviced-of.*CriminalEvent≡Criminal
>
> Person⊓∃is-member-of.TerroristOrganisation≡ Terrorist
>
> Person⊓∃ *accused-of.*CriminalEvent≡SuspectedCriminal
>
> Person⊓∃ *accused-of.*(CriminalEvent⊓∃ *used-in.*WeponOfMassDestruction)≡Terrorist

We can also provide semantics for interpreting the relations for infrastructure elements:

> Infrastructure ⊓ (∃ *is-headquaters-of.*MilitaryOrganisation ⊔
>
> ∃ *is-headquarters-of.*GovernmentOrganisation) ≡ CriticalInfrastructure

3 Transformation of Semantic Model to Complex Network

Until now, scientists have tried to construct theoretical models describing the behavior of real systems, which is the main reason of Complex Networks applications. The main aim of research in this area is to uncover the mechanisms hidden in the structure of complex systems, which can further lead to the discovery of real networks characteristics and their explanation. Apparently, networks derived from real data (most often spontaneously growing) have "six degree of separation", power low degree distributions, hubs occurring and many other interesting features. Complex Networks have *Scale Free*, *Small Word* and *Clustering* properties that make them accurate models of networks such as social networks, in particular, a terrorist organization with features mentioned above. The *Scale Free* and *Small World* networks, while being fault tolerant, are still very prone to acts of terrorism. The *Scale Free* feature distinguish immunity against random attacks (it is hard to hit a hub). The *Small World* feature can dramatically affect communication among network nodes. Thus both concepts and underlying theories are highly pertaining to the presenting idea subject and objectives [22]. For the purpose of this work we have developed a transformation from a created semantic network into a set of Complex Networks.

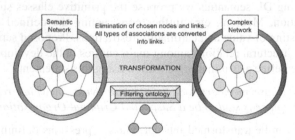

Fig. 2. The transition between a semantic network and a complex network using ontology filtering

First we have to choose the ontology, which is most significant from the analysis point of view. It leads to leave only a subset of nodes and edges connecting them. At this moment we have produced graph with uniform node and edge type. As a result of transformation, one of possible Complex Networks has been generated. In order to find representation of a terrorist organization as a complex network, we should apply the algorithm presented in the Fig.2.

Formally, we can write transformation T of semantic network S_1 into complex network S_2 as follows: $T : S_1 \xrightarrow{FO} S_2$, where FO describes filtering ontology from Fig.2,

where: $S_1 = \left\langle G_1 = \langle N_1, A_1 \rangle, \{f_{i1}(n)\}_{\substack{i \in \{1,...,LF_1\} \\ n \in N_1}}, \{h_{j1}(a)\}_{\substack{j \in \{1,...,LH_1\} \\ a \in A_1}} \right\rangle$, N_1, A_1 – sets of graph's nodes and arcs, respectively, $f_{i1} : N_1 \rightarrow Z_{i1}$ – the i-th function described on the graph's nodes, $i = 1,...LF_1$, (LF_1 – number of node's functions); $h_{j1} : A_1 \rightarrow Z_{j1}$ – the j-th function described on the graph's arcs, $j = 1,...LH_1$ (LH_1–number of arc's functions), Z_* – any set (e.g. types of vertices); S_2 – defined by analogy but it has a single function described on the nodes and arcs: $S_2 = \left\langle G_2 = \langle N_2, A_2 \rangle, \{f_2(n)\}_{n \in N_2}, \{h_2(a)\}_{a \in A_2} \right\rangle$.

4   Models and Methods for Complex Networks Analysis

4.1   Standard Centrality Measures

Turning to the analysis of network, we start by introducing centrality measures, which are the most fundamental and frequently used measures of network structure. The central vertices in Complex Networks are of particular interest because they might play the role of organization hubs. Centrality measures address the question "Who (what) is the most important or central person (node) in the given network?". Here, we consider five centrality measures: degree, radius, closeness, betweenness and eigenvector.

Let us accept the following notations: k_i - the degree of a vertex v_i, d_{ij} - the shortest path length between nodes i and j in a graph (number of links in it), A_{ij} – an element of graph adjacency matrix, $A_{ij}=1$, if there is an edge (arc) between vertices i and j, and 0 otherwise, V – set of graph vertices, $n = |V|$, m – number of arcs.

The degree centrality measure gives the highest score of influence to the vertex with the largest number of first-neighbors. This agrees with the intuitive way to estimate someone's influence from the size of his immediate environment: $k_i = \sum_{j=1}^{n} A_{ij}$. In a network of n vertices, the degree centrality of vertex i, is defined as:

$$Center_i^{Degree} = \frac{k_i}{n-1} \qquad (1)$$

If we need to find influential nodes in an area modeled by the network it is quite natural to use the radius centrality measures [11], which choose the vertex using the pessimist's criterion. So the radius centrality of vertex i, can be defined as:

$$Center_i^{Radius} = \frac{1}{\max_j d_{ij}} \qquad (2)$$

This notion of centrality focuses on the idea of communications between different vertices. Since these measures are defined as 'closeness', the inverse of the mean distance of a vertex from all others is used [18]. Hence, if d_{ij} is the shortest distance between vertices i and j in terms of edge steps:

$$Center_i^{Closeness} = \left[\frac{\sum_{j \in V} d_{ij}}{n-1} \right]^{-1} = \frac{n-1}{\sum_{j \in V} d_{ij}} \qquad (3)$$

Betweenness (load) Centrality

If (v_x, v_i, v_y) is the set of all shortest paths between vertices v_x and v_y passing through vertex v_i and (v_x, v_y) is the set of all shortest paths between vertices v_x and v_y, then [8]:

$$Center_i^{Load} = \frac{\sum_{v_x \in V} \sum_{v_y \neq v_x \in V} \frac{|(v_x, v_i, v_y)|}{|(v_x, v_y)|}}{(n-1) \cdot (n-2)} \qquad (4)$$

This definition of centrality explores the ability of a vertex to be 'irreplaceable' in the communication of two random vertices. It is of particular interest in the study of network attacks, because at any given time the removal of the maximum betweenness vertex seems to cause maximum damage in terms of connectivity and mean distance in the network.

Where degree centrality gives a simple count of the number of connection a vertex has, eigenvector centrality [4] acknowledges that not all connections are equal. In general, connections to people who are themselves influential will give a person more influence than connections to less influencial people. If we denote the centrality of vertex i by e_i, then we can allow for this effect by making e_i proportional to the average of the centralities of the i's network neighbors.

$$\lambda - const., \quad e_i = \frac{1}{\lambda} \sum_{j=1}^{N} A_{ij} e_j, \quad \vec{e} = \frac{1}{\lambda} A \vec{e} \Rightarrow A \vec{e} = \lambda \vec{e} \qquad (5)$$

So we have $A \vec{e} - \lambda I \vec{e} = 0$ and the λ value we can calculate using $\det(A - \lambda I) = 0$. Hence, we see that $\vec{e}$ is an eigenvector of adjacency matrix with eigenvalue λ.

4.2 Multicriteria Approach to Weighted Graphs Similarity

In some cases the use of a graph to represent the Complex Network does not provide a complete description of the real-world systems under investigation. For instance, in a terrorist organization net represented as a simple graph, we only know whether terrorists are connected, but we cannot model the strength of these connections or the rank of individual terrorists. Moreover, single vertices measures presented in the previous section can be unsatisfactory for such an analysis. Therefore, we defined the multicriteria weighted graph similarity decision problem (MWGSP), which is based on our previous works [20], [21]. The presented idea is a significant improvement of a social network analysis and we will show how to use it in an anti-terrorism application. Let us define the weighted graph WG as follows:

$$WG = \left\langle G, \{f_i(n)\}_{\substack{i\in\{1,...,LF\} \\ n\in N_G}}, \{h_j(a)\}_{\substack{j\in\{1,...,LH\} \\ a\in A_G}} \right\rangle, \text{ where: } G - \text{Berge's graph, } G = \langle N_G, A_G \rangle,$$

N_G, A_G – sets of graph's nodes and arcs, $A_G \subset \{\langle n, n' \rangle : n, n' \in N_G\}$, $f_i : N_G \to R^n$ – the i-th function described on the graph's nodes, $i = 1,...LF$, (LF – number of node's functions); $h_j : A_G \to R^n$ – the j-th function described on the graph's arcs, $j = 1,...LH$ (LH – number of arc's functions).

Having two weighted graphs G_A and G_B we propose to calculate two types of similarities of the G_A and G_B: structural and non-structural (quantitative).

To calculate the structural similarity between G_A and G_B it is proposed to use the approach defined in [3]. We will obtain structural similarity matrix $S(G_A, G_B)$ between nodes of graphs G_A and G_B as follows:

$$S(G_A, G_B) = [s_{ij}]_{n_B \times n_A} \tag{6}$$

We can also use other structural similarity measures presented at the beginning of section 1 instead of the matrix distance measure. Having matrix $S(G_A, G_B)$, we can formulate and solve the optimal assignment problem (using e.g. Hungarian algorithm) to find the best allocation matrix $X = [x_{ij}]_{n_B \times n_A}$ of nodes from graph describing G_A, G_B

(see details in [20], [21]). The $d_S(G_A, G_B) = \sum_{i=1}^{n_B} \sum_{j=1}^{n_A} s_{ij} \cdot x_{ij} \to \max$ describes the value of

the *structural similarity measure* of G_A and G_B. Let us note that we can easily adopt centrality measures from sect.4.1 to use them or their combinations instead of s_{ij}.

To calculate non-structural (quantitative) similarity between G_A and G_B we should consider the similarity between function values of the node and arc (*nodes and arcs quantitative similarity*). To compute nodes quantitative similarity we propose to create a vector $v(G_A, G_B) = \langle V_1, ..., V_{LF} \rangle$ of matrices, where $V_k = [v_{ij}(k)]_{n_B \times n_A}$, $k = 1, ..., LF$, describing the similarity matrix between nodes of G_A and G_B from the point of view of the k-th node's function ($f_k^A : N_{G_A} \to R^n$ for G_A and $f_k^B : N_{G_B} \to R^n$ for G_B) and $v_{ij}(k) = \|f_k^B(i) - f_k^A(j)\|$ describes "distance" between the i-th node of G_B and the j-th node of G_A from the point of view of f_k^B and f_k^A, respectively. We can apply the norm with parameter $p \ge 1$ as the distance measure:

$$\|f_k^B(i) - f_k^A(j)\| = \|f_k^B(i) - f_k^A(j)\|_p = \left(\sum_{r=1}^{n} |f_{k,r}^B(i) - f_{k,r}^A(j)|^p \right)^{1/p} \tag{7}$$

where $f_{k,r}^A(\cdot)$, $f_{k,r}^B(\cdot)$ describe the r-th component of the vector being the value of f_k^A and f_k^B, respectively. Next, we computed for each $k = 1, ..., LF$ a normalized matrix $V_k^* = [v_{ij}^*(k)]_{n_B \times n_A}$, where $v_{ij}^*(k) = v_{ij}(k) / \|V_k\|_F$. This procedure guarantees that each

$v_{ij}^*(k) \in [0,1]$. Finally, we compute total quantitative similarity between the i-th node of G_B and the j-th node of G_A as follows:

$$\bar{v}_{ij} = \sum_{k=1}^{LF} \lambda_k \cdot v_{ij}^*(k), \quad \sum_{k=1}^{LF} \lambda_k = 1, \quad \underset{k=1,...,LF}{\forall} \lambda_k \in [0,1] \tag{8}$$

The $d_{QN}(G_A, G_B)$ *nodes quantitative similarity measure* of G_A and G_B we compute solving the previous assignment problem substituting $-\bar{v}_{ij}$ for s_{ij} (because the smaller value of $\bar{v}_{ij}$ the better) and $d_{QN}(G_A, G_B)$ for $d_S(G_A, G_B)$ (see details in [20], [21]). We can obtain *arcs quantitative similarity measure* $d_{QA}(G_A, G_B)$ by analogy to $d_{QN}(G_A, G_B)$ [20].

Let $SG = \{G_1, G_2, ..., G_M\}$ be given as a set of weighted graphs defining some objects. Moreover, we have weighted graph P that defines some pattern objects. The problem is to find such a graph G^o from SG that is most similar to P. We define this problem as the multicriteria weighted graphs similarity problem (*MWGSP*), that is the multicriteria optimization problem in the space SG with relation R_D:

$$MWGSP = (SG, F, R_D) \tag{9}$$

where $F: SG \rightarrow R^3$,

$$F(G) = \left(d_S(P,G), d_{QN}(P,G), d_{QA}(P,G)\right) \tag{10}$$

and

$$R_D = \left\{ \begin{array}{c} (Y,Z) \in SG \times SG : d_S(P,Y) \geq d_S(P,Z) \wedge \\ d_{QN}(P,Y) \leq d_{QN}(P,Z) \wedge \\ d_{QA}(P,Y) \leq d_{QA}(P,Z) \end{array} \right\} \tag{11}$$

There are many methods for solving the problem (9). One of the methods, which can be applied, is the scalar function $H(G): SG \rightarrow R$ as a weighted sum of objectives:

$$H(G) = \alpha_1 \cdot d_S(P,G) + \alpha_2 \cdot \left(-d_{QN}(P,G)\right) + \alpha_3 \cdot \left(-d_{QA}(P,G)\right)$$
$$\alpha_1, \alpha_2, \alpha_3 \geq 0, \quad \alpha_1 + \alpha_2 + \alpha_3 = 1 \tag{12}$$

We can also use other methods for solving the *MWGSP* [20]: hierarchical optimization (the idea is to formulate a sequence of scalar optimization problems with respect to the individual objective functions subject to bounds on previously computed optimal values), the trade-off method (one objective is selected by the user and the other ones are considered as constraints with respect to individual minima), the method of distance functions in L_p-norm ($p \geq 1$) and others.

Taking into account (12) the problem of finding the most matched G^o to P can be formulated as follows: to determine such a $G^o \in SG$, that $H(G^o) = \max_{G \in SG} H(G)$.

5   An Experimental Analysis of a Terrorist Network

Resistance of a terrorist network on any losses of their members is most important for network's survival. There are at least three indicators of network's destruction: (1) the information flow through the network is seriously reduced; (2) the network, as a decision body, can no longer reach consensus; (3) the network, as an organization, loses the ability to effectively perform its task.

Robustness analysis of a terrorist organization [10] with *Scale Free* [1] and *Small World* [16], [19], [23] feature that prepared and executed the September 11, 2001 attacks show that this network is exceptionally resilient to the loss of a random node. This is a significant fact, if we remember that most of the nodes in a *Scale Free* network may have a small degree. To effectively destroy a *Scale Free* network, one must simultaneously remove at least 5% of the nodes in the central position. Apparently, terrorist networks have many other interesting features like "six degree of separation", power low degree distributions, hubs occurring and others. This kind of network presents a perfect example of a complex network [17].

According to Krebs' analysis [12] the network has 62 members, of which 19 were hijackers, and 43 supporters. The average degree of nodes is 4.9. Degree distribution of nodes is particularly interesting (degree from 1 to 4=>37 nodes, degree from 50 to 8=>19 nodes; degree from 9 to 12=>4 nodes, degree>=13 =>2 nodes). The degree of most nodes is low, while only few nodes have a high degree. It is easy to show, that degrees of nodes have an exponential distribution. This property characterizes *Scale Free* networks, rising spontaneously, without a particular plan or intervention of the architect. Nodes that are members of the network for a longer time are better connected with other nodes, and these nodes are more significant for the network's functionality and are also more "visible" to new members.

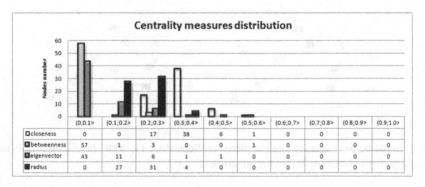

Fig. 3. Centrality measures distributions for the terrorists network that prepared and executed September 11, 2001 attacks

The average radius centrality of nodes is 0.23 (Fig.3). According to this the measure node is central, if it is the closest to the farthest node (min-max criterion). Four nodes (with the highest radius centrality value – 0.33) need three steps at most to get to the farthest node in the network. If it is critical that all members of the organization must share information in the shortest time, nodes with the highest radius centrality are the best start points for the propagation process. These nodes are Mohamed Atta, Zacarias Moussaoui, Ramzi Bin Al-Shibh and Imad Eddin Barakat Yarkas.

The average closeness of nodes is 0.33 (Fig.3). This centrality measure postulates that a node is more central, if it is closer on average to all other nodes. Many of nodes in terrorist network have a high value of closeness centrality e.g. in the range from 0.3 to 0.4. Seven nodes have even a higher value, especially one (Mohamed Atta) with closeness centrality value equal 0.59.

The average betweenness centrality of nodes (Fig.3) is 0.035 indicating very high average redundancy. Values of betweenness centrality for 57 are in the range from 0 to 0.1 (in fact even less than 0.01!). Only 5 nodes have betweenness centrality higher than 0.1. These 5 nodes are extremely critical for information flow in the analyzed network, especially the one (Mohamed Atta) with betweenness of almost 0.6, meaning that almost 60% of shortest path between nodes pass through this central node.

The average eigenvector centrality of nodes is 0.085 (Fig. 3). It assumes that a node is central, if it has many central neighbors. Two nodes have the highest eigenvector value 0.45 (Mohamed Atta) and 0.37 (Marwan Al-Shehhi).

In Fig. 4 we have a subnet of a terrorist network with two cases: a long time before the airplane hijackings (a) and a short time before the airplane hijackings (b). When we use the network from Fig. 4a as some "normal" communication between terrorists then we can use *MWGSP* problem solving to recognize the threat situation (structural similarity between the net from Fig. 4a and Fig. 4a (self-similarity) is equal 0,990 and between the net from Fig. 4a and Fig. 4b is equal 0,880: this difference can indicate a threat situation; we can also set some threshold values for similarity changes, which in experts' opinion indicates a threat situation). We can also take into account some quantitative characteristics of a terrorist network (e.g. frequency of communication between terrorists) to evaluate weighted terrorist graphs similarity (details in [21]).

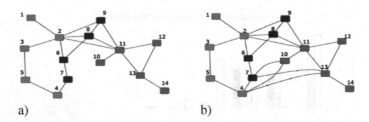

a)   b)

Fig. 4. Terrorist network that hijacked airplanes on September 11, 2001: (a) a long time before the hijacking; (b) a short time before the hijacking

6 Summary

The presented idea is an original attempt at integrating theories and practices from many areas, in particular: semantic models, social networks, graph and network theory, decision theory, data mining and security, as well as multicriteria optimization. It utilizes the theoretical basis for a very practical purpose of growing importance and demand: widely understood countering terrorism. Moreover, the presented approach combines well-known structural and rarely considered non-structural (quantitative) similarity between graphs as models of some objects and can significantly improve social network analysis.

Our models and methods of networks analysis have been used in the criminal justice domain to search large datasets for associations between crime entities in order to facilitate crime investigation. However indirect association and link analysis still faces many challenging problems, such as information overload, high search complexity, and heavy reliance on domain knowledge.

Acknowledgements

This work was partially supported by projects: N^{oo} PBZ-MNiSW-DBO-01/I/2007 and MNiSW OR00005006 titled "Integration of command and control systems".

References

1. Barabási, A., Albert, R.: Emergency of Scaling in Random Networks. Science 286, 509–512 (1999)
2. Barthelemy, M., Chow, E., Eliassi-Rad, T.: Knowledge Representation Issues In Semantic Graphs for Relationship detection, UCRL-CONF-209845. In: Proceedings of the 2005 AAAI Spring Symposium on AI Technologies for Homeland Security, Palo Alto, CA (US), March 21-23 (2005)
3. Blondel, V., Gajardo, A., Heymans, M., Senellart, P., Van Dooren, P.: A Measure of Similarity Between Graph Vertices: Applications To Synonym Extraction and Web Searching. SIAM Review 46(4), 647–666 (2004)
4. Bonacich, P.: Factoring and Weighting Approaches to Status Scores and Clique Identification. Journal of Mathematical Sociology 2, 113–120 (1972)
5. Brandes, U.: A Faster Algorithm for Betweenness Centrality. Journal of Mathematical Sociology 25, 163–177 (2001)
6. Chmielewski, M., Gałka, A., Krasowski, K., Jarema, P., Kosiński, P.: Semantic Knowledge Representation in Terrorist Threat Analysis for Crisis Management Systems. In: Nguyen, N.T., Kowalczyk, R., Chen, S.-M. (eds.) ICCCI 2009. LNCS (LNAI), vol. 5796, pp. 460–471. Springer, Heidelberg (2009)
7. Davies, J., Fensel, D., Harmelen, F.: Towards the Semantic Web: Ontology-driven Knowledge management, HPL-2003-173. John Wiley & Sons, Chichester (2003)
8. Freeman, L.: A set of Measures of Centrality Based on Betweenness. Sociometry 40, 35–41 (1977)
9. Friedman-Hill, E.: Jess in Action: Java Rule-Based Systems, Manning (2003) ISBN 1930110898
10. Golbeck, J., Mannes, A., Hendler, J.: Semantic Web Technologies for Terrorist Network Analysis. IEEE Press, Los Alamitos (2005)
11. Harary, F., Hage, P.: Eccentricity and centrality in networks. Social Networks 17, 57–63 (1995)
12. Krebs, V.: Mapping Networks of Terrorist Cells. Connections 24(3), 43–52 (2002)
13. Najgebauer, A., Antkiewicz, R., Chmielewski, M., Kasprzyk, R.: The Prediction of Terrorist Threat on The Basis of Semantic Associations Acquisition and Complex Network Evolution. In: Proceedings of the Military Communications and Information Systems Conference MCC 2007, Bonn, Germany (2007)
14. Najgebauer, A., Antkiewicz, R., Kulas, W., Pierzchała, D., Rulka, J., Tarapata, Z., Chmielewski, M.: A concept of simulation based diagnostic support tool for terrorism threat awareness. In: Proceedings of NATO Modelling and Simulation Group Conference, Koblenz, Germany, October 07-08 (2004)

15. National Consortium for the Study of Terrorism and Responses to Terrorism, Global Terrorism Database, http://www.start.umd.edu/start/
16. Newman Mark, E.J.: Models of the small world: A review. J. Stat. Phys. 101, 819–841 (2000)
17. Newman Mark, E.J.: The structure and function of complex networks. SIMA Review 45(2), 167–256 (2003)
18. Sabidussi, G.: The Centrality Index of a Graph. Psychometrica 31, 581–603 (1966)
19. Strogatz, S.H.: Exploring complex networks. Nature 410, 268–276 (2001)
20. Tarapata, Z.: Multicriteria weighted graphs similarity and its application for decision situation pattern matching problem. In: Proceedings of the 13th IEEE/IFAC International Conference on Methods and Models in Automation and Robotics, Szczecin, Poland, August 27-30, pp. 1149–1155 (2007)
21. Tarapata, Z., Kasprzyk, R.: An application of multicriteria weighted graph similarity method to social networks analyzing. In: Proceedings of the 2009 International Conference on Advances in Social Networks Analysis and Mining ASONAM 2009, Athens, Greece, July 20–22, pp. 366–368. IEEE Computer Society, Los Alamitos (2009)
22. Wang, X., Chen, G.: Complex Networks: Small-world, scale-free and beyond. IEEE Circuits and Systems Magazine 3(1), 6–20 (2003)
23. Watts, D.J., Strogatz, S.H.: Collective dynamics of small-world networks. Nature 393, 440–442 (1998)

A Real Time Player Tracking System for Broadcast Tennis Video

Bao Dang[1], An Tran[1], Tien Dinh[1], and Thang Dinh[2]

[1] Faculty of Information Technology, University of Science, VNU-HCMC,
227 Nguyen Van Cu, Ho Chi Minh City, Vietnam
{dangvinhbao,tranminhan.my}@gmail.com,
dbtien@fit.hcmus.edu.vn
[2] Computer Science Department, University of Southern California, USA
thangdin@usc.edu

Abstract. This paper proposes a novel framework for tennis player detection and tracking. The algorithm is built on (1) a powerful court-line pixel detection method utilizing intensity and texture pattern, (2) a fast RANSAC-based line parameter estimation which also determines line extents, and (3) a player segmentation and tracking algorithm exploiting knowledge of tennis court model. The content of the video is then explored at a highly semantic level. The framework was tested extensively on numerous challenging video sequences with various court environments and lighting conditions. The results show the robustness and the promising direction of our algorithm.

Keywords: Object detection, object tracking, sports analysis, feature extraction, camera calibration.

1 Introduction

Automatic analysis of sport videos is an interesting application of content analysis, since it allows people to query information of their specific interest easily. Since each group of users has their own preferences and requirements, applications in sport video analysis are very broad [14]. To analyze a tennis video at a higher semantic level such as activity recognition or content classification, a robust detection and tracking algorithm is required. This paper presents a novel content analysis framework based on an automatic players and court detection and tracking method in tennis domain.

Due to the highly dynamic nature of sport videos, existing state of the art methods for single object tracking [3, 4, 7] cannot be applied effectively. Camera motion, player motion, player-articulated poses and scene changes are just some of the difficulties. Therefore, a successful sport analysis system should exploit game specific information rather than apply a generic solution. In court games such as tennis and soccer, typical and useful information is the uniform court color and white court-lines. Moreover, players should only be in the playing field or close to the surrounding boundary. In [9, 10], Hough transform is adopted to detect court-lines, which are subsequently used for calibration. However, the approach is slow and not robust because of the computation complexity and inaccuracy of Hough transformation. In

N.T. Nguyen, M.T. Le, and J. Świątek (Eds.): ACIIDS 2010, Part II, LNAI 5991, pp. 105–113, 2010.

[13], a court-line detection algorithm is described using straight-line detection method to build up a court model. However, it does not provide good results when the court is partially occluded.

In this paper, we propose a real-time and robust framework for developing a fully automatic tennis player tracking system. The flowchart of our system is presented in Fig. 1. First, a court detector is applied to determine precise position of the court in the current frame. This is done by a simple white pixel detection procedure and a RANSAC-based line detection algorithm which also helps to determine the extent of the lines. Some heuristic pruning techniques are also applied to reduce the number of possible candidates for the actual court position. Second, the positions of players are detected in the image by background subtraction. In this step, court model information is used to exclude distracters. It enhances the robustness of our method on broadcast tennis videos impressively. Finally, the positions of players with respect to the court are obtained by combining the position information of the court, the players, and the court model.

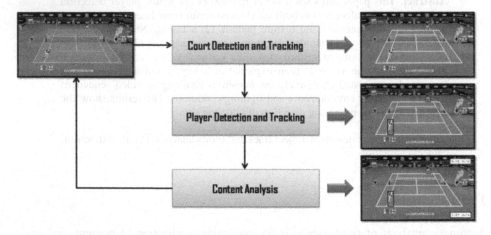

Fig. 1. The overview of proposed system

The rest of this paper is organized as follows. The details of the court detector are presented in Section 2. Player detection and tracking algorithm is described in Section 3. Experimental results on various video sequences and some applications are shown in Section 4, followed by conclusions and future work in Section 5.

2 Court Detection and Tracking

2.1 White Pixel Extraction

Based on the observation that the color of court-lines is always white, [10] suggested a luminance threshold method to extract white pixels with an additional constraint to exclude large white areas from the detection result. In order to eliminate large areas of white pixels, which usually do not belong to court-lines, darker pixels are checked if they can be found at a horizontal or vertical distance τ from the candidate pixel. In which, τ is the approximate line width. This means white court-line pixels must be enclosed either horizontally or vertically by darker pixels.

2.2 Line Parameter Estimation

From the set of obtained white pixels, court-line candidates will be detected by RANSAC algorithm [6]. The most dominant lines are determined and removed from the dataset in an iterative process, until we get enough line candidates. Dominant lines are defined by the start and the end positions of line segments. Those segments are further linked together to make longer segments if possible and classified as horizontal or vertical lines. These steps will be explained clearly in the following sections.

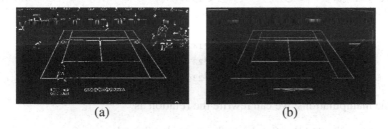

(a)   (b)

Fig. 2. (a) White-pixels image and (b) Detected lines by RANSAC

2.2.1 RANSAC Based Dominant Line Detection

RANSAC is a randomized algorithm that hypothesizes a set of model parameters and evaluates the quality of the parameters. After several hypotheses have been evaluated, the best one is chosen. Specifically, a line is hypothesized by randomly selecting two court-line pixels. Then, the line parameter g is computed. An evaluation score is defined as

$$s(g) = \sum_{(x,y) \in P} \max(\tau - d(g, x, y), 0), \tag{1}$$

where $d(g, x, y)$ is the distance of the pixel (x, y) to line g, P is the set of court-line pixels, and τ is the line width from Section 2.1 which has default value of 4. This score effectively indicates the support of a hypothesized line as the number of white pixels close to the line and also determine two end points of the line. The process is repeated until about 25 hypothesized lines are generated randomly [6]. Finally, the hypothesized line with the highest score is selected, as illustrated in Fig. 2.

2.2.2 Line Segment Refinement and Classification

The RANSAC based line detector produces some neighboring short-line segments that actually belong to one court-line. This misalignment happens because of poor quality of the video or the court itself. To address this issue, we merge two neighboring segments l_1 and l_2 if $(\widehat{l_1, l_2}) < 0.75°$ and $d(l_1, l_2) < 5$. Obtained lines are also classified in advance as horizontal lines or vertical lines based on two end point coordinates. A candidate line having two end points (x_1, y_1) and (x_2, y_2) is considered as a horizontal line if $|y_1 - y_2| < 20$. Furthermore, horizontal lines are sorted from the top to the bottom; whereas, vertical lines are sorted from the left to the right. Both steps aim to reduce the number of possible hypotheses in calibration parameters searching space and hence improve the performance of the algorithm. All above parameters are set empirically according to the resolution of the input video.

2.3 Model Fitting

The model fitting step determines the correspondences between the detected lines and the lines in the standard court model. Once these correspondences are known, we can construct the homography between real-world coordinates and image coordinates. According to [12], an image coordinate (x, y) is related to a court model coordinate (α, β) by a homography matrix H

$$\begin{pmatrix} x \\ y \\ 1 \end{pmatrix} = \begin{pmatrix} h_{11} & h_{12} & h_{13} \\ h_{21} & h_{22} & h_{23} \\ h_{31} & h_{22} & h_{23} \end{pmatrix} \begin{pmatrix} \alpha \\ \beta \\ 1 \end{pmatrix} = H \times \begin{pmatrix} \alpha \\ \beta \\ 1 \end{pmatrix}. \tag{2}$$

With some manipulations, we can rewrite the relation as

$$h_{11}\alpha + h_{12}\beta + h_{13} - h_{31}\alpha x - h_{32}\beta x - h_{33}x = 0,$$
$$h_{21}\alpha + h_{22}\beta + h_{23} - h_{31}\alpha y - h_{32}\beta y - h_{33}y = 0. \tag{3}$$

These two linear equations have nine unknown parameters $h_{11} \dots h_{33}$ of the linear mapping matrix H. If dividing each term in the equations by the element h_{33}, we are left with only eight unknown parameters. Matching two candidate lines with two model lines, we acquire a complete system of linear equations, which can be solved by the Gaussian elimination method.

In order to find the best homography that fits the dataset, [10] proposed a matching score approach. After all transformation matrices have been evaluated, the matrix with the highest score is selected as the ultimate solution. In this paper, the Extreme Fast Line Algorithm [1] is used for the scoring step. Although the computation for each calibration setting can be done fast, the evaluation of the model support is computationally intensive. Therefore, some simple and quick tests to reject physically impossible calibration parameters are applied.

In addition, another evaluation step is proposed to confirm the calibration settings. Let (μ_R, μ_G, μ_B) and $(\delta_R, \delta_G, \delta_B)$ be defined as the means and the standard deviations of red, green and blue channel of non-white pixel set inside the court, respectively. Due to the uniform of the court color, calibration parameters are invalid if $\max(\delta_R, \delta_G, \delta_B) > \theta$, where θ is a threshold. In other words, the detection result is unreasonable when the color inside the court varies too much.

2.4 Model Tracking

The previous detection algorithm only has to be applied once in the bootstrapping process at the first frame of every new playing shot. Given that the change in camera speed is small, the camera parameters of subsequent frames are predictable. Similar to [9], the detected court-lines in the previous frame are extended to obtain a search region for the current frame, shown in Fig. 3. The same detection technique is executed, but now within the local search area, to track the model. The obvious benefits are that the court-lines are estimated more accurately and that the calibration setting can be determined rapidly by matching each court-line with the closest predicted model line. Our court detection and tracking method has been experimented in different types of court under various lighting conditions. In case of court occlusion or camera in motion, only two precise court-lines are required to reconstruct the court successfully.

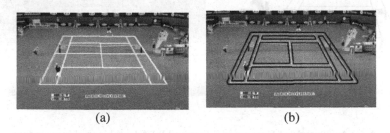

(a)   (b)

Fig. 3. (a) Previous court detection result and (b) Local search area for the current frame

3 Player Detection and Tracking

This section explains the process to obtain the real-world positions of the players, starting with the pixel-level detection, up to the final trajectory computation at the object-level. In general, the moving area of a player is limited to the playing field inside the court and partially the surrounding area. Moreover, the color of the playing field and the surrounding area is almost uniform. These features allow us to separately construct background models for the field inside the court and the surrounding area, instead of creating a complete background for the whole image [9]. This approach has two advantages. First, the background image cannot be influenced by any camera motions. Second, only color and spatial information are considered when constructing the background models, which avoids complex motion estimation.

The RGB color space is used for modeling the background. The stored background model for the playing field inside the court is $[\mu_R, \mu_G, \mu_B, \delta_R, \delta_G, \delta_B]$, defined in Section 2.3. A pixel inside the court is marked as foreground if

$$|r - \mu_R| > \alpha \, \delta_R \ \lor \ |g - \mu_G| > \alpha \, \delta_G \ \lor \ |b - \mu_B| > \alpha \, \delta_B, \tag{4}$$

where α is a constant, depending on the color variance of the court, and r, g, b denote the red, green, blue values of the current pixel, respectively. The same process is applied to all pixels in the area surrounding the court. This step produces a foreground image as shown in Fig. 4a. Here, the background model is assumed to be persistent during a playing shot; therefore, only the first frame background is modeled.

Exploiting the knowledge of the game court, two searching windows in the foreground image, above and below the tennis net line are used. By assuming that the player sizes in the video do not change much, the player boundary for each searching window is defined as a fixed-size rectangle which encloses the maximum foreground pixels (Fig. 4b). The initial windows are fixed for the first frame of every playing shot, whereas the windows in subsequent frames are adjusted, surrounding the previous locations of the players [11]. The reference point of the player's position in the image plane is defined as the lowest vertical point on the boundary rectangle, which is transformed using Eq. (2) to obtain the real-world location of the player.

(a)   (b)

Fig. 4. (a) An example of foreground image and (b) Adaptive searching windows

Unfortunately, the player is not the only foreground object in the video frame. As shown in Fig. 5a, other objects, such as referees and ball boys, can affect the tracking result. This issue has not been mentioned in any recent works. Thus, we propose a heuristic method to solve the problem. Obviously, the positions of distracters are usually fixed on the court during a playing shot. We take advantage of this fact to detect and track those objects and enhance the player tracking process, even when the camera is moving. After the players are detected in the first frame, all remaining medium-sized foreground blobs are considered as the distracters. Their image coordinates are then converted to the real-world model by using the calibration parameters. In successive frames, these coordinates are transformed back to the image domain by the new camera setting to locate all of those distracters in order to remove them from the foreground image. The proposed method can significantly eliminate the false detection (Fig. 5b).

(a)   (b)

Fig. 5. (a) False player detection problem and (b) Our proposed method

4 Experimental Results and Applications

In this section, experimental results of our system on four video sequences recorded from Grand Slam tournaments [2], which are the most important tennis events of the year, are provided. All sequences have the resolution of 480 by 270 pixels, the frame rate of 25 fps, and the length of 9000-11000 frames. The system is implemented in C++ and runs at 75 fps on a Intel Core 2 Duo 2.0 GHz processor. This efficiency performance is due to the combination of a specialized court-line pixel detector, a RANSAC-based line detector and exploiting the knowledge of the game court. Hence there is room for further processing steps can be applied to extract high level semantic in content analysis applications which required real-time performance.

Table 1. Court detection results

	Playing frames	Detected	Correct	Miss	False
Australian Open	10135	10123	10118 (99%)	17	5
French Open	5026	4985	4970 (98%)	56	15
Wimbledon	7667	7413	7396 (96%)	361	107
US Open	8327	8282	8279 (99%)	48	3

Table 2. Player detection results

	Playing frames	Player 1	Player 2
Australian Open	10135	9127 (90%)	9336 (92%)
French Open	5026	4786 (95%)	4415 (88%)
Wimbledon	7667	5478 (71%)	5622 (73%)
US Open	8327	7863 (94%)	7521 (90%)

The system achieves an impressive 96-99% court detection rate (Table 1). Our algorithm is very robust to occlusion, partial court views, poor lighting, and camera motion as shown in Fig. 7 since it requires only two precise court-lines to track the court model. This result outperforms the algorithms in [5, 13], which only handle full court view frames. The player detection accuracy is about 87% (Table 2), where the criterion is that at least 70% of the body of the player is included in the detection window. Fig. 6 illustrates some visual results, where the top and bottom players are bounded with red boxes and blue boxes, respectively.

According to [8], the tracking results are analyzed to extract higher semantic information, such as:

— **Instant speed of the player:** The speed of each player is definitely an important factor to reveal the current status of a player (running or still). It also indicates the intensity of the match. The speed is estimated based on the court model size and player positions during the shot, as shown in Fig. 6.
— **Speed change of the player:** Acceleration and deceleration of a player occurs during changes in action behavior.

- **Relative position of the player on the court model:** This position is the main source for the recognition of those events that are characterized by a particular arrangement of players on the playing field. **Fig. 7** shows our event detection results of three typical events, which are service, baseline rally and net approach.
- **Temporal relations among each event:** In some sports games like tennis and baseball, there are strong temporal correlations among key events. For example, in a tennis video, service is always at the beginning of a playing event, while the baseline rally may interlace with net approaches.

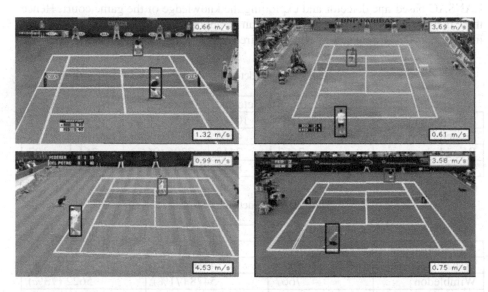

Fig. 6. Visual experimental results (the court is indicated by yellow lines, the red and blue rectangles represent the players) in different court types, lighting conditions, and camera views

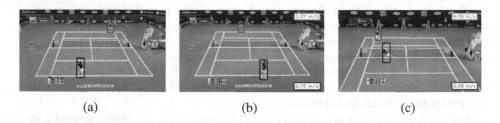

     (a)   (b)   (c)

Fig. 7. Some examples of event detection: (a) Service, (b) Baseline Rally and (c) Net Approach

5   Conclusions and Future Work

We have presented a novel framework for tennis sports tracking and content analysis. The algorithm includes the playing field detection using court-line detection and tracking, player segmentation. The system provides a high-level semantic content analysis in tennis video with a limited human-interaction at the beginning. Possible future directions of the system would be upgrading of homography matrix into full camera matrix, including the curved line segments into the court model, or employing the popular HMM model to describing the dynamic of the video events.

Acknowledgements

We would like to acknowledge the support of AI Lab of the University of Science, VNU-HCMC for providing facilities for this research.

This work is a part of the KC.01/06-10 project supported by the Ministry of Science and Technology, 2009-2010.

References

1. Extremely Fast Line Algorithm, http://www.edepot.com/linee.html
2. Roger Federer Highlights Video, http://www.rogerfedererpoints.com
3. Arulampalam, S., Maskell, S., Gordon, N., Clapp, T.: A Tutorial on Particle Filters for On-line Non-linear/Non-Gaussian Bayesian Tracking. In: IEEE Transactions on Signal Processing, pp. 174--188 (2002)
4. Avidan, S.: Ensemble Tracking. In: IEEE Computer Society Conference on Computer Vision and Pattern Recognition, San Diego, USA (2005)
5. Calvo, C., Micarelli, A., Sangineto, E.: Automatic Annotation of Tennis Video Sequences. In: Proceedings of the 24th DAGM Symposium on Pattern Recognition, pp 540--547. London, UK (2002)
6. Clarke, J., Carlsson, S., Zisserman, A.: Detecting and tracking linear features efficiently. In: Proc 7th British Machine Vision Conference, pp. 415--424. Edinburgh (1996)
7. Collins, R., Liu, Y., Leordeanu, M.: Online Selection of Discriminative Tracking Features. In: IEEE Transactions on Pattern Analysis and Machine Intelligence, pp. 1631--1643 (2005)
8. Farin, D., Han, J., de With, P.H.N.: Multi-level analysis of sports video sequences. In: SPIE Conference on Multimedia Content Analysis, Management, and Retrieval, pp. 1--12. San Jose, USA (2006)
9. Farin, D., Han, J., de With, P.H.N., Lao, W.: Automatic Tracking Method for Sports Video Analysis. In: 26th Symposium on Information Theory, pp. 309--316. Brussels, Belgium (2005)
10. Farin, D., Krabbe, S., de With, P.H.N., Effelsberg, W.: Robust Camera Calibration for Sport Videos Using Court Models. In: SPIE Storage and Retrieval Methods and Applications for Multimedia, pp. 80--91 (2004)
11. Jiang, Y. C., Lai, K.T., Hsieh, C.H., Lai, M.F.: Player Detection and Tracking in Broadcast Tennis Video. In: Proceedings of the 3rd Pacific Rim Symposium on Advances in Image and Video Technology, pp. 759--770. Tokyo, Japan (2009)
12. Segarra, D. L.: Tennis Stroke Detection in Video Sequences. In: Master Degree Project. Stockholm, Sweden (2005)
13. Sudhir, G., Lee, C., Jain, K.: Automatic Classification of Tennis Video for High-level Content-based Retrieval. In: Proceedings of International Workshop on Content-Based Access of Image and Video Databases, pp. 81--90. Washington DC, USA (1998)
14. Wang, J. R., Parameswaran, N.: Survey of sports video analysis research issues and applications. In: Workshop on Visual Information Processing, pp. 87--90, Sydney, Australia (2004)

Twittering for Earth: A Study on the Impact of Microblogging Activism on Earth Hour 2009 in Australia

Marc Cheong and Vincent Lee

Clayton School of Information Technology,
Monash University, Clayton
Victoria, 3800
Australia
marc.cheong@infotech.monash.edu.au,
vincent.lee@infotech.monash.edu.au

Abstract. The role of Twitter - a form of microblogging - as both influencer and reflector of real-world events is fast emerging in today's world of Web 2.0 and social media. In this investigation, we survey how the use of Twitter in Australia is linked to the real-world success of the Earth Hour 2009 campaign. The results of this research will give us an idea of the emergence of microblogging as a new medium of influencing human behavior and providing a source of collective intelligence in planning and decision making, specifically in the Australian context. We found that, from our observations, there is a correlation between the inter-state total energy reduction during this campaign with the amount of inter-state online Twitter discussion. We also identified a link between the Twitter discussion frequency and the total real-life population of the locale in which the chatter takes place, which could be used as a yardstick to analyze the reach of online technologies in the real world.

1 Introduction

Twitter [26], a microblogging site, has become one of the fastest growing trends on the Internet, with an exponentially-growing user base [19]. Users participate by expressing their thoughts or feelings within 140 characters, publishing their messages (or 'tweets') for friends, family, and the rest of the world to see. Of late, Twitter has been an effective method for information dissemination and interpersonal communication, and can be used as a gauge to measure the current *zeitgeist* on certain topics or issues. The shift from microblogging of daily trivialities to the usage of Twitter for dissemination of information can be seen in the transformation of the original Twitter motto – from the original *"what are you doing?"* to the current *"what's happening?"* [26].

In this research, our aim is to perform a case study to observe how online activity on Twitter which is part of the "social web" [6] can be reflected in a real-world social system. Such studies allow us to gain insight as to how social media in a virtual online setting can be linked to real-world human behavior, which

N.T. Nguyen, M.T. Le, and J. Świątek (Eds.): ACIIDS 2010, Part II, LNAI 5991, pp. 114–123, 2010.

supplements existing studies on online memetics [1,27], information-sharing behavior [5], and social participation and dynamics.

Specifically, in this study we investigate how Twitter activity in a collective action campaign can be a reflective indicator of real-world sentiment on real-world events – in this case, the microblogging pattern of Australian Earth Hour 2009 participants on Twitter. We also intend to measure the success of the Australian Earth Hour campaign on a state-by-state basis in terms of energy savings recorded and attempt to determine a link, if any, between activism on Twitter and the efficacy of the Earth Hour campaign. This is interesting in the fact that Earth Hour 2009 not only focused on real-world campaigning and activism, but also had a strong online presence [29]. Finally, we determine any possible correlation between population levels and the level of participation in the Twitter microblogging platform, which allows us to see how adoption rates of online microblogging and social networking technologies can be linked to the size of the population.

The motivation behind our work is that no prior work has been done in this domain, and the potential results obtained from such a study could give us an insight into how an online microblogging platform (with elements of social networking and communication [6,9], collaborative applications [9], and information dissemination [5,12,16]) could effectively 'mirror' a real-world social system.

2   Background and Prior Work

2.1   An Overview of Twitter

There exists limited prior literature on Twitter and microblogging research. Java et al. [12] surveyed the topological, geographical and statistical properties of the Twitter social network. In their paper, they conducted an analysis on the taxonomy of "user intentions" of microblogging via Twitter, and found that users on Twitter not only answers the basic question of "what [they] are doing" at the moment. Users are inclined to share information, news, and sentiments [12] – this is also described by O'Reilly and Milstein [19] in their Twitter book. A high-level overview of the trending topics (top ten topics discussed on Twitter) by Cheong [6] came to a similar conclusion. From a social science and humanities perspective, Mischaud [16] stated that users on Twitter use it as an "expression of themselves" by using it as a medium for information and news sharing.

Krishnamurthy et al. [14] and Huberman et al. [10] studied the growth pattern of the existing Twitter network, user connections based on network theory, geographical presence based on time zone, and Twitter usage ('tweeting') habits versus geographic location. They found that Twitter updates roughly correlate to day-to-day activities of the user base, and that Twitter is predominantly used in the West but is quickly gaining popularity in other countries.

Honeycutt & Herring [9] studied the communication aspect of Twitter by studying the trends of Twitter replies, or "@ messages" and observed that Twitter messages also constitute a form of interpersonal communication amidst the flurry of chatter generated by users.

Cheong and Lee's research on Twitter Trends and user demography [6] have found that such 'Trends' can be used as efficient indicators of the current sentiment of a section of Twitter users. This is corroborated by their findings that Twitter users contributing to a particular topic can be clustered according to their demographics and that such information can be used for policy-planning and decision making.

2.2 Social Information Spread

Social information spread is also relevant in discussing our case study. Arbesman [1] and Wasik [27] in their experiments with memetics and flash mobs respectively have shown that information dissemination via electronic media have been proven effective in triggering a 'viral' spread of information. This in turn translates in a real-world cultural phenomenon where people act upon the viral information that is spread online; an example would be people participating in real-world 'flash mobs' as a consequence of Wasik's email experiment [27].

Relating to Twitter, Cheong's work on surveying the commonly found text strings on Twitter Trends [5] found that 'activism campaigns' on Twitter are commonplace and can generate large amounts of discussion on Twitter due to increased "worldwide awareness and conversation".

2.3 Twitter and OSNs as 'Facilitators'

Relating to real world deployment issues, Hughes and Palen [11] have studied the usage patterns of Twitter in real-world scenarios of emergency response and mass convergence. Jungherr [13] and Goolsby [7] also discussed about how Twitter is being used as a facilitator to express political dissent.

Twitter has also been given much discussion in terms of it being a catalyst for activism and citizen journalism. Examples include an overview of the extensive role of Twitter in Barack Obama's 2008 US presidential campaign [8], in political activism [13,17,21], and also reporting terrorist threats and disaster events as they unfold [3,4,25]. In the context of computer-based media art, Patel [20] explains that "context-specific [Twitter] feeds reveal a rich vein of social possibilities to mine ... in activism".

Studies in activism among users of *de facto* online social networks (OSNs) have also been given much attention, particularly in the domain of social sciences and the humanities. There is a study by Song [23] in feminist cyber-activism with Facebook, which showed how Facebook – an OSN – can be leveraged to enhance activism. Mankoff [15] has also performed a study of how such OSNs can be used in eco-activism by encouraging members on OSNs to reduce their ecological footprint in real life.

Relating to our case study, there was prior work done by Solomon [22] on analyzing the energy drop recorded during the 2007 Earth Hour conducted Sydney-wide from an economic perspective, which revealed findings that users "overstate their participation in the Earth Hour project", as observed from the total energy drop registered during the 2007 Earth Hour. For the sake of comparison, this case study will also address the issue of whether the energy drop registered during the 2009 edition of Earth Hour has actually become significant compared to past years due to the increasing popularity and media coverage given to it by the organizers [29].

3   Methodology

3.1   Analyzing Twitter Chatter during Earth Hour in Australia

For the Earth Hour 2009 campaign, users on Twitter are encouraged by the organizers to publish Twitter messages in such a format to express their support for the Earth Hour campaign:

... use the hashtags #earthhour or #voteearth along with your #location to get the word out. [28]

'Hashtags' are words preceded with a hash [#] symbol to tag Twitter messages, indicating a "use of social tagging to categorize posts to allow ease of communication and searching for related posts" [5].

Using this notation, we are able to create a search query to seek out Twitter messages from Australia in support of Earth Hour. The Twitter official API [26], however, is ill-suited for this purpose as it only allows the retrieval of the past 1500 messages, limited to a period of approximately one month (whichever limit is reached first) as discovered in prior research [6].

Therefore for the purposes of this project, we use the Hashtags.org website [2], a Twitter API-based website which automatically tracks users with hashtags and has a backdated, browsable archive for the hashtags '#earthhour' or '#voteearth'.

For the location, the list of Earth Hour messages obtained are then scanned through to identify names of Australian state capitals, major cities, and their abbreviations. The messages are then collated according to state (with case-insensitive matching), as per the hashtag keywords in Table 1.

From the corpus of the filtered Twitter messages, Twitter activity will be expressed as total Twitter messages per each of the aforementioned states. A

Table 1. Hashtag keywords found in Earth Hour Twitter messages Australia-wide, grouped on a state-by-state basis

State	String describing state name, abbreviation, state capital, or major cities
NSW	Nsw, NewSouthWales, Sydney
QLD	Qld, Queensland, Brisbane
SA	SouthAustralia, Adelaide
TAS	Tasmania, Hobart
VIC	Victoria, Melbourne, Bendigo

scatter-plot of Twitter activity to percentage of power savings will be generated to identify if there is any link between the two variables. The R-squared coefficient of determination will be obtained from the graph to statistically validate our findings.

3.2 Measuring Power Consumption during Earth Hour 2009 in Australia

The Australian National Electricity Market Management Company Limited (NEMMCO) [18][1] publishes electricity market (supply and demand) data, updated on a half-hourly basis, for five Australian states - New South Wales, Queensland, South Australia, Tasmania, and Victoria. By using the data for Earth Hour 2009, we are able to come up with an authoritative measure of how much energy is saved during the year's observance of Earth Hour (8.30pm to 9.30pm local time for each state, on the 28th of March 2009).

Power consumption data in megawatts (MW) for the Earth Hour time period (two half-hour periods, 8.30pm–9.00pm and 9.00pm–9.30pm) for a period of one week centered on the 28th of March 2009 (25th of March – 31st of March inclusive) is gathered for each of the states mentioned above.

The average non-Earth Hour power consumption (on a state-by-state basis) is calculated by averaging the wattage for the 3 days before and 3 days after the Earth Hour event. Hence, the energy reduction during Earth Hour could be expressed as a percentage of the average consumption.

To test the significance of the energy drop, a paired Student's t-Test for statistical significance is performed (using an α value of 0.05) with the aid of a spreadsheet package on the entire set of wattage data. This gives us an idea of whether the energy drop is statistically significant. For visualization and data tabulation, the energy consumption data will be formatted in bar graph and table form.

3.3 Investigating Twitter Activity versus State Population

As a secondary part of our investigation, we compare the Twitter usage rate as above to the population of each of the specified Australian states. This ratio provides us an insight into the adoption rate of microblogging (expressed by the proportion of people participating in the Earth Hour Twitter campaign) relative to the size of each state in terms of its population.

The latest population data on a state-by-state basis, dated December 2008, is obtained from the Australian Bureau of Statistics [24], as it is an authoritative source of population and demographical statistics. A scatterplot of the interstate Twitter usage rate versus state population will be generated and the R-squared value obtained to statistically validate our findings.

---

[1] As of July 2009, NEMMCO's operations have been taken over by the Australian Energy Market Operator (AEMO).

4   Results

By calculating the average state-by-state consumption of power for two half-hourly periods – Period 1: 8.30pm–9.00pm and Period 2: 9.00pm–9.30pm in each state's respective time zone – three days before and after Earth Hour, and also the actual consumption recorded during the same half-hourly periods on Earth Hour day (28th March 2009), we obtain the following data in Table 2. The number following the state abbreviation represents the half-hourly period.

Table 2. State-by-state energy consumption for both half-hourly periods during Earth Hour and the corresponding average consumption during non-Earth Hour days (expressed in MW to 2 decimals)

State & period	Average consumption	Earth Hour consumption	Energy drop
NSW:1	8688.59	7729.56	959.03
NSW:2	8373.54	7607.68	765.86
QLD:1	6454.31	6001.81	452.50
QLD:2	6330.06	5789.71	540.35
SA:1	1496.56	1370.52	126.04
SA:2	1450.41	1372.22	78.19
TAS:1	1071.12	1031.49	39.63
TAS:2	1030.07	1008.60	21.47
VIC:1	5806.23	5317.49	488.74
VIC:2	5659.26	5323.03	336.23

Table 3. Percent reduction of energy use during Earth Hour, to 2 decimal places; and tabulation of Twitter messages observed

State	Percentage of energy reduction during Earth Hour	Number of Twitter messages observed
NSW	20.18%	48
QLD	15.55%	21
SA	13.81%	6
TAS	5.78%	3
VIC	14.36%	4

Using a spreadsheet package, we calculate the t-probability using Student's two-tailed, paired t-Test for the sets of average consumption data and the Earth Hour consumption data. Using an α value of 0.05, and 18 degrees of freedom, we obtain a t-value of 3.24 (which is greater than the required t-value of 2.10 obtained from the given parameters). This indicates a statistically significant drop of energy consumption during Earth Hour.

We then obtain the net reduction of energy for each of the states expressed as a percentage value, and the aggregated Twitter usage count on a state-by-state basis in Table 3.

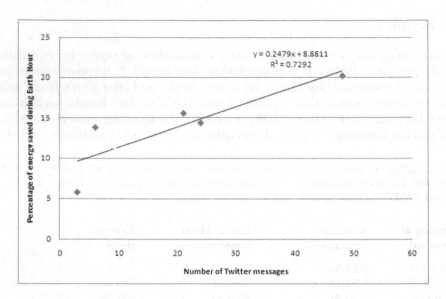

Fig. 1. Scatter-plot of energy savings (expressed as percentages) versus count of Twitter messages for the Australian states

Table 4. State-by-state Australian population data (in millions)

State	State population (millions)	Number of Twitter messages observed
NSW	7.0414	48
QLD	4.3495	21
SA	1.6120	6
TAS	0.5003	3
VIC	5.3648	4

A scatter-plot of the percentage of energy savings and Twitter usage count for the 5 states is then created (Figure 1).

Population data for each of the states is acquired from the website of the Australian Bureau of Statistics (ABS) as of the end of 2008 [24], and tabulated in Table 4. The scatter-plot for Twitter message activity versus population data is per Figure 2.

5   Discussion

Firstly, based on the analysis performed on total energy savings, the drop of energy consumption suggests that the Earth Hour campaign 2009 showed statistically significant results in promoting energy conservation. Although this study is confined to five major Australian states - NSW, QLD, SA, TAS and VIC; it does suggest an efficacy on the part of the organizers of such campaigns in promoting awareness on the need for energy conservation.

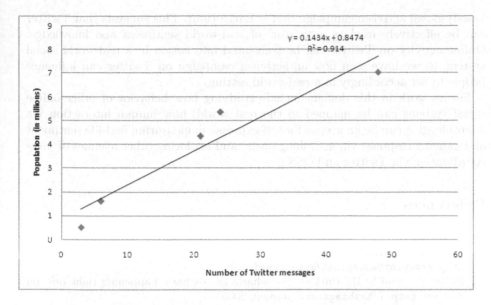

Fig. 2. Scatter-plot of state population (millions) and the Twitter messages for the Australian states

The results seen here are clearly better than the 2007 edition of the same campaign [22]. Among the chief reasons for this is the awareness generated among the populace by a successful marketing campaign which involves commitments from local governments worldwide; and also the engagement of the social media (namely blogs, social networks, and microblogging sites), corroborating research which illustrates the efficacy of social media in activism (aforementioned examples include [15,23]), and also the 'viral' spread of information online as discussed in [1,27].

Secondly, by analyzing the scatter-plot comparing Twitter message activity with the net energy reduction for the 5 states, we obtain a correlation coefficient (R-squared value) of 0.729. This suggests a relation between the two parameters mentioned, implying that the frequency of Twitter message activity might be influential to the percentage of energy savings in the Australian states.

On another note, comparing the Twitter message activity with state population gives us a high correlation coefficient of 0.9140. This suggests that there is a strong correlation between Twitter activity per state with the state's total population in real life. An interpretation of this is that the usage rate of Twitter and such microblogging technologies depend on how populated a particular locale is. This has potential as a basis for future work to measure the penetration rate of social media and microblogging in Australia.

6   Conclusion

To conclude this study, we prove the existence of a link between Australian Twitter usage patterns and the efficacy of a Twitter-based (and social media-based) global activism campaign that is Earth Hour. This suggests that Twitter can be effectively used as a "mirror" of real-world sentiment and knowledge. Online activity on Twitter can be translated into action in a real-world social system, as we have seen how an activism campaign on Twitter can influence people to act accordingly in a real-world setting.

Future work in this domain include studying how behavior of other online social systems can be mapped to the real world; how human interaction via microblogging can be an avenue for self-expression; measuring real-life sentiment and gauging response via microblog posts; and exploring other avenues of mass coordination via Twitter and OSNs.

References

1. Arbesman, S.: The Memespread Project: An initial analysis of the contagious nature of information in social networks (2004),
   http://www.arbesman.net/memespread.pdf
2. Bailey, C., Smith, B., Burkert, B.: #hashtags - what's happening right now on Twitter, http://hashtags.org (June 9, 2009)
3. Beaumont, C.: Mumbai attacks: Twitter and Flickr used to break news. The Daily Telegraph (November 27, 2008)
4. Cashmore, P.: Mashable: Jakarta bombings – Twitter user first on the scene, http://mashable.com/2009/07/16/jakarta-bombings-twitter/ (July 16, 2009)
5. Cheong, M.: What are you Tweeting about?: A survey of Trending Topics within the Twitter community. Tech. Rep. 2009/251, Clayton School of Information Technology, Monash University (2009)
6. Cheong, M., Lee, V.: Integrating web-based intelligence retrieval and decision-making from the Twitter Trends knowledge base. In: Proc. CIKM 2009 Co-Located Workshops: SWSM 2009, pp. 1–8 (2009)
7. Goolsby, R.: Lifting elephants: Twitter and blogging in global perspective. In: Liu, H. (ed.) Social Computing and Behavioral Modeling, pp. 1–7. Springer, Heidelberg (2009)
8. Harris, M.: Barack to the future. Engineering & Technology 3(20), 25 (2008)
9. Honeycutt, C., Herring, S.: Beyond microblogging: Conversation and collaboration via Twitter. In: Proc. 42nd Hawaii International Conference on System Sciences, pp. 1–10 (2009)
10. Huberman, B.A., Romero, D.M., Wu, F.: Social networks that matter: Twitter under the microscope (2008), http://ssrn.com/abstract=1313405
11. Hughes, A.L., Palen, L.: Twitter adoption and use in mass convergence and emergency events. In: Proc. 6th International ISCRAM Conference (2009)
12. Java, A., Song, X., Finin, T., Tsen, B.: Why we Twitter: An analysis of a microblogging community. In: Proc. 9th WebKDD and 1st SNA-KDD 2007 workshop on Web mining and social network analysis, pp. 118–138. Springer, Heidelberg (2009)
13. Jungherr, A.: The DigiActive guide to Twitter for activism (2009),
    http://www.digiactive.org/wp-content/uploads/
    digiactive_twitter_guide_v1-0.pdf
14. Krishnamurthy, B., Gill, P., Arlitt, M.: A few chirps about Twitter. In: Proc. WOSN 2008, pp. 19–24 (2008)

15. Mankoff, J., Matthews, D., Fussell, S., Johnson, M.: Leveraging social networks to motivate individuals to reduce their ecological footprints. In: Proc. HICSS-40, pp. 1–10 (2007)
16. Mischaud, E.: Twitter: Expressions of the Whole Self. Master's thesis, London School of Economics and Political Science (2007)
17. Mungiu-Pippidi, A., Munteanu, I.: Moldova's "Twitter Revolution". Journal of Democracy 20(3), 136–142 (2009)
18. NEMMCO Melbourne/Australian Energy Market Operator: Nemmco, http://www.nemmco.com.au/ (April 8, 2009)
19. O'Reilly, T., Milstein, S.: The Twitter Book. O'Reilly Media, Inc., Sebastopol (2009)
20. Patel, K.: *@reality*: A proposal exploring communication between physical and virtual realms through location based Twitter projections (2008)
21. Simon, M.: Student 'twitters' his way out of Egyptian jail. CNN, http://www.cnn.com/2008/TECH/04/25/twitter.buck/ (April 25, 2008)
22. Solomon, D.: How effective are individual lifestyle changes in reducing electricity consumption? - measuring the impact of earth hour. Tech. rep., University of Chicago, Graduate School of Business (2008)
23. Song, M.: The new Face(book) of Malaysian cyberfeminist activism. Honors thesis, School of Arts & Social Sciences: Monash University Malaysia (2008)
24. of Statistics, A.B.: 3101.0 - Australian Demographic Statistics (December 2008), National Statistics, http://www.abs.gov.au/ausstats/abs.nsf/mf/3101.0/ (June 9, 2009)
25. Terdiman, D.: Photo of Hudson River plane crash downs TwitPic. CNET News (2009), http://news.cnet.com/8301-1023_3-10143736-93.html
26. Twitter Inc.: *Twitter* (2009), http://www.twitter.com
27. Wasik, B.: And Then There's This: How Stories Live and Die in Viral Culture, Penguin Group (USA), New York, NY (2009)
28. William-Ross, L.: LAist: Lights out, Los Angeles: Earth Hour is tonight, http://laist.com/2009/03/28/lights_out_los_angeles_earth_hour_i.php (March 28, 2009)
29. WWF: World Wide Fund for Nature: Earth Hour (2009), http://www.earthhour.org/

Student Courses Recommendation
Using Ant Colony Optimization

Janusz Sobecki and Jakub M. Tomczak

Institute of Informatics, Wroclaw University of Technology
Wyb.Wyspianskiego 27, 50-370 Wroclaw, Poland
Janusz.Sobecki@pwr.wroc.pl, Jakub.Tomczak@pwr.wroc.pl

Abstract. In the paper we present recommendation of student courses using Ant Colony Optimization (ACO). ACO is proved to be effective in solving many optimization problems, here we show that ACO also in the problem of prediction of final grades students receives on completing university courses is able to deliver good solutions. To apply ACO in any recommender system we need special problem representation in form of a graph, where each node represents a decision in the problem domain.

1 Introduction

The main goal of the recommendation systems (RS) is to deliver customized information to a great variety of users of ever increasing web-based systems population. RS are used to solve problems coming from different domains, such as [1,6]: net-news filtering, web recommender, personalized newspaper, sharing news, movie recommender, document recommender, information recommender, e-commerce, purchase, travel and store recommender, e-mail filtering, music recommender and music list recommender. Here we will present quite new application of RS in the student course recommendation using Ant Colony Optimization (ACO).

ACO is a natural computation algorithm that mimics the behaviour of living ants. ACO was proposed by Italian scholar M. Dorigo [2]. In this paper ACO is applied as an information filtering method in RS. We can distinguish the following information filtering methods [6]: demographic (DF), collaborative (CF), content-based (CBF) and hybrid (HA). Other authors [7] present also case-based reasoning (CBR) and rule-based filtering (RBF) [10]. ACO enables to introduce a new type of HA method which is called *integrated* HA.

In this paper we concentrate on application of Ant Colony Optimization (ACO) for recommendation of student courses. In the following section we present the short recommendation systems overview and the area of application. In the section 3 the ACO as well as its application to hybrid filtering are presented. The section 4 presents the experiment carried out on real-life dataset. The final section presents conclusions and future work in the area of the user interface recommendation.

N.T. Nguyen, M.T. Le, and J. Świątek (Eds.): ACIIDS 2010, Part II, LNAI 5991, pp. 124–133, 2010.
© Springer-Verlag Berlin Heidelberg 2010

2 University Courses Recommendation

The today's web-based information systems are successful because of delivering of the customized information for their users. The systems with functionality of that kind are often called recommendation systems (RS) [6]. There are many different applications of recommendation systems. One of the most popular of their applications is movie or video recommendation, such as: Hollywood Video, jimmys.tv, MovieLens, Everyone's a Critic and Film Affinity. The other very popular applications are following: information retrieval, browsing, interface agent and different shopping recommender systems [6].

In this paper we would like to present quite novel application for courses recommendation for university students. This is common in most of the universities all over the world that their students may choose smaller or greater part of the courses to study in each semester. Sometimes they can also choose the teacher or the terms and place. In the increasing number of universities the process of enrolment is managed using computer information systems, mainly web-based ones. This is also the case at Wroclaw University of Technology, where since several semesters students use computer system called EdukacjaCL, which goal is to computerize the whole information system concerning education process. So the process of enrolment is only a part of the system.

The main idea of courses recommendation is to deliver for each student prediction of grade for all the courses so that he or she has an opportunity to enroll. This is obvious that each student would prefer to enroll only for the courses which give more chances to be completed with the positive result.

The recommendation is performed using integrated hybrid approach (HA) information filtering that is based on ACO. We can distinguish three basic types of filtering: demographic (DF), content-based (CBF) and collaborative (CF). DF uses stereotype reasoning [5] and is based on the information stored in the user profile that contains different demographic features [6]. According to [5], stereotype reasoning is a classification problem that is aimed at generating initial predictions about the user. CBF takes descriptions of the content of the previously evaluated items to learn the relationship between a single user and the description of the new items [6]. CF delivers predictions (filtering) about the recommended items by collecting and using information about tastes of other users. Most of CF based recommender systems use user item rating matrix applied for both: identifying similar users and recommend items highly ranked by those users. All the basic methods have some disadvantages, for example DF delivers too general predictions and does not provide any adaptation to user interests changing over time; CBF depends on so called an objective description of the recommended items and it tends to overspecialize its recommendations; and finally CF gives poor prediction when the number of other similar users is small.

The disadvantages of each of the above mentioned recommendation approaches could be overcome by applying HA. For example the disadvantage of the insufficient number of the similar users at the early stages of the system operation using CF may be overcome by the application of the demographic stereotype reasoning. In [10] the following HA methods were enumerated: weighted, switching, mixed, feature combination, cascade, feature augmentation and meta-level. In this paper we present a new integrated HA based on ACO that was applied in the area of user interface recommendation and movie recommendation [9].

3 ACO Filtering Method

Ant Colony Optimization (ACO) is a new natural computation algorithm from mimic the behaviours of ants colony, and proposed by Italian scholar M. Dorigo [2]. The original intention of ACO is to solve the complicated combination optimization problems, such as traveling salesman problem (TSP). ACO is based on observations about ant colonies and their mutual behaviour. The behaviour of a single ant is quite simple, but in the case of the whole ant colony it is more complicated and therefore ACO is able to complete complicated task [3].

The main idea of using ant colony metaphor is that living ants, despite their being almost blind, are able to find their way from the nest to food and back. They are able to do it because ant colony communicates through of a chemical substance called *pheromone*. As the ant moves from the nest to food, the pheromone is released on its path. Ants are able to perceive this pheromone and recognize its density. These ants have a larger probability to move along the path that has the greater hormone density [3].

In the recommendation system the determination of the prediction may be solved as a finding the best path in the oriented acyclic weighted graph $G=(V,E,\tau,\eta)$, where V is the set of vertices, E is the set of edges and functions $\tau,\eta:E\rightarrow[0,1]$ are called trail and visibility, respectively. The best path is defined as a path from the start node to the finish node that is determined according the following procedure [2]:

1. All ants remember already visited vertices using an ordered list called tabu list (TL) (graph is oriented and acyclic, so it guaranties that each vertex is visited only once).
2. At every step the ant chooses, using a probabilistic rule, a vertex to move to among those not in the TL.
3. After a tour has been completed the ant u_k lays a trail in form of tuple (u_k,τ_{ij}) on edge $e(i,j)$ used (trail is the analog of pheromone left by the ant u_k) and clears its tabu list.

The probabilistic rule used for selection the vertex j in the second step is based on the selection rule presented in [2]:

$$P(j)=\frac{\tau_{ij}^{\alpha}\cdot\eta_{ij}^{\beta}}{\sum_{l\in TL}\tau_{il}^{\alpha}\cdot\eta_{il}^{\beta}},$$

where τ_{ij} is the amount of trail on edge $e(i,j)$ and η_{ij} is called visibility of a node j from the node i, $\alpha \in [0,1]$, and $\beta \in [0,1]$ describe importance of trail and visibility in calculating probability.

The procedure mentioned above is rather general. To apply it in the area of prediction of student grades we should first define the problem space (see fig. 1). At the Wroclaw University of Technology students may receive the following seven final grades from the highest (excellent) to the lowest (unsatisfactory): 5.5, 5.0, 4.5, 4.0, 3.5, 3.0 and 2.0.

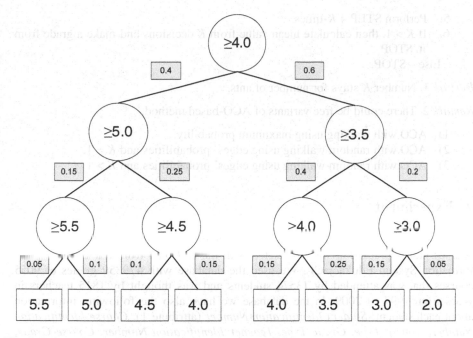

Fig. 1. Balanced tree of grades, in the circle nodes conditions are placed and in the rectangles – decisions (grades). In small rectangles exemplary probabilities are shown.

Procedure of ACO-based recommendation method

ACO filtering method used for grades' recommendation is as follows:

1. Define balanced tree of grades.
2. Calculate trail:
   a. Count empirical probability for each grade, p_n, using M neighbours.
   b. On the edges, starting from the leaves, put trail equals to the p_n for respective grades.
   c. On the edges, going from the leaves to the root, put trail equals to cumulative probability of grades below.
3. Calculate visibility:
   a. Count empirical probability for each grade, p_n, using user's content.
   b. On the edges from leaves put trail equals to the p_n for respective grades.
   c. On the edges, going up to the root, put trail equals to cumulative probability of grades below.
4. Search paths from root to the leaf using one ant due to the following steps:
   a. Calculate probability of moving from i-th node to j-th node:

$$P_{ij} = \frac{\eta_{i,j}^{\alpha} \tau_{i,j}^{\beta}}{\sum_j \eta_{i,j}^{\alpha} \tau_{i,j}^{\beta}}$$

   b. Due to the calculated probability go to next node.

5.   Perform STEP 4 K-times.
6.   If $K > 1$, then calculate mean value from K decisions and make a grade from it. STOP
   Else – STOP.

Remark 1. Number K stays for number of ants.

Remark 2. There could be free variants of ACO-based method:

1)   ACO with walking using maximum probability.
2)   ACO with random walking using edges' probabilities and $K = 1$.
3)   ACO with random walking using edges' probabilities and $K > 1$.

4 Experiment

The verification of the ACO-based recommendation method (see previous section) of courses grades prediction was conducted with the real data from the University Information System EdukacjaCL. We used the database with 479359 grades of 9055 courses that was attended by 13574 students and was thought by 1815 teachers in years from 1997 to 2009. In the database we have also the following information about each student: *Student Identification Number* (attribute 1), *Course Identification Number*, *Course Type*, *Grade Type*, *Teacher Identification Number*, *Course Grade*, *Semester* (att. 8), *Studies Specialization* (att. 9), *Studies Type* (att. 10), *Recruitment Grade* (att. 11), *Recruitment Number* (att. 12). From these data we can calculate the following mean values for each type of course and type of grade concerning each student: *grade from lectures with final exam* (att. 2), *grade from lectures with pass exam* (att. 3), *grade from exercises* (att. 4), *grade from laboratories* (att. 5), *grade from projects* (att. 6), *grade from seminars* (att. 7).

As a recommendation task the grade from the lecture with final exam was chosen. We used three algorithms based on ACO:

1.   With choosing maximal probability.
2.   With one ant and random walking using edges' probabilities.
3.   With K ants and random walking using edges' probabilities.

ACO methods were compared with following filtering methods:

1.   Random recommender (Random) – recommendation (a grade) is given randomly with uniform probability density function.
2.   Content-Based recommender (CBF) – recommendation is given as a mean value of other items in the student's profile (attributes 3-7).
3.   Collaborative Filtering using demographic information recommender (CF1) – recommendation is given as a mean value of grades from N neighbours (neighbourhood is calculated using attributes 8-12 by weighted 0-1 similarity measure).
4.   Collaborative Filtering using Euclidean metric recommender (CF2) - recommendation is given as a mean value of grades from N neighbours (neighbourhood is calculated using attributes 3-7 with Euclidean similarity measure).

5. Collaborative Filtering recommender (CF3) - recommendation is given as a mean value of grades from N neighbours (neighbourhood is calculated using attributes 3-7 as in CF2 and 8-12 as in CF1).

We applied the following measures for comparison of the above mentioned methods [4]: *Mean Absolute Error* (MAE), *Normalized Mean Absolute Error* (NMAE), *Prediction Accuracy* (PA), *Mean Squared Error* (MSE), *Root Mean Squared Error* (RMSE), *Standard Error Variance* (SEV), *Classification Accuracy* (classAcc) that measures how many predictions were exactly the same as actual grades in the test set.

All methods were implemented in Matlab® environment. The program was launched 10 times for each method with different training set (70% of original dataset) and test set (30% of original dataset). Dataset was divided into training and test sets randomly. All values of performance measures are mean values from 10 program launches.

Values of parameters were as follows: $\alpha = 0.4$, $\beta = 0.4$, number of neighbours was set as 20. After parameter tuning it turned out that for all methods the same values were the best. Moreover, number of neighbours more than 20 gave no better results. Further, for ACO, more than 100 ants gave no better results. The results are presented in Table 1 and 2, and in the following Figures 2 – 5.

Table 1. Results for three variants of ACO and Random as a comparison

Measure	Random	ACO, max prob.	ACO, 1 ant	ACO, 100 ants
MAE	1.25	0.52	0.76	0.41
NMAE	0.36	0.15	0.22	0.12
PA	0.64	0.85	0.78	0.88
MSE	2.45	0.58	1.095	0.38
RMSE	1.56	0.76	1.05	0.61
SEV	0.94	0.56	0.72	0.45
classAcc	0.03	0.375	0.28	0.42

Table 2. Results of PA and classAcc for ACO, CBF, CF1, CF2, CF3

Measure	ACO, 100 ants	CBF	CF1	CF2	CF3
PA	0.88	0.82	0.87	0.87	0.88
classAcc	0.42	0.08	0.11	0.11	0.13

Presented results show that the best method is ACO with 100 ants and with random walk. However, it is slightly better than CF3, CF1 or CF2 in PA measure, but other measures, and especially classAcc, prove the best performance of ACO-based method.

A little bit more attention should be paid to classAcc. This measure shows how many grades were predicted exactly the same as actual grades from the test set. Result for ANT with maximum probability (around 35%, see Table 1) and ANT with 100

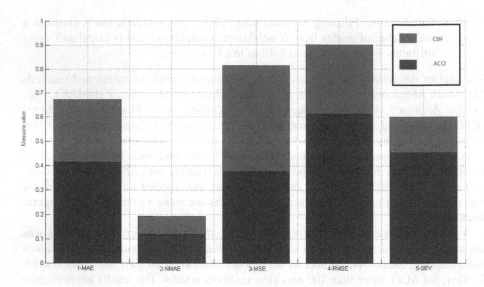

Fig. 2. Measures' (MAE, NMAE, MSE, RMSE, SEV) values for CBF recommender and Ant-Colony-Based with random probability walk with 100 ants

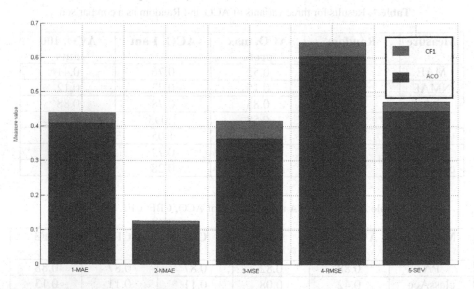

Fig. 3. Measures' (MAE, NMAE, MSE, RMSE, SEV) values for CF1 recommender and Ant-Colony-Based with random probability walk with 100 ants

ants and random walk (around 42%, see Table 1 and 2) are extremely high in such difficult task as grade recommendation. Especially that in our experiment student is described only by few attributes. Thus interesting is if classAcc could have higher value for more detailed student description. However, this issue has to be postpone for further research.

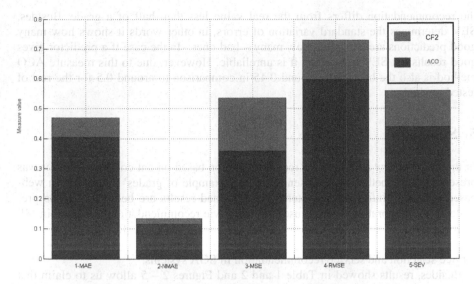

Fig. 4. Measures' (MAE, NMAE, MSE, RMSE, SEV) values for CF2 recommender and Ant-Colony-Based with random probability walk with 100 ants

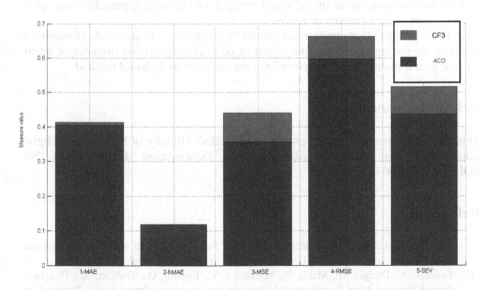

Fig. 5. Measures' (MAE, NMAE, MSE, RMSE, SEV) values for CF3 recommender and Ant-Colony-Based with random probability walk with 100 ants

Also comparison of the other measures: MAE, NMAE, MSE, RMSE, SEV, shows that ACO method was the best. All of the presented measures show that ACO method behaves more stable than any other recommender. First of all, MAE characterizes what is the difference between real a value and predicted one. Value of that measure for ACO-based methods is around 0.4 (see Table 1), which means that in mean sense the recommendation differs from the real value less than half of a grade. Besides, SEV determines the standard variation of errors, in other words it shows how many good predictions are made and how many – bad ones. In the case if a predictor gives good results but SEV is big then it is unreliable. However, due to this measure ACO method is also the best (with around 0.45 in comparison to around 0.5 for the rest of tested methods).

5 Summary

In this paper a new approach for recommendation based on ant colony metaphor was presented. This method was presented on the example of grades' prediction in web-based education information system. The obtained results are better that those presented for collaborative filtering used in the movie recommendation presented in [8]. This approach was also used in other domains such as user interface and movie recommendation [9], so this approach could be also used in other applications, such as: service selection and service recommendation in SOA systems.

Besides, results showed in Table 1 and 2 and Figures 2 – 5 allow us to claim that ACO-based method is promising. However, more research should be carried out with different datasets and for different applications, but we say that ACO-based method enables to overcome most of the disadvantages of classical approaches and gives higher values of performance measures.

In further works other applications should be taken into consideration. Moreover, it could be interesting to examine the aspect of description precision (number of attributes), e.g. of a student, for the performance measures of ACO-based method.

Acknowledgements

This work has been partially supported by the Polish Ministry of Science and Higher Education within the European Regional Development Fund, Grant No. POIG.01.03.01-00-008/08.

References

[1] Chen Q., Aickelin U, Movie Recommendation Systems using an artificial immune system. Poster Proceedings of ACDM 2004, Engineers' House, Bristol, UK (2004)

[2] Colorni, A., Dorigo, M., Maoli, F., Maniezzo, V., Righini, G., Trubian, M.: Heuristics from Nature for Hard Combinatorial Optimization Problems. International Transactions in Operational Research 3(1), 1–21 (1996)

[3] Gao, W.: Study on Immunized Ant Colony Optimization. In: Third International Conference on Natural Computation (ICNC 2007), vol. 4, pp. 792–796 (2007)

[4] Herlocker, J.L., Konstan, J.A., Terveen, L.G., Riedl, J.: Evaluating Collaborative Filtering Recommender Systems. ACM Transactions on Information Systems 22, 5–53 (2004)

[5] Kobsa, A., Koenemann, J., Pohl, W.: Personalized Hypermedia Presentation Techniques for Improving Online Customer Relationships. Knowledge Eng. Rev. 16(2), 111–155 (2001)

[6] Montaner, M., Lopez, B., de la Rosa, J.P.: A Taxonomy of Recommender Agents on the Internet. Artificial Intelligence Review 19, 285–330 (2003)

[7] Nguyen, N.T., Sobecki, J.: Using Consensus Methods to Construct Adaptive Interfaces in Multimodal Web-based Systems. Universal Access in Inf. Society 2(4), 342–358 (2003)

[8] Sarwar, B., Konstan, J., Borchers, A., Herlocker, J., Miller, B., Riedl, J.: Using Filtering Agents to Improve Prediction Quality in the GroupLens Research Collaborative Filtering System. In: CSCW 1998, Seattle Washington, USA, pp. 1–10 (1998)

[9] Sobecki, J.: Ant colony metaphor applied in user interface recommendation. New Generation Computing 26(3), 277–293 (2008)

[10] van Setten, M.: Supporting People in Finding Information. Hybrid Recommender Systems And Goal-Based Structuring, Enschede, The Netherlands, Telematica Instituut Fundamental Research Series 016 (2005)

Web Ontology Building System for Novice Users: A Step-by-Step Approach

Shotaro Yasunaga[1], Mitsunori Nakatsuka[1], and Kazuhiro Kuwabara[2]

[1] Graduate School of Science and Engineering, Ritsumeikan University
[2] College of Information Science and Engineering, Ritsumeikan University
1-1-1 Noji-Higashi, Kusatsu-shi, Shiga-ken, 525-8577, Japan
{cc012057,cc008057}@ed.ritsumei.ac.jp,
kuwabara@is.ritsumei.ac.jp

Abstract. This paper proposes a system for building Web ontologies for novice users. Web ontology plays a central role in Semantic Web applications, and is utilized for various purposes. Systems for building ontologies have been extensively studied in diverse ways. However, Web ontology is not widely exploited in many applications, in part because ontology construction requires expert knowledge. To allow novice users to create ontologies, we propose a system for building Web ontologies. In the proposed system, the process of building ontologies is divided into several steps. A user is provided with step-by-step instructions, and enters information either in the form of a web page or by choosing from a list of words. In this paper, we describe the architecture of the proposed system, and show how Web ontology building can be assisted in an example application of a disaster mitigation system.

Keywords: Web ontology, ontology construction, Semantic Web.

1 Introduction

These days, there are many web sites available on the Internet providing a wide range of information. It would be desirable to search different web sites at the same time using the same query. However, there are two major problems with retrieving information from different web sites. One is that we cannot conduct a search with the semantics of the data taken into consideration. The other is that when we integrate information from multiple web sites, the difference in words used to express the same thing needs to be resolved. In order to solve these problems, we use Semantic Web technologies [1], especially OWL, Web ontology language [2], to represent relationships among words used in different web sites.

There are many studies on building Web ontology, such as automated ontology construction using machine learning techniques [3], and ontology editing tools [4] to facilitate ontology development by human users. While an ontology editor is an indispensable tool for building ontologies, the use of the ontology editor generally requires extensive knowledge of ontology, and is not suitable for novice users. However, constructing ontologies should not be limited to experts. Various domain experts need to

N.T. Nguyen, M.T. Le, and J. Świątek (Eds.): ACIIDS 2010, Part II, LNAI 5991, pp. 134–143, 2010.

participate in ontology construction. In order to allow users from different backgrounds to build ontologies, we propose a Web ontology building system especially targeted for novice users.

For beginners in ontology construction, the following problems can be pointed out.

1. They do not know specific steps to follow in building ontologies.
2. They do not know what type of class and property can be defined in OWL.
3. The OWL format is complex, making it difficult to understand how an ontology is defined.
4. The area of vocabularies is so broad that users cannot build them alone.
5. They do not know what RDF [5] triples are to be generated by inference with OWL.

These problems prevent many people who want to incorporate the Web ontology technologies from building vocabularies. A simpler system for building ontology is needed for beginners. In order to solve these problems, we propose an ontology building system that includes the following functions.

1. The process of building an ontology is broken into small and simple steps, and the proposed system provides users with step-by-step instructions.
2. The proposed system restricts each step in the building process to a simple task such as entering the information into the form or choosing a word from a list of vocabularies.
3. The ontology file is converted into an RDF graph and presented visually to users.
4. By implementing as a web application, users can use a web browser to build an ontology. Thus, distributed and cooperative development can easily be achieved.
5. The proposed system provides a checking function that infers data with built-in OWL definitions.

The rest of the paper is structured as follows. The next section describes related work, while sections 3 and 4 present the design of the proposed system, and an example usage of the system. The final section concludes the paper with discussion on future work.

2 Related Works

In order to help construct Web ontologies, several ontology editors have been developed. Protégé [6], written in Java and developed at Stanford University, allows users to create classes, properties, or instances easily. The users can also edit metadata about the ontology, such as version information or other imported ontologies. Its interface is grouped into tabs for each function. In addition, Protégé has a testing facility to find logical errors. By using Protégé, users can build an ontology properly according to their desires. OntoStudio [7], developed by Ontoprise, uses a unique method, called On-To-Knowledge, to build ontologies. It uses F-Logic as an inference engine to translate axioms. Furthermore, we can import or export ontologies written in OWL. WebODE [8] is a workbench that supports the whole lifecycle of building ontologies. The MINDSWAP project at the University of Maryland is also developing an ontology editor called Swoop [9]. Swoop has an interface modeled after a web browser. Users can open an editing window for each item to be edited by clicking a button. Hozo [10] focuses on the role and relationship of ontologies.

These tools are very useful for users who are familiar with Web ontology. However, they are very difficult to use for people who do not know much about Web ontology. For ontology novices, a system is needed that allows ontology building by simple operations without expert knowledge.

3   Ontology Building Environment for Novices

3.1   Features of the Proposed System

The proposed system is implemented as a web application that can be used on a web browser. This system presents each step to take in making ontologies. Users can complete the ontology building by following the steps, each of which consists of simple processes like filling in a form on the web pages.

The description languages of ontologies in the system are RDFS and OWL. It has the function to export files in these formats. Although RDFS and OWL languages are XML-based, users do not need to know XML when they use the system. A created ontology is checked for any logical inconsistency. The system visualizes the ontology and displays it in an RDF graph.

In many cases, all of the ontology files are being built by a system developer. That imposes a burden on the developer, making it harder to use ontologies. If many people can create the same ontology collaboratively, vocabularies can possibly be defined more widely and extensively. Not only a system that can build an ontology easily, but also a system that makes distributed and collaborative development possible is needed so that a lot of people can build ontologies collaboratively. The proposed system is composed as a web application. Only a web browser and the environment of the Internet are needed to use the proposed system.

There is also a problem in that novice users do not understand derived data from the definitions in RDF and OWL rules. When they finish building ontologies, it is difficult to confirm whether the built ontologies are adequate. The proposed system incorporates a function to confirm that unintended data will not be derived when a query is executed by presenting both the inferred data and the original for comparison.

3.2   Ontology Building Flow

It is crucial for making operations easier and more efficient to divide building operations into several stages, and show users how to follow each step. We describe the stages of ontology building below.

3.2.1   Deciding the Aim of the Ontology and Creating Project

Building ontology is started from clarifying the ontology's target domain, target use, and target user. In the proposed system, ontologies are managed by projects. As the first step, users must give ontology a proper name.

3.2.2 Searching Existing Ontologies

When building a new ontology, users must search existing ontologies for possible reuse. If an existing Web ontology is found, which can be used as a basis for the new ontology, the ontology found is imported and a new ontology is to be built by adding new vocabulary or links to it.

3.2.3 Defining Classes

Defining classes is one of the important stages in ontology building. Users must consider the domain of their vocabulary, and define classes, class expressions or class axioms.

In the stage of defining classes, there are two main operations. Users can begin to define classes from any one of the operations, which are:

1. Defining new classes,
2. Adding OWL class expressions, class axioms, or property restriction to existing or new classes.

When users want to define new classes, the URI of each new class is needed. The proposed system shows users a text form for inputting the URI, and names the class by the URI.

As for the case of adding OWL class expressions or class axioms to existing classes, the first step is selecting a class to which users want to add new class elements, from a list of URIs of existing classes (the list is made by the system using loaded ontology file(s)). When a class is selected, a list of class elements will be shown. Users can select a class element and input requested information using input forms designed for each class element. In this way, adding class expressions, such as class hierarchy, which is important to use for ontology inferences, is made easier for ontology novices.

3.2.4 Defining Properties

Properties are defined to express references between things or properties. It is important to define what resources are linked by a property. Similar to the case of classes, defining properties involves two main operations:

1. Defining new properties,
2. Adding OWL property axioms to existing or new properties.

If users want to define a new property, they can define the property's type by selecting owl:ObjectProperty or owl:DatatypeProperty (in OWL, there are other types of properties, annotation property and ontology property, which we will discuss later). When a new property is made, users must define the domain and range of the property. The proposed system shows a list of existing classes from the loaded ontology and users can define it only by choosing classes as the domain or range from this list.

If users want to add a property axiom to an existing property, they can select the target property from a list of existing properties in the loaded ontology. Users then select a property axiom, and go to the editing page designed for each axiom.

In this stage, users can easily create properties considering their hierarchy, equivalence, restrictions, or logic.

3.2.5 Checking Built Ontologies

The next stage of ontology building is to check the ontology developed. If there is inconsistency in the semantics of the ontology, or inference rule violation, applications that use the ontology will not work correctly. Creating a correct vocabulary and accurate ontology will increase the reliability of applications using the ontology.

In addition, it is important to confirm that the ontology is constructed as the user intends. Unfortunately, OWL files are written in XML, and tend to be very large in size in many cases. Moreover, in reading XML files, it is difficult to understand where and/or which vocabulary is written. Presenting a visual graph representation of the ontology makes it easier to build or confirm the ontology. Figure 1 shows the OWL file in the XML file format (a) and RDF graph format (b). The proposed system converts an OWL file into graphical data type.

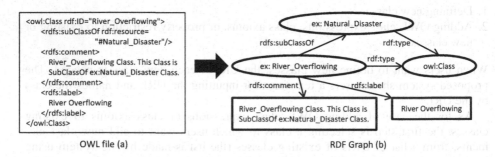

Fig. 1. OWL file and graphical representation

3.2.6 Checking Derived Data

The final stage of building ontologies is to check if the intended data can be inferred. In this stage, users first upload an RDF file(s) that contains checking data from which new data is to be inferred (Fig. 2 (a)). In the next step, new data are derived from the uploaded data with the ontologies built (Fig. 2 (b)). Finally, users compare inferred data (Fig. 2 (c)) with the original, and examine difference between them. Users can confirm whether they can obtain correct results by inference. By doing this, novice users can understand the role of ontologies, and make inferences with ontologies.

Figure 3 shows a screenshot of the proposed system for checking derived data. Uploaded and inferred RDF data are shown in the N3 format, which is easier to read than the XML format for novice users.

3.3 Distributed and Collaborative Development

In the proposed system, with a web browser and Internet connection, users can participate in building an ontology from anywhere. Distributed and collaborative development is encouraged to help with building large-scale ontologies.

For distributed development, an import function is needed. The import function uses the owl:imports property, which is an ontology property on the ontology header, and imports an external ontology and adds it to the current ontology. Using this function and connecting ontologies that are built in a distributed environment, we can make large ontologies more quickly.

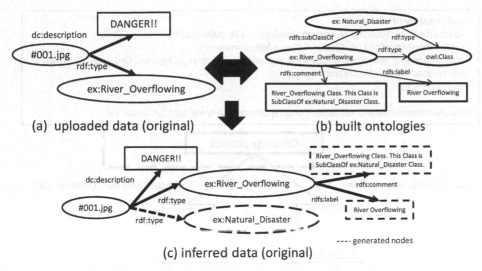

Fig. 2. Checking derived data

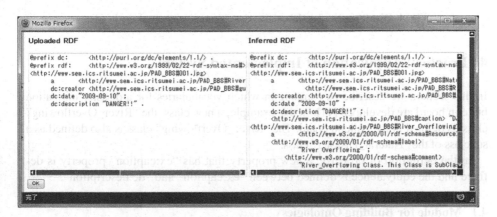

Fig. 3. Screenshot of proposed system for checking derived data

When we develop an ontology collaboratively, we need a version management function. The information on the version is described using owl:versionInfo, which is the annotation type property. The described content on the ontology header is the date of creation, user's name, modified point, and so on. Figure 4 shows an example of distributed and collaborative development and an ontology header. In this example, Users A and B build ontologies collaboratively. In the ontology header, there is information about the version, date of creation, user's name, modified point, and so on.

We are implementing an ontology building support system as a web application based on the above steps.

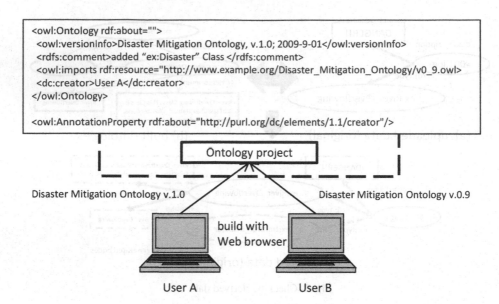

```
<owl:Ontology rdf:about="">
   <owl:versionInfo>Disaster Mitigation Ontology, v.1.0; 2009-9-01</owl:versionInfo>
   <rdfs:comment>added "ex:Disaster" Class </rdfs:comment>
   <owl:imports rdf:resource="http://www.example.org/Disaster_Mitigation_Ontology/v0_9.owl>
   <dc:creator>User A</dc:creator>
</owl:Ontology>

<owl:AnnotationProperty rdf:about="http://purl.org/dc/elements/1.1/creator"/>
```

Ontology project

Disaster Mitigation Ontology v.1.0   Disaster Mitigation Ontology v.0.9

build with
Web browser

User A   User B

Fig. 4. Distributed and collaborative development and ontology header

4 Practical Application of Implemented System

In this section, we describe an example in which vocabularies for a disaster mitigation bulletin board are developed [11]. In this example, a new class, the "River_Overflowing" class, is defined in the bulletin board. The "River_Overflowing" class is also defined as a subclass of the "Natural_Disaster" class.

In the definition of property, a new property that has "ex:caption" property is defined and the equivalence is defined between "ex:caption" and "dc:description".

4.1 Module for Building Ontologies

In this example, new vocabularies are added to the existing ontologies. We describe an example of building flow below.

4.1.1 Making a Project

At the start of ontology construction, a user makes a decision about the area of ontology, and names a project. In this example, the defined ontology has information about disaster mitigation, and the project is named "Disaster_Mitigation_Ontology".

In addition, when users start to define new vocabularies, they must enter the prefix of namespace. The prefix is defined to be "ex:" which is shorthand for "http://www.example.org/terms/".

4.1.2 Uploading an Existing Ontology

If users need to reuse an existing ontology, they can upload an ontology file. Uploaded files are managed in the same project. When an ontology file is exported, uploaded files are incorporated into a main file with the "owl:imports" property in the ontology header (Fig. 4).

4.1.3 Defining Classes

The next step is to define a new class. In this example, the "ex:River_Overflowing" class and the "ex:Natural_Disaster" class are to be defined. In this step, necessary information is the class's URI. Figure 5 shows a screenshot of the proposed system. In Fig. 5 (a), users can define a new class and its expanded information. Users choose a prefix "ex:" and enter the URI in the text form. In addition, users can enter information that has a label and comments in a natural language. These comments are not used internally, but for a human user. They are important for understanding ontology.

Next in this example, a semantic hierarchy of the terms used in the disaster mitigation bulletin boards is defined. The "River_Overflowing" class is defined as a subclass of the "Natural_Disaster" class. Figure 5 (b) shows that users can add a new class vocabulary to the existing classes. In this example, a user selects a class, "River_Overflowing", and adds a new class vocabulary using, "rdfs:subClassOf". In the page of the defined "rdfs:subClassOf" in Fig. 5 (c), users choose necessary data, the super-class's URI, from the list of class URIs. With this information, the proposed system adds a new class vocabulary.

4.1.4 Defining Properties

Defining properties is similar to defining classes. Users need to select the prefix "ex:" and enter the URI of property, "caption". Moreover, to define the equivalent relationship, users choose the property, "owl:equivalentProperty", from the list of vocabularies and specify that the new property is equivalent to the property "dc:description".

4.1.5 Checking Ontology

If users finish defining the ontology, the proposed system converts an ontology file into graphical data type, RDF graphs (Fig. 1).

Finally, the proposed system provides users with a checking tool. In this example, first, users upload RDF files that have image and meta-information for disaster mitigation (Fig. 2 (a)). Second, users define a semantic hierarchy of the terms used in the bulletin boards. Specifically, "ex:River_Overflowing" is defined as a subclass of "ex:Natural_Disaster" (Fig. 2 (b)). When a user makes a query about "ex:Natural_Disaster," the data about "ex:River_Overflowing" can also be retrieved according to the subclass relationship. For the subclass relationship, a new node (in this example, "ex:Natural_Disaster" for #001.jpg) is inferred to be added so that a user can make a query with a term of broader meaning (Fig. 2 (c)).

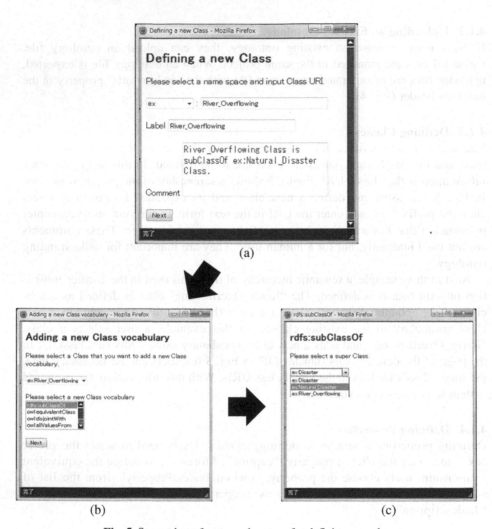

Fig. 5. Screenshot of proposed system for defining new classes

5 Conclusion and Future Work

In this paper, we have proposed a step-by-step ontology building approach implemented as a web application. By providing specific steps of creation, we can support
users who do not have expert knowledge on ontology in making ontologies. In this
way, we can provide opportunity for users to build and use Web ontology. This will
contribute to wider usage of Semantic Web technologies. In ontology building, users
perform only simple tasks like inputting information into a form on a web page and
choosing something from the presented list. The visualization of ontology and file-
making is supported at the system side. As a result, users do not need to think about
the format of ontology files, and this makes it easier to build them. In addition, users

are provided with a function to check inferences of built ontologies, so they can understand what data is generated by inference.

In order to complete ontology building, it is also necessary to support the process of enumerating all the vocabulary used in the domain, which is the work of a lower layer. Future work will deal with supporting all processes of ontology building.

References

1. Berners-Lee, T., Hendler, J., Lassila, O.: The Semantic Web. Scientific American 284(5), 34–43 (2001)
2. Bechhofer, S., Harmelen, F.V., Hendler, J., Horrocks, I., McGuinness, D.L., Patel-Schneider, P.F., Stein, L.A.: OWL Web Ontology Language Reference, http://www.w3.org/TR/owl-ref/ (accessed September 16, 2009)
3. Maedche, A., Staab, S.: Ontology Learning for the Semantic Web. IEEE Intelligent Systems 16(2), 72–79 (2001)
4. Kozaki, K., Mizoguchi, R.: A Present State of Ontology Development Tools. Journal of Japanese Society for Artificial Intelligence 20(6), 707–714 (2005)
5. Resource Description Framework (RDF), http://www.w3.org/RDF/ (accessed September 16, 2009)
6. Noy, N.F., Sintek, M., Decker, S., Crubezy, M., Fergerson, R.W., Musen, M.A.: Creating Semantic Web contents with Protege-2000. IEEE Intelligent Systems 16(2), 60–71 (2001)
7. Davies, J., Grobelnik, M., Mladeni, D.: OntoSTUDIO® as a Ontology Engineering Environment. In: Semantic Knowledge Management, pp. 51–60. Springer, Heidelberg (2008)
8. Arpírez, J.C., Corcho, O., Fernández-López, M., Gómez-Pérez, A.: WebODE: a Scalable Workbench for Ontological Engineering. In: Proceedings of the 1st International Conference on Knowledge Capture, pp. 6–13 (2001)
9. Kalyanpur, A., Parsia, B., Sirin, E., Grau, B.C., Hendler, J.: Swoop: A Web Ontology Editing Browser, Web Semantics - Science Services and Agents on the World Wide Web 4, 144–154 (2007)
10. Kozaki, K., Kitamura, Y.: Hozo: An Environment for Building/Using Ontologies Based on a Fundamental Consideration of "Role" and "Relationship". In: Gómez-Pérez, A., Benjamins, V.R. (eds.) EKAW 2002. LNCS (LNAI), vol. 2473, pp. 155–163. Springer, Heidelberg (2002)
11. Yasunaga, S., Kuwabara, K.: Information Retrieval from Multiple Regional Bulletin Boards using Semantic Web Technology. In: Proceedings of the Second Asian Joint Workshop on Information Technologies (Joint Workshop of Ritsumeikan University and Dalian University of Technology), p. 54 (2009)

Automatic Lexical Annotation Applied to the SCARLET Ontology Matcher

Laura Po and Sonia Bergamaschi

DII, University of Modena and Reggio Emilia, Italy
name.surname@unimore.it

Abstract. This paper proposes lexical annotation as an effective method to solve the ambiguity problems that affect ontology matchers. Lexical annotation associates to each ontology element a set of meanings belonging to a semantic resource. Performing lexical annotation on the ontologies involved in the matching process allows to detect false positive mappings and to enrich matching results by adding new mappings (i.e. lexical relationships between elements on the basis of the semantic relationships holding among meanings).

The paper will go through the explanation of how to apply lexical annotation on the results obtained by a matcher. In particular, the paper shows an application on the SCARLET matcher.

We adopt an experimental approach on two test cases, where SCARLET was previously tested, to investigate the potential of lexical annotation. Experiments yielded promising results, showing that lexical annotation improves the precision of the matcher.

Keywords: Ontology matching, lexical annotation, mapping discovery.

1 Introduction

Finding correspondences between heterogeneous conceptual structures is inherent to all systems that combine multiple information sources. In information integration and ontology engineering communities, the task of identified matching is a core task.

Many different matching solutions have been proposed in literature [4]. They take advantage of various properties of ontologies, e.g., structures, data instances, semantics, or labels, and use techniques from different fields, e.g., statistics and data analysis, machine learning, automated reasoning, and linguistics. Some approaches have been proposed for validating mappings with respect to the semantics of the involved ontologies. Example of works in this direction are the S-Match system [6] and the theoretical study proposed in [17]. Although the approaches are important for the validation of mappings, they do not discern elements with different meanings. Some other tools incorporate the linguistic features inside the matcher (we find some example in H-MATCH [3], Cupid [9] and Falcon-AO [8]). Differently from these matchers that include a linguistic component, lexical annotation is able to disambiguate elements, to enable an effective comparison of them from other online ontologies or thesauri.

N.T. Nguyen, M.T. Le, and J. Świątek (Eds.): ACIIDS 2010, Part II, LNAI 5991, pp. 144–153, 2010.
© Springer-Verlag Berlin Heidelberg 2010

Until now, we developed automatic lexical annotation techniques that allow us to extract lexical knowledge from structured and semi-structured data sources detecting mappings useful for the data integration process [2]. These techniques have been developed within the MOMIS data integration system [1] and the evaluation performed on real data sets has shown good performance.

In this paper, we apply automatic lexical annotation on the elements involved in the mappings discovered by the a matcher. In particular, we will show the application of lexical annotation on the SCARLET matcher[1], but lexical annotation can be applied in general to the output of different matchers.

The SCARLET matcher has been selected as a candidate matcher because it belongs to a new generation of ontology matchers that focused on exploiting the increasing amount of online semantic data available on the Web. These applications handle the high semantic heterogeneity introduced by the increasing number of available online ontologies (different domains, different points of view; different conceptualisations). These matching algorithms exhibit very good performance, but they rely on merely syntactical techniques to anchor the terms to be matched to those found on the Semantic Web. As a result, their precision can be affected by ambiguous terms. A critical issue is to solve these ambiguity problems by introducing lexical annotation techniques, which validate the mappings by exploring the semantics of the elements involved in the matching process. In addition, lexical annotation allows the discovery of new mappings (derived from the lexical knowledge), thus enriching the results of the matcher.

Automatic lexical annotation is obtained by the application of a set of Word Sense Disambiguation (WSD) algorithms. We make use of the tool ALA (Automatic Lexical Annotator) [2] to detect the correct annotations for each ontology concept. Then, we apply rules to detect the false positive mappings discovered by the matcher and to discover new mapping among concepts.

The evaluation has been done on two test cases and compared with other WSD techniques previously tested on the SCARLET output [7].

The paper is organized as follows: Section 2 describes the SCARLET matcher, section 3 focus on the application of lexical annotation techniques on the matcher output. In section 4 the evaluation of lexical annotation techniques is shown on two different test cases. Conclusion are sketched in section 5.

2 SCARLET Matcher

SCARLET[2] [11,15] is a technique for discovering relationships between two concepts by making use of online available ontologies. Developed in the context of the NeOn[3] and OpenKnowledge[4] projects, SCARLET has been primarily used to support tasks such as ontology matching and enrichment.

---

[1] http://scarlet.open.ac.uk/
[2] http://scarlet.open.ac.uk/
[3] http://www.neon-project.org/web-content/
[4] http://www.openk.org/

SCARLET discovers semantic relationships between concepts by using the entire Semantic Web as a source of background knowledge: by using semantic search engines (Swoogle [5] and WATSON [13]), it finds online ontologies containing concepts with the same names as the candidate concepts and then it derives mappings from the relationships in the online ontologies.

Scarlet is able to identify disjoint relations, subsumption relations, and correspondences [12,14]. All relations are obtained by using derivation rules which explore not only direct relations but also relations deduced by applying subsumption reasoning within a given ontology.

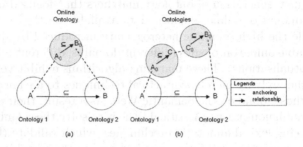

Fig. 1. SCARLET strategy using a single online ontology (a) or more ontologies (b)

Figure 1 illustrates the idea of SCARLET. A and B are the concepts to relate, and the first step is to find online ontologies containing concepts A_0 and B_0 equivalent to A and B. This process is called *anchoring* and A_0 and B_0 are called the *anchor terms* (or *anchor concepts*). Based on the relationships that link A_0 and B_0 in the retrieved ontologies, a mapping is then derived between A and B. In Figure 1 (a), the strategy assumes that a semantic relationship between the candidate concepts can be discovered in a single ontology. However, some relationships could be distributed over several ontologies. Therefore, SCARLET develops a recursive strategy to combine knowledge contained in several ontologies, and thus derives mappings from two (or more) ontologies, as shown in Figure 1 (b).

2.1   Limitations of SCARLET

As depicted in [12], the SCARLET paradigm is feasible. A baseline implementation of the SCARLET technique applied on a large-scale, real life data set has led to a precision value of 70% which correlates with the performance of other background knowledge based matchers. An analysis of the causes of false positive mappings revealed that more than half of them were due to an incorrect anchoring caused by ambiguities: elements of the source ontology have been anchored to online ontologies on the basis of the syntax. Therefore, we can affirm that the major limitation of SCARLET prototype remains its simple, string comparison based anchoring which generated more than half of the false mappings.

SCARLET is not able to take advantage of the ontological context in which a concept appears. Instead, lexical annotation techniques, exploiting the context, can define the meaning for the concept itself. By identifying a meaning (or a set of meanings) for a concept it is possible to, more accurately, compare the concept with the concepts that appear in online ontologies. For example, if we look for a online ontology that contains the term "star" we retrieve 14 results[5]. Some of these ontologies use the word "star" as a famous actor/actress, some other assume the meaning of a celestial body. Because the SCARLET anchoring ground on all the retrieve documents, it can potentially derived false positive relationships.

On the SCARLET matcher some disambiguation techniques have already been applied. In [7] two different techniques of WSD are investigated to improve the SCARLET results, by detecting and solving the ambiguity problems inherent to the use of heterogeneous sources of knowledge. The experiments carried out confirmed that precision can be improved by using the semantic techniques. However, both the techniques proposed (semantic similarity measures) have an important limitation: they need some training set to detect an accurate threshold under which two terms are not consider synonyms. Unlike these techniques, the method proposed in this paper offers a definite answer regarding the detection of synonym relationships.

3   Lexical Annotation Applied to the SCARLET Matcher

An Annotation is a piece of information added in a book, document, online record, video, or other data. Lexical Annotation is a particular kind of Annotation that refers to a semantic resource. Each lexical annotation has the property to own one or more lexical descriptions. Lexical annotation of an ontology class is the explicit assignment of its meaning w.r.t. a semantic resource (i.e. entries in a thesaurus, dictionary or semantic network). Lexical Annotation differs from the Ontology-based Annotation where the annotation is performed w.r.t. an ontology and it is not mandatory that an ontology class has a lexical description.

Lexical annotation leads to several improvements in the matching process:

- it improves the precision of the matcher by detecting the false positive mappings;
- it enriches the matching results by discovering new mappings based on the lexical relationships among meanings;
- it is able to identify synonymous and more general classes of a concept, giving the matcher the possibility to widen the search among online ontologies.

The third improvement has not been developed yet, it will be the focus of our future work.

---

[5] Data obtained on 21st June 2009 looking in WATSON for a word match of class entity.

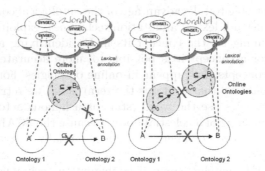

Fig. 2. Lexical annotation improvements: detection of false positive mappings

Detection of false positive mappings: The idea is to apply a combination of WSD algorithms to the source and background ontologies involved in the matching process. After the annotation of these ontologies, we examined the concepts involved in the anchoring. If a concept and its anchoring concept have disregarding meanings (i.e. if they do not have the same list of meanings), the anchoring is discharged. Lexical annotation can thus filter out wrong anchoring (with a good precision) and so, it can improve the efficiency of the matcher. Figure 2 shows how lexical annotation influences the anchoring. Let us focus on the (a) subfigure, after the annotation of all the concepts involved in the anchoring (A, B, A_0, B_0), it is possible to compare the meanings of a concept with the meanings of its anchoring concept. The anchoring between A and A_0 is preserved because the concepts have the same meanings. Instead, the anchoring between B and B_0 is discharged because the concepts have different meanings. As a consequence, the mapping among A and B is detected as false positive mapping. Also anchoring across online ontologies benefit from the lexical annotation, as shown in figure 2 (b).

New mapping discovery: Lexical annotation can also enrich the matching results by discovering new mappings. In Figure 3, the process of identifying new mappings among elements is shown. First, the lexical annotation of the

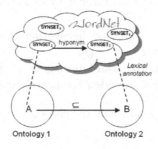

Fig. 3. Lexical annotation improvements: new mapping discovery

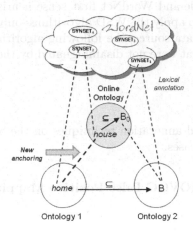

Fig. 4. Lexical annotation improvements: identification of synonymous and more general concepts

source and target elements is performed, then, the WordNet network is explored looking for lexical relationships between the selected meanings. For any relationships found, a mapping is inserted between the corresponding source and target elements.

Identification of synonymous and more general concepts: Lexical annotation of an ontology leads important consequences. Identifying a meaning for a ontology class means that we are able to detect synonymous (terms that share a meaning) and more general concepts of the given class. Synonymous and more general concepts are new terms that can be used by the matcher to widen the search in online ontologies (see Figure 4).

3.1 WSD Techniques

To perform lexical annotation we use a combination of WSD algorithms of different natures. Ensemble methods are becoming more and more popular as they allow one to overcome the weaknesses of single approaches [10]. Different strategies can be applied such as majority voting, probability mixture, rank-based combination or maximum entropy combination. We chose to combine algorithms in a sequential composition based on their reliability: each algorithm has been previously tested on several scenarios and its precision has been calculated; then, this precision value has been used to define the reliability of the algorithm.

We employ the ALA tool [2] to perform lexical annotation of the onologies involved in the matching process (source ontologies and online ontologies). With ALA we combine the output of four WSD algorithms (Structural Disambiguation algorithm, WordNet Domains Disambiguation algorithm, Gloss Similarity algorithm and Iterative Gloss Similarity algorithm) and two heuristic rules

(Monosemic heuristic rule and WordNet first sense heuristic rule). We select a sequential composition to apply the WSD algorithms: only the first algorithm is executed on the entire data source, the following algorithms are executed only on the set of concepts that were not disambiguated by the previous ones.

4   Evaluation

The application of lexical annotation techniques on the SCARLET results has been tested on two test cases.

4.1   NALT and AGROVOC False Positive Mappings Evaluation

The first test case was composed of real life thesauri [15]: the United Nations Food and Agriculture Organization (FAO)'s **AGROVOC** thesaurus, and the United States National Agricultural Library (NAL) Agricultural thesaurus **NALT**. On this scenario, a sample of 1000 mappings obtained by SCARLET has been manually validated, resulting in a promising 70% precision. Our evaluation has been performed on the 217 false positive mappings (the detection of them was previously done by a domain expert that knows the ontologies and their characteristics and is able to select correct mappings). After the lexical annotation of concepts involved in the anchoring, we discovered if the meanings are consistently linked or not, then we detected the false positive mappings.

We performed the automatic lexical annotation on each sub-ontology involved in the mapping and then, evaluated the results of the annotation on the anchoring. As previously explained in section 3, it was sufficient that the lexical annotation reveal that a concept has a meaning different from its anchoring, so that the anchoring is discharged and the mapping is revealed not valid. Thanks to the lexical annotation of the concepts, 12 out of 14 mappings have been recognized as false positive.

Unfortunately, not all the background ontologies were still available when we performed the test and some of them were not correctly written. At the end, the test case was not meaningful for the lexical annotation evaluation because it examines only 14 mappings.

4.2   OAEI Evaluation

The second test case is based on the OAEI 2006 benchmark; this was the test case where SCARLET and another disambiguation method have been previously evaluated [15]. The benchmark[6] is bibliographic domain, the bibliographic ontologies we took into account are the reference ontology and the Karlsruhe ontology.

On this test case, we tested the lexical annotation techniques over both correct and incorrect anchoring to evaluate not only which wrong mappings are

---

[6] Available at http://oaei.ontologymatching.org/2006/benchmarks/

discharged after lexical annotation, but even, which true negative mappings are lost due to lexical annotation. For each matching found by SCARLET, we compared the meanings of the terms on the source ontologies with the meaning of the correspondent anchoring terms in the background ontologies. If both the couples have converging meanings, the anchoring is confirmed. If one couple has disregarding meanings, the anchoring is discharged. After lexical annotation the obtained results have been compared with the manual evaluation done by an expert on the entire set of matching.

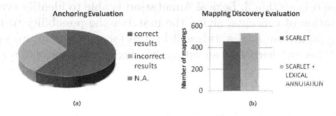

Fig. 5. Lexical annotation evaluation on OAEI 2006 benchmark: detection of incorrect anchoring (a) and new mapping discovery (b)

We examined a set of 109 mappings. The evaluation found agreement with the manual evaluation of the anchoring results in 65 cases (62 true positive anchoring and 3 true negative anchoring). The disagreement with the manual evaluation has been found in 34 cases (in these cases our algorithm retrieved 25 false positive and 9 false negative). 10 cases were impossible to disambiguate. A graphical representation is shown in figure 5 (a).

Moreover, we compared our results with a multiontology disambiguation method [18] that has been applied on the SCARLET matcher [7] and evaluated on the OAEI test case   [15]. Because the multiontology disambiguation method retrieves similarity measures, the comparison of these two disambiguation methods permitted to evaluate some possible threshold on the similarity measures (we retrieved a threshold for 0.19).

On the OAEI scenario we also evaluated how lexical annotation can enrich the matcher results proving new relationships. After the lexical annotation of the OAEI sources each concept has one or more meanings associated. Exploring the WordNet network, we computed a mapping between two concepts, if a relationships exists between their meanings in WordNet. Some of these mappings confirmed the relationships found by SCARLET (we retrieved 18 mappings that confirm the SCARLET results), and some other were new mappings that enrich the matcher results (we retrieved 77 new mappings, with a precision of 0.75%). Figure 5 (b) reports the improvements yielded by the lexical annotation.

5   Conclusion and Future Work

In this paper we described and experimentally investigated the application of automatic lexical annotation techniques in order to solve the ambiguity problems and to improve the results obtained by a matcher. The method has been applied on the SCARLET matcher, a semantic web based matcher which discovers mappings between two concepts by making use of online ontologies. Nevertheless, the method could be coped with any matcher.

We adopted an experimental approach on two test cases where SCARLET was previously tested, to investigate the potentiality of our method. The results confirmed our initial hypothesis (the precision is increased by solving ambiguity problems and new relationships are discovered), thus proving the value of the approach.

As mentioned in section 3, Lexical Annotation is able to identify synonymous and generalization of concepts, giving the matcher the possibility to widen the search among online ontologies, this will be the focus of our future work. This paper constitutes the kernel of an automatic lexical annotator for real world schemata/ontologies. To cope with complex schemata/ontologies, our method needs to be extended by including the treatment of compound terms and abbreviations [16].

Acknowledgements

The research reported in the paper was developed during a research period at KMI (Knowledge and Media Institute at Open University, Milton Keynes UK). We thank Marta Sabau, Enrico Motta, Jorge Gracia for their help with the evaluation reported in this paper. This work was partially supported by MUR FIRB Network Peer for Business project (http://www.dbgroup.unimo.it/nep4b) and by the IST FP6 STREP project 2006 STASIS (http://www.dbgroup.unimo.it/stasis).

References

1. Beneventano, D., Bergamaschi, S., Guerra, F., Vincini, M.: Synthesizing an integrated ontology. In: IEEE Internet Computing, September-October 2003, pp. 42–51 (2003)
2. Bergamaschi, S., Po, L., Sorrentino, S., Corni, A.: Dealing with uncertainty in lexical annotation. In: ER Demo Sessions (2009)
3. Castano, S., Ferrara, A., Montanelli, S.: Matching ontologies in open networked systems: Techniques and applications, pp. 25–63 (2006)
4. Choi, N., Song, I.-Y., Han, H.: A survey on ontology mapping. SIGMOD Record 35(3), 34–41 (2006)
5. Ding, L., Pan, R., Finin, T.W., Joshi, A., Peng, Y., Kolari, P.: Finding and ranking knowledge on the semantic web. In: Gil, Y., Motta, E., Benjamins, V.R., Musen, M.A. (eds.) ISWC 2005. LNCS, vol. 3729, pp. 156–170. Springer, Heidelberg (2005)
6. Giunchiglia, F., Shvaiko, P., Yatskevich, M.: Semantic schema matching. In: Meersman, R., Tari, Z., Hacid, M.-S., Mylopoulos, J., Pernici, B., Babaoglu, Ö., Jacobsen, H.-A., Loyall, J.P., Kifer, M., Spaccapietra, S. (eds.) OTM 2005. LNCS, vol. 3760, pp. 347–365. Springer, Heidelberg (2005)

7. Gracia, J., Lopez, V., d'Aquin, M., Sabou, M., Motta, E., Mena, E.: Solving semantic ambiguity to improve semantic web based ontology matching. In: Shvaiko, P., Euzenat, J., Giunchiglia, F., He, B. (eds.) OM. CEUR Workshop Proceedings, CEUR-WS.org, vol. 304 (2007)
8. Jian, N., Hu, W., Cheng, G., Qu, Y.: Aligning ontologies with falcon. In: Ashpole, B., Ehrig, M., Euzenat, J., Stuckenschmidt, H. (eds.) Integrating Ontologies. CEUR Workshop Proceedings, CEUR-WS.org, vol. 156 (2005)
9. Madhavan, J., Bernstein, P.A., Rahm, E.: Generic schema matching with cupid. In: Apers, P.M.G., Atzeni, P., Ceri, S., Paraboschi, S., Ramamohanarao, K., Snodgrass, R.T. (eds.) VLDB, pp. 49–58. Morgan Kaufmann, San Francisco (2001)
10. Navigli, R.: Word sense disambiguation: A survey. ACM Comput. Surv. 41(2) (2009)
11. Sabou, M., d'Aquin, M., Motta, E.: Using the semantic web as background knowledge for ontology mapping. In: Shvaiko, P., Euzenat, J., Noy, N.F., Stuckenschmidt, H., Benjamins, V.R., Uschold, M. (eds.) Ontology Matching. CEUR Workshop Proceedings, CEUR-WS.org, vol. 225 (2006)
12. Sabou, M., d'Aquin, M., Motta, E.: Exploring the semantic web as background knowledge for ontology matching. J. Data Semantics 11, 156–190 (2008)
13. Sabou, M., Dzbor, M., Baldassarre, C., Angeletou, S., Motta, E.: Watson: A gateway for the semantic web. Poster session of the European Semantic Web Conference, ESWC (2007)
14. Sabou, M., Gracia, J.: Spider: bringing non-equivalence mappings to oaei. In: Third International Workshop On Ontology Matching (OM 2008) (October 2008)
15. Sabou, M., Gracia, J., Angeletou, S., d'Aquin, M., Motta, E.: Evaluating the semantic web: A task-based approach. In: Aberer, K., Choi, K.-S., Noy, N., Allemang, D., Lee, K.-I., Nixon, L.J.B., Golbeck, J., Mika, P., Maynard, D., Mizoguchi, R., Schreiber, G., Cudré-Mauroux, P. (eds.) ASWC 2007 and ISWC 2007. LNCS, vol. 4825, pp. 423–437. Springer, Heidelberg (2007)
16. Sorrentino, S., Bergamaschi, S., Gawinecki, M., Po, L.: Schema normalization for improving schema matching. In: Laender, A.H.F., Castano, S., Dayal, U., Casati, F., de Oliveira, J.P.M. (eds.) ER 2009. LNCS, vol. 5829, pp. 280–293. Springer, Heidelberg (2009)
17. Stuckenschmidt, H., Serafini, L., Wache, H.: Reasoning about ontology mappings. Technical Report, ITC-IRST, Trento (2005)
18. Trillo, R., Gracia, J., Espinoza, M., Mena, E.: Discovering the semantics of user keywords. J. UCS 13(12), 1908–1935 (2007)

State of the Art of Semantic Business Process Management: An Investigation on Approaches for Business-to-Business Integration

Hanh Huu Hoang[1], Phuong-Chi Thi Tran[2], and Thanh Manh Le[1]

[1] Hue University
3 Le Loi Street, Hue City, Vietnam
{hhhanh,lmthanh}@hueuni.edu.vn
[2] Phu Xuan Private University
176 Tran Phu Street, Hue City, Vietnam
phuongchi0910@gmail.com

Abstract. Based on an in-depth analysis of existing approaches in applying semantic technologies into Business Process Management (BPM) research in perspective of business-to-business integration (B2Bi), we analyse, discuss and compare the methodologies, applications and best practices of the surveyed approaches. This paper identifies various relevant research directions in semantic-based BPM or Semantic BPM (SBPM). Based on the result of our investigation we summarise the state of the art of SBPM and address areas and directions of further research activities.

1 Introduction

Business Process Management (BPM)'s efforts are to bring business and IT communities together to solve the problem of the integration and collaboration cross-enterprises. BPM standards such as BPMN [1], XPDL [2], BPEL [3] and their combination [4] have not fulfilled the expectation of two communities in takling the integration challenge. The gap is still there: how enterprises can make the cross collaboration each other; and how a process based on the graphical notation can be fully mapped into the appropriate executable process without its semantics. Semantics is applied into the BPM to bridge the gap between the business world and information systems, especially in the context of B2B integration. This addresses the ever enduring need of new weaponry in struggle for survival in the e-business environment.

In this paper, we sketch out the current state of the research on the fusion of semantic technologies and BPM leading to new research directions to deal with mentioned issues. This survey is based on an investigation of approximately 30 selected publications on the mentioned issues and e-business standards. The material used for this paper originates from various important publications databases and citation indexes. Furthermore, the approaches are analysed in this paper are very recently high-rated.

N.T. Nguyen, M.T. Le, and J. Świątek (Eds.): ACIIDS 2010, Part II, LNAI 5991, pp. 154–165, 2010.

From the data gathered from this survey and based on similarities of research goals, relevant research directions in Semantic BPM are identified. While the classifications sometimes do not differ much in methodology, they seem sufficiently separate and logical with regards to research goals. Besides research directions, this paper also analyses the literature for the common methodologies and gives a short discussion about mentioned issues.

1.1 Scope of Survey

The goal of this analytical survey paper is aimed at a full investigation of current approaches in semantics-based BPM, so-called Semantic Business Process Management (SBPM), and its research efforts in the B2B integration, or called as cross-enterprise collaboration. The focused issues in this investigation include:

- BPM in the context of the cross-enterprise collaboration;
- SBPM standards analysis and evaluation;
- The role of semantics and its relevant technologies applied into BPM to tackle the mentioned concerns.

From the analysis, an evaluation and a roadmap of applying SBPM into the e-business environment will be drawn for further steps ahead.

1.2 Survey Criteria

Main objectives of our investigation are described according to the following criteria:

- *BPM Standards for the BP collaboration*
In this criterion, we are going to evaluate on applying BPM standards in the cross-enterprise collaboration perspective. We show missing paces of current BPM standards and methodologies in the collaboration context.

- *B2B Standards for cross-enterprise collaboration*
A throughout consideration into B2B standards will be carried out in the perspective of the integration of virtual enterprises. Therefore, the challenges for B2B integration will be addressed.

- *Semantic Web and BPM*
With this criterion, we would like to investigate the fusion of semantic technologies and BPM in order to tackle the difficulties in the e-business environment. The investigation is an evaluation of Semantic Web-based approaches to develop semantics-augmented BPM standards for the B2B integration. It will cover from the modeling techniques at high level to the transformation to execution tasks with semantics of process remained;

To start over, a brief evaluation on current BPM standards will be presented in the following section as a preface for our investigation's objectives.

2   Evaluation on Standards in B2B Integration Perspective

2.1   BPM Standards Overview

There are many BPM standards, and so we attempted to classify them primarily into four groups: (1) Graphical (2) Execution and (3) Interchange standards [5]. As shown in Fig. 1, graphical standards are currently the highest level of expressing business processes (i.e. most natural to humans) while the lowest (i.e. most technical) level is the execution standards. While graphical standards diagrammatically express contemporary business processes, the execution standards aim to automate business processes via computers.

The graphical standards are graph-oriented representations while the execution standards mainly belong to the syntax-based block-oriented representation. Block-oriented languages define a business process' control flow via nesting different elements and attributes of their syntax, while the graph-oriented languages specify control flow by using different kinds of nodes and arcs. Arcs connect nodes with one another, and these connections intrinsically represent temporal and logical connections [6]. Hence, transformations from the graph-oriented graphical standards to the block-oriented execution standards and vice versa are problematic and often experience a loss of information [6]. This created a need for interchange standards, the middle level of Fig. 1.

Fig. 1. The BPM Standards Categories

Interchange standards act as the middle ground between the two contrasting modes of representation. Standardization groups such as OMG which pioneered interchange standards often claim their creations as the missing link between the business analyst and the IT specialist. This is only half the story as there are still many aspects of business process modelling that current standards fail to address (e.g. goals, context, role definitions, decomposition from high-level goals to automated business processes, etc.) [5]. In fact, it is more accurate to say that interchange standards are the non-contextual translator between graphical standards and execution standards.

2.2   BPM Approaches for Cross-Enterprise Collaboration

Oh et al [7] tried to tackle difficulties in the B2B integration by layering the integration process into different layers. An integrated process is designed with BPMN[1],

---

[1] Business Process Modeling Notation (OMG/BPMI), http://www.bpmn.org/

divided into the three types of processes, and automatically transformed to XPDL[2] and BPEL[3] for process enactment. Specifically, collaborative processes are mapped to corresponding actual partners through virtual partners. However they failed to define what the virtual partner is; and ignored the semantics in the integration process.

In another approach, [8] wanted to solve the integration issues using the approach of formulating collaborative processes in an certain "abstract" level. However they ignore the method to define the level of the abstraction. Furthermore, the approach failed to point out the root of problems in the B2B integration is the semantics of enterprises' business processes.

2.3 B2B Standards for Cross-Enterprise Collaboration

B2B integration or cross-enterprise collaboration is about the integration of collaborative business processes of business partners. Although there are B2B standards issued recently such as IDEF Family [9], ebXML BPSS [10], RosettaNet PIPs [11], and UBL [12] in order to unify the information exchange between enterprises, however the compatibility of the information semantics in the transaction is an unsolved problem in the B2B integration. According to [13], the reason behind addressed problems is the semantic difference in using business terms in business documents, transaction protocols and message interfaces. In general, the ultimate goal of B2Bi standards such as ebXML, RosettaNet is to provide complete components for cross-enterprise transaction using XML. In this section we focused on RosettaNet and ebXML standards because they are widely accepted in the industry.

RosettaNet is a vertical standard; therefore, it clearly defines collaborative business processes using Partner Interface Processes (PIPs). Business partners have to obey the standardized processes and the transactional semantics defined by PIPs whenever they want to make collaborations in the framework of RosettaNet. Meanwhile, ebXML is a horizontal standard and has not defined business processes especially. Instead, ebXML provides business partners methodologies and tools to specify business processes according their usages.

ebXML uses its Business Process Specification Schema (BPSS) as the core component to define collaborative business processes. The concepts of business collaboration, choreography and business document workflow for defining collaborative business processes are similar to RosettaNet PIPs. In short, RosettaNet defines PIPs to specify business workflows between partners, and ebXML use BPSS for the same task. The basic differences of these standards are clarified in Table 1.

As described in Table 1, B2B standards provide frameworks for business collaboration standardized by business documents definitions, data-formatted messages in the B2B integration. These definitions are attached with the semantics explanation in theses standards' vocabularies. However, these documents have not modelled by any formal description language; as the result, they cannot be processed automatically and semantically by applications and machines [14].

Specification documents for business transaction in RosettaNet have been mainly modelled using DTD and XML schema, and as we have known, they are lack of semantic presentations for constraints and semantics of business documents [15].

---

[2] XML Process Definition Language (WfMC), http://www.wfmc.org/xpdl.html
[3] Business Process Execution Language for Web Services (OASIS).

Table 1. ebXML and RosettaNet comparison

Criteria	ebXML	RosettaNet
Type	Horizontal	Vertical
Collaborative business process definition	Business Process Specification Schema (BPSS)	Partner Interface Processes (PIPs)
Vocabulary	BPSS and Core Components (CC)	RosettaNet Dictionary: business (RNBD) and technical (RNTD)
Execution framework	Message Services	Execution framework (RNIF)
Repository	support	N/A
Partners agreement	Collaborative Protocol Profile (CPP) and Collaborative Protocol Agreement (CPA)	N/A

ebXML defines components' semantics by two ways: BPs are defined using BPSS; and CC is used for modelling with the approach for building the business vocabulary. However, ebXML's CC does not meet all requirements specification despite its architecture has been designed to solve the semantic difference in transactions by associating structured information and contexts.

2.4 Missing Pace

Collaborative business processes are public business processes that cross enterprises' borders [16]. There is currently an increasingly popular group of standards which facilitate the B2B integration for globalised supply chains. These standards are technically B2B information exchange standards and they include: the IDEF Family [9], WS-CDL [17], ebXML BPSS [10], RosettaNet PIPs [11], and newly emerging UBL [12]. Though they facilitate businesses' entry into e-commerce instead of the usual traditional methods, these standards merely standardize the information exchange and do not address the real needs of a dynamic business process collaborations like those discussed in [18, 19]. These approaches focus on the metadata in common for the information exchange between business partners. However, these standards ignore the most importance of the enterprise integration is to understand the meaning of each other [16].

In summary, although B2B standards have created advantages for B2B e-commerce transactions, they are only the standards for the information exchange in B2B e-commerce applications. However, they do not reflect the dynamic aspects of the business processes collaboration. Current BPM and B2B standards focus on metamodels for the information exchange between businesses and ignore the most important issue in the business integration that is the semantics of information and contexts during transactions; they help businesses 'understand' each other. Therefore, the B2B integration is still a challenge for current industrialized B2Bi standards.

3   Semantic BPM: Semantic Approaches for B2B Integration

3.1   SBPM vs. BPM

BPM has gained significant attention in both research and industry, and a range of BPM tools are available. However, the degree of mechanization in BPM is currently very limited. The major obstacle preventing a coherent view on business processes is that the business processes are not accessible to machine reasoning. Additionally, businesses cannot query their process space by logical expressions, e.g. in order to identify activities relevant to comply with regulations.

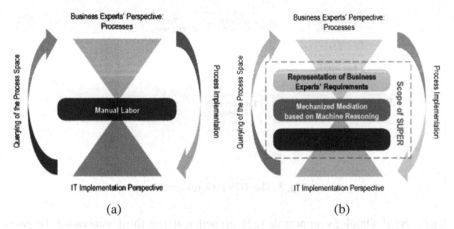

Fig. 2. BPM in business and IT perspectives (a) and the semantic-based approach in the SUPER project (b) which tends to bridge the gaps using semantic technologies such as ontologies, and Semantic Web Services

Founded on ontologies Semantic Web technology provides scalable methods and tools for the machine-readable representation of knowledge. Semantic Web Services (SWS) make use of Semantic Web technology to support the automated discovery, substitution, composition, and execution of software components (Web Services) [20]. BPM is a natural application for the Semantic Web and SWS technologies, because the latter provide large-scale, standardized knowledge representation techniques for executable artefacts [21].

The Semantic Web and, in particular, SWS technology offer the promise of integrating applications at the semantic level. By combining them with BPM, and developing one consolidated technology, we would create horizontal ontologies which describe business processes and vertical telecommunications oriented ontologies to support domain-specific annotation (Fig. 2).

3.2   Semantic Approaches for Cross-Enterprise Collaboration and Integration

Since the failure of the BPM and B2B approaches as mentioned above, research efforts have been emerged from the motivation of knowledge management and applying

Semantic Web technologies into the BPM research to bring the administrative side and IT side together.

The first research effort in the semantic approach was the TOVE project at Toronto University. The TOVE project (TOronto Virtual Enterprise project) aimed at developing and ontology framework for business integration in early 1990s by Mark S. Fox [22]. The TOVE project's main objective is to build a set of ontologies (Fig. 3) for modelling the integration and trading between businesses. However, the implantation of TOVE was not successful because there was a gap between the modelling and the execution of processes. In another way, TOVE focused on the IT aspect and ignore the execution of business processes.

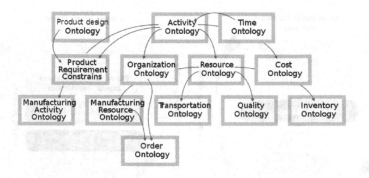

Fig. 3. The TOVE Ontologies

Jenz's BPM Ontology approach [23] argued that the third generation business process management is different in that it provides an integrated view on business processes. According to Jenz's, the business oriented view has a counter piece in the form of the IT view, and both must be on an equal footing. The business view can be segmented into three layers: core business ontology layer; industry-specific ontology layer; and organization-specific ontology layer. The IT view is not segmented into layers and is completely organization-specific.

SUPER [20] addresses the ever enduring need of new weaponry in struggle for survival in optimistic business environment where profit margins dramatically drop while competitiveness reaches the new sky high limits. This project answers the two most urgent issues emerging from BPM:

- shift in control of processes from IT professionals to business natives;
- carrying up business process management to a new complexity level.

The major objective of the SUPER project is to raise BPM to the business level, where it belongs, from the IT level where it mostly resides now [20]. This objective requires that BPM is accessible at the level of semantics of business experts. SUPER's approach has tried to transform existing BPMN and BPEL standards into a semantics-enriched form, respectively called sBPMN (so-called BPMO – Business Process Modeling Ontology) and sBPEL [24, 25] in the attempt to realize their goals.

In the same line, the SemBiz project[4] aims at bridging the gap between the business level perspective and the technical implementation level in Business Process Management (BPM) by semantic descriptions of business processes along with respective tool support. This approach takes emerging frameworks for Semantic Web Services, namely the Web Service Modeling Ontology (WSMO)[5] as a basis for defining an exhaustive semantic description framework for business processes. On basis of this, novel functionalities for BPM on the business level can be supported by inference-based techniques that work on semantic process descriptions.

These three approaches try to bring semantic technologies into their approaches to tackle the challenge in B2B integration. However, SUPER delivers more noticeable deliverables for research communities.

The Enterprise Ontology [26], as developed by Dietz, is the starting point for profoundly understanding the organization of an enterprise and subsequently for analyzing, (re)designing, and (re)engineering it. The approach covers numerous issues in an integrated way: business processes, in- and outsourcing, information systems, management control, staffing etc.

The analysis on these approaches are summarised in the following table:

Table 2. Comparisons on the semantic BPM approaches

Approaches	Specification	Acces-sibility	Ontology levels consis-	Ontology modelling language	Usage/ Execu-tion
SUPER and SemBiz	Ontologies stack	+	+	WSMO	+/SWS
Jenz's BPM Ontology	core business ontology; industry-specific ontology; organization-specific ontology	+	+	OWL	-/WS
Enterprise Ontology	Activity, Organisation, Strategy, Marketing, Time	+/-	+	Informal (text), semi-formal (Ontolingua)	-
TOVE	Ontology set (Fig. 3)	-	-	F- Logic	-

As described in Table 2, we can see the approaches of SUPER and SemBiz are to apply Semantic Web ontologies to deal with the challenge in the business collaboration/integration. Furthermore, SUPER has attracted the research communities due to its advantages in compared to others, and they are as follows:

-    Proposed ontologies are associated to create a stack that make them consistent and a tight linkage between them;
-    Using WSMO as the modelling foundation, WSMO is a flexible ontology language with dynamic reasoning features, supports repositories and the execution based-on Web services as well;

---

[4] SemBiz Project, http://www.sembiz.org/
[5] Web Service Modeling Ontology, http://www.wsmo.org/

- SUPER framework has been successfully applied in the scenario of tele-communication [20].

Through the evaluation and comparison of these approaches, we can see that the fusion of BPM and the Semantic Web or ontology-based techniques become promising research direction in the domain. This research approach can bring new opportunities, new prospects and useful tools for e-business and B2B integration especially.

We summarise BPM and B2B standards, SBPM approaches and related issues in the following *landscape* (Fig. 4) in order to draw an overview picture of what we have discussed.

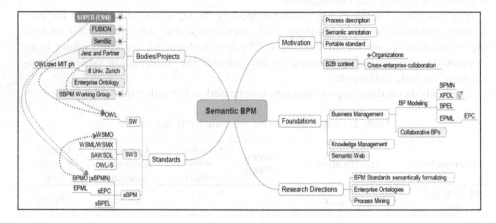

Fig. 4. Landscape of Semantic Business Process Management

3.3 BP Repositories: Process Description and Reuse

The goal of the MIT Process Handbook project [27] is to develop rich online libraries for sharing and managing the business knowledge. For example, using the libraries parties can find interesting case studies, best practices, generate innovative ideas about new business possibilities, and develop new computer programs. Starting in 1991, they have developed one such library. It is called the Process Handbook – an extensive online knowledge base including entries for over 5000 business activities and a set of software tools for managing this knowledge [28].

The Process Handbook can help people to (1) redesign existing (business) processes, (2) invent new process, especially those that take advantage of information technology, novel coordination structures, or exception handling approaches, (3) organize and share knowledge about organizational practices, and (4) automatically, or semi-automatically, be used to generate software to support or analyze business processes [27]. In the past year the project's activities have been mainly focused on consolidating the effort by publishing an edited volume about the subject.

At the same time at University of Zurich, an OWLized version of the MIT process handbook has been developed including an ontology and approximately 8000 business processes [29].

In the previous work of us [16], we have taken advantages from the fusion of BPM standards and Semantic Web technologies to propose an approach called BizKB that can assist enterprise to establish dynamic partnerships through a semantics-enriched BP repository. In BizKB, BPs are modelled using BPMO with their instances with different levels of its decomposition (a.k.a business rules). With the assistance of BizKB's querier, enterprises can look up their needed BP process patterns as well as BP instances for objective collaborations.

4 Discussion

Business Process Management's efforts are to bring business and IT communities together to solve the problem of the integration and collaboration cross-enterprises. BPM's approach is to execute business tasks with the IT support in the business expert perspective rather than technical one. BPM is challenged with issues of the automation of dynamic business collaboration and the integration/collaboration of business process in B2Bi. There are tools and technologies to support the BP modelling based on graphs such as UML's Activity Diagrams, BPMN and EPC to the query and execution languages including BPQL and BPEL. However, the level of the automation of BPM is limited by the lack of formal representation so that machines can *understand* and process themselves.

Semantic Web technologies with ontologies bring in a promising approach in providing methodologies and tools for representing the knowledge of businesses that is *machine-readable*. Accompanying with the high-level modelling, Semantic Web Services (SWS) support the execution level by discovering and composing web services to fulfil the goals of high-level modelled BP, and can execute the collaborative processes automatically. The combination of BPMO at the high level and SWS will realize the objectives of SBPM in the business knowledge formalisation to the execution of the result processes [21].

The Semantic Web with SWS opens a new prospect for the integration applications at semantics-enriched levels in order to automatically perform the enactment of collaborative business processes. The fusion of the Semantic Web, Semantic Web Services and BPM create a unified solution based on business semantics, and it is called Semantic Business Process Management or SBPM [21, 30]. The ultimate goal of the SBPM is to augment the automatic level in the BPM lifecycle [20]. Though this analytical survey on SBPM methodologies and approaches, it shows how semantics, particularly Semantic Web technologies, is becoming one of answers for the B2B integration challenge discussed in Section 2.

5 Conclusion

A number of common patterns can be detected in the approaches described in this paper. On the technical level, it can be concluded that in the working context of semantic models, quite many of the used common methodologies are to bring the semantics in BPM to solve the problems in B2B e-commerce, with current approaches using the Semantic Web and its relevant technologies such as Semantic Web Services.

We have found out that the trend of taking advantages of service-oriented methodologies in BPM and the fusion of BPM and the Semantic Web that have reflected of ontology-based solutions and Semantic Web Services, with the execution level based-on the web services infrastructure - the SOA.

Besides, the competition in modeling language is also interesting. Taking focus on the SBPM, we can see the three main lines: using native logics such as F-logics, tradition OWL with OWL-S, and newly emerging BPMO with WSMO. Last both approaches are based on Web services execution platforms.

Through this investigation, we also see the open issues in research, and it needs our efforts to continually make the promising SBPM approach happen and provide a complete framework that support the dynamic semantics-based B2B integration in the new e-business environment.

References

1. OMG: Business Process Modeling Notation (BPMN). vol. 2007 (2007)
2. WfMC: XML Process Definition Language (XPDL), ver 2.1. (June 2008)
3. OASIS: Web Services Business Process Execution Language (WSBPEL). vol. 2007 (2007)
4. Nathaniel, P.: Understanding the BPMN-XPDL-BPEL Value Chain. Business Integration Journal, 54–55 (November/December 2006)
5. Ko, R.K.L.: BPM Standards: A Survey (Technical Report). Nanyang Technological University, Singapore (2007)
6. Koskela, M., Haajanen, J.: Business Process Modeling and Execution: Tools and Technologies Report for the SOAMeS Project. VTT Research Notes 2407. VTT Technical Research Centre of Finland (2007)
7. Oh, J.Y., Jung, J.-y., Cho, N.W., Kim, H., Kang, S.-H.: Integrated Process Modeling for Dynamic B2B Collaboration. In: Khosla, R., Howlett, R.J., Jain, L.C. (eds.) KES 2005. LNCS (LNAI), vol. 3683, pp. 602–608. Springer, Heidelberg (2005)
8. Mous, K., Ko, K., Lee, S., Tan, P., Lee, E.: High-level Business Processes for Agile B2B Collaboration. In: Proceedings of International MATADOR Conference, pp. 169–172 (2007)
9. Mayer, R.J., Painter, M.K., de Witte, P.S. (Knowledge Based Systems): IDEF Family of Methods for Concurrent Engineering and Business Re-engineering Applications. Knowledge Based Systems (1994)
10. UN/CEFACT, OASIS: ebXml, Xmlbpss Version 1.01 Technical Specification (2002)
11. RosettaNet: RosettaNet Implementation Framework: Core Specification (1998)
12. Meadows, B., Seaburg, L.: Universal Business Language 1.0. Organization for the Advancement of Structured Information Standards (OASIS), Committee Draft (September 2004)
13. Nurmilaakso, J.M.: EDI, XML and e-business frameworks: A survey. Computers in Industry (2007)
14. Gessa, N.: An ontology-based approach to define and manage B2B interoperability. OASIS (2007)
15. Kotinurmi, P., Vitvar, T., Haller, A., Richardson, R., Boran, A.: Semantic web services enabled B2B integration. In: Lee, J., Shim, J., Lee, S.-g., Bussler, C.J., Shim, S. (eds.) DEECS 2006. LNCS, vol. 4055, pp. 209–223. Springer, Heidelberg (2006)

16. Hoang, H., Le, T.: BizKB: A Conceptual Framework for Dynamic Cross-Enterprise Collaboration, p. 401. Springer, Heidelberg (2009)
17. Kavantzas, N., Olsson, G., Mischkinsky, J., Chapman, M.: Web Service Choreography Description Language (WSCDL) 1.0. W3C Working Draft (2004)
18. Tan, P.S., Goh, A.E.S., Lee, S.S.G., Lee, E.W.: Issues and Approaches to Dynamic, Service-oriented Multi-enterprise Collaboration. In: 2006 IEEE International Conference on Industrial Informatics, pp. 399–404 (2006)
19. Wombacher, A., Fankhauser, P., Mahleko, B., Neuhold, E.: Matchmaking for business processes, pp. 7–11. IEEE Comput. Soc., Newport Beach (2003)
20. Born, M., Drumm, C., Markovic, I., Weber, I.: SUPER - Raising Business Process Management Back to the Business Level. ERCIM News 70, 43–44 (2007)
21. Wetzstein, B., Ma, Z., Filipowska, A., Kaczmarek, M., Bhiri, S., Losada, S., Lopez-Cob, J.-M., Cicurel, L.: Semantic Business Process Management: A Lifecycle Based Requirements Analysis. In: Workshop on Semantic Business Process and Product Lifecycle Management (SBPM 2007), Innsbruck, Austria. CEUR-WS, vol. 251 (2007)
22. Fox, M.S.: The TOVE Project: towards a common sense model of the enterprise. In: Enterprise Integration. MIT Press, Cambridge (1992)
23. Jenz, D.E.: Ontology-Based Business Process Management: The Vision Statement. Strategic White Paper. Jenz & Partner GmbH, Erlensee, Germany (2003)
24. Dimitrov, M., Simov, A., Stein, S., Konstantinov, M.: A BPMO Based Semantic Business Process Modelling Environment. In: Workshop on Semantic Business Process and Product Lifecycle Management (SBPM 2007), Innsbruck, Austria. CEUR-WS, vol. 251 (2007)
25. Yan, Z., Cimpian, E., Zaremba, M., Mazzara, M.: BPMO: Semantic Business Process Modeling and WSMO Extension. In: IEEE International Conference on Web Services, 2007. ICWS 2007, pp. 1185–1186 (2007)
26. Dietz, J.L.G.: Enterprise ontology: theory and methodology. Springer, Heidelberg (2006)
27. MIT: The MIT Process Handbook Project (2003)
28. Malone, T.W., Crowston, K., Herman, G.A.: Organizing Business Knowledge: The Mit Process Handbook. MIT Press, Cambridge (2003)
29. IFI-UZH: OWL MIT Process Handbook. Dynamic & Distributed Information Systems Group - University of Zürich (2005)
30. Hepp, M., Leymann, F., Domingue, J., Wahler, A., Fensel, D.: Semantic business process management: a vision towards using semantic Web services for business process management. In: IEEE International Conference on e-Business Engineering. ICEBE 2005, pp. 535–540 (2005)

Evolving Concurrent Petri Net Models of Epistasis

Michael Mayo[1] and Lorenzo Beretta[2]

[1] Dept. of Computer Science, University of Waikato, New Zealand
mmayo@cs.waikato.ac.nz
[2] Referral Center for Systemic Autoimmune Diseases,
Fondazione IRCCS Ospedale Maggiore Policlinico di Milano, Italy
lorberimm@hotmail.com

Abstract. A genetic algorithm is used to learn a non-deterministic Petri net-based model of non-linear gene interactions, or statistical epistasis. Petri nets are computational models of concurrent processes. However, often certain global assumptions (e.g. transition priorities) are required in order to convert a non-deterministic Petri net into a simpler deterministic model for easier analysis and evaluation. We show, by converting a Petri net into a set of state trees, that it is possible to both retain Petri net non-determinism (i.e. allowing local interactions only, thereby making the model more realistic), whilst also learning useful Petri nets with practical applications. Our Petri nets produce predictions of genetic disease risk assessments derived from clinical data that match with over 92% accuracy.

Keywords: Petri net, genetic algorithm, epistasis, concurrency, systemic schlerosis, digital ulcers.

1 Introduction

Petri nets [13] are widely used abstract computational models of concurrent processes. Recently, they have found application as useful modeling tools in biochemistry, genetics and medicine (e.g. [2,6]).

They are best described as executable graphs with two different types of node: places and transitions. In a biochemical modeling situation, a place usually represents a substance and a transition stands for a reaction or process in which one or more input substances are transformed over time into one or more output substances. Petri nets have potential to realistically model what could be happening in real world situations because they are inherently concurrent. For example, in a net, two pathways of multiple transitions may fire simultaneously, thus simulating two concurrent processes.

Figure 1 depicts a simple Petri net with three places and two transitions. The places, P_0, P_1 and P_2, represent three different chemical substances, and the transitions, T_0 and T_1, represent two different reactions that can occur between them. Petri nets represent the concentration of a substance at a particular point in time by "marking" each place with an integer number of tokens. These tokens move around the net as the transitions fire.

N.T. Nguyen, M.T. Le, and J. Świątek (Eds.): ACIIDS 2010, Part II, LNAI 5991, pp. 166–175, 2010.
© Springer-Verlag Berlin Heidelberg 2010

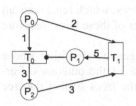

Fig. 1. Example of a Petri Net

For example, suppose in Figure 1 that there are 10 tokens at P_0, and no tokens at P_1 and P_2. The overall marking of the entire net is the vector <10,0,0>. The arcs indicate either transition inputs or outputs, depending on the directionality. They are labeled with a quantity of tokens consumed or produced. T_0, for example, represents a chemical process in which P_0 is being converted into P_2, with one unit of P_0 being consumed for every three units of P_2 being produced. If T_0 fires once, the marking of the net will become <9,0,3>. If it fires twice, it will become <8,0,6>. T_1, on the other hand, represents an entirely different reaction with P_0 and P_2 as inputs, and P_1 as output. Because T_1 requires three units of P_2 as an input, it cannot fire until T_0 has fired at least once. If this happens, the marking will change from <10,0,0> to <9,0,3> (after T_0 fires) and then to <7,5,0> (after T_1 fires).

Transitions can only fire if there are sufficient input tokens available (i.e. the number of tokens at an input place cannot fall below zero) and if they are not inhibited. An example of an inhibitor in Figure 1 is the arc from P_1 to T_0: if ever P_1 has a non-zero quantity of tokens present, then T_0 is effectively turned off.

The only other time that a transition cannot fire is if one of its output places has insufficient capacity. For example, suppose the maximum capacity of all places in Figure 1 is 10 tokens, and the current marking is <7,6,9>. Although T_1 has sufficient inputs available at P_0 and P_2, there is insufficient capacity at the output place P_2, so T_1 cannot fire.

It should be evident by now that Petri nets are concurrent and non-deterministic models. Transitions may fire in any order, and if they do not share common inputs or outputs, they can fire concurrently.

Non-determinism does have some issues when models are to be executed on serial computers. If there are two or more transitions enabled, which one should fire first? The simplest answer to this question is to enforce an arbitrary priority amongst the transitions [13]. For example, in Figure 1, T_0 may have a higher priority and therefore always fire before T_1, if they are both enabled at the same time. This strategy simplifies a non-deterministic Petri net into a deterministic model.

An alternative answer is to make the transitions fire stochastically. Of those that are enabled, one of them is selected to fire at random; and sometimes, in order to give all enabled transitions a fair chance of firing, those that have recently fired are not permitted to fire again until a certain amount of time has elapsed.

A significant issue with both of these solutions is that they require global coordination. In other words, in order to select the next transition to fire, all transitions must be examined globally. Nature, however, is unlikely to employ this level of global coordination; natural systems are more likely to evolve gradually with many local, concurrent interactions. The issue is therefore how to relax the requirement of global coordination from our Petri net models in order to make them more realistic and therefore more interesting.

In this paper, we address this specific problem in the context of modeling disease-causing epistatic interactions between genes. Our solution is to convert the Petri net model into an alternative representation called a set of state trees, which represents all possible orderings in which the transitions can fire. The leaves of these trees therefore represent all possible final outcomes.

We show that it is possible to evolve a Petri net using a genetic algorithm whose state tree outcomes match clinical observations in over 92% of the outcomes. Furthermore, it is also possible to limit the depth and size of the trees so that the tree remains relatively small, thereby permitting inspection.

This new approach eliminates the need for global coordination of the transition firings in the Petri net. Instead, transitions can fire in any order, and the Petri net therefore exhibits only more realistic local interactions.

2 Method

We describe firstly our Petri net models of non-linear gene interaction, and then discuss the conversion of a Petri net to a multiple state trees. Finally we describe the specific genetic algorithm that we employed to learn our Petri net-based models.

2.1 Petri Net Models of Epistasis

Epistasis [11] refers to the phenomenon of non-linear gene interaction. In the context of genetic disease, it manifests when no single genetic cause for a disease can be isolated; instead, scientists determine that it is the curious interaction between multiple genes that causes the disease. The main question is how this interaction could be happening, and Petri nets are useful as a means of hinting at a hypothesis explaining the interaction.

In biological reality, each gene is actually a sequence comprising hundreds of thousands of nucleotides. Mutations to these sequences may occur in many ways, but one of the most common is a change to a single nucleotide, known as a Single Nucleotide Polymorphism (SNP). A single SNP may completely alter the behavior of a gene. In this paper, we will refer to the value of an SNP as A (the original, wild-type) or a (its mutant form). In an individual, nucleotides come in unordered pairs (alleles); so therefore an individual has three possible genotypes per SNP: AA, Aa, or aa[1].

For modeling purpose, the nucleotide level of detail is far too complex. We therefore model entire genes as "gene units" within our Petri nets. Each gene unit is assumed to vary *only* by a single SNP; that is, all nucleotides except for one are assumed constant. This representation is depicted in Figure 2.

As Figure 2 illustrates, a gene is modeled as two places and a transition. The first place is called the "activating place" (AP) and represents the substance that activates or turns on the gene; the second substance is the "product place" (PP), and represents the output of the gene. There is also an optional "inhibitory place" (I) that can turn the gene unit off completely. The key

Fig. 2. A gene unit

---

[1] This is a convention we use in this paper for readability by non-geneticists. To be technically correct, we should use nucleotide notation, e.g. CC/CG/GG.

point is that the rate of production of the gene unit, the value g, is controlled by a genotype varying only by a single SNP.

Following biological investigations [3,5,12], it is assumed that the SNP's mutant form a causes an over-production of the gene's output substance at some fixed ratio. The values of g in Figure 2, therefore, have been set to 3 for genotype AA; 6 for genotype Aa; and 9 for genotype aa.

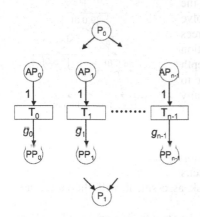

Figure 3 shows an overall Petri net-based architecture comprising several gene units. It should evident that whenever n genes are being modeled, there must be up to 3^n different genotypes involved.

There are two additional places in this architecture: P_0 and P_1. P_0 denotes the initial source of tokens in the network, or from a biological point of view, it is the trigger event that initiates the chain of reactions leading up to the disease. P_1 represents the output of this process; it is the toxic disease-causing substance. Following previous studies [1,7,9], we use a threshold to determine whether the toxic substance is in such abundance as to cause a high risk of the disease. In all of our simulations, this threshold is set to 50% of the maximum capacity of P_1. Thus, if the threshold at P_1 is

Fig 3. A generalized Petri net architecture comprising n gene units

reached or exceeded, it is assumed that the current genotype leads to a high risk of disease; otherwise, there is only a low risk.

Besides the gene units and P_0 and P_1, we also assume the existence of an arbitrary additional number of places and transitions. These are places and transitions not forming the parts of any specific gene unit, but they do have significant influence because they connect to the gene unit's APs and PPs.

In all of our simulations the maximum place capacity and arc weights are set to 10.

2.2 From Petri Nets to Sets of State Trees

Petri nets are inherently non-deterministic, concurrent computational models. That is, transitions that are co-enabled can fire in any order, as long as one of the transitions does not disable the other, and transitions that do not share common inputs and outputs may fire concurrently. In order to evaluate the behaviour of such a model it is necessary to "unroll" its non-deterministic aspects into a deterministic form that can be properly assessed.

We propose a tree representation that we call a state tree as the deterministic form of a Petri net. A state tree is an alternative representation of a Petri net in which nodes represent markings, and arcs represent transitions. A path from the root of the state tree to a leaf represents, therefore, a single execution of the Petri net from start state to final state. Figure 4 depicts a state tree for the very simple Petri net depicted in Figure 1.

In Figure 4, the starting state is <10,0,0>, indicating 10 tokens at P_0 and no tokens anywhere else. Only transition T_0 is enabled initially, but after it fires once, both T_0 and T_1 are thereafter enabled. The final states that are reached, which depend on the ordering of transition firings, are <7,0,9>, <6,5,3>, or <7,5,0>.

Clearly, for a Petri net of significant size or complexity, the state tree can be very large. Furthermore, if a state is visited more than once, then at that point the state tree can have effectively infinite depth. To resolve these problems, we limited the depth of our state trees to 10 and automatically excluded from consideration any Petri nets whose state tree exceeded this depth limit. We also limited the number of leaves per tree to 100 or less, again excluding from consideration any trees that did not conform. Finally, we also made use of a domain-specific heuristic to further trim the tree. Since P_1 only ever accumulates tokens and is never the input place for another transition, it is possible to stop growing the state tree as soon as the number of tokens

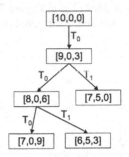

Fig 4. A state tree for the Petri net depicted in Figure 1

at P_1 exceeds the threshold of 50%, since the risk assessment will thereafter not change.

These measures for the most part kept the size of the state trees manageable, whilst still being practical for solving the problem of non-linear gene modeling.

As Figure 3 shows, there are n variables $g_0, g_1, ..., g_{n-1}$, that are genotype dependent within each net. As each gene has 3 different possible values (AA/3, Aa/6 or aa/9), this means that there are 3^n possible genotypes. Now, each genotype will produce a different Petri net execution dynamics, and therefore a different state tree must be constructed for each and every genotype. Thus, in our problem domain, every single Petri net is converted into not one but a set of 3^n state trees.

2.3 Evolving Petri Nets

We propose the use of a genetic algorithm to learn a Petri net model of the observed non-linear gene interactions. Genetic algorithms [4] use random mutations and crossover operators to gradually optimize solutions to problems. In the specific field of gene interaction modeling, Moore and Hahn [9], Mayo [7], Mayo and Beretta [14], and Beretta et al. [1] all apply genetic algorithms to learn Petri nets. The key difference between those previous works and our current work presented here is that previously, deterministic Petri nets were used, whereas now we are concerned with relaxing the determinism criteria and instead learning Petri nets that may execute non-deterministically (i.e. the transitions may fire in any order) whilst still remaining a highly accurate model of the interaction.

In our case, we have a set of 3^n genotypes, each genotype being labeled either "high risk" or "low risk", and we want a Petri net that, after all transitions have fired, always reaches or exceeds the threshold at P_1 for high risk genotypes, but never exceeds the threshold at P_1 for low risk genotypes. Construction of a state tree for each genotype, therefore, is essential in order to assess all possible outcomes. The model should show how the genes activate, produce, and interact in all situations in order to produce the correct desired behavior.

Our representation of a Petri net for the genetic algorithm is as follows: we fix the number of places to $2n+2$ and the number of transitions to $n+10$, where n is the number of genes, and model each net as a list of directed arcs. Arcs can be either weighted or inhibitory. Our genetic algorithm randomly constructs its initial Petri nets, putting a random arc with random weight between a place and transition with probability 0.2. Of those arcs, 10% of them are chosen randomly to be inhibitors.

Our genetic algorithm has a population size of 2,000 individuals. From the random initial population and for each subsequent generation, the top 5% of individuals are retained for the following generation. The rest are created via either the mutation operator or the crossover operator. The mutation operator either (i) adds one, two or three random arcs to the net; or, (ii) deletes a random arc, or (iii) modifies an existing arc by changing its weight or type, with equal probability. The crossover operator merges the arc lists of two parent nets, while maintaining the criteria that there is no more than one arc between any pair of nodes.

In our initial testing, we found that the mutation operator was far more effective than the crossover operator, and so set the probability of crossover to 5% and the probability of mutation to 95%. Parent nets are selected stochastically with probability proportionate to fitness. The genetic algorithm continues to iterate until 2,000 generations pass without any gains in fitness. At that point, the search is complete and the best net is returned.

In our non-linear gene modeling scenario, there are 3^n genotypes, and therefore 3^n state trees per net. To compute the fitness of each net, we iterate over the genotypes and generate for each genotype its corresponding state tree. For example, if $n=3$, then the genotypes will be *AA-AA-AA*, *AA-AA-Aa*, *AA-AA-aa*, ..., *aa-aa-aa*, where *AA* corresponds to arc weight 9, *Aa* to weight 6, and *aa* to weight 3.

Since each genotype will have a risk assessment (either high or low), we examine the leaves of its state tree and compute the proportion of leaves with the correct predicted assessment. This is what we term the accuracy of the state tree. The overall fitness is then the average accuracy across all genotypes, with a small bias against net size subtracted. During testing, we also found that squaring this fitness value tended to give marginally better results than not squaring it, and so the final result is squared. In mathematical terms, the fitness function is given by the equation below, where r is a genotype.

$$fitness(net) = \left[\frac{\sum_{r=0}^{3^n-1} accuracy(net,r)}{3^n} - 0.01 \times size(net) \right]^2$$

The fitness function ranges in value between 0.0 and 1.0, with a greater value indicating a better solution. The size component of the function is determined by dividing the actual number of arcs by the maximum possible number.

Figure 5 illustrates the computation of the fitness value for the very simple Petri net from Figure 1, assuming that the P_0-T_0-P_2 portion of the net is now a single gene unit. Since g can take three possible values, specifically 3,6 or 9, there are three possible state trees. If the AA genotype is low risk (P_1 must be less than 50% of maximum capacity) whilst the Aa and aa genotypes are high risk (P_1 must be greater than or equal to 50% capacity, which is 5 tokens), then Figure 5 shows that this net is only 33.3% accurate when $g=3$, but 100% accurate when $g=6$ or 9. Overall, then, the fitness of this net is $(0.33+1.0+1.0)/3.0-0.01(6/6)\approx0.77$.

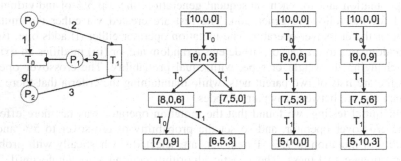

Fig. 5. (a) A Petri net with a single gene unit and (b) its corresponding state trees for $g=3$, 6 and 9 respectively

3   Evaluation

3.1   Non-linear Gene Interaction Model of Digital Ulcers

A recently discovered disease-causing non-linear gene interaction is used as a test-bed for our method [1]. This model, depicted in Figure 6, describes the risk of developing digital ulcers in a population of 200 Italian systemic sclerosis patients and was built using the Multifactor Dimensionality Reduction (MDR) kernel [10]. The model concerns two SNPs (IL-2 C-330G SNP and IL-6 G-174C SNP, hereafter referred to as IL2 and IL6), and one non-SNP mutation (HLA-B35, hereafter referred to B35). Due to the complexity of B35, only the presence or absence of a particular mutant allele (HLA-B*3501) is recorded; we refer to the absence of this allele as AA, and its presence as Aa/aa.

In each cell of Figure 6, there are two bars. The left bars indicate the frequency of patients (cases) with digital ulcers, and the right bars indicate the frequency of patients without digital ulcers (the controls). If the ratio of cases to controls exceeds a certain threshold, patients are labeled as high risk (which are the dark-shaded cells), otherwise they are low risk (the light-shaded cells).

We want to use our genetic algorithm to learn a Petri net model corresponding to the architecture in Figure 3 that shows how the various genotypes could lead to either a high risk or low risk of the disease, for each of the 18 genotypes in the matrix.

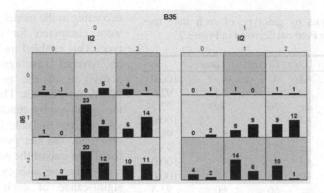

Fig. 6. Multifactor dimensionality reduction (MDR) model of non-linear gene-gene interaction. Key: For IL2 and IL6, cell indices 0, 1 and 2 denote genotypes *AA*, *Aa* and *aa* respectively. For B35, cell index 1 indicates *Aa/aa* and index 0 indicates genotype *AA*.

3.2   Results and Analysis

We performed 32 runs of our genetic algorithm. The maximum fitness value after a run obtained was 0.85, and the minimum was 0.64. The mean best fitness value was 0.70. We examined the Petri net, depicted in Figure 7, with the maximum fitness of 0.85. This net required 5,558 generations to learn, and it has 40 arcs. Rather than showing the net graphically, which would be difficult to interpret, we present it instead as a list of transitions.

$$
\begin{array}{lll}
T_{B35}: & AP_{B35}(1) & \Rightarrow & PP_{B35}(g_{B35}) \\
T_{IL6}: & AP_{IL6}(1) & \Rightarrow & PP_{IL6}(g_{IL6}) \\
T_{IL2}: & AP_{IL2}(1),P_1(i) & \Rightarrow & PP_{IL2}(g_{IL2}) \\
T_3: & PP_{B35}(8),PP_{IL6}(9) & \Rightarrow & P_1(10) \\
T_4: & PP_{IL2}(7) & \Rightarrow & P_1(3),AP_{IL2}(9) \\
T_5: & PP_{IL6}(3) & \Rightarrow & P_1(4),AP_{B35}(4),AP_{IL2}(8),AP_{IL6}(9) \\
T_6: & PP_{IL2}(7),PP_{IL6}(i) & \Rightarrow & P_1(4) \\
T_7: & PP_{IL6}(6) & \Rightarrow & AP_{B35}(3),AP_{IL2}(2) \\
T_8: & PP_{IL2}(7) & \Rightarrow & P_1(2),AP_{IL6}(7) \\
T_9: & P_0(4) & \Rightarrow & AP_{B35}(9),AP_{IL6}(1) \\
T_{10}: & PP_{IL2}(3) & \Rightarrow & P_1(3)\ AP_{IL6}(7) \\
T_{11}: & PP_{IL2}(7) & \Rightarrow & P_1(2),AP_{IL6}(2) \\
T_{12}: & PP_{IL2}(3),PP_{IL6}(i) & \Rightarrow & P_1(6),AP_{IL2}(10)
\end{array}
$$

Fig. 7. Best Petri net obtained with our genetic algorithm. In the figure, P_0 is the initial token source for the net, and P_1 is the toxic output that is thresholded to determine the risk assessment. The LHS of each transition is a list of inputs and weights (or i if the input is an inhibitor), and the RHS is a list of outputs and weights. AP_k and PP_k are the activating and product places of gene k, and g_k denotes the genotype-controlled weight for gene k.

We generated all 18 state trees (each state tree being derived from one genotype, as described in Section 2), and calculated the number of different outcomes (leaf nodes) for each tree. Of those, the number that gave the correct risk assessment (high or low

Table 1. Analysis by genotype of each state tree derived from the Petri net depicted in Figure 7

B35	IL2	IL6	Correct	Total	%
AA	AA	AA	4	4	100.0
AA	AA	Aa	50	54	92.6
AA	AA	aa	36	39	92.3
AA	Aa	AA	4	4	100.0
AA	Aa	Aa	95	99	96.0
AA	Aa	aa	57	61	93.4
AA	aa	AA	4	4	100.0
AA	aa	Aa	68	68	100.0
AA	aa	aa	56	60	93.3
Aa/aa	AA	AA	2	2	100.0
Aa/aa	AA	Aa	2	2	100.0
Aa/aa	AA	aa	0	2	0.0
Aa/aa	Aa	AA	2	2	100.0
Aa/aa	Aa	Aa	2	2	100.0
Aa/aa	Aa	aa	2	2	100.0
Aa/aa	aa	AA	2	2	100.0
Aa/aa	aa	Aa	2	2	100.0
Aa/aa	aa	aa	2	2	100.0

according to the model in Figure 6) were computed for each genotype. This enabled us to compute an overall average accuracy across all of the genotypes of 92.6% for this net. The results of the genotype-by-genotype analysis are given in Table 1. As can be observed, the net performs well for all but one of the genotypes.

Finally, we also examined the significance of each individual transition in the network. Taking the Petri net depicted in Figure 7, and iteratively deleting each transition and all arcs incident on it achieved this. We then recomputed the value of the network without the transition, before replacing the transition and its arcs. The recomputed values give an indication of each transition's significance, and can be compared to the original value of 0.85. Table 2 lists, for each transition, these recomputed values. The lower the value of the transition, the greater its significance on the Petri net's dynamics.

As can be observed in Table 2, transitions T_{B35}, T_{IL6} and T_9 are the most significant: remove them, and the network fails almost completely. This is clearly because the source of the initial tokens (P_0) is only accessible via T_9, which feeds into B35 and IL6 gene units. All of the remaining transitions except for the final three have different degrees of significance. Interestingly, the final three transitions, T_4, T_8, and T_{11}, have almost no significant impact on the Petri net's behaviour. They could, therefore, be entirely removed, thereby reducing the size of the net from 40 arcs to 31 arcs. This transition analysis could be employed during the genetic algorithm search process itself in order to reduce the size of Petri nets without relying on the random mutation operator.

Table 2. Recomputed values of the Petri Net by transition

Transition	Value
T_{B35}	0.25
T_{IL6}	0.25
T_9	0.25
T_{IL2}	0.42
T_7	0.56
T_3	0.58
T_6	0.66
T_{10}	0.67
T_5	0.67
T_{12}	0.76
T_4	0.84
T_8	0.84
T_{11}	0.84

4 Conclusion

We have shown how to construct a Petri net-based model of a set of concurrent processes. Unlike previous approaches to Petri net learning that require global coordination (in the form of transition priorities or randomised transition firing), our approach "unrolls" the non-determinism by converting a Petri net into a set of state trees, and evaluates the trees rather than the net. This gives an indication of the net's behavior when the transitions are allowed to fire concurrently or in any order whatsoever – a situation most suitable for modeling real world processes.

We have applied this approach to the modeling of non-linear gene interactions, and shown that not only is this approach computationally feasible for a practical application, but also that the analysis of the Petri net's learned using this method may lead to useful insight into the problem being modeled.

References

1. Beretta, L., Santaniello, A., Mayo, M., Cappiello, F., Marchini, M., Scorza, R.: Genetic and Biological Models of Epistasis to Predict Digital Ulcer Occurrence in Italian Systemic Sclerosis Patients. Article In Submission to Annals of Human Genetics (2009)
2. Cheng, S., Yeh, H., Lin, Y., Lin, S., Soo, V.: Inferring Gene Regulatory Networks from Microarray Data Based on Transcription Factor Analysis and Conditional Independency. In: BIOCOMP 2007, pp. 65–71 (2007)
3. Fishman, D., Faulds, G., Jeffery, R., et al.: The effect of novel polymorphisms in the inter-leukin-6 (IL-6) gene on IL-6 transcription and plasma IL-6 levels, and an association with systemic-onset juvenile chronic arthritis. J. Clin. Invest. 102, 1369–1376 (1998)
4. Goldberg, D.: Genetic Algorithms in Search, Optimization and Machine Learning. Addison-Wesley, Reading (1989)
5. Hoffmann, S.C., Stanley, E.M., Darrin Cox, E., et al.: Association of cytokine polymorphic inheritance and in vitro cytokine production in anti-CD3/CD28-stimulated peripheral blood lymphocytes. Transplantation 72, 1444–1450 (2001)
6. Lin, Y., Yeh, H., Cheng, S., Soo, V.: Comparing Cancer and Normal Gene Regulatory Networks Based on Microarray Data and Transcription Factor Analysis. In: Proc. of the 7th IEEE International Conference on Bioinformatics and Bioengineering, BIBE 2007, pp. 151–157 (2007)
7. Mayo, M.: Learning Petri net models of non-linear gene interactions. BioSystems 82(1), 74–82 (2005)
8. McGarry, K., Loutfi, M., Moscardini, A.: Stochastic Simulation of the Regulatory Pathways involved in Diabetes using Petri-nets. In: Proc. of the International Conference on Computer Theory and Applications (ICCTA 2007), Alexandria, Egypt (2007)
9. Moore, J., Hahn, L.: Petri net modelling of high-order genetic systems using grammatical evolution. BioSystems 72, 177–186 (2003)
10. Moore, J.: Computational analysis of gene-gene interactions using multifactor dimensionality reduction. Expert Review of Molecular Diagnostics 4(6), 795–803 (2004)
11. Phillips, P.: Epistasis–the essential role of gene interactions in the structure and evolution of genetic systems. Nat. Rev. Genet. 9, 855–867 (2008)
12. Pociot, F., Molvig, J., Wogensen, L., Worsaae, H., Nerup, J.: A TaqI polymorphism in the human interleukin-1β (IL-1β) gene correlates with IL-1β secretion in vitro. Eur. J. Clin. Invest. 22, 396–402 (1995)
13. Reisig, W.: Petri nets: an introduction. In: EATCS Monographs on Theoretical Computer Science. Springer, Heidelberg (1985)
14. Mayo, M., Beretta, L.: Modelling Epistasis in Genetic Disease using Petri Nets, Evolutionary Computation and Frequent Itemset Mining. Submission, Expert Systems with Applications: An International Journal (2009)

Partial Orderings for Ranking Help Functions

Sylvia Encheva[1] and Sharil Tumin[2]

[1] Stord/Haugesund University College, Bjørnsonsg. 45, 5528 Haugesund, Norway
sbe@hsh.no
[2] University of Bergen, IT-Dept., P.O. Box 7800, 5020 Bergen, Norway
edpst@it.uib.no

Abstract. In this paper we propose ranking help functions in an intelligent tutoring system with respect to their usefulness. A help function is regarded as useful to a student if the student has succeeded to solve a problem after using it. Methods from the theory of partial orderings are further applied facilitating an automated process of suggesting individualised advises on how to proceed in order to solve a particular problem.

Keywords: Decision support services, uncertainty management, partial orderings.

1 Introduction

The level of usefulness of a service is usually established by sending users questionnaires and summarising the obtained responses. Naturally new users lack in-depth understanding of the subject they study and cannot really judge to which extend a particular help function is facilitating the learning process. What they actually express is their overall likings of the tool. Among other factors that influence their responses are friends' opinions on the matter, student's degree of interest in that subject, honesty, i.e. are their responses anonymous or a sender can be tracked down and so on.

This work is intended to facilitate automated provision of help functions via an intelligent tutoring system, applying mathematical methods from partially ordered sets. Our approach to evaluate help functions will avoid the influence of the above mentioned subjective factors since a help function in this work is regarded as useful to a student if the student has succeeded to solve a problem after using it.

The rest of the paper is organised as follows. Section 2 contains definitions of terms used later on. Section 3 explains how to rank help functions according to personal responses and Section 4 is devoted to a system description. Section 5 contains the conclusion of this work.

2 Methodology

2.1 Ordered Sets

Determining a consensus from a group of orderings and making statistically significant statements about orderings have been discussed in [3].

N.T. Nguyen, M.T. Le, and J. Świątek (Eds.): ACIIDS 2010, Part II, LNAI 5991, pp. 176–184, 2010.

A relation I is an *indifference* relation when given AIB neither $A > B$ nor $A <$ B has place in the componentwise ordering. A partial ordering whose indifference relation is transitive is called a *weak ordering*.

Let w_1, w_2, w_3 be weak orderings. Then w_2 is between w_1 and w_3 if each decision made by w_2 is made by either w_1 or w_3 and any decision made by both w_1 and w_3 is made by w_2, i.e. $w_1 \cap w_3 \subseteq w_2 \subseteq w_1 \cup w_3$.

The distance $d(w_1, w_3)$ is defined as $d(w_1, w_2) + d(w_2, w_3) = d(w_1, w_3)$. The distance is a metric in the usual sense, it is invariant under permutation of alternatives, and the minimum positive distance is 1.

If given a choice between two alternatives, a person will choose one. Thus obtained partial orderings appeared to be all linear orders since each pair of alternatives is compared. A social welfare function acts on k-tuples representing either weak or linear orderings of k individuals for m alternatives. The representation of an individual's ordering can be thought of as a column vector in which an integer in position i represents the preference level the individual assigns to alternative i. Thus k individuals presented with m alternatives can illustrated by k-tuples of orderings in a $m \times k$ matrix of integers of preference levels.

2.2 Related Works

"Hints are an important ingredient of natural language tutorial dialogues. Existing models of hints, however, are limited in capturing their various underlying functions, since hints are typically treated as a unit directly associated with some problem solving script or discourse situation, [10]." The taxonomy developed by the authors has five dimensions: domain knowledge reference, inferential role, elicitation status, discourse dynamics, and problem solving perspective.

Hinting is a general and effective tutoring tactic in one-on-on tutoring when the student has trouble solving a problem or answering a question, [12]. Among other things the article addresses the problem of effectiveness of a tutoring system depending mainly on hinting for facilitating a learning process.

The role of specific and unspecific tasks for learning declarative knowledge and skills with a web based learning system is considered in [4]. According to the obtained results in this study learners with specific tasks where better. It is further concluded that instructions should be selected carefully in correspondence with desired learning goals.

An investigation of whether a cognitive tutor can be made more effective by extending it to help students acquire help-seeking skills can be found in [8]. The authors have performed an investigate whether a cognitive tutor can be made more effective by extending it to help students acquire help-seeking skills. A preliminary model involves 57 production rules and registers productive and unproductive help-seeking attempts.

A proliferation of hint abuse (e.g., using hints to find answers rather than trying to understand) was found in [1] and [8]. However, evidence that when used appropriately, on-demand help can have a positive impact on learning was found in [9].

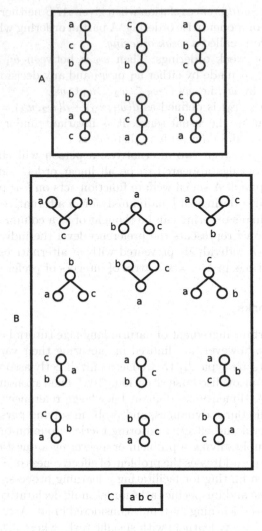

Fig. 1. Orderings of three elements

3   Rankings

In the process teaching students how to of solve new problems is always difficult to estimate the amount of assistance they should be offered. We are focussing on modelling a system that can provide assistance tailored to individual needs. We assume that a particular type of assistance is the one needed by an individual when that individual can solve a given problem after been presented with that type of assistance (we will refer to it as 'element').

I) Suppose the system offers three elements. The following orderings are to be considered:

- the three elements are ordered consecutively
- one of the elements is placed higher than the other two elements and no preference is shown with respect to those two elements
- one of the elements is placed lower than the other two elements and no preference is shown with respect to these two elements
- only two of the elements are considered

Representatives of these posets are summarized in Fig. 1. They are placed in three sets A, B, C where all members of A signify that the three elements have been ranked, all members of B signify that elements have been ranked as higher and lower, membership in C signify that none of the elements has been addressed.

II) The process of ranking four elements is more laborious. Their meaning can be described as follows.

- the four elements are not ranked
- only two elements are ranked
- two couples of elements are ranked
- three couples of elements are ranked
- four couples of elements are ranked
- one element is ranked higher than the other three
- three elements are ranked higher than the forth element
- three elements are ranked linearly and the forth element is not ranked
- three elements are ranked linearly and the forth element is ranked higher than the last element
- three elements are ranked linearly and the forth element is ranked below than the top element
- one element is compared to the other three elements and another element ranked higher than the rest
- one element is compared to the other three elements and two elements are ranked higher than the rest
- one element is compared to the other three elements and two elements are ranked higher than the rest
- every element is compared to two of the other elements
- all elements are ranked linearly

Representatives of these posets are summarized in Fig. 2. They are placed in four sets A, B, C, D where all members of A signify that the four elements have been ranked, all members of B signify that three of elements have been ranked only, all members of C signify that two of elements have been ranked only, all members of D signify that none of the elements has been addressed.

III) The case with five elements can be worked out in a way similar to the case with four elements. Due to space constrain we will restrict go and list all the sixty three posets of five elements. These posets can be arranged in five sets where the first set contains not ranked elements, the second set contains all posets where only two of the elements are ranked and so on.

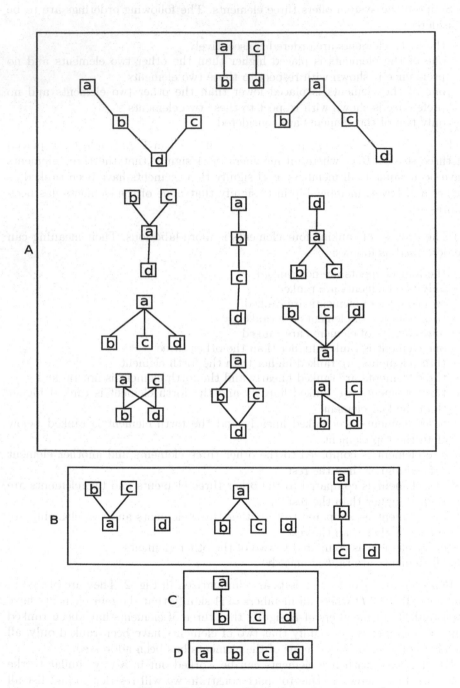

Fig. 2. Orderings of four elements

3.1 Discussion

A natural question here will be if an original system is build for facilitating
three elements could be extended to working with four elements. Technically
this could be done by simply mapping the three elements ordering into the
four elements ordering and add the fourth element whenever needed. Problems
may arise if some weights have been assigned in the first version and some other
decision principals are applied for the second version of the system. For example,
if distances between two posets in Fig. 1 are seriously involved in the decision
process for the first version than the meaning of this information cannot be 'just'
transferred in the second version. New decision principles should be considered.

3.2 Aggregation of Preferences

One of the very important questions related to such processes is how to incorporate aggregation of preferences. According to [6] accumulation of preferences
has been of research interest for more than two centuries. Most of the time the
focus has been on working out a consensus ranking that satisfies a predetermined
criterion. The authors develop an axiomatic structure related to the concept of
distance between rankings. Adopting the median ranking as a form of consensus,
it is shown that this ranking can be determined, in the case of complete rankings,
by solving a certain assignment problem.

For the purpose of our study we suggest two possible sums, cardinal and
ordinal. Suppose we operate with two structures

$$\mathcal{A} = (X, R) \ \text{and} \ \mathcal{B} = (Y, S),$$

where X and Y are sets and R and S are binary relations on them. The cardinal
and ordinal sums are

$$\mathcal{A} + \mathcal{B} = (X \cup Y, R \cup S)$$

and

$$\mathcal{A} \oplus \mathcal{B} = (X \cup Y, R \cup S \cup (X \times Y)).$$

X and Y are assumed to be disjoint. In case they are not they can be replaced
by isomorphic copies whose universes are disjoint, [2]. In cardinal sums the order
relation of each set is preserved and an element in the first set is never comparable
with an element in the second set. In ordinal sums the order relation of each set
is preserved and each element in the first set is less than any element in the
second set.

3.3 Weights

The problem of aggregating a set of numerical readings in order to obtain a mean
value is addressed in [11].

If $x_1, x_2, ..., x_n$ is a set of readings then the aggregating process is denoted as $Agg(x_1, x_2, ..., x_n) = a$. The aggregation operator has the following properties

1. $Agg(a) = a$, (a natural boundary). It means that in the case of a single reading the aggregated value is taken to be that single reading.

2. If $Agg(x_1, x_2, ..., x_n) = a$ then

$$Agg(x_1, x_2, ..., x_n, a) = Agg(x_1, x_2, ..., x_n) = a,$$

(self-identity). It implies that adding an element equal to an already existing value does not change the aggregation value.

3. $Agg(x_1, x_2, ..., x_n) \geq Agg(y_1, y_2, ..., y_n), x_i \geq y_i, \forall 1 \leq i \leq n$, monotonicity

Agg is idempotent ($Agg(x_1, x_2, ..., x_n) = a$ if $x_i = a \forall 1 \leq i \leq n$) if (1) and (2) hold [11].

If $Agg(x_1, x_2, ..., x_n)$ has a natural boundary and self-identity, is monotonic and idempotent, then

$$a \leq Agg(x_1, x_2, ..., x_n) \leq b$$

where

$$a = Min[x_i] \text{ and } b = Max[x_i].$$

Theorem 1. *[11] If Agg is a monotonic and self-identity operator then*

$$Agg(x_1, ..., x_n, K) \geq Agg(x_1, ..., x_n), \text{ if } K > Agg(x_1, ..., x_n),$$

and

$$Agg(x_1, ..., x_n, K) \leq Agg(x_1, ..., x_n), \text{ if } K < Agg(x_1, ..., x_n).$$

Theorem 1 implies that addition of new elements with values greater than the current mean value results in an increased mean value and addition of new elements with values smaller than the current mean value results in an decreased mean value.

Theorem 2. *[11] If Agg is an idempotent and monotonic operator, not equal to the Max or Min operators then Agg is not associative.*

Theorem 2 implies that if Agg is idempotent and monotonic its extension to a greater cardinality does not have to be done in a consistent way. At the same time if Agg has a natural boundary and self-identity than by Theorem 1 certain requirements for consistency have to be satisfied while extending its cardinality.

The weights must satisfy the following conditions

$$\sum_{j=1}^{n} w_{nj} = 1 \ \forall \ n \text{ and } w_{nj} \geq 0 \ \forall \ n, j$$

For the case with n arguments $Agg(x_1, ..., x_n) = \sum_{j=1}^{n} w_{nj} x_n$. The weights w_{nj} are uniquely determined by the formula

$$w_{ni} = w_{n1} \left(\frac{1}{w_{i,1}} - \frac{1}{w_{i-1,1}} \right), \quad 2 \leq i \leq n$$

and the ratio $\frac{w_{n1}}{w_{n-1,1}}$ between the first element in the current iteration and the first element in the previous iteration, [11].

Another approach for calculating the weights w_{nj} involves the Lukasiewicsz t-conorm $S(x_1, ..., x_n)$ [7],

$$S(x_1, ..., x_n) = Min\left[1, \sum_{i=1}^{n} x_i\right].$$

Thus

$$w_{nj} = \frac{S_j - S_{j-1}}{S_n},$$

where

if $S_j < 1$ then $w_{nj} = \frac{x_j}{S_n}$

if $S_{j-1} \geq 1$ then $w_{nj} = 0$

if $S_j = 1, S_{j-1} < 1$ then $w_{nj} = 1 - S_{j-1}$.

If factors like importances and frequencies are to be involved in the decision proses one may use the method of incorporating quantitative weights into aggregation, [5]. In statistics the weighted mean is presented as

$$W(x_1, ..., x_n) = \frac{\sum_i n_i x_i}{\sum_i n_i}$$

where n_i are integer weights and x_i are frequencies of observations. W is further generalized to

$$W(x_1, ..., x_n) = \sum_i w_i x_i, \quad w_i \geq 0$$

where the weight w_i corresponds to the ith input x_i. Quantitative weights can be assigned to different entries reflecting the sets' belonging as presented in Fig. 2.

4   The System

A Web-based application using Apache HTTP server [13], mod_python module [14] and SQLite database [15] can be considered for building a system prototype. The mod_python module provides programmable runtime support to the HTTP server using Python programming language. The whole application components are

1. Web-based users interface,
2. application logic and application interfaces written in Python, and
3. relational database.

5   Conclusion

Evaluating help functions based on students' responses is more accurate than using questionnaires. Such an approach is not affected by subjective opinions and provides useful feedback to content developers.

The presented approach is a response to the the increased demand for the necessity of developing effective learning tools that can be smoothly integrated in the educational process.

References

1. Aleven, V., Koedinger, K.R.: Limitations of Student Control: Do Student Know when they need help? In: Gauthier, G., VanLehn, K., Frasson, C. (eds.) ITS 2000. LNCS, vol. 1839, pp. 292–303. Springer, Heidelberg (2000)
2. Arrow, K.J.: Social Choice and Individual Values. Wiley, New York (1951); 2nd ed. (1963) ISBN 0-300-01364-7
3. Bogart, K.P.: Some social sciences applications of ordered sets. In: Rival, I. (ed.) Ordered Sets, pp. 759–787. Reidel, Dordrecht (1982)
4. Brunstein, A., Krems, J.: Helps and Hints for Learning with Web Based Learning Systems: The Role of Instructions. In: Lester, J.C., Vicari, R.M., Paraguaçu, F. (eds.) ITS 2004. LNCS, vol. 3220, pp. 794–796. Springer, Heidelberg (2004)
5. Calvo, T., Mesiar, R., Yager, R.: Quantitative weights and aggregation. IEEE T. Fuzzy Systems 12(1), 62–69 (2004)
6. Cook, W.D., Seiford, M.L.: Priority Ranking and Consensus Formation. Management Science 24(16), 1721–1732 (1978)
7. Dubois, D., Prade, H.: A review of fuzzy sets aggregation connectives. Information Science 36, 85–121 (1985)
8. Aleven, V., McLaren, B.M., Roll, I., Koedinger, K.R.: Toward tutoring help seeking. In: Lester, J.C., Vicari, R.M., Paraguaçu, F. (eds.) ITS 2004. LNCS, vol. 3220, pp. 227–239. Springer, Heidelberg (2004)
9. Schworm, S., Renkl, A.: Learning by solved example problems: Instructional explanations reduce self-explanation activity. In: Gray, W.D., Schunn, C.D. (eds.) Proceeding of the 24th Annual Conference of the Cognitive Science Society, pp. 816–821. Erlbaum, Mahwah (2002)
10. Tsovaltzi, D., Fiedler, A., Horacek, H.: A Multi-dimensional Taxonomy for Automating Hinting. In: Lester, J.C., Vicari, R.M., Paraguaçu, F. (eds.) ITS 2004. LNCS, vol. 3220, pp. 772–781. Springer, Heidelberg (2004)
11. Yager, R.R., Rybalov, A.: Noncommutative self-identity aggregation. Fuzzy Sets and Systems 85, 73–82 (1997)
12. Zhou, Y., Freedman, R., Glass, M., Michael, J.A., Allen, A., Rovick, A.A., Evens, M.W.: Delivering hints in a dialogue-based intelligent tutoring system. In: Proceedings of the sixteenth national conference on Artificial intelligence and the eleventh Innovative applications of artificial intelligence conference innovative applications of artificial intelligence, pp. 128–134 (1999)
13. Apache HTTP Server Project, http://httpd.apache.org/
14. Python Programming Language, http://www.python.org/
15. SQLite, http://www.sqlite.org/

Robust Prediction with ANNBFIS System

Robert Czabanski[1], Michal Jezewski[1], Krzysztof Horoba[2], Janusz Jezewski[2], and Janusz Wrobel[2]

[1] Silesian University of Technology, Institute of Electronics, ul Akademicka 16,
44-101 Gliwice, Poland
robert.czabanski@polsl.pl
[2] Department of Biomedical Informatics, Institute of Medical Technology and
Equipment, ul. Roosevelta 118, 41-800 Zabrze, Poland

Abstract. In this paper a learning method of Artificial Neural Network Based on Fuzzy Inference System (ANNBFIS) is presented. It is based on deterministic annealing, ε-insensitive learning by solving a system of linear inequalities, and robust fuzzy c-means clustering. To find the unknown number of fuzzy if-then rules we proposed the procedure of robust clusters merging. The performance of the learning method was demonstrated through the benchmark sunspot prediction problem.

1 Introduction

The concepts of fuzzy sets and fuzzy logic [11] have been used in many areas of science. The basic problem when designing a fuzzy system is the determination of its fuzzy rule base, representing the knowledge of the phenomena under consideration. There is no explicit method of expert knowledge acquisition, therefore the automated methods of rules extraction are actively investigated. A set of fuzzy conditional statements may be obtained from numerical data describing input/output system characteristics using learning capabilities of artificial neural networks. In this work a learning procedure for Artificial Neural Network Based on Fuzzy Inference System (ANNBFIS) [5] is presented. The main goal of the proposed algorithm is to improve the learning efficiency and the same the quality of the extracted fuzzy if-then rules.

ANNBFIS is a neuro-fuzzy system with parameterized consequents that generates inference results based on fuzzy if-then rules:

$$\underset{i=1,2,\ldots,I}{\forall} R^{(i)} : \textbf{if and} \overset{t}{\underset{j=1}{}} \left(x_{0j} \text{ is } A_j^{(i)} \right) \text{ then } Y \text{ is } B^{(i)}\left(\theta\right), \qquad (1)$$

where I denotes a number of rules, t is the number of inputs, x_{0j} is the j-th element of the input vector $\mathbf{x}_0 = [x_{01}, x_{02}, \ldots, x_{0t}]^T$, Y is the output linguistic variable, $A_j^{(i)}$ and $B^{(i)}\left(\theta\right)$ are fuzzy sets representing the linguistic values (terms) in antecedents and consequents respectively, and θ is a set of parameters defining the fuzzy set in the (parameterized) consequence.

N.T. Nguyen, M.T. Le, and J. Świątek (Eds.): ACIIDS 2010, Part II, LNAI 5991, pp. 185–194, 2010.

In ANNBFIS system the fuzzy sets of linguistic values in rule antecedents have Gaussian membership function and the explicit connective "**and**" of multi-input rule predicates is defined as an algebraic product. Accordingly, the firing strength (the degree of activation) of the rule can be evaluated as:

$$\underset{i=1,2,\ldots,I}{\forall} F^{(i)}(\mathbf{x}_0) = \prod_{j=1}^{t} A_j^{(i)}(x_{0j}) = \exp\left[-\frac{1}{2}\sum_{j=1}^{t}\left(\frac{x_{0j} - c_j^{(i)}}{s_j^{(i)}}\right)^2\right], \quad (2)$$

where $c_j^{(i)}$ and $s_j^{(i)}$, $\forall i = 1, 2, \ldots, I, \forall j = 1, 2, \ldots, t$, are membership function parameters, center and dispersion, respectively. The inference process depends on a method of if-then rules interpretation. In the ANNBFIS the Larsen product is applied:

$$B^{(i)\prime}(\theta) = F^{(i)}(\mathbf{x}_0) \cdot B^{(i)}(\theta), \quad (3)$$

where $B^{(i)\prime}(\theta)$ is the conclusion after inference but before aggregation. The consequents have symmetric triangular membership functions, defined by width of the triangle base $w^{(i)}$, and center of gravity location $y^{(i)}(\mathbf{x}_0)$ determined by linear combinations of fuzzy system inputs:

$$y^{(i)}(\mathbf{x}_0) = p_0^{(i)} + p_1^{(i)}x_{01} + \ldots + p_t^{(i)}x_{0t} = \mathbf{p}^{(i)T}\mathbf{x}_0'. \quad (4)$$

Finally, the membership function of fuzzy set in the consequence is a function of $w^{(i)}$ and $\mathbf{x}_0$:

$$B^{(i)}(\theta) = B^{(i)}\left(\mathbf{x}_0, w^{(i)}\right). \quad (5)$$

The above dependency formulates so called parameterized consequent [5]. The output fuzzy set is evaluated in the aggregation process using a normalized arithmetic mean. To evaluate the crisp output of the fuzzy system the defuzzification process is performed. In the ANNBFIS system the modified indexed center of gravity defuzzifier is used. Finally, the crisp output value of the ANNBFIS system can be evaluated from the following formula:

$$y_0 = \sum_{i=1}^{I} G^{(i)}(\mathbf{x}_0) y^{(i)}(\mathbf{x}_0) = \sum_{i=1}^{I} \frac{g\left(F^{(i)}(\mathbf{x}_0), w^{(i)}\right)}{\sum_{k=1}^{I} g\left(F^{(k)}(\mathbf{x}_0), w^{(k)}\right)} y^{(i)}(\mathbf{x}_0). \quad (6)$$

Function $g\left(F^{(i)}(\mathbf{x}_0), w^{(i)}\right)$ is the area under a curve defined by membership function of conclusion:

$$g\left(F^{(i)}(\mathbf{x}_0), w^{(i)}\right) = \frac{1}{2} w^{(i)} F^{(i)}(\mathbf{x}_0). \quad (7)$$

To determine the rule base of the ANNBFIS the set of unknown parameters $\forall_{1 \leq i \leq I} \zeta^{(i)} = \left[\mathbf{c}^{(i)T}, \mathbf{s}^{(i)T}, w^{(i)}, \mathbf{p}^{(i)T}\right]^T$ have to be estimated. The number of rules I is unknown and the number of antecedents t is directly defined by the size of input training vector. The fuzzy system with parameterized consequents is equivalent to the radial basis function neural networks [5]. Therefore, the unknown neuro-fuzzy system parameters can be calculated using learning algorithms of neural networks. In this work we applied the deterministic annealing and ε-insensitive approach for learning.

2   Deterministic Annealing

The final goal of learning is an extraction of the set of fuzzy if-then rules that leads to the best fuzzy system performance. The extraction process consists in an estimation of membership function parameters of antecedents as well as consequents being a result of a squared-error (cost) function minimization:

$$E = \sum_{n=1}^{N} E_n = \sum_{n=1}^{N} \frac{1}{2} \left(t_0\left(n\right) - y_0\left(n\right) \right)^2, \tag{8}$$

where N is the number of elements in the training set and $t_0\left(n\right) \in \mathbb{R}$ is the desired output. In order to increase the ability to avoid many local minimums that trap descent methods, the deterministic annealing (DA) [8] can be used. We modified the DA algorithm to apply it as learning method of fuzzy system with parameterized consequents [5], [3]. Deterministic annealing is based on the simulated annealing framework but it replaces computationally intensive stochastic simulation by straightforward deterministic optimization of modeled system error energy. The DA algorithm consists in minimization of the cost function while simultaneously controlling the entropy level of a solution. The constrained optimization of the cost E with imposed level of entropy S is equivalent to unconstrained minimization of the Lagrangian:

$$L = E - T \cdot S, \tag{9}$$

where T is the Lagrange multiplier and S is the Shannon entropy:

$$S = -\sum_{n=1}^{N} \sum_{i=1}^{I} G^{(i)}\left(\mathbf{x}_0\right) \log G^{(i)}\left(\mathbf{x}_0\right). \tag{10}$$

The minimization procedure involves series of iterations while the randomness level is reduced gradually. To achieve the global minimum of the cost function the simulated annealing method framework is used. At high level of temperatures T, the minimization of the Lagrangian L is unambiguous with the entropy maximization. In other words, we seek a set of fuzzy rules that are equally associated with each input data point and cooperate to work out a desired output. As temperature is lowered more emphasis is placed on reducing the square-error, which also leads to a decrease in entropy. We get more and more competitive fuzzy rules each associated with given data more closely. We cross gradually from cooperation to competition. Finally, with $T = 0$ the minimization is conducted regardless of the entropy level. The temperature reduction procedure is defined by 'annealing schedule'. In our numerical experiments we assume an exponential schedule function:

$$T(k+1) = q\,T(k), \tag{11}$$

where $q \in (0, 1)$ is a preset parameter and k denotes the iteration index.

For each value of temperature we minimize the Lagrangian iteratively. The parameters of neuro-fuzzy system are given as:

$$\zeta(k+1) = \zeta(k) - \frac{\eta_{\text{ini}}}{\sqrt{\sum_{i=1}^{n_i}\left(\frac{\partial L}{\partial \zeta_i}\right)^2}} \left.\frac{\partial L}{\partial \zeta}\right|_{\zeta(k+1)=\zeta(k)}, \tag{12}$$

where η_{ini} denotes initial learning step and n_i is the number of optimized parameters. For parameters of membership function of fuzzy sets in antecedents n_i is equal to $2 \cdot I \cdot t$, for parameters of linear function in consequents $n_i = I \cdot (t+1)$, and for triangle base widths $n_i = I$.

The parameters of linear equations from consequents of fuzzy conditional rules can be estimated using the least square (LS) method as well [5]. The application of LS allows the learning convergence to accelerate. The other method for estimating parameters of rules consequents - the ε-insensitive learning - increases the learning quality of the ANNBFIS neuro-fuzzy system in aspect of generalization ability increase [7]. To improve the ANNBFIS parameters optimization with DA procedure we integrated it with ε-insensitive learning [4].

3 ε-Insensitive Learning

The neuro-fuzzy system with parameterized consequents as well as many other neuro-fuzzy systems have intrinsic inconsistency: they may perform approximate reasoning but simultaneously its learning methods are not tolerant of imprecision. Usually the neuro-fuzzy learning algorithms use the quadratic error function to control the performance of fuzzy modeling process. Therefore, only perfect match of the fuzzy model and the modeled phenomenon results in zero error value. Other approach to neuro-fuzzy modeling was based on the premise that human learning and reasoning is tolerant of imprecision. A limiting value of tolerance for imprecision was denoted as ε (epsilon). If the error value is less than ε then the zero loss is obtained. The learning methods based on ε loss function are called ε-insensitive learning. The ε-insensitive learning for the ANNBFIS can be formulated as the equation system [7]:

$$\begin{cases} \mathbf{p}^{(i)} = \left(\mathbf{X}_{0e}'^{T}\mathbf{D}_e\mathbf{X}_{0e}' + \frac{\tau}{2}\tilde{\mathbf{I}}\right)^{-1}\mathbf{X}_{0e}'^{T}\mathbf{D}_e\left(\mathbf{t}_{0e} + \mathbf{b}\right), \\ \mathbf{e} = \mathbf{X}_{0e}'\mathbf{p}^{(i)} - \mathbf{t}_{0e} - \mathbf{b} = \mathbf{0}. \end{cases} \tag{13}$$

where $\mathbf{X}_{0e}' = [\mathbf{x}_0'(1), \mathbf{x}_0'(2), \ldots, \mathbf{x}_0'(N), -\mathbf{x}_0'(1), -\mathbf{x}_0'(2), \ldots, -\mathbf{x}_0'(N)]^T$ is an extended input matrix, $\mathbf{t}_{0e} = [t_0(1) - \varepsilon, t_0(2) - \varepsilon, \ldots, t_0(N) - \varepsilon, -t_0(1) - \varepsilon, -t_0(2) - \varepsilon, \ldots, -t_0(N) - \varepsilon]^T$ is an extended output vector, $\tilde{\mathbf{I}} = \text{diag}([0, \mathbb{1}_{t\times 1}^T])$, $\mathbb{1}_{t\times 1}$ is a $(t \times 1)$ dimensional vector with all entries equal to 1 and $\mathbf{D}_e$ denotes a diagonal weights matrix, $\mathbf{D}_e = \text{diag}(G^{(i)}(\mathbf{x}_0(1))/|e_1|, G^{(i)}(\mathbf{x}_0(2))/|e_2|, \ldots, G^{(i)}(\mathbf{x}_0(N))/|e_N|, G^{(i)}(\mathbf{x}_0(1))/|e_{N+1}|, \ldots, G^{(i)}(\mathbf{x}_0(N))/|e_{2N}|)$ where e_i is the i-th component of the error vector $\mathbf{e}$. The regularization parameter $\tau \geq 0$ is responsible for the model generalization ability. A larger τ results in an increase of the model generalization ability. The vector $\mathbf{b}$ is called margin vector because its components determine the distance between data and insensitivity region. The solution of the (13) is called ε-insensitive Learning by Solving a System of Linear Inequalities (εLSSLI). In this method, the margin vector components are modified by corresponding error, only if the change results in an increase of the margin vector components:

$$\mathbf{b}(k+1) = \mathbf{b}(k) + \rho\left(\mathbf{e}(k) + |\mathbf{e}(k)|\right), \tag{14}$$

where $\rho > 0$ is a parameter and k denotes the iteration index. It was proven [7], that for $0 < \rho < 1$ the above algorithm is convergent for any matrix $\mathbf{D}_e$.

Another problem is the initialization of the learning and the determination of the fuzzy if-then rules number. In the subsequent sections we proposed the solution of these problems based on the robust clustering.

4   Robust Fuzzy c-Means Algorithm

Let us consider a dataset containing N feature vectors $\mathbf{X} = \{\mathbf{x}_1, \mathbf{x}_2, ..., \mathbf{x}_N\}$, from t-dimensional Euclidean space. Clustering is based on a partition of the feature vectors into c classes represented by prototypes $\mathbf{v}_i \in \mathbf{V} \subset \mathbb{R}^t \; \forall i = 1, 2, ..., c$. The certainty of the assignment of the n-th sample in the i-th class is measured by grade of membership u_{in}. The elements of the $(c \times N)$ - dimensional partition matrix $\mathbf{U} = [u_{in}]$ are defined as:

$$u_{in} \in [0, 1], \quad \mathop{\forall}_{1 \leq n \leq N} \sum_{i=1}^{c} u_{in} = 1, \quad \mathop{\forall}_{1 \leq i \leq c} 0 < \sum_{n=1}^{N} u_{in} < N. \tag{15}$$

The RFCM method minimizes the criterion function given by:

$$J_r(\mathbf{U}, \mathbf{V}) = \sum_{i=1}^{c} \sum_{n=1}^{N} (u_{in})^r \rho\left(d_{in}/\gamma\right), \tag{16}$$

where $r \in [1, \infty)$ is the weighted exponent, γ is a scaling constant, $d_{in} = \|\mathbf{x}_n - \mathbf{v}_i\|_2$ is an Euclidean distance between prototype and data element, and $\rho(x)$ is a Huber function:

$$\rho(x) = \begin{cases} 0.5x^2, & \text{if } |x| \leq 1, \\ |x| - 0.5, & \text{if } |x| > 1, \end{cases} \tag{17}$$

which is used to reduce the influence of outliers.

The iterative scheme of clustering algorithm, leading to either a local minimum or a saddle point of the objective function (15), involves series of commutative modifications of the partition matrix as well as prototypes. If we fix values of the parameters r and c then using technique of the Lagrange multipliers we can get the following updating equations [6]:

$$\mathop{\forall}_{1 \leq i \leq c} \mathop{\forall}_{1 \leq n \leq N} \quad u_{in} = \left[\sum_{j=1}^{c} \left(\frac{\rho(d_{in}/\gamma)}{\rho(d_{jn}/\gamma)} \right)^{\frac{1}{r-1}} \right]^{-1}, \tag{18}$$

and

$$\underset{1 \leq i \leq c}{\forall} \quad \mathbf{v}_i = \frac{\sum\limits_{n=1}^{N} (u_{in})^r \, w \, (d_{in}/\gamma) \, x_{in}}{\sum\limits_{n=1}^{N} (u_{in})^r \, w \, (d_{in}/\gamma)}, \qquad (19)$$

where $w(x) = \rho'(x)/x$ are Huber weights. Using Huber function we can reduce the influence of data points that are in a long distance from cluster centers thereby making the algorithm robust to outliers. The scaling constant can be estimated using the median absolute deviation about the median (MAD) [6]. Applying $\gamma = 3 \cdot \text{MAD}$ we get as a result the quadratic influence on objective function as well as partition matrix of data points with a distance $d_{in} \leq 3 \cdot \text{MAD}$ and linear otherwise. The schema of RFCM algorithm is very similar to well known fuzzy c-means method [2] (FCM), with the exception for the calculation of scaling constant and the application of Huber function.

5   Determination the Number of Rules

One of the main disadvantages of ANNBFIS learning is the necessity of determining the optimal number of fuzzy rules prior to learning execution. Nevertheless, the number of rules equals to number of clusters identified with the initialization partition procedure. To find the unknown number of clusters and ipso facto the number of rules, we can apply the idea of robust compactness and separation [1]. We can start the clustering algorithm with the maximal initial number of clusters c_{ini} and merge the clusters for which the separation measure between cluster centers is small in relation to their compactness (radius). The robust compactness of a cluster $\mathbf{v}_i$ is the ratio of the fuzzy robust variation σ_i and the fuzzy cluster cardinality n_i:

$$\pi_i = \frac{\sigma_i}{n_i} = \frac{\sum\limits_{n=1}^{N} (u_{in})^r \, \rho \, (d_{in}/\gamma)}{\sum\limits_{n=1}^{N} (u_{in})^r}. \qquad (20)$$

To calculate the separation measure between two clusters $\mathbf{v}_p$ and $\mathbf{v}_q$ the Euclidean distance can be used:

$$\ell_{pq} = \|\mathbf{v}_p - \mathbf{v}_q\|_2 \, .$$

The merging criterion is given by the robust merge ratio:

$$\varpi_{pq} = \frac{\rho \, (\ell_{pq}/\gamma)}{\pi_p} \, .$$

The merge ratio is calculated for every cluster after each partition iteration. The clusters $\mathbf{v}_p$ and $\mathbf{v}_q$ are merged if the $\varpi_{pq} \leq \alpha$, where α is constant. If we define $\alpha < 1$ then we join two clusters for which the distance $\rho \, (\ell_{pq}/\gamma)$ is less than cluster compactness. In [1] the range $\alpha \in [0.1, 0.3]$ was provided. After decision to merge the clusters, a new cluster center is placed closer to the one with the larger fuzzy cardinality:

$$\mathbf{v}_n = \frac{n_p}{n_p + n_q}\mathbf{v}_p + \frac{n_q}{n_p + n_q}\mathbf{v}_q,$$

where $\mathbf{v}_n$ is the center of the new cluster. The membership values for new cluster structure are calculated during the subsequent iteration of clustering algorithm.

6 Learning Algorithm

To improve the learning quality of ANNBFIS system we integrated the least squares method as well as ε-insensitive learning with deterministic annealing procedure. As a result, the parameters of fuzzy sets from premises and consequents of fuzzy rules are evaluated separately. Premise parameters $c_j^{(i)}, s_j^{(i)}$ as well as triangle base widths $w^{(i)}$ of fuzzy sets in consequents were estimated by means of deterministic annealing method. Parameters of linear equations from consequents $\mathbf{p}^{(i)}$ were calculated using the least squares algorithm or the ε-insensitive learning. To initialize the learning procedure we applied the result of robust clustering using the following formulas:

$$c_j^{(i)} = \frac{\sum\limits_{n=1}^{N} (u_{in})^r x_{0j}(n)}{\sum\limits_{n=1}^{N} (u_{in})^r}, \quad s_j^{(i)} = \sqrt{\frac{\sum\limits_{n=1}^{N} (u_{in})^r \left(x_{0j}(n) - c_j^{(i)}\right)^2}{\sum\limits_{n=1}^{N} (u_{in})^r}}. \tag{21}$$

Since the algorithm might lead to a local minimum of the objective function (15), the calculations were repeated for various random realizations of initial partition matrix. To evaluate the clusters quality we applied the Xie-Beni validity index from [10]. For comparison we used fuzzy c-means algorithm as an initialization procedure as well. Consequently, we got four learning methods: DALS (deterministic annealing with least squares method) and εDA (deterministic annealing with ε-insensitive learning), both initialized with RFCM or FCM procedure.

7 Numerical Experiments

To validate the introduced method of ANNBFIS learning we performed a numerical experiment using benchmark database concerning the problem of sunspots prediction [9]. The dataset consists of 280 samples $x(n)$ of sunspots activity measured within a one-year period from 1700 to 1979 A.D. The goal is the prediction of a number of sunspots (output value) $y(n) = x(n)$ using past values combined in the embedded input vector $[x(n-1) \, x(n-2) \ldots x(k-12)]^T$. We divided the samples into the training set, containing the first 100 input-output data, and the testing set including the remaining 168 pairs. As a result of learning we chose the set of fuzzy rules parameters leading to the best generalization ability (the lowest mean square error MSE for testing data). The initial clustering procedures were stopped if the maximum number of 500 iterations was achieved or when in sequential iterations the change of the criterion function was less than 10^{-5}. We set the value of the weighted exponent to $r = 2$. We repeated the partition process 25 times for different random initialization of the partition matrix. Results characterized by the minimal value of Xie-Beni validity index were chosen for initialization of the ANNBFIS learning. There is no method which could provide values for parameters of ε-insensitive learning leading to the best prediction quality. According to [7] we set: $\mathbf{b}^{[1]} = 10^{-6}$, $\rho = 0.98$, $\kappa = 10^{-4}$ and $k_{\varepsilon \max} = 1000$. To determine τ and ε we performed the ε-insensitive learning separately, changing the parameter values from 0.01 to 0.1 with step 0.01. The results characterized by the highest generalization ability of prediction were used in the ANNBFIS learning. For the deterministic annealing based methods we applied the following settings: $\eta_{\text{ini}} = 0.01$, $q = 0.95$, $k_{\max} = 10$ and $T_{\min} = 10^{-5}T_{\max}$. As there was no method for automatic estimation of initial temperature $T_{\max}$, we were changing its value from 10^{+5} to 10^{-5} with the common ratio equal to 0.1.

To determine the number of if-then rules we applied the RFCM algorithm with clusters merging procedure. We were changing the maximal number of clusters (rules) c_{ini} from 6 to 10 and α from 0.1 to 0.3 with the step 0.5. The obtained results (Table 1) provided the range for the number of if-then rules from 2 to 7. The results indicated also the optimal number of fuzzy rules $I = 2$ or $I = 3$. Nevertheless, the verification based on ANNBFIS learning was needed. To predict the number of sunspots we performed learning of ANNBFIS using DALS and εDA algorithm initialized with RFCM or FCM clustering. The prediction results are shown in Table 2 and 3 respectively. For comparison we provided the results of original ANNBFIS learning based on a combination of the steepest descent optimization and least squares method (SDLS) as well. The best results are marked.

Clearly, in all the examples, the ε-insensitive learning based method demonstrates consistent improvement of the prediction quality. However, it must be noted, that the computational complexity of the εLSSLI is approximately 3 times

Table 1. Determination of the rules number with cluster merging procedure

α	c_{ini}				
	6	7	8	9	10
0.10	3	3	3	7	7
0.15	3	3	3	6	7
0.20	3	3	3	3	4
0.25	2	3	3	3	3
0.30	2	3	2	2	3

Table 2. RMSE of the prediction with FCM initialization

			εDA		DALS		SDLS
I	ε	τ	T_{max}	MSE $\cdot 10^3$	T_{max}	MSE $\cdot 10^3$	MSE $\cdot 10^3$
2	0.09	0.10	10^{-1}	**6.6008**	10^{-1}	**7.2967**	**7.7950**
3	0.09	0.05	$10^{\ 0}$	6.8348	10^{-3}	7.4432	8.1231
4	0.03	0.06	10^{+3}	7.0882	10^{-3}	9.9788	12.226
5	0.80	0.01	10^{+5}	6.6769	10^{+5}	11.834	12.565
6	0.01	0.10	10^{+4}	6.6023	10^{-2}	13.569	20.393
7	0.02	0.03	10^{+3}	6.2954	10^{+5}	15.045	33.388

Table 3. RMSE of the prediction with RFCM initialization

			εDA		DALS		SDLS
I	ε	τ	T_{max}	MSE $\cdot 10^3$	T_{max}	MSE $\cdot 10^3$	MSE $\cdot 10^3$
2	0.09	0.10	10^{-2}	**6.5256**	10^{-2}	**7.5768**	**7.7477**
3	0.10	0.06	10^{+1}	6.8411	10^{+5}	8.2737	9.8536
4	0.05	0.03	10^{+3}	7.0008	10^{-5}	9.6698	9.6410
5	0.05	0.02	10^{+4}	6.8182	10^{+5}	9.9530	13.409
6	0.03	0.01	10^{+2}	6.8306	10^{+3}	14.281	20.287
7	0.03	0.01	10^{+2}	6.4740	10^{+3}	20.149	33.287

greater comparing to the LS method. Nevertheless, we can not give the clear answer if the application of RFCM clustering improves the learning results.

The true advantages of RFCM clustering can be shown during testing the robustness of outliers of the proposed ANNBFIS learning. For this purpose we added one outlier to the training set: the minimal output sample $y(1)$ equal to 0 was set to double value of maximal output sample $2y(67)$ equal to 1.6150. The learning results are presented in Table 4.

Table 4. Results of test for robustness of outliers

	FCM		RFCM	
	I	MSE $\cdot 10^3$	I	MSE $\cdot 10^3$
εDA	3	7.6643	3	7.4995
DALS	2	15.847	3	13.618
SDLS	2	19.909	3	9.8536

The explicit increase of the prediction quality with RFCM algorithm can be noticed. The εDA approach improves the generalization ability for the sunspots prediction problem in the presence of outliers in the training set over the reference algorithms as well.

8   Conclusions

In this paper the learning algorithms of ANNBFIS neuro-fuzzy system based on deterministic annealing, ε-insensitive learning by solving a system of linear inequalities and robust fuzzy c-means clustering were presented. The unknown number of fuzzy if-then rules was determined using the robust clusters merging procedure. The experiments show the usefulness of the presented solutions and indicates the learning procedure which integrates the deterministic annealing algorithm with ε-insensitive learning as the most efficient. The obtained results show also the consistent improvement of outliers robustness when robust fuzzy c-means initialization is used.

Acknowledgments. This work was supported in part by the Ministry of Sciences and Higher Education resources in 2007-2010 under Research Project NN518 335935.

References

1. Aranda, T.G.: Robust fuzzy n-means clustering. A Research Paper Presented to the Faculty of the Division of Mathematical Sciences Midwestern State University, pp. 1–25 (2000)
2. Bezdek, J.C.: Pattern Recognition with Fuzzy Objective Function Algorithms. Plenum Press, New York (1982)
3. Czabanski, R.: Neuro-fuzzy modeling based on a deterministic annealing approach. Int. J. Appl. Math. Comput. Sci. 15(4), 561–575 (2005)
4. Czabanski, R.: Extraction of fuzzy rules using deterministic annealing integrated with ε-insensitive learning. Int. J. Appl. Math. Comput. Sci. 16(3), 357–372 (2006)
5. Czogala, E., Leski, J.: Fuzzy and Neuro-Fuzzy Intelligent Systems. Physica-Verlag, Springer-Verlag Comp., Heidelberg (2000)
6. Kersten, P.R., Lee, R.R.-Y., Verdi, J.S., Yankovich, S.P., Carvalho, R.M.: Segmentation sar images using fuzzy clustering. In: Proc. 19th Inter. Conf. of North American Fuzzy Information Processing Society NAFIPS, pp. 105–108 (2000)
7. Leski, J.: Neuro-fuzzy system with learning tolerant to imprecision. Fuzzy Sets and Systems 138(2), 427–439 (2003)
8. Rose, K.: Deterministic annealing for clustering, compression, classification, regression and related optimization problems. Proc. IEEE 86(11), 2210–2239 (1998)
9. Weigend, A.S., Huberman, B.A., Rumelhart, D.E.: Predicting the future: a connectionist approach. Int. J. Neural Syst. 1(3), 193–209 (1990)
10. Xie, X.L., Beni, G.: A validity measure for fuzzy clustering. IEEE Trans. Pattern Analysis and Machine Intelligence 13(8), 841–847 (1991)
11. Zadeh, L.A.: Fuzzy sets. Information and Control 8(4), 338–353 (1965)

An Unsupervised Learning and Statistical Approach for Vietnamese Word Recognition and Segmentation

Hieu Le Trung[1], Vu Le Anh[2], and Kien Le Trung[3]

[1] St. Petersburg State University, Saint Petersburg, Russia
[2] Hoa Sen University, 8. Nguyen Van Trang, Q1, Ho Chi Minh City, Vietnam
[3] Institue of Mathematics, Arndt University, Germany

Abstract. There are two main topics in this paper: (i) Vietnamese words are recognized and sentences are segmented into words by using probabilistic models; (ii) the optimum probabilistic model is constructed by an unsupervised learning processing. For each probabilistic model, new words are recognized and their syllables are linked together. The syllable-linking process improves the accuracy of statistical functions which improves contrarily the new words recognition. Hence, the probabilistic model will converge to the optimum one.

Our experimented corpus is generated from about 250.000 online news articles, which consist of about 19.000.000 sentences. The accuracy of the segmented algorithm is over 90%. Our Vietnamese word and phrase dictionary contains more than 150.000 elements.

1 Introduction

Word recognition and segmentation of a given sentence into words are important steps in many applications of natural language processing such as text mining, text searching and document classification. These problems are not difficult in Occidental languages since words are determined by space characters. In some Oriental languages such as Vietnamese, Chinese and Japanese, they become much more difficult. Word, *a meaningful linguistic unit*, can be one syllable, or a combination of two or more syllables. Vietnamese word recognition and segmentation problems *can not be solved completely* due to the following two reasons:

There does not exist an algorithm that segments a given Vietnamese sentence into words exactly according to its meaning if the sentence is considered isolated.
Let us consider the following sentence: *"Cái bàn là của tôi"*. This sentence has two quite different meanings depending on the different word segmentations: (i) *"It is my iron"* for the word segmentation *"Cái | bàn là | của | tôi"*, and (ii) *"The table is mine"* for the word segmentation *"Cái | bàn | là | của | tôi"*. Clearly, no word segmentation algorithm works on this input sentence. The explanation is that each syllable can be a component of different words. Moreover, a Vietnamese sentence is written as a sequence of syllables, not a sequence of words, and its meaning can not be determined without the context.

N.T. Nguyen, M.T. Le, and J. Świątek (Eds.): ACIIDS 2010, Part II, LNAI 5991, pp. 195–204, 2010.

There is no official definition of word and complete dictionary in Vietnamese.
Nowadays, Vietnamese linguists still discuss and do not agree with each other
about *"What is the word definition in Vietnamese language?"* [4,2]. For examples, *"máy tính xách tay" (laptop), "máy bay lên thẳng" (helicopter), "xe gắn máy"*
(motorcycle), etc. have no final official definition that they are single words or
combinations of two words. Moreover, most of new words in Vietnamese online
documents, which are from foreign languages (*"avatar", "sms", ...*) or commonly
used by teenagers (*"mún", "xì tin", "chảnh",...*) are not in any Vietnamese dictionaries. According to [5], the biggest Vietnamese dictionaries contain less than
33.000 words while *The Second Edition of the Oxford English Dictionary* contains
over 250.000 words. Furthermore, as we know, there is no complete Vietnamese
dictionary of proper names and names of places and organizations.

Our work intends to address and solve two problems: (i) Recognizing words
under probability viewpoint; (ii) Constructing the optimum probabilistic model
of huge corpus using an unsupervised learning process.

Our approach for the first problem is as follows. We obverse the corpus, which
is a huge set of syllable sequences, and decide *which pair of syllables, (α, β),
is probably a word or an infix of word.* (α, β) is chosen if it is *confident* and
supported. The support S is defined as the number of occurrence of event, E, in
which $\alpha\beta$ is infix of some sentence. $\mathcal{H}$ is the hypothesis in which (α, β) is neither
a word nor an infix of word. We use a probabilistic model with assumption $\mathcal{H}$,
and estimate S by S'. $\mathcal{H}$ is probably wrong if there is a big difference between
S' and S. The confidence of (α, β) (*about $\mathcal{H}$ is wrong*) is proportional to the
popularity of event E (S) and the ratio, $\frac{S}{S'}$. Obviously, if (α, β) is supported
and confident enough, (α, β) is probably a word or an infix of word.

The optimum probabilistic model is constructed by an unsupervised learning
processing. The initial corpus is a huge set of sentences generated from online
documents on the internet. For each learning iteration, we shall do the following
steps: (i) Finding only the *local maximum confident* sequences of syllables in sentences; (ii) Linking the local maximum confident sequences of syllables together
to be new syllables; (iii) Recomputing all probability values of new corpus and
return to step (i). Basing on confident functions, we build a recognition function
in which each pair of syllables can be determined whether they are infix of some
words or undecidable. Local maximum confident sequences of syllables, which
are determined by the recognition function and by comparing neighbor pairs of
syllables, are probably infix of some words. Let us see following example:

Sentence S is *"Công việc của chúng tôi đã thành công".* The considered pairs
of syllables of the first iteration are *"công việc", "việc của", "của chúng", "chúng
tôi", "tôi đã", "đã thành", "thành công".* Recognition function shows that *"chúng
tôi"* and *"thành công"* are infix of some words. Moreover, the confident of their
neighbors is quite lower than theirs. S is rewritten as *"Công việc của chúng_ tôi
đã thành_ công"* with two new syllables *"chúng_ tôi"* and *"thành_ công".* The
considered pairs of syllables of the second iteration are *"công việc", "việc của",
"của chúng_ tôi", "chúng_ tôi đã", "đã thành_ công".* Suppose *"công việc"* is the

local maximum confident pair. S is rewritten as *"Công_ việc của chúng_ tôi đã thành_ công"*.

By replacing local maximum confident sequence of syllables by the new sylla-bles, the confusion of syllables and words is reduced and the statistical functions are more precise. Contrarily, precise probability values will improve finding the local maximum confident sequences. Therefore, the quality of the probabilistic model is improved by each iteration.

Our contribution. We introduce and study a new algorithm for recognizing the new Vietnamese words in huge corpus based on statistics. The unsupervised learning process for building an optimum probabilistic model and corpus is also introduced and discussed. With the experimental corpus generated from over 250.034 online news, a new Vietnamese dictionary and the optimum corpus will be introduced and used in public.

Section 1 is the introduction. Section 2 is the related works. The probabilistic model, confident functions and basic concepts are studied in section 3. The learning process is discussed in section 4. Section 5 is the experiments. Section 6 concludes the paper.

2   Related Works

As we known, we are the first group, who study the Vietnamese word recognition based on statistical methods. Vietnamese word segmentation has be studied by several groups [3,7,8,6,9]. There are two main approach: *manual corpus based approach* [3,7,8] and *unsupervised statistical approach* [6,9].

The solutions of the first group are built around the theory of supervised learning machines. Dinh [3] is based on the WFST model and Neural Network. Nguyen [8] is based on CRF (conditional random fields) and SVM (support vector machines). Le [7] is based on hybrid algorithms with *maximal-matching method* concept. Their learning machines learn from manual dictionaries or man-ual annotated corpora, which are limited by human resource. [3] used 34.000-word dictionary, [8] used about 1.400 annotated news, [7] ignored the new words. They claimed the accuracy of their methods are over 90% but only for very small manual annotated corpora.

In second approach, Ha [6] applied the maximum probability of tri-gram in a given chunk of syllables over huge corpus. Thanh [9] used Mutual Information (MI) formulas for n-gram combined with Genetic Algorithm. Their works and ours have three big different points: (i) They did not have learning process to improve the accuracy of statistical information (ii) In our work, the relationship of syllables in same word are generalized by confidence concept using different probability formulas, not only MI formula [6] or maximum probability of n-gram [9] (iii) Their corpus is quite smaller than us. Therefore, the accuracies of their algorithms are 50% [6] and 80% [9], which are lower than us (90%). Moreover, our approach can apply for proper names, names of places or organizations and phrases recognition.

3   Probabilistic Model

3.1   Basic Concepts

Syllable is an original syllables (such as *"của"*, *"đã"*) or a linking syllables (such as *"công_ việc"*, *"chúng_ tôi"*, *"thành_ công"*). Given a syllable β, we denote $\alpha \in Pre(\beta)$ $(\alpha \in Suf(\beta))$ if α is a prefix (suffix) of β. For example, *"công"* $\in Pre(công\_ việc)$, *"việc"* $\in Suf(công\_ việc)$, and *"của"* $\in Pre(của) \cap Suf(của)$. *Sentence* is a sequence of syllables. *"Công_ việc của chúng_ tôi đã thành_ công"* sentence is denoted by $S = \alpha_1\alpha_2\ldots\alpha_5$ in which $\alpha_1=$*"công_ việc"*, $\alpha_2=$*"của"*, ..., and $\alpha_5=$*"thành_ công"*. $\beta_1\beta_2\ldots\beta_l$ is an infix of sentence $S = \alpha_1\alpha_2\ldots\alpha_k$ $(1 \leq l \leq k)$ if $\exists 1 \leq i \leq k - l + 1 : \beta_j = \alpha_{i+j-1} \forall j = 1, \ldots, l$.

A probabilistic model $\mathcal{P}$ is defined as a triple $(\mathcal{C}, \Sigma_C, F_C)$.

Corpus, $\mathcal{C}=\{s_1, s_2, \ldots, s_n\}$, is a finite set of sentences. $\pm_C$ is the set of syllables, which are infix of some sentence s_i of $\mathcal{C}$. $\mathcal{F}_C$ is *the set of statistical functions*. A probabilistic function $f_C \in \mathcal{F}_C$ is a map from $\pm_C^*$ to $\mathcal{R}$. It can be a constant $(\emptyset \mapsto \mathcal{R})$, a function of syllable $(\pm_C \mapsto \mathcal{R})$ or a function of pair of syllables $(\Sigma_C^2 \mapsto \mathcal{R})$ and so on. Here are basic statistical functions used in our work:

Suppose $\alpha, \beta \in \pm_C$. $N(\alpha)$ is denoted for the number of occurrence of α in $\mathcal{C}$. We define: $N_p(\alpha) = \sum_{\beta:\alpha \in Pre(\beta)} N(\beta)$; $N_s(\alpha) = \sum_{\beta:\alpha \in Suf(\beta)} N(\beta)$; $N_1 = \sum_{\alpha \in \Sigma_C} N(\alpha)$. The probabilities of the events that α occurs in $\mathcal{C}$ as independent syllable or prefix, suffix of some syllable are estimated respectively as follows: $P(\alpha) = \frac{N(\alpha)}{N_1}$; $P_p(\alpha) = \frac{N_p(\alpha)}{N_1}$ and $P_s(\alpha) = \frac{N_s(\alpha)}{N_1}$. Similarly, $N(\alpha\beta)$ is denoted for the number of occurrence of $\alpha\beta$ in some sentence of $\mathcal{C}$, $N_2 = \sum_{\alpha,\beta \in \Sigma_C} N(\alpha\beta)$. The probability of event $\alpha\beta$ occurs in $\mathcal{C}$ is estimated by $P(\alpha\beta) = \frac{N(\alpha\beta)}{N_2}$.

3.2   Confident Functions and Word Recognition

The *optimum corpus* is the one in which each sentence is segmented into sequence of words exactly according to its meaning. Each syllable in optimum sentence is a word. We have shown in the introduction section that there is no algorithm to construct the optimum if each sentence is considered isolated. However, words are recognized with the help of confident functions.

Confident functions are statistical functions of pair of syllables, which measure how probably the given ordered pair of syllables is an infix of word. Suppose $\mathcal{H}$ is the hypothesis that $\alpha\beta$ is not infix of any word. Each confident function $f_{C,M}(\alpha, \beta)$ is based on a probabilistic model, $\mathcal{M}$, in which: (i) $\mathcal{H}$ is assumed to be true (ii) the probability of event, E, $\alpha\beta$ occurs in $\mathcal{C}$, is estimated as $P'(\alpha\beta)$. $f_{C,M}(\alpha, \beta)$ is proportional with the popularity of E, and the ratio $\frac{P(\alpha\beta)}{P'(\alpha\beta)}$.

Definition 1. *Suppose $c \in \mathcal{R}$ is a constant. The confident function $f_{C,M}(\alpha, \beta)$: $\pm_C^2 \mapsto \mathcal{R}$ over probabilistic model , $\mathcal{M}$, and corpus, $\mathcal{C}$, is defined as:*

$$f_{C,M}(\alpha, \beta) = c * \frac{P(\alpha\beta)^2}{P'(\alpha\beta)}$$

We choose randomly two neighbor syllables $x_1 x_2$ in some sentence, and A is the event $x_1 = \alpha$ is suffix of some words in the optimum sentence; B is the event in which $x_2 = \beta$ is prefix of some word in the optimum sentence. $\mathcal{H}$ implies that for each occurrence $\alpha\beta$ in $\mathcal{C}$: (i) α must be suffix of some words and (ii) β must be prefix of some words. Hence, $P'(\alpha\beta) = P(A \cap B) \equiv P(AB)$. Here are different models for $P(AB)$ estimation:

Model 1: Assumption: A, B are independent events, $P(A) = c_1 P(\alpha)$, $P(B) = c_2 P(\beta)$ (c_1, c_2 const.), and $P(AB) = c_1 c_2 P(\alpha) P(\beta)$. Hence for $c = \frac{1}{c_1 c_2}$:

$$f_{\mathcal{C},1}(\alpha, \beta) = \frac{P(\alpha\beta)^2}{P(\alpha)P(\beta)}$$

In reality, $P(\alpha)P(\beta)$ is much more smaller than $P(AB)$ since in natural language α, β never stand by each other purely random. We suggest $P(AB) = P(\alpha)^{\varphi_s(\alpha)} P(\beta)^{\varphi_p(\beta)}$. The experiments shows that $\varphi_s(\alpha) = 1 - \frac{H(X_{s,\alpha})}{\log N_s(\alpha)}$ ($\varphi_p(\beta) = 1 - \frac{H(Y_{p,\beta})}{\log N_p(\beta)}$) is the good estimation in which $X_{s,\alpha}$ ($Y_{p,\beta}$) is a variable presenting syllables appear before (after) the syllable α (β), and $H(\cdot)$ is an *entropy operator*. Because of the limit of this paper, we will study the construction and the properties of φ_s, φ_p functions in another work.

Model 2: Assumption: $P(AB) = c_1 P(\alpha)^{\varphi_s(\alpha)} P(\beta)^{\varphi_p(\beta)}$ (c_1 const.). Let's $c = \frac{1}{c_1}$:

$$f_{\mathcal{C},2}(\alpha, \beta) = \frac{P(\alpha\beta)^2}{P(\alpha)^{\varphi_s(\alpha)} P(\beta)^{\varphi_p(\beta)}}$$

Model 3: Assumption: A, B are independent events. Obviously, $P(A) \simeq P_s(\alpha)$ and $P(B) \simeq P_p(\beta)$. $P(AB)$ is estimated by $P_s(\alpha)P_p(\beta)$. Here, we take $c = 1$:

$$f_{\mathcal{C},3}(\alpha, \beta) = \frac{P(\alpha\beta)^2}{P_s(\alpha)P_p(\beta)}$$

Connector words (such as *"và"* (and), *"thì"* (then), *"là"* (is), *"của"* (of), etc.) are important factors in Vietnamese. The occurrences of these words are very high comparing to normal ones. There is a famous assumption [1] about Vietnamese word recognition which says that $\alpha\beta$ is a word in given sentence if and only if we can not place any connector word between them that not change the meaning of the sentence. Suppose W is the set of connector words. $N_W(\alpha\beta)$ is denoted for the number of occurrence of event, E, $\delta\gamma\eta$ is an infix of some sentence of $\mathcal{C}$ in which $\gamma \in W$, $\alpha \in Suf(\delta)$ and $\beta \in Pre(\eta)$. $N_3 = \sum_{\alpha,\beta \in \Sigma_C} N_W(\alpha\beta)$. The probability of event E is estimated by $P_W(\alpha\beta) = \frac{N_W(\alpha\beta)}{N_3}$. The number of occurrence of event AB is proportional to the number of occurrent E.

Model 4: Assumption: $P(AB) = c_1 P_W(\alpha\beta)$. Hence for $c = \frac{1}{c_1}$:

$$f_{\mathcal{C},4}(\alpha, \beta) = \frac{P(\alpha\beta)^2}{P_W(\alpha\beta)}.$$

The extent version of the confident function $f_{C,M}$, $f_C^* : \pm c^+ \mapsto \mathcal{R}$, is defined as follows: $f_C^*(w) = P(w)$ if $w \in \pm c$. Otherwise, $f_C^*(w) = \frac{P^2(w)}{P^*(w)}$ in which $P^*(w) = Max_{w=uv} P'(u,v)$ where u, v are prefix, suffix of w and $w = uv$. $P'(u,v)$ is the estimated probability of the event that w is not word and is segmented into $u|v$ using M and necessary statistical values.

$\mathcal{P} = (\mathcal{C}, \Sigma_C, F_C)$ is a probabilistic model. $m_{sup}, M_{sup}, m_{con}, M_{con} \in \mathcal{F}_C$ are constant functions, in which $0 < m_{sup} \leq M_{sup}$ and $0 < m_{con} \leq M_{con}$. $f_C \in \mathcal{F}_C$ is a confident function. Word recognition function is defined as follows:

Definition 2. $f_R : \Sigma_C^2 \mapsto \{-1, 0, 1\}$ *is the word recognition function of* f_C *over* $\mathcal{P}$ *with parameters* $(m_{con}, M_{con}, m_{sup}, M_{sup})$ *in which:*

$$f_R(\alpha, \beta) = \begin{cases} 1 & \text{if } (f_C(\alpha, \beta) \geq M_{con}) \wedge (N(\alpha\beta) \geq M_{sup}); \\ -1 & \text{if } (f_C(\alpha, \beta) < m_{con}) \vee (N(\alpha\beta) < m_{sup}); \\ -0 & \text{otherwise.} \end{cases}$$

If return value of recognition word function is 1, the input pair of syllables are supported and confident and it is probably infix of some word. If return value is -1, the input pair of syllables belongs two different words. We have no decision in the case the return value is 0. Obviously, if $m_{sup} = M_{sup}$ and $m_{con} = M_{con}$, the return values can not be 0 and there does not exist undecidable case. In the case, we have different confident functions and different recognition word functions. We can combine them by some fuzzy rules to be only one *universal word recognition function*, f_R^*.

4   Learning Process and Main Results

4.1   Learning Rules and Learning Process

Suppose $\mathcal{P} = (\mathcal{C}, \Sigma_C, F_C)$ is a probabilistic model; $f_C, f_R^* \in \mathcal{F}_C$ is a confident function and the universal word recognition function respectively; $D_{con} \in \mathcal{F}_C$ is a positive constant; $s = \alpha_1\alpha_2 \ldots \alpha_k \in \mathcal{C}$ is a sentence in corpus, and $w = \alpha_l\alpha_{l+1} \ldots \alpha_{l+m}$ is an infix of s $(1 \leq l < k, 0 < m \leq k - l)$.

Definition 3. w *is a* local maximum confident sequence *(LMC for short) of* s *over* $\mathcal{P}$ *with* f_C, f_R^* *and* D_{con}, *if it satisfies following conditions:*

(i) $\forall i = l, \ldots, m - 1 : f_R^*(\alpha_i, \alpha_{i+1}) = 1;$
(ii) if $l > 1$: $f_R^*(\alpha_{l-1}, \alpha_l) = -1$ *or*
$f_R^*(\alpha_{l-1}, \alpha_l) = 0 \wedge f_C(\alpha_l, \alpha_{l+1}) > f_C(\alpha_{l-1}, \alpha_l) + D_{con}$
(iii) if $l + m < k$: $f_R^*(\alpha_{l+m}, \alpha_{l+m+1}) = -1$ *or*
$f_R^*(\alpha_{l+m}, \alpha_{l+m+1}) = 0 \wedge f_C(\alpha_{l+m-1}, \alpha_{l+m}) > f_C(\alpha_{l+m}, \alpha_{l+m+1}) + D_{con}$

Condition (i) guarantees that all pairs of neighbor syllables of w are infixes of some words. In condition (ii) and (iii), the neighbors of w, $(\alpha_{l-1}, \alpha_l)(l > 1)$ and $(\alpha_{l+m}, \alpha_{l+m+1})$ $(l + m < k)$, are considered. They guarantee that the neighbors do not effect to w under confident viewpoint. Therefore, w is a sequence of words.

Suppose $w = \beta_1\beta_2\ldots\beta_l$ occurs $T(w)$ times as a LMC in some sentence. Here are learning rules sorted by the priority.

Rule 0. If $Link(w) \in \pm_C$: Replace w by $Link(w)$. $Link(w) \in \pm_C$ implies that in the past we have learned that w is infix of some word.

Rule 1. If $m = 1$: Replace w by $Link(w)$. w is a two syllables word.

Rule 2. Sorting the neighbor pairs of w by the confident value. If (β_i, β_{i+1}) is the first one and the difference of the confident values of first- and second pairs more than D_{con}: Replace $\beta_i\beta_{i+1}$ by $Link(\beta_i\beta_{i+1})$. The difference of the confident values guarantees $\beta_i\beta_{i+1}$ does not belong two different words probably.

Rule 3. If $m = 3, 4$ and $T_w \geq M_{sup}$ and $f_C^*(w) \geq M_{con}$: Replace ω by $Link(\omega)$.

Rule 0 is always considered as it is the nature of learning. Rule 1 and 2 find pairs of syllables which are infix of some word. The priority of Rule 1 is higher than Rule 2 since Rule 1 helps us founding out the "two-syllables" words. Rule 3 have the lowest priority since it needs many statistical values and computing resources. The case in which $m > 4$ are ignored since there does not exist 5-syllable word in Vietnamese. Learning process for Rule 0, 1 and 2 is as follows:

```
1.  Learning-Process-1                       9.   until No linking pair is found
2.  repeat                                    10.  for each sentence s
3.    repeat                                  11.    for each w is LMC of s
4.      for each sentence s                   12.      if infix αβ of w satisfies Rule2 then
5.        for each w is LMC of s              13.        s is rewritten by replacing αβ
6.          if w satisfies Rule 0 or 1 then                with Link(αβ)
7.            s is rewritten by replacing      14.      Update, create new necessary
              w with Link(w)                              statistical values
8.            Update, create new necessary     15.  until No linking pair of Rule 2 is found
              statistical values
```

Learning Strategy. The system of parameters (*SP for short*), $(m_{con}, M_{con}, m_{sup}, \ldots)$, decides the quality of learning process. For example, as higher as M_{con}, the number of new words is smaller but the quality of learning is higher. Our proposed strategy of learning process is *"Slowly but Surely"*. Suppose $(m_{con}^*, M_{con}^*, m_{sup}^*, \ldots)$ is the desired SP for the optimum corpus. We increase the quality of learning process by generating sequences of SPs: $(m_{1,con}, M_{1,con}, m_{1,sup}, \ldots)$, $\ldots$, $(m_{n,con}, M_{n,con}, m_{n,sup}, \ldots) = (m_{n,con}^*, M_{n,con}^*, m_{n,sup}^*, \ldots)$ in which the quality of learning of system i-th is higher than system $(i + 1)$-th: $m_{i+1,con} \leq m_{i,con}$, $M_{i+1,con} \leq M_{i,con}$, $\ldots$ Here is the proposed learning process for Rule 0, 1, 2 and 3 and desired SP $(m_{n,con}^*, M_{n,con}^*, m_{n,sup}^*, \ldots)$:

```
1.  Learning-Process                          6.  for each sentence s
2.    Computing necessary statistical values  7.    for each local maximum confident
        of syllable and pairs of syllables           sequence w of s
3.  for i = 1 to n do                          8.      if w satisfies Rule3 then
4.    Setting the parameters to                9.        s is rewritten by replacing w
        (m_{i,con}, M_{i,con}, m_{i,sup}, ...)            with Link(w)
5.    Learning-Process-1
```

The **Learning-Process-1** loops n-times (line 3-5) with n systems of parameters, which converge to the desired one. The way we choose these systems guarantees the quality of learning process. Rule 3 is considered (line 6-9) which guarantees all 3-4 syllables words are recognized.

4.2 Optimum Segmentation Algorithm, Dictionary

Suppose $\mathcal{P}^* = (\mathcal{C}^*, \Sigma_C^*, F_C^*)$ is the probabilistic model which is the result of the learning process, $M_{sup}^* \in \mathcal{F}_C^*$ is the minimum support of words. The dictionary of words, $\mathcal{D}$, generated from $\mathcal{P}^*$ is defined as: $\mathcal{D} = \{w \in \Sigma_C^* | N_1(w) \geq M_{sup}^*\}$. Experiments show that if desired parameters is chosen exactly, we can extract not only the dictionary of Vietnamese words but also phrases, names of organizations and so on.

Different word segmentation algorithms introduced in another works [3,7,8] can use the optimum corpus as the annotated corpus. Learning process itself is an word segmentation algorithm. The input sentence is segmented by the learning algorithm introduced in previous subsection. It uses the statistical values which are produced by the optimum probabilistic model. Experiments shows that the accuracy our algorithm is about 90%.

5   Experiments

Corpus. Our corpus is generated from 250.034 articles in the *Tuoi Tre* (The Youth) online newspaper. After applying the data normalization (fix the code font and repair spelling mistake of syllables) and sentences segmentation (using punctuation mark, comma, question mark, semicolon, etc.), the initial corpus has 18.676.877 sentences whose total length is 131.318.974 syllables.

Model. Because of the limit of this paper, we present only the experiments using Model 2 with confident function $f_{C,2}$ in which the formula of SP is ($m_{con}, M_{con}, m_{sup}, M_{sup}, D_{con}$). The results of other models, the comparison and combination of different models will be represented in the extent version of this paper.

Algorithm. Our learning process applied the strategy *"Slowly but Surely"*. There were 9 learning iterations with 9 different SPs. In 1-4 iterations, M_{con} and M_{sup} are very high since we want recognize the most common two-syllable words,

Table 1. System of parameters and number of recognized words in learning iterations. $\log(SP)$ is denoted of ($\log m_{con}, \log M_{con}, \log m_{sup}, \log M_{sup}, \log D_{con}$).

$\log(SP)$	No. Linking	1-Syl.	2-Syl.	3-Syl.	4-Syl.	(> 4)-Syl.	Sum
(-5.0,5.0,10,200,2.0)	9.331.392	11.107	1.990	0	0	0	**13.097**
(-4.5,3.0,10,100,1.5)	21.150.384	10.415	7.556	9	0	0	**17.980**
(-4.5,2.0,10,80,1.5)	27.545.021	10.019	14.761	108	0	0	**24.888**
(-4.0,1.0,10,50,1.0)	32.273.932	9.594	30.479	582	6	0	**40.661**
(-4.0,0.0,10,40,1.0)	36.547.210	9.191	42.531	2.159	80	25	**53.986**
(-3.5,-1.0,10,30,0.5)	39.023.942	8.598	51.681	6.187	587	123	**67.176**
(-3.5,-1.5,10,20,0.5)	40.246.564	8.413	66.394	9.985	1069	261	**86.122**
(-3.0,-2.0,10,20,0.0)	41.763.245	8.180	85.500	12.985	1.947	514	**109.126**
(-3.0,-3.0,10,20,0.0)	45.516.469	7.676	106.696	32.573	5.835	1.788	**154.568**

Table 2. The confident function's values of some pairs of syllables, of which *"cá"* is the first syllable. *"reg."* is the short of *recognized word.*

Syllables	St. 1	St. 2	St. 3	St. 4	Syllables	St. 1	St. 2	St. 3	St. 4
cá nhân	3.58	reg.	reg.	reg.	cá cảnh	-2.55	-2.01	-0.66	1.02
cá cược	3.60	reg.	reg.	reg.	cá biển	-2.85	-2.60	-0.8	1.1
cá độ	1.86	3.03	reg.	reg.	cá lóc	0.88	1.5	2.51	reg.
cá heo	0.51	1.31	2.01	reg.	cá bỏ	-7.86	-7.49	-6.83	-6.05

whose average occurrence is about 30% in Vietnamese sentences. Hence linking their syllables will improve effectively the quality of statistical values. In 5-7 iterations, M_{con} is decreased slowly so most of words are recognized in this time. In 8-9 iterations, all parameters are change slowly so that all proper nouns, phrases, etc are recognized. The results of the iterations are shown in Table 1. Our dictionary contains about 60.000 words, over 30.000 phrases, and 20.000 proper names, names of place, foreign words.

Accuracy. The accuracy of word recognition and segmentation algorithms is measured by choosing randomly some recognized words or segmented document and counting the mistakes. After checking many times and computing the average of mistake, the accuracy of our dictionary and segmented algorithm is 95% and 90% respectively. They are depended strongly on the confident function. Table 2 shows the values of the confident function of 8 pair of syllables, which are words except *"cá bỏ"* for 1-4 iterations. The confident values are increased by each iteration, and these words are recognized step by step except *"cá bỏ"*.

Scalar. According to the number of words and phrases, our dictionary is the biggest public Vietnamese dictionary. Table 3 shows the list of words, proper names, names of places, phrases which are in our dictionary and rarely found in another ones. However, our dictionary is incomplete. All missing words are rarely used in modern Vietnamese or in the professional language for newspapers. Our corpus is expanded easily with our web crawler. Currently, we have downloaded

Table 3. Some special phrases, names of place, proper names, and new words in our dictionary

Phrases	Place names	Place names	Proper names	New words
ách tắc giao thông	Bắc Triều Tiên	Quận Ninh Kiều	Alfred Riedl	acid béo
áp thấp nhiệt đới	Bờ Biển Ngà	Quận Đống Đa	Alex Ferguson	mobile phone
bất phân thắng bại	Bồ Đào Nha	Quận Ba Đình	Barack Obama	bản photocopy
bật đèn xanh	Ch Czech	Quận Bình Tân	Bảo Đại	bánh heroin
bất khả xâm phạm	Ch Ireland	Quận Bình Thạnh	Nguyễn Minh Triết	băng cassette
bặt vô âm tính	Trung Quốc	Quận Bình Thủy	Nguyễn Tấn Dũng	bánh pizza
cân đo đong đếm	Triều Tiên	Quận Gò Vấp	Phạm Ngọc Thạch	máy in Laser
lở mồm long móng	Nhật Bản	Quận Hai Bà Trưng	David Beckham	quả penalty
càng sớm càng tốt	Thủy Điển	Quận Hải Châu	Leonardo Da Vinci	nhạc rock
chất độc màu da cam	Ấn Độ	Quận Hoàn Kiếm	Công nương Diana	Windows Vista

about 9.000 Vietnamese online-books at the Web http://vnthuquan.net and the news of the most popular Vietnamese online newspapers. The next version of our dictionary is more complete and the optimum corpus is more precise, too.

6 Conclusion

We have proposed a Vietnamese word recognizing algorithm based on statistic. The algorithm works well on different corpora and can extract the name of persons, places or organizations. The experiments show that the complete Vietnamese dictionary can be built with this approach.

We have also studied an unsupervised learning iterations to construct the optimum probabilistic model and perfect word segmentation algorithm. There are two main factors effect to the learning: (i) Linking the local maximum confident pairs of syllables in sentences together as new syllables; (ii) probability values of the corpus. Two factors effect to each others, and both are improved by learning process. The output of the learning iterations is the optimum probabilistic model. The experiments show that by using the optimum probabilistic model generated from our corpus, the accuracy of our word segmentation is over 90%.

The Vietnamese language is not explained and described well by grammar rules. One of our research direction is: Finding the most common formulas of Vietnamese sentences based on statistic. We believe that with computer and huge corpus, we can solve many problems of Vietnamese language processing based on statistic.

References

1. Cao, X.H.: Vietnamese - Some Questions on Phonetics, Syntax and Semantics. Nxb Giao duc, Hanoi (2000)
2. Chu, M. N., Nghieu, V. Đ, Phien, H. T. : Cơ sở ngôn ngữ học và tiếng Việt. Nxb Giáo dục, Hanoi, pp. 142–152 (1997)
3. Dien, D., Kiem, H., Toan, N.V.: Vietnamese Word Segmentation. In: The Sixth Natural Language Processing Pacific Rim Symposium, Tokyo, Japan, pp. 749–756 (2001)
4. Giap, N. T.: Từ vựng học tiếng Việt. H., Nxb Giao duc (2003)
5. Thu, C. B., Hien, P.:Về một xu hướng mới của từ điển giải thích (2007) http://ngonngu.net/index.php?p=319
6. Ha, L.A.: A method for word segmentation in Vietnamese. In: Proceedings of Corpus Linguistics 2003, Lancaster, UK (2003)
7. Le, H.P., Nguyen, T.M.H., Roussanaly, A., Ho, T.V.: A hybrid approach to word segmentation of Vietnamese texts. In: Martín-Vide, C., Otto, F., Fernau, H. (eds.) LATA 2008. LNCS, vol. 5196, pp. 240–249. Springer, Heidelberg (2008)
8. Nguyen, C.T., Nguyen, T.K., Phan, X.H., Nguyen, L.M., Ha, Q.T.: Vietnamese word segmentationwith CRFs and SVMs: An investigation. In: Proceedings of the 20th Pacific Asia Conference on Language, Information and Computation (PACLIC 2006), Wuhan, CH (2006)
9. Nguyen, T.V., Tran, H.K., Nguyen, T.T.T., Nguyen, H.: Word segmentation for Vietnamese text categorization: an online corpus approach. In: Research, Innovation and Vision for the Future, The 4th International Conference on Computer Sciences (2006)

Named Entity Recognition for Vietnamese

Dat Ba Nguyen, Son Huu Hoang, Son Bao Pham, and Thai Phuong Nguyen

Faculty of Information Technology
University of Engineering and Technology
Vietnam National University, Hanoi
{datnb,sonhh,sonpb,thainp}@vnu.edu.vn

Abstract. Named Entity Recognition is an important task but is still relatively new for Vietnamese. It is partly due to the lack of a large annotated corpus. In this paper, we present a systematic approach in building a named entity annotated corpus while at the same time building rules to recognize Vietnamese named entities. The resulting open source system achieves an F-measure of 83%, which is better compared to existing Vietnamese NER systems.

1 Introduction

Automatically identifying and classifying named entities is an important task for many natural language processing tasks including information extraction, information retrieval and machine translation. Specific named entity recognition (NER) tasks organized at Message Understanding Conference have attracted wide participation and improved the performance of NER systems. However, the majority of existing NER systems relies on a reasonably large annotated corpus, which mainly covers popular languages such as English, French and German. This is clearly a bottleneck for under-resourced languages as building a NER annotated corpus is rather expensive.

In this paper, we present a rule-based system for Vietnamese NER where rules are incrementally created while annotating a named entity corpus. To the best of our knowledge, this is the first publicly available open-sourced project for building a named entities annotated corpus and an NER system.

In section 2 we will present related work on named entities recognition system. Section 3 covers our take on corpus development while the system is described in section 4. Section 5 presents our experimental results and error analysis. Finally, section 6 gives concluding remarks and pointers to future work.

2 Related Work

Existing approaches to building NER systems can be classified into three categories: manually built rule-based systems [9], [4], statistically based systems [8] and hybrid approaches [5].

N.T. Nguyen, M.T. Le, and J. Świątek (Eds.): ACIIDS 2010, Part II, LNAI 5991, pp. 205–214, 2010.

Rule-based system consists of a set of manually created rules for specific purposes. For NER tasks, a rule utilizes morphological information (e.g. upper case, lower case characters), syntactic information (e.g. part of speech) or contextual information [3]. For example, in the following sentence:

"President Bush said that it's time to leave Iraq."

"President Bush" is recognized as a person name since "Bush" stands after "President". MUSE (Multi source entity finder) is a rule based NER system for various domains built on the Gate framework [9]. Jape grammars in the Gate framework have been used widely in specifying rules for NER tasks in many domains and languages [13], [10]. For Vietnamese, VN-KIM IE uses Jape grammars to recognize entities of various types (Organization, Location, Person, Date, Time, Money and Percent) with an F-measure of 81% [4].

Various machine learning algorithms have been applied for recognizing named entities such as hidden markov model [1], maximum entropy [2] and support vector machine [17], [8]. Conditional random field is used to recognize Vietnamese named entities of the following types: person, location, organization, percent, time, number and currency [11]. This system reports an F-measure of 81%. Pham described a system using SVM for Vietnamese NER in the health service domain with an F-measure of around 83%. [14].

Hybrid systems exploit advantages of both rule-based and corpus-based approaches but have not been applied for Vietnamese NER. Hybrid systems have been shown to achieve promising results for languages other than English such as Chinese [16], [5].

3 Corpus Development

Consulting NER tasks for various languages [7], we decided to recognize the following types for Vietnamese: Person, Organization, Facility, Location, Nationality, and Religion. Other entity types such as Percent, Money or Date can be easily recognized by regular expression and are not ambiguous and hence won't be considered in our work.

We have a rule for entity recognition in documents, that is: entities cannot overlap each other. A new entity will be recognized only when the previous one had been finished. In case they overlap each other, the longest entity will be recognized. For example:

"Phòng Giáo dục huyện Mỹ Đức đang họp."

"The Education and Training Department of My Duc District is having a meeting."

In this sentence, we only recognize "Phòng Giáo d•c huy•n M• •c" (The Education and Training Department of My Duc District) is an entity of Organization, and we skip "M• •c" as a entity of Location.

The corpus annotation process is carried out together with the process of rules creation as shown in figure 1. Initially we build a basic Jape-based NER system for Vietnamese. Then, we use this system to annotate the documents automatically. These annotated documents will then be manually checked and corrected for any remaining errors. Two independent teams do the annotation of every document and the result will be crosschecked and committed only when both teams agree. Interleaving the two processes by the same people has the following advantages:

- Having the rule-based system ready will bootstrap the annotation process and the corpus is already pre-annotated.
- Correcting errors in the pre-annotated corpus bring about experience in how to modify and improve the rules.

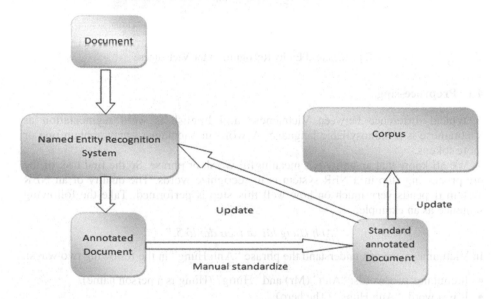

Fig. 1. Development process of the system

4 System Development

Our NER system is built as plugins in Gate framework with the architecture shown in Figure 2. The system comprises of four pipelined processing resources:

- Word segmentation
- Part of speech Tagger
- Gazetteer
- Transducer

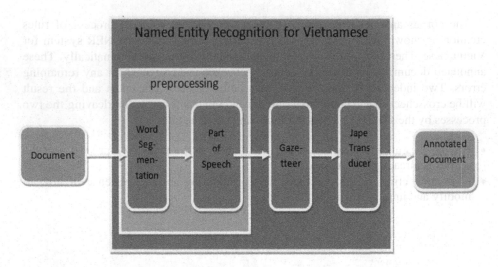

Fig. 2. Named Entity Recognition for Vietnamese

4.1 Preprocessing

A typical difference between Vietnamese and English is word segmentation as Vietnamese is a monosyllabic language. A word in Vietnamese may contain one or more tokens.

We all know that an entity is a meaningful word, or phrase. So the first task of the pre-processing step in a NER system is to recognize words. The quality of an NER system depends very much on how well this step is performed. Take the following sentence as an example:

"Anh Hùng lái xe trên đại lộ 5."

In Vietnamese, we can understand the phrase "Anh Hùng" in the following two ways:

- It contains two words: "Anh" (Mr) and "Hùng" (Hùng is a person name).
- It is a word: "Anh Hùng" (The hero).

If the word segmentation module works well, the result should be:

"Anh Hùng lái_xe trên đại_lộ 5."

"Mr. Hung drove in highway no 5."

"Hùng" standing after the prefix "Anh" will be recognized as a person accurately. However, if the word segmentation module doesn't work well, we will receive a wrong result:

"Anh_Hùng lái_xe trên đại_lộ 5."

"The hero drove in highway no 5."

The problem is that the set of words doesn't contain "Hùng" as a word by itself and it is almost impossible to recognize it as an entity.

We package the word segmentation module and part of speech tagger of Pham [15] as plugins in Gate so that their results will be represented as *Word* annotations with *pos* feature.

4.2 Gazetteer

The Gazetteer module consists of several dictionaries or gazetteers that will be used to create annotations Lookup over words with specific semantics to be used in writing rules at later stages. Each dictionary contains words with the same meaning, such as person names, organization entities, or phrases that signal the type of surrounding named entities. The list of gazetteers was formed and updated during the process of manually annotating the corpus. They can be grouped as followed:

- Gazetteers that contain potential named entities such as person, organization, location and nationality.
- Gazetteers containing phrases used in contextual rules such as name prefix or verbs that likely follow a person name.
- Gazetteer of potential ambiguous named entities.

4.3 Transducer

The transducer module is a cascade of Jape grammars or rules. A Jape grammar allows one to specify regular expression patterns over semantic annotations. Thus, results of previous modules including word segmentation, part of speech tagging and gazetteer in the form of annotations can be used to recognize and classify named entities.

The cascade of Jape grammars consists of the following components in order:

- Preprocess: Remove incorrect Lookup annotations and identify potential named entities.
- Recognizing Organization and Facility entities
- Recognizing Location and Nationality entities
- Recognizing Religion entities
- Recognizing Person entities
- Resolving ambiguity improve quality using contextual rules.

The pre-processing removes Lookup annotations that are only a partial part of a word. For example the word "tr••ng" (school) is a clue for recognizing Organazation but this word can also be just part of another word with totally different meaning. Take the following sentence:

"Thị trường Việt Nam thật ảm đạm trong thời kỳ khủng hoảng."

"The market of Vietnam was so dull during the crisis."

"tr••ng" is not a word but part of the word "Th• tr••ng"(market).

The pre-processing steps is also responsible for recognizing potential named entities so that later steps focus on typing them as well as disambiguating ambiguous cases. Apart from those named entities recognized by corresponding gazetteers, we also create annotations NamePhrase over consecutive words with their first letters capitalized.

Different named entities types are recognized together in one step such as Organization and Facility, Location and Nationality because they are somewhat ambiguous and hard to distinguish without using the context.

Person names are particularly difficult to recognize due to the fact that almost any word can be used as a person name. Following is an example of a contextual Jape grammar that recognizes a person name:

```
Rule: PersonAfterTitle
(
{Lookup.majorTupe == "titleperson"}({NamePhrase}):name
)-->
:name.Person = {kind = "Person", sure = "90", rule = "PersonAfterTitle"}
```

It is worth noting that for each rule, we assign a level of sureness signaling the level of confidence of recognizing corresponding named entities.

Apart from grammars that use local sentential context, we also look at the document context to resolve ambiguity. In case when multiple instances of a named entity phrase appear in a document, we will apply the type of the most reliable one having the level of sureness over a certain threshold to the remaining instances. For example, in the following paragraph:

"Bà Nùng vừa hút tẩu thuốc vừa kể: "Đời thằng A Lưới khổ lắm. Nhà chẳng còn ai, lao động quần quật cả năm mà vẫn không đủ ăn. Không biết đời nó bao giờ mới có vợ." Ấy thế mà niềm vui bất ngờ đã đến với A Lưới, anh gặp Hoa, cô giáo miền xuôi mới lên bản."

"A L••i" appears twice but only the first instance is recognized as a person name with a high level of confidence due to the title prefix "th•ng" (Mr). Therefore the second instance of that phrase is also typed as a person name.

5 Experiment and Error Analysis

For our experiment, we use a corpus taken from the Vietnamese Treebank corpus [12], which has not been annotated with named entities. Our corpus is randomly divided into two parts with the first part containing 53 documents (2814 sentences) and the second containing 20 documents (1125) sentences. We manually annotate part one (called training data) first before annotating the second part of the corpus. However we only update the Gazetteer and Transducer modules while annotating part one. Part two of the corpus (called test data) is not used for updating the grammars, as it will be kept for testing the performance of the built grammars. As mentioned in section 4, both the training data and test data are annotated and crosschecked by two annotators.

5.1 Evaluation Metrics

We evaluate the performance of the built system on the test data using the standard Precision, Recall and F-measure metrics using two criteria:

- Strict criteria: an entity is recognized correctly when both the span and the type are the same as in the annotated corpus.
- Lenient criteria: an entity is recognized correctly when the type is correct and the span partially overlaps with the one in the annotated corpus.

5.2 Experimental Result

Table 1 and Table 2 show the performance of the system on the training and test data respectively.

Table 1. Performance on the training data using strict criteria

Type	No. of entity	No. of recognized entity	No. of correctly recognized entity	Precision	Recall	F-measure
Person	1096	940	914	0.97	0.83	0.89
Organization	200	170	160	0.94	0.80	0.86
Location	544	510	488	0.95	0.89	0.91
Nationality	174	168	144	0.85	0.82	0.83
Facility	156	152	138	0.90	0.88	0.88
Religion	16	16	16	1.00	1.00	1.00
All	2186	1956	1860	0.95	0.85	0.89

Table 2. Performance on test data using strict criteria

Type	No. of entity	No. of recognized entity	No. of correctly recognized entity	Precision	Recall	F-measure
Person	370	285	273	0.96	0.74	0.84
Organization	48	33	25	0.75	0.52	0.61
Location	260	239	222	0.92	0.85	0.88
Nationality	82	68	56	0.82	0.68	0.74
Facility	40	33	29	0.87	0.72	0.78
Religion	10	9	9	1.00	0.89	0.94
All	810	667	614	0.92	0.76	0.83

It can be seen from table 2 that the system achieves a respected overall F-measure of 0.83. Given the performance of the system on the training data reaches only 0.89, we believe there is still a lot of room for improvement. While Religion and Location entities are well recognized, Organization entities appear to be a challenge. A reason is that the names of Vietnam's organizations sometimes are quite long, and hard to recognize, especially when they are not capitalized. For example:

"Công ty Hợp tác lao động nước ngoài - LOD , thuộc Bộ GTVT."
"Overseas Labour Cooperation Company - LOD, under the Ministry of Transportation"

The above sentence consists of 3 entities of the organization type namely: "Công ty Hợp tác lao động nước ngoài", "LOD" and "Bộ GTVT". However, the system only recognizes "Công ty Hợp tác" (wrong recognition) and "Bộ GTVT" (correct recognition). This problem can be alleviated if we do not penalize partial recognition. Table 3 shows result when evaluating the system using lenient criteria.

Table 3. Performance on the test data using lenient criteria

Type	No. of entity	No. of recognized entity	No. of correctly recognized entity	Precision	Recall	F-measure
Person	370	285	276	0.96	0.75	0.84
Organization	48	33	30	0.9	0.62	0.73
Location	260	239	224	0.93	0.86	0.89
Nationality	82	68	57	0.83	0.69	0.75
Facility	40	33	31	0.93	0.77	0.84
Religion	10	9	9	1.00	0.89	0.94
All	810	667	627	0.94	0.77	0.85

Clearly, while almost results of recognizing other entities do not increase much, the result of recognizing the organization entities increases significantly (F-measure from 0.61 to 0.73).

Table 4 shows the comparison of our system with existing Vietnamese NER systems. Although we do not make comparison on the same data, our simple and systematic approach already results in a system with reasonable performance with room for improvement. The open-sourced nature of system also facilitates future improvement from the research community.

Table 4. Comparison of our system with existing NER system

	Precision	Recall	F-measure
NER system using CRF [11]	0.81	0.79	0.80
VN-KIM IE System [4]	0.81	0.81	0.81
Our system Coltech.NER	0.92	0.76	0.83

5.3 Analyzing Errors

Looking at the errors of the system, we notice a number of ambiguous cases in distinguishing between Location and Nationality entities. Take the following sentence for example:

> *"Sau khi con tàu đã có một bản " khai sinh " khác thì bọn chúng tìm mối tiêu thụ và giao tàu tận nơi, Philippines là địa điểm mà chúng thường đến ..."*
> *"After getting a new "birth" profile for the ship, they seek buyers and deliver the ship, the Philippines is the location they usually go to ..."*

The system incorrectly recognizes "Philippines" as a Nationality entity instead of Location; however, it is very difficult to get right in this case. A possible solution is to use the semantics of the surrounding context with the phrase "là •a i•m" (is the location).

Person entities are also responsible for many ambiguous cases due to the fact that any word in Vietnamese can be used as a person name.

In addition, a large number of errors are due to incorrect results of pre-processing module such as the word segmentation module. For example the sentence:

> *"Linh cảm thấy buồn."*
> *"Linh felt sad."*

This sentence was word-segmented as "Linh_c•m th•y bu•n." (The feeling is sad); hence "Linh" is not considered a word and will not be recognized as a person entity.

6 Conclusion

We have successfully built an open source system on the Gate framework for recognizing Vietnamese named entities[1]. We presented a systematic approach in building a rule-based system together with processes in annotating a corpus for the task of Vietnamese NER. This can be applied to other under-resourced languages.

Our system achieves an F-measure of 83% which is comparable or even better than existing Vietnamese NER systems. Given the performance of system on the training data reaching an F-measure of 89%, we believe that the system can be further improved.

In the future, we plan to apply machine learning approaches utilizing the built annotated corpus and investigate ways of combining with our rule-based approach into a hybrid method.

Acknowledgement

This work is partly supported by the research fund from University of Engineering and Technology, Vietnam National University, Hanoi.

---

[1] The system can be downloaded from www.coltech.vnu.edu.vn/~sonpb/projects/VNER.html

References

1. Bikel, D., Miller, S., Schwartz, R., Weischedel, R.: A High- Performance Learning Name-finder. In: Proceedings of the Fifth Conference on Applied Natural Language Processing, pp. 194–201 (1998)
2. Borthwick, A., Sterling, J., Agichtein, E., Grishman, R.: Exploiting Diverse Knowledge Sources via Maximum Entropy in Named Entity Recognition. In: Proceedings of the Sixth Workshop on Very Large Corpora, Montreal, Canada (1998)
3. Budi, I., Bressan, S.: Association Rules Mining for Name Entity Recognition. In: Proceedings of the Fourth International Conference on Web Information Systems Engineering (2003)
4. Nguyen, V.T.T., Cao, T.H.: Automatic Extraction of Vietnamese Named-Entities on the Web. In: Proceedings of the Journal of New Generation Computing. Ohmsha, Ltd. Springer (2007)
5. Fang, X., Sheng, H.: A Hybrid Approach for Chinese Named Entity Recognition. In: Proceedings of the Fifth International Conference on Discovery Science (2002)
6. Kim, J., Kang, I., Choi, K.: Unsupervised Named Entity Classification Models and their Ensembles. In: Proceedings of the Nineteenth International Conference on Computational Linguistics (2002)
7. Mai, T.D.: Named Entity Guideline for Vietnamese. Bachelor thesis, College of Technology, Vietnam National University, Hanoi (2008)
8. Mansouri, A., Affendey, L., Mamat, A.: Named Entity Recognition Using a New Fuzzy Support Vector Machine. Proceedings of the International Journal of Computer Science and Network Security, IJCSNS 8(2), 320–325 (2008)
9. Maynard, D., Tablan, V., Ursu, C., Cunningham, H., Wilks, Y.: Named Entity Recognition from Diverse Text Types. In: Proceedings Recent Advances in Natural Language Processing 2001 Conference, Tzigov Chark, Bulgaria (2001)
10. Maynard, D., Bontcheva, K., Cunningham, H.: Towards a Semantic Extraction of Named Entities. In: Proceedings Recent Advances in Natural Language Processing, Borovets, Bulgaria (2003)
11. Nguyen, T.C., Tran, O.T., Phan, H.X., Ha, T.Q.: Named Entity Recognition in Vietnamese Free-Text and Web Documents Using Conditional Random Fields. In: Proceedings of the Eighth Conference on Some Selection Problems of Information Technology and Telecommunication, Hai Phong, Viet Nam (2005)
12. Nguyen, T.P., Vu, L.X., Nguyen, H.M.T., Nguyen, H.V., Le, P.H.: Building a Large Syntactically-Annotated Corpus of Vietnamese. In: Proceedings of the Third Linguistic Annotation Workshop (LAW) at ACL-IJCNLP 2009 (2009)
13. Pastra, K., Maynard, D., Hamza, O., Cunningham, H., Wilks, Y.: How Feasible is the Reuse of Grammars for Named Entity Recognition? In: Proceedings of the Conference on Language Resources and Evaluation, LREC 2002 (2002)
14. Pham, T., Kawazoe, A., Dinh, D., Collier, N.: Construction of Vietnamese Corpora for Named Entity Recognition. In: Conference RIAO 2007, Pittsburgh PA, U.S.A, May 30-June 1 (2007)
15. Pham, D.D., Tran, G.B., Pham, S.B.: Vietnamese Word Segmentation Using Part of Speech tags. In: Proceedings of the First International Conference on Knowledge and Systems Engineering, Hanoi, Vietnam (2009)
16. Srihari, R., Niu, C., Li, W.: A Hybrid Approach for Named Entity and Sub-Type Tagging. In: Proceedings of the Sixth Conference on Applied Natural Language Processing (2000)
17. Wu, Y., Fan, T., Lee, Y., Yen, S.: Extracting Named Entities Using Support Vector Machines. In: Bremer, E.G., Hakenberg, J., Han, E.-H(S.), Berrar, D., Dubitzky, W. (eds.) KDLL 2006. LNCS (LNBI), vol. 3886, pp. 91–103. Springer, Heidelberg (2006)

Task Allocation in Mesh Connected Processors with Local Search Meta-heuristic Algorithms

Wojciech Kmiecik, Marek Wojcikowski, Leszek Koszalka, and Andrzej Kasprzak

Dept. of Systems and Computer Networks,
Wroclaw University of Technology1 Wroclaw, Poland
leszek.koszalka@pwr.wroc.pl

Abstract. This article contains a short analysis of applying three metaheuristic local search algorithms to solve the problem of allocating two-dimensional tasks on a two dimensional processor mesh in a period of time. The primary goal is to maximize the level of mesh utilization. To achieve this task we adapted three algorithms: Tabu Search, Simulated Annealing and Random Search, as well as created a helper algorithm Dumb Fit and adapted another helper algorithm – First Fit. To measure the algorithms' efficiency we introduced our own evaluating function Cumulative Effectiveness and a derivative Utilization Factor. Finally, we implemented an experimentation system to test these algorithms on different sets of tasks to allocate. In this article there is a short analysis of series of experiments conducted on three different classes of task sets: small tasks, mixed tasks and large tasks.

Keywords: Network structure, task allocation, Tabu Search, Simmulated Annealing, experimentation system.

1 Introduction

Recently, processing with many parallel units is gaining on popularity very rapidly. It is applied in various environments, ranging from multimedia home devices to very complex machine clusters used in research institutions. In all these cases, success depends on a wise task allocation [1], enabling the user to utilize the power of a highly parallel system. Research has shown, that in most cases, parallel processing units give only a fraction of their theoretical computing power [2] (which is a multiplication of the potential of a single unit used in the system). One of the reasons for this is high complexity of task allocation on parallel units.

Meta-heuristic algorithms have been invented to solve a subset of problems, for which finding an optimal solution is impossible or far too complex for contemporary computers. Algorithms like Tabu Search [3], or Simulated Annealing [4], [5], [6] are among the most popular. They are capable of finding near-optimum solutions for a very wide range of problems in a time incomparably shorter than the time that it would take to find the best solution [7].

We decided to adapt three ideas for solving the allocation problem: the aforementioned Tabu Search and Simulated Annealing as well as a simplified local search

N.T. Nguyen, M.T. Le, and J. Świątek (Eds.): ACIIDS 2010, Part II, LNAI 5991, pp. 215–224, 2010.

meta-heuristic Random Search used for comparison. In our implementation, of these algorithms we also applied the First Fit algorithm [8]. The difference from typical approach is that it is not used as a solution by itself, but only as a local algorithm to help evaluating the results obtained in any iteration by the main algorithm. Moreover, we use two incarnations of the First Fit algorithm: one which is here named Dumb Fit as the simplest form of First Fit and one which is here named First Fit as actually a richer form of the classical First Fit. Our FF enables some reorganization of the task set and is also used to generate results that are to be used as reference when examining the efficiency of the main algorithms. Moreover, we have introduced the indices of performing for evaluating algorithms: (i) the Cumulative Effectiveness, and (ii) Utilization. The investigations of efficiency of solutions found by the examined algorithms in various conditions (mesh sizes, task sizes, task processing times, etc.) is based on results of simulations made with our own experimentation system. This system gives many possibilities to generate task lists, conduct series of experiments and to process their results.

Next sections of the paper contain what follows: Section 2 states the problem to be solved, Section 3 describes used algorithms and their roles, and Section 4 presents the experimentation system. Section 5 contains an analysis of results of series of experiments made for three categories of tasks: (i) small tasks, (ii) mixed tasks, and (iii) large tasks. Finally, Section 6 contains conclusions and sums up the article.

2 Problem Statement

Basic Terms. The basic terms used in the problem are defined as follows:

A *node* is the most basic element which represents a processor in a processor mesh. It is a unit of the dimensions of a mesh, submesh or task. Such node can be busy or free.

A *processor mesh*, which thereafter will be simply referred to as 'mesh', is a 2-D rectangular structure of nodes distributed regularly on a grid. It can be denoted as $M(w, h, t)$, where w and h are the width and height of the mesh and t is the time of mesh's life, respectively. The value of t may be zero or non-zero. A zero value means that the mesh will be active until the last task from the queue is processed.

A *position* (x, y) within a mesh M refers to the node positioned in the column x and row y of the mesh, counting from left to right and top to bottom, starting with 1.

A *submesh* S is a rectangular segment of a mesh M – a group of nodes, defined in a certain moment of time, denoted as $S(a, b, e, j)$ with its top left node in the position (a,b) in the mesh M, and of width e and height j. This entity, as a separate being, has only symbolic value (it is used in this paper to describe various conditions). If a submesh is occupied, it means that all its nodes are busy.

A *mesh* in a certain moment of time $M(w, h, t1)$, can be depicted as a matrix of integers, where each number corresponds to a node. Zero can be denoted as a dot (.) and it means a free node. Non-zero number (same for a submesh processing one allocated task) indicates a busy node, its value is the time left to execute the task. Such depiction is portrayed in Fig. 1 (four various tasks allocated on a small mesh).

A *task,* denoted $T(p, q, s)$, is stored in a list. All the contents of the list are known before the allocation. Tasks from the list are to be allocated on a mesh. There, they occupy a submesh S of width p height q for s units of time (s is task processing time).

Fig. 1. A sample depiction of a mesh with 4 allocated tasks in a moment of time

Efficiency metrics. The main evaluating function invented by us is the *Cumulative Effectiveness* (CE) expressed by (1). In (1) p_i q_i and s_i are width, height and processing time of the i-th of n processed tasks, respectively. CE is used when there is a non-zero time of life defined for a mesh. Knowing it and the parameters of used mesh we can count a more self-descriptive factor: the Usage Factor (UF) expressed by (2). In (2) w, h, t are width, height and time of life of the used mesh, respectively.

$$CE = \sum_{i=1}^{n} \left(p_i \cdot q_i \cdot s_i \right) \tag{1}$$

$$U = \frac{CE}{w \cdot h \cdot t} \cdot 100\% \tag{2}$$

The CE function can be interpreted as the cumulative volume of all allocated tasks and UF is the percentage of mesh's volume used by the processed tasks. It allows us to easily and objectively determine how much of the mesh's potential was "wasted" and how much was utilized. As the evaluating function can also be taken the Time of Completion denoted as T. The time T is the moment when the last of all tasks has been fully executed. This factor can only be used for comparing algorithms, not for objectively evaluating their efficiency.

3 Algorithms

Basic Concept. We have implemented three main metaheurstic local search algorithms: SA – Simulated Annealing explained in [4], [5], [6], and [7], TS – Tabu Search explained in [3] and [7], RS – Random Search (not to be confused with simple evaluating of a random solution), explained in [7]. All of them work for a number of iterations. In each of iterations, they operate on a single solution and its neighborhood and evaluate the results. A *solution* is defined here as a permutation of tasks to be allocated, stored in a list. Such permutation is evaluated by performing on it a simulation, using one of two atomic algorithms (First Fit and Dumb Fit) and, basing on its result, counting one of the evaluation functions, explained above. There are also various kinds of neighborhood to be explored by the main algorithms. We implemented two of them:

insert and *swap*. In case of the first one, a neighboring solution is found by taking one element of the permutation and putting it in some other position. In case of the second kind, two elements are taken and their positions are swapped (hence the name). Success level for each of the main algorithms depends on the instance of the problem (mesh size, tasks sizes and life/processing times), and on internal parameters of algorithms e.g. the atomic algorithms used to initiate them.

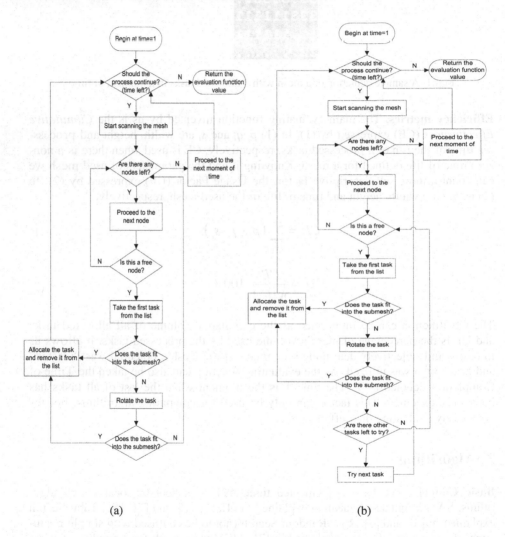

(a)   (b)

Fig. 2. Block diagrams for the Dumb Fit (a) and First Fit (b) algorithms

Random Search Algorithm (RS). The RS is the simplest local search algorithm. The algorithm starts from a solution and in each iteration; it finds and evaluates a new solution from the neighborhood of the current one. In the next iteration, the new one becomes the current one and the process continues. In RS there are no additional parameters except for the number of iterations. This algorithm is highly resistant to local minimums but it does not improve certain solutions too precisely.

Simulated Annealing Algorithm (SA). The SA works in a more complex way. Its main parameters are starting and ending temperatures. During the course of its work the temperature drops (logarithmically or geometrically). In each iteration, a random solution from the neighborhood of the current one is found and evaluated. When the temperature is high there is high probability to accept the new solution as the current one, even if it is worse. When the temperature is low only these solutions are accepted as new current ones which are better. Such approach makes this algorithm resistant to local minimums in the beginning and precisely improving a current solution in the end, going down to the nearest minimum.

Tabu Search (TS). Our implementation of the TS algorithm is similar to the SA algorithm with low temperatures, except for the fact that it does not accept a new solution as the current one, if the same solution is found in the taboo list. Whenever a new current solution is set it is added to the taboo list. The taboo list has limited length which is the main parameter of the algorithm. This algorithm is forced to leave the vicinity of a local minimum. This vicinity is limited by the length of the taboo list. At the same time TS tries to precisely improve a current solution. It also returns the best solution found during its work.

Atomic Algorithms. The atomic algorithms implemented here are: Dumb Fit (DF) and First Fit (FF) - block diagrams for both are shown in Fig. 2. The DF works as follow: It takes a solution (task permutation) and does not modify its order. In each discrete moment of time it scans through the mesh from top to bottom, left to right. If it encounters a free node, it checks whether there is enough other free nodes right from it and below it, to allocate the first task from the list. If this try fails, it rotates the first task from the list by ninety degrees and tries to fit it in the hole again. If it succeeds the task is taken from the list, a corresponding submesh is allocated for the appropriate amount of time and the algorithm keeps scanning the mesh and tries to allocate next tasks until the mesh ends. This process is repeated in each moment of mesh's lifetime or until the last task is allocated (if the mesh's lifetime is set to 0). The FF algorithm works very similarly to DF, except that upon encountering a free submesh, it does not limit itself to trying to allocate only the first task from the list, but tries each of them. Therefore, it can modify the permutation by removing elements from its middle or end. When DF or FF is working, the appropriate evaluating function value is counted and then it is returned to the main algorithm that is currently working.

4 Experimentation System

We tried to design as versatile simulation environment as possible, to be able to evaluate all combinations of parameters for various problem instances. As a result we have developed a console application (to be run under Microsoft Windows OS), written in the C++ language, with various abilities to read parameters for the experiments and to write their results. It has two main modes of operation – menu mode and command line mode.

Inputs. Generally, in all modes of operation, our software allows the user to set certain input parameters. First group of them defines the problem. It allows the user to choose ranges of dimensions (p, q) and processing times (s) for the tasks and the task-list length. The user can also define the size and lifetime of the mesh: w, h, t. All the parameters from the first group allow the program to randomly create a task-list and define a mesh, which, together, form a problem instance. The other group of parameters varies and consists of specific parameters of the chosen algorithm, like: number of iterations, starting and ending temperatures for SA, temperature profile for SA, tabu-list length for TS, etc. Specifying both groups of parameters makes it possible to solve a predefined problem with a chosen, custom configured algorithm.

Menu Mode. When the software is run without parameters in command line, it goes into the menu mode. It features a main menu consisting of options giving three main kinds of operation. First of them involves tracing a single simulation for a given task-list and a given mesh, using a chosen atomic algorithm. It shows the simulation graphically, step by step and shows the evaluation function's value, which helps to understand the way of evaluating a solution with a given algorithm. The second way of using the program in menu mode is providing a file with predefined test series. The third, and main, way to use the menu mode, is performing a single experiment with a chosen algorithm. It allows the user to specify all the parameters easily without using an input file and watch the algorithm work (it's progress and current evaluation function value is shown). It would not be suitable for performing a series of experiments for a major research but is quite convenient for calibrating the parameters for a test series being in design.

Command Line Mode. This mode is the preferred one for running a series of experiments for a certain research. It allows the user to specify, as command line parameter, a file with a designed test series. Such file begins with a set of parameters defining the problem instance. The task-list is generated only once and same problem instance is used for the whole test series defined in the file. Also the number of repetitions for each test can be specified. These parameters are followed by any number of lines with parameters defining each test (used algorithm and its specific options). Therefore, any number of any kinds of tests for a certain problem instance can be defined in an input file. When using the command line mode, the user can create a batch file (*.bat*) for running a series of series of tests (a series of program executions for more than one input file).

Outputs. For each experiment (single execution of a main algorithm), in both program execution modes, an output file can be specified. It contains some specific input parameters set for the chosen algorithm and, most importantly, lines showing the current and best evaluation function value for any iteration. Such data can later be analyzed with appropriate software.

5 Investigations

The three complex experiments for the three problem instances were carried out. The distinct categories of instances were: (i) tasks relatively small (as compared with the mesh size), (ii) tasks relatively large, and (iii) mixed tasks (mixed small and large).

For each instance the used algorithms behaved differently, so we designed three corresponding test series that are analyzed in the following subsections. In each test, the result of FF is used as a level of reference. Evaluating algorithm means the algorithm used for each solution evaluation by the main algorithms. Initiating algorithm means the algorithm used to generate the initial solution. In Table 1 the experiment design (input values for all three complex experiments) is presented.

Table 1. Experiment Design

Parameter	Test		
	Mixed tasks	*Small tasks*	*Large tasks*
p	2÷12	2÷10	4÷12
q	2÷12	2÷10	4÷12
s	2÷12	2÷10	4÷12
w	12	50	12
h	12	50	12
t (task-list length when t=0)	1000	0 (1000 tasks)	1000
tested algorithms	SA, TS, RS	SA, TS, RS	SA, TS, RS
evaluating algorithms	DF, FF	DF, FF	DF, FF
initiating algorithms	DF, FF	DF, FF	DF, FF
evaluating function	CE	T	CE
neigborhood	*swap, insert*	*swap, insert*	*swap, insert*

Mixed Tasks. In this case almost all tests were performed for 20000 iterations for all main algorithms for the same task set. Only for tests in which FF was the evaluating algorithm, 5000 iterations were tested. The aim was to keep all algorithms running for about 100 seconds. Each test was repeated 3 times and means values were calculated.

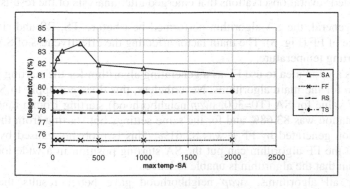

Fig. 3. SA algorithm for various starting temperatures vs. other algorithms

The results shown in Fig. 3 were obtained for various starting temperatures in SA and for the best found configuration: the starting permutation being random, DF used, *swap* neighborhood, 0.01 final temperature and geometrical temperature profile. It may be observed that for this configuration and problem instance, the best starting temperature of SA is around 300. It makes achieving over 83% of mesh usage against almost 80% for TS, the second one in the contest. In general the best value of the UF achieved by the SA for this instance (20000 iterations) was of 83.68%.

The winning algorithm outperformed the FF result in its 4522nd iteration. The values of current and best results for any iteration, for the best SA passage, are shown in Fig. 4. It may be seen, that in the beginning, the current result tends to be lower than the best one, but then it starts to "stick" to the best result as with the temperature fall, the algorithm acts more like Descending Search and less like Random Search. It is also visible when the best and current results surpass the value found by FF.

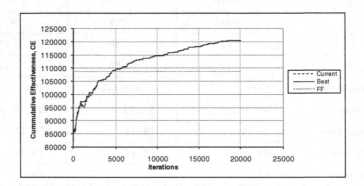

Fig. 4. The current and the best results for the winning SA (Mixed Tasks)

The main results for Mixed Tasks case may be specified as:

- The best result: SA, *swap*, evaluation function DF, starting temp. 300, geometrical profile: U=83.68%,
- The difference between the best result and FF: 10% of the result by FF.

There are listed several observations that emerged after analysis of the results:

- In general, the SA algorithm performed best before TS, RS and single passage of FF (Fig. 3). The main factor affecting the effectiveness of SA was the starting temperature.
- It is a good idea to use FF as a generating algorithm for the starting permutation for the main algorithms. Nevertheless, this does not apply to SA: mean *UF* value for SA (T0=300, *swap* neighborhood) starting from a random permutation was 83.68% and for the same settings but starting from the permutation generated by FF it was 82.61%. This is probably caused by the fact that the FF algorithm can put the SA starting position in a wide local minimum that the algorithm is unable to leave.
- For all algorithms, *swap* neighborhood gave better results than *insert* neighborhood (about 3% difference in average for the UF).

- TS algorithm with tabu list of length around 1000 performed slightly better than for lengths around 100 (*UF* was of 79.8% and of 78.8%, respectively).
- If the FF algorithm is used during the work of the main algorithms then they perform well but the processing time increases significantly. After lowering the iteration count it gives comparable results to the configurations with 20000 iterations and DF algorithm.

Small Tasks. In this case we decided to use the second evaluating factor – *T*. It is less objective than the first one but still allows comparing the algorithms and gives much better ability to spare experimentation time. It is so, because for a large mesh and small tasks it would be needed to process a huge list of tasks, so as not to run out of them in e.g. 1000 units of mesh's lifetime. This would make a series of experiments very long to conduct in our conditions. These experiments also spawned a conclusion, namely one, saying that using metaheuristic algorithms for allocating small tasks, does not make sense. Tasks are here small enough, in comparison to the size of the mesh that the FF algorithm manages to fit a task from the list into almost every free submesh. Therefore, even metaheuristics basing on the result given by FF can not achieve any better result (see e.g. Fig. 5). What is more, due to semi-random characteristics of tested metaheurstics, ones that started from a random solution and did not use FF, gave even worse results, e.g. SA, in such case, gave a result of *T*=124.

The main results for Small Tasks case may be specified as:

- The best result: SA/TS/RS: *T*=98,
- The difference between the best result and FF: 0.

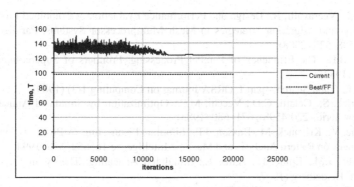

Fig. 5. The current and the best results for SA (Small Tasks)

Large Tasks. In this case, the achieved results and algorithm's behavior were very similar to the general case of mixed tasks (we also used the same scheme of testing as then). The achieved result of the winning algorithm was even better, but only by a tiny margin. Also, as in the case of mixed tasks, the **SA** algorithm was the best and the same parameters as previously caused maximum performance.

The main results for Large Tasks case may be specified as:

- The best result: SA, *swap*, evaluation function DF, starting temp. 300, geometrical profile: U=83.98%,
- The difference between the best result and FF: 13.3% of the result by FF.

6 Conclusions and Perspectives

We have adapted three metaheuristic local search algorithms for the problem of task allocation on a processor mesh, created an experimentation environment for testing them and performed research with it.

The experiments showed, that in general, local search metaheuristic algorithms perform well in solving the considered problem. Only for allocating small tasks on a large mesh, it is needless to use them as they do not achieve better results than the basic ΓΓ algorithm which itself performs well, due to the easiness of fitting small tasks into free submeshes. The unquestioned leader of all our tests was the SA algorithm. It defeated all others for tasks of mixed and large sizes. It also achieved reasonable results of over 83% of mesh usage, which we find quite satisfactory, even though it was achieved in a relatively small number of iterations.

We noticed that it is very important to design the experiment well. We did that following our concept described in [9], what allowed to check many possible combinations of parameters.

It definitely can not be said that the problem has been fully explored and researched. We are planning in further research (i) to adapt ideas presented in [10] performing future experiments, (ii) to construct a versatile testing environment, and (iii) to implement new allocation algorithms based on the Genetic Algorithm or ACO.

References

1. Goh, L.K., Veeravalli, B.: Design and Performance Evaluation of Combined First-Fit Task Allocation and Migration Strategies in Mesh Multiprocessor Systems. Parallel Computing 34, 508–520 (2008)
2. Buzbee, B.L.: The Efficiency of Parallel Processing. Frontiers of Supercomputing, Los Alamos (1983)
3. Glover, F.: Tabu Search – part I. ORSA Journal on Computing 1(3) (1989)
4. Kirkpatrick, S., Gelatti, C.D., Vecchi, M.P.: Optimization by Simulated Annealing. Science New Series 220(4598), 671–680 (1983)
5. Granville, V., Krivanek, M., Rasson, J.P.: Simulated Annealing: A Proof of Convergence. IEEE Trans. on Pattern Analysis and Machine Intelligence 16, 652–656 (1994)
6. Laarhoven, J.M., Emile, H., Aarts, L.: Simulated Annealing: Theory and Applications. Springer, Heidelberg (1987)
7. Glover, F., Kochenberger, G.A.: Handbook of Metaheuristics. Springer, Heidelberg (2002)
8. Byung, S., Das, C.R.: A Fast and Efficient Processor Allocation Scheme for Mesh Connected Multi-computers. IEEE Trans. on Computers 1, 46–59 (2002)
9. Koszalka, L.: Simulation-based Evaluation of Distributed Mesh Allocation Algorithms. In: Thulasiraman, P., He, X., Xu, T.L., Denko, M.K., Thulasiram, R.K., Yang, L.T. (eds.) ISPA Workshops 2007. LNCS, vol. 4743, pp. 335–344. Springer, Heidelberg (2007)
10. Koszalka, L., Lisowski, D., Pozniak-Koszalka, I.: Comparison of Allocation Algorithms for Mesh Networks with Multistage Experiment. In: Gavrilova, M.L., Gervasi, O., Kumar, V., Tan, C.J.K., Taniar, D., Laganá, A., Mun, Y., Choo, H. (eds.) ICCSA 2006. LNCS, vol. 3984, pp. 58–67. Springer, Heidelberg (2006)

Computer System for Making Efficiency Analysis of Meta-heuristic Algorithms to Solving Nesting Problem

Pawel Bogalinski, Leszek Koszalka, Iwona Pozniak-Koszalka, and Andrzej Kasprzak

Dept. of Systems and Computer Networks, Faculty of Electronics,
Wroclaw University of Technology, 50-370 Wroclaw, Poland
iwona.pozniak-koszalka@pwr.wroc.pl

Abstract. The paper concerns the strip nesting problem. This kind of problems is NP-hard even fort rectangular shapes and plate with no rotation allowed. In this paper computer system for making efficiency comparison of heuristic algorithms is proposed. The evaluation of tabu search, ACO and simulated annealing algorithms with raster approach to the geometry of problem is described. The evaluation was made on the basis of the results of simulations made using the experimentation system designed and implemented by authors.

Keywords: Strip nesting problem, tabu search, ant algorithm, simulated annealing, quantization, computer system.

1 Introduction

Nesting belongs to the *cutting and packing* class of problems and usually refers to the problem of placing a number of shapes within the bounds of some container (material), such that no pair of shapes overlaps. In Fig. 1 an example is given.

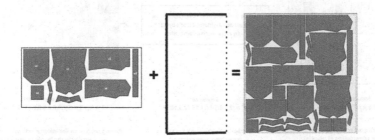

Fig. 1. A nesting problem example – tight packing parts of clothes on the material

This figure shows two-dimensional nesting problem which arise in the clothing production when cutting fixed parts of clothes from a roll of fabric. In this case, the goal is minimisation the waste of fabric that is equivalently to maximisation the

N.T. Nguyen, M.T. Le, and J. Świątek (Eds.): ACIIDS 2010, Part II, LNAI 5991, pp. 225–236, 2010.

utilization of the fabric necessary to cut all involved parts of clothes. It is a typical *strip nesting problem*, where given are: set of shapes and width of rectangular region, and the aim of optimisation is minimizing length of region containing all shapes. In the last years, a multitude of approaches solving nesting problem was developed. Solution ideas range from simple heuristics to more complicated methods like evolutionary approaches [7, 11], meta-heuristic algorithms [2, 9], while using different geometric approaches to the problem [1]. Nesting is computationally NP-hard so it is necessary to use intricate optimisation techniques to find solution in satisfying time. This therefore is motivated to find approaches reducing time computing and bringing good solutions.

The main subject of this paper is the designed and implemented computer experimentation system for making efficiency analysis of algorithms for solving nesting problem. The system allows (i) providing shape and problem conversion to the mathematical model by quantization and shape coding, (ii) finding solutions of strip nesting problem along with own implementation of algorithms based on simulated annealing, tabu search and ant colony, (iii) making comparison of results. The system may use the problem instances accorded by *ESICUP*[1]. Moreover, it uses raster model to resolve geometric aspect of the problem.

The rest of paper is organized as follow. Section 2 states the problem of nesting. Section 3 describes implemented approaches. In Section 4 the experimentation system is presented. Section 5 shows results of investigations. Section 6 contains final conclusions and perspectives for further research in the area.

2 Problem Statement

We concentrate on the *strip nesting problem,* i.e. problem of packing the shapes within the rectangular strip of material such that length of used material is minimized.

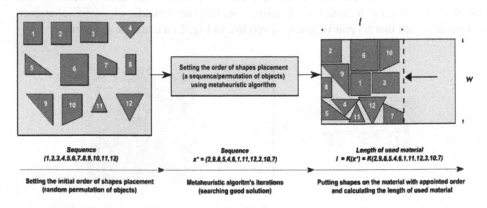

Fig. 2. The idea of solving the strip nesting problem

---

[1] EURO Special Interest Group on Packing and Cutting – ESICUP's website: *http://paginas.fe.up.pt/~esicup/*

Assumptions: *(i)* rotations of shapes are not allowed during the process of searching solution, *(ii)* each shape is identified by unique index: *1...n,* (*n* - total number of shapes), *(iii)* shapes are placed on the material along with the *bottom left (BL)* placement rule [12] (this rule iteratively moves each piece horizontally as far to the left as possible and then vertically until it is able to move horizontally again or touches another piece or bottom of the stock sheet), *(iv)* each solution x^* is represented by a sequence of shapes described as a *n*-element permutation of indices.

Given: $S = \{s_1,...s_n\}$ – set of shapes, w – material width, X – set of solutions *i.e.* *n*-elements permutations, $K: X \rightarrow R$ – the cost function defined as the length l.

Constraints: Overlaps between each pair of shapes are not allowed.

Objective. To find a permutation $x^* \in X$, such that the cost function K is a minimum:

$$K(x^*) = \min_{x \in X} K(x)$$

3 Solution to the Problem – Algorithms

The proposed solution to the nesting problem is composed of three stages, including

Quantization - changing of the graphic representation of shapes and material to the maps of pixels – making a mathematical model of a given problem instance. Each shape is characterized by array of bits and next is coded.

Optimisation – all implemented methods run following scheme:

   Step 1. Determination of shapes (sequence) order (at random or by shape sorting).
   Step 2. Tightly placing shapes on the material along with BL and the defined order.
   Step 3. Calculation of the length of the used material.

After step 3 it stops his work or repeats procedure starting from step 2 after change of sequence order of shapes – after identifying so-called neighbourhood solution.

Visualisation – changing of the mathematical model of solution to the graphics.

We implemented three algorithms basing on the meta-heuristic ideas.

Simulated annealing (SA)

There are known implementations of SA in literature (e.g. [9] describes SA method with NFP approach to the geometry). In this paper, SA was implemented for the raster model - see scheme of SA implementation.

```
begin
    generate(x⁰);   {random initial solution}
    initialize(T₀, s); {T₀ - start temp., s - iterations}
    i := 0;
    x:= x⁰;   {x - current solutions}
repeat
    for m := 1 to s do
    begin
        x' := disorder(x); {x' - solution get by swap(i,j) move of
        permutation for randomi,j)}
        if (K(x') ≤ K(x) then x := x';
        {K(x') - length l of used material for x' perm.}
        else if random(0,1) < e ⁻ᐞ/ᵀᴵ then x := x';
        {Δ = K(x') - K(x) - difference between length of used material for
        x' and x solutions}
    end;
    updateTemperature(Tᵢ);
    i := i +1;
until stopCriteria;
end;
```

Disturbed solution. Neighbourhood solution is get by swap (in the current permutation) two elements with randomly chosen indexes i and j.

Cooling schedule. It is defined by following parameters: T_o and T_k as an initial and terminating temperature, respectively, *Decrement temperature* determined by using one of three schemas: (*i*) Linear: $T_{i+1} = T_i - \lambda, \lambda > 0$, (*ii*) Geometric: $T_{i+1} = T_i \lambda , 1 > \lambda > 0$, (*iii*) Logarithmic: $T_{i+1} = T_i/(1+\lambda T_i), \lambda > 0$.

Stop criteria. Algorithm stops when current temperature t is equal to T_k or when counter i reach the limit of iterations defined by the user.

To choose the most suitable parameters of SA for the nesting problem (Tab. 2) we made 1350 tuning tests for three problem instances with various parameters (Tab. 1).

Table 1. Range of variability of parameters during SA method tuning process

SA parameters		From	Step	To
Initial temp. T_0		100	100	500
Terminating temp. T_k		0	-	0
Iterations number		20	20	100
Stop criteria (iterations) s		10	20	110
Cooling schedule	Linear - λ	1	4	9
	Geometric - λ	0,1	0,4	0,9
	Logarithmic - λ	0,1	10	30

Table 2. The best parameters for the considered problem instances

Problem instance	T_0	T_k	s	stop	schema	λ
Blaz	400	0	80	110	logarithmic	20,1
Han	300	0	100	90	logarithmic	20,1
Shapes0	300	0	80	110	logarithmic	20,1
THE BEST	**350**	**0**	**90**	**100**	**logarithmic**	**20,1**

Tabu search (TS)

The TS implementation to the nesting problem was presented by authors in [3]. The improved implementation for the problem described in Section 2 is presented below.

```
begin
define(Restarts);  {Restarts - number of alg. restarts} ITER = 0;
   repeat
      generate(x⁰);   {random initial solution}
      x*:= x⁰;  {x⁰ - the best solution at now}
      K*:= K(x*);  {K* - length l of used material for x*}
      initializeTabuList(); {blank list}
      i := 0;
      repeat
         Mᵢ := numberOfElements(N(xᵢ));
         {N(xⁱ) - swap neighborhood of xⁱ permutation}
         K' := ∞;
         {K' - value of the best solution in N(xⁱ) space}
         for j := 1 to Mᵢ do
         begin
            x:=nextElement(N(xⁱ));
            if (K(x) < K') AND
            (move xⁱ→ x is not on tabu list) OR
            (move xⁱ→ x is on tabu list but K(x) is
            better than ever before)
               then K' := K(x); xⁱ⁺¹ := x;
         end;
         if K' < K* then x* := xˣ⁺¹; K* = K';
         updateTabuList();
         i := i+1;
      until stopCriteria; ITER := ITER +1;
   until (ITER < Restarts)
end;
```

Initial solution. A random permutation of numbers (indices).

Neighbourhood. Neighbourhood of permutation type *swap*, has been got by all possible swap moves (swap positions) of two elements in the permutation.

Tabu list. There are stored pairs of numbers i and j that all permutations where element with number i is before element with number j are forbidden.

Stop Criteria. Algorithm stops when counter i is equal to the number of iterations determined by the user.

Restart. This parameter determines how many times algorithm starts from different initial solutions. It allows checking wider solution space during searching process.

To choose appropriate parameters of TS method we made 900 tuning tests for three problem instances and various parameters (Tab. 3). After analysis of tests results we adjusted the parameters of TS by determining their values such that TS gets the best values of length of the used material (Tab. 4).

Table 3. Range of variability of parameters during TS method tuning process

TS parameters	From	Step	To
Restarts	1	2	20
Iterations	1	5	50
Length of tabu list	4	2	20

Table 4. The best TS parameters for each problem instance

Problem instance	Restarts	Iterations	Length of tabu list
Blaz	9	26	10
Han	11	46	10
Shapes0	19	36	10
THE BEST	13	36	10

Ant colony optimisation (ACO)

The implementation of ant colony approach is presented below.

```
begin
    calculate(η_ij);{calc. the visibility for all shapes}
    initialize(m, w, α, β, ITER);
        {m - number of ants,
         w - pheromone evaporation factor,
         α, β- parameters,
         ITER-number of algorithm iterations-stop criteria}
    x*:= 0; {x* - the best solution}
    K*:= ∞; {K* - length of used material for x* perm.}
    distribute Randomly m Number of Ants();
    t := 0;
    repeat {loop to realize the single route of m ants}
        for i := 1 to m do
        begin
            ant(i) make a solution x^i;
            if K* > K(x^i) then K* := K(x^i); x* := x^i;
        end
        update τ_ij(t); {update pheromone - trial intensity}
        t := t+1;
    until (t < ITER);
end;
```

Visibility (η_{ij}). This parameter describes how the shape just placed fits to the next item to be placed. It takes values from range (0,1> and it is used to calculate transition probability (the detailed formula for calculating this probability is given in [3]). The illustration of the way to calculate this parameter is shown in Fig. 3.

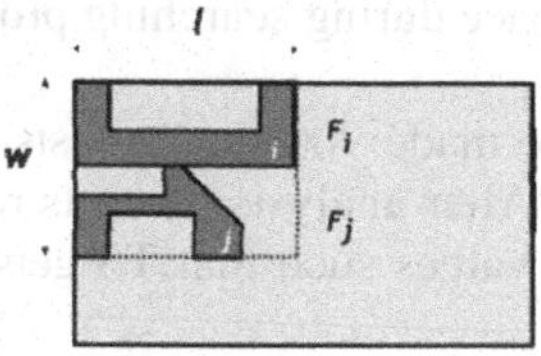

$$\eta_{ij} = \frac{Total Area_{ij}}{Shape Area_i + Shape Area_j} = \frac{l \cdot w}{F_i + F_j}$$

Fig. 3. Calculating visibility of shape i and shape j

Evaporation factor (w). It describes how quickly pheromone evaporates on ant paths – it is used to calculate and update pheromone trial intensity (see [3]).

α, β – parameters. These parameters allow adjusting *visibility* of shapes and *popularity* of route among ants. Great value of *α* cause that ants choose mainly pheromone paths. Low *α* can approximate ant searching to greedy algorithm.

Stop criteria. Algorithm stops after performing *ITER* iterations defined by the user. To choose the best parameter we made 3750 tests for three problem instances and parameter ranges specified in Tab. 5. The adjusted values are given in Tab. 6.

Table 5. Range of variability of parameters during ACO method tuning process

ACO parameters	From	Step	To
Number of units m	20	20	100
Evaporation factor w	0,1	0,2	0,9
Alpha parameter α	1	6	25
Beta Parameter β	1	6	25
Number of iterations ITER	10	20	110

Table 6. The best ACO parameters for each problem instance

Problem instance	m	w	α	β	ITER
Blaz	60	0,3	19	19	50
Han	60	0,1	13	1	30
Shapes0	80	0,3	13	19	90
THE BEST	**70**	**0,3**	**15**	**13**	**65**

ACO tuning process was very complicated, in particular with choosing appropriate values of *α* and *β*. The parameters were strongly dependent on shapes geometry properties. E.g. for more complicated problems like *Han* (many of different shapes) the best result of ACO was found for α much greater than β.

4 Computer Experimentation System

The created computer system called *KSBAN* consists of four main modules:

Problem Generator Module (PGenM) – for problem instances reading from files and creating new problem instances.

Geometry Module (GeoM) – for graphic transformation, it is based on methods for mathematical model creation, methods for shapes description, and methods for shapes collisions detection and shapes placement on the material. *Optimisation Module (OM)* – for searching the best sequence (order) of shapes. *Presentation Module (PM).*

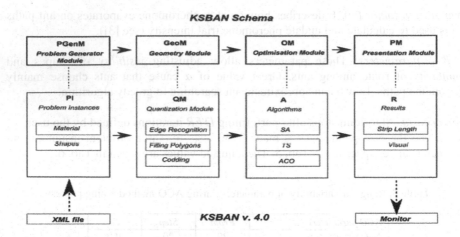

Fig. 4. Scheme of computer experimentation system

Fig. 4. contains also modules implemented in the newest version of the system:

Problem instances (PI) – the module which can read problem instances from XML files by *ESICUP* (test data can be downloaded from website Data Sets Section).
Quantization Module (QM) – responsible for handling the problem geometry (Fig. 5).

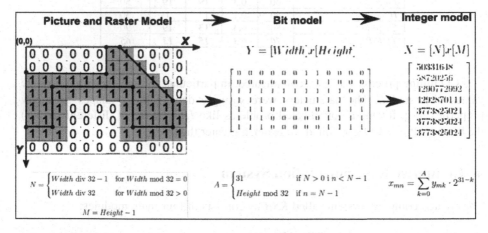

Fig. 5. *QM module* - creating a mathematical representation of a shape

The raster model contains: *(1) Edge recognition.* Edge detection and quantization. The Bresenham's line algorithm to make a raster model of shapes edges is applied, *(2) Filling polygon.* Scan line filling algorithm to fill shapes and make theirs raster models is used, *(3) Coding.* After raster/pixel model of shapes creation (using edge rec. and filling polygon) this model is transformed to array of bits and next to array of integer values to reduce the memory needed to cache mathematical models of shapes.

Algorithms (A). The algorithms SA, TS and ACO are implemented (see Section 3).

Results (R). This module is responsible for comparison of the results produced by algorithms. The main indices of performance are: the length of the used material and the usage of material. The visual representation of shapes placement is available.

5 Investigation

The evaluating factors were (A) the length l of the used material, and (B) the computing time. Experiments were made for ten different *ESICUP's* instances – their characteristics are in column *Problem* (Tab. 7), in brackets, including the number of different shapes, the number of all shapes, and the width of material, successively.

Table 7. Investigation results – Min, Avg. and Max of 100 solutions

Problem	Alg.	Length [px]			Usage [%]			Time computing [s]		
		Min	Avg.	Max	Min	Avg.	Max	Min	Avg.	Max
Blaz (7,28,15)	SA	41	41,7	43	79,65	82,15	83,54	9,72	10,86	18,45
	TS	41	42,83	44	77,84	79,98	83,54	12,33	12,96	13,64
	ACO	42	44,58	46	74,46	76,84	81,55	4,45	4,54	4,63
Dagli (10,30,60)	SA	72	74,95	80	75,43	80,54	83,81	55,31	65,63	72,92
	TS	74	75,97	77	78,37	79,44	81,55	86,55	93,28	109,5
	ACO	77	79,68	82	73,59	75,75	78,37	17	18,02	9,39
Fu (12,12,38)	SA	40	41,24	44	77,1	82,28	84,81	6,7	7,76	9,39
	TS	41	42,11	44	77,1	80,6	82,74	1,39	1,48	1,58
	ACO	42	44,8	47	72,18	75,79	80,77	2,33	2,57	4,16
Han (20,23,58)	SA	52	54,38	62	67,22	76,72	80,15	30,25	35,05	39,88
	TS	53	55,02	56	74,42	75,76	78,64	29,22	32,06	34,72
	ACO	55	58,74	62	67,22	70,99	75,78	10,19	10,78	11,59
Jakobs1 (25,25,40)	SA	17	18,34	20	75,24	82,12	88,52	6,28	7,43	9,02
	TS	18	18,62	19	79,2	80,88	83,6	6,42	6,8	7,28
	ACO	18	19,41	21	71,66	77,59	83,6	3,03	3,17	3,28
Jakobs2 (25,25,70)	SA	31	33,22	36	69,68	75,56	80,92	20,94	23,9	26,75
	TS	32	33,61	35	71,67	74,65	78,39	22,88	24,5	26,5
	ACO	34	35,67	37	67,8	70,35	73,78	7,39	7,86	8,75
Marques (8,24,104)	SA	87	91,59	97	79	83,69	88,08	77,14	97,01	117,83
	TS	90	92,85	95	80,66	82,54	85,14	78,5	87,85	95,17
	ACO	93	97,99	103	74,4	78,24	82,4	24,06	26,05	28,81
Poly1a (15,15,40)	SA	20	21,17	25	58,05	68,73	72,56	4,41	5,08	5,78
	TS	21	21,84	23	63,1	66,48	69,11	1,41	1,5	1,58
	ACO	21	22,95	24	60,47	63,3	69,11	1,7	1,78	1,89
Shapes0 (4,43,40)	SA	80	83,2	87	66,13	69,18	71,92	75,41	81,9	92,84
	TABU	83	84,4	86	66,9	68,21	69,32	265	285,8	319,5
	ACO	85	87,4	90	63,93	65,84	67,69	25,42	26,05	27,05
Trousers (17,64,79)	SA	284	291	300	80,95	83,47	85,51	725,8	840	968,3
	TABU	289	290	291	83,46	83,75	84,04	6283	6370	6458
	ACO	292	301	307	79,11	80,74	83,17	211,8	220,6	229,8

For each of the selected instances we executed 100 running tests for all implemented algorithms. Algorithms parameters were adjusted after tuning process (as described in Section 3). The experiments were performed on PC with Pentium 4 3,09 GHz.

Experiment (**A**). The *Minimum*, *Average*, and *Maximum* lengths are shown in Fig. 7.

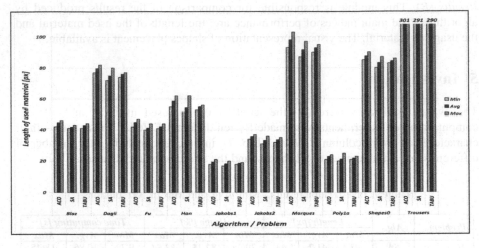

Fig. 7. Min, Avg. and Max of length of the used material for 100 solutions for all algorithms

It can be seen in Fig. 7 that the best solutions were obtained when using SA. For all considered instances, this algorithm got the best results (minimal length of the used material) and for most instances got the best values of the average length. Second-quality solutions returned TS algorithm. The average results were very similar to those produced by SA and the results by TS were of very similar mean values. Low variance of lengths can be explained by the deterministic nature of TS method.

Experiment (**B**). The *Minimum*, *Average*, and *Maximum* times are shown in Fig. 8.

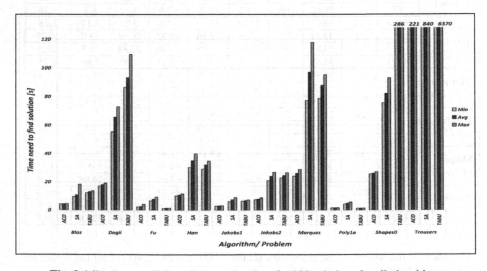

Fig. 8. Min, Avg. and Max of computing time for 100 solutions for all algorithms

It can be observed in Fig. 8 that: for small instances SA needs more time than other methods, for large instances TS needs a much more time than SA and ACO, and that for all instances ACO return solutions in the shortest time.

The best solutions to some instances of problems obtained with KSBAN are in Fig. 9.

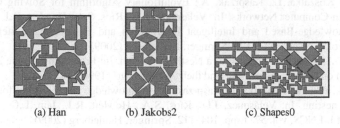

(a) Han   (b) Jakobs2   (c) Shapes0

Fig. 9. Visualisation of the best solutions found by *KSBAN* during experiments

6 Conclusions

The main objective of this work was implementation of simulation system with raster model and meta-heuristic algorithms such as SA, TS, ACO for solving strip nesting problem. The KSBAN system allows tuning algorithms and comparing their efficiencies measured by the length of the used material and the computing time. The investigations made for various instances of problems show that algorithm based on SA found the shortest lengths but that based on ACO was the fastest one.

Future development of the KSBAN system will focus on implementing shape rotation and implementing additional algorithms e.g. based on genetic ideas as well as preparing new modules to allow designing multistage experiments in automatic way.

References

1. Bennell, J.A., Oliveira, J.F.: The geometry of nesting problems: A tutorial. European Journal of Operational Research 184(2), 397–415 (2008)
2. Blazewicz, J., Hawryluk, P., Walkowiak, R.: Using a tabu search approach for solving the two-dimensional irregular cutting problem. Ann. Oper. Res. 41(1-4), 313–325 (1993)
3. Bogalinski, P., Pozniak-Koszalka, I., Koszalka, L., Kasprzak, A.: Algorithms to Solving Strip Nesting Problem based on Tabu Search and ACO. In: Proc. to 23rd IAR Conf., pp. 240–245 (2008)
4. Budzynska, L., Kominek, P.: The application of GLS algorithm to 2 dimension irregular-shape cutting problem. In: Bubak, M., van Albada, G.D., Sloot, P.M.A., Dongarra, J. (eds.) ICCS 2004. LNCS, vol. 3038, pp. 1241–1248. Springer, Heidelberg (2004)
5. Chmaj, G., Koszalka, L.: Quantization Applied to 2D Shape Arrangement. Acta Mosis 112 (2006)
6. Kendall, G., Burke, E.: Applying evolutionary algorithms and the no fit polygon to the nesting problem. In: IC-AI 1999: The 1999 International Conference on Artificial Intelligence (1999)
7. Kendall, G., Burke, E.: Applying ant algorithms and the no fit polygon to the nesting problem. In: Foo, N.Y. (ed.) AI 1999. LNCS, vol. 1747. Springer, Heidelberg (1999)

8. Egeblad, J., Nielsen, B.K., Odgaard, A.: Fast neighbourhood search for two-and three-dimensional nesting problems. European Journal of Operational Research 183(3), 1249–1266 (2007)

9. Oliveira, J.F., Ferreira, J.S.: Algorithms for nesting problems. In: Applied Simulated Annealing. LNEMS, pp. 255–273 (1993)

10. Ohia, D., Koszalka, L., Kasprzak, A.: Evolutionary Algorithm for Solving Congestion Problem in Computer Networks. In: Velásquez, J.D., Ríos, S.A., Howlett, R.J., Jain, L.C. (eds.) Knowledge-Based and Intelligent Information and Engineering Systems. LNCS (LNAI), vol. 5711, pp. 112–121. Springer, Heidelberg (2009)

11. Rintala, T.A.: Genetic Approach to a Nesting Problem. In: Proceedings to the 2nd Nordic Workshop on Genetic Algorithms and their Applications (1996)

12. Chmaj, G., Pozniak-Koszalka, I., Kasprzak, A.: A knowledge based system for minimum rectangle nesting. In: Velásquez, J.D., Ríos, S.A., Howlett, R.J., Jain, L.C. (eds.) KES 2009, Part I. LNCS, vol. 5711, pp. 104–112. Springer, Heidelberg (2009)

Towards Collaborative Library Marketing System for Improving Patron Satisfaction

Toshiro Minami

Kyushu Institute of Information Sciences, 6-3-1 Saifu, Dazaifu, Fukuoka 818-0117 Japan and
Kyushu University Library, 6-10-1 Hakozaki, Higashi, Fukuoka 812-8581 Japan
minami@kiis.ac.jp, minami@lib.kyushu-u.ac.jp

Abstract. Patron satisfaction is the biggest issue for libraries, which are the organizations that aim to provide information services to public. Libraries have been providing various services so far; usually planned based on the librarians' intuition. In these one to two decades, our life style has been changing very rapidly due to the development of information technology such as the Internet, personal computers, mobile phones, and so on. It is now quite difficult to provide the library patrons with services that match their needs based only on the librarians' intuition any more. So we are interested in the methods of finding services based on the various data that libraries are able to collect, which we call the library marketing. In this paper we propose some new ideas of collaborative library marketing system. In this framework we put special interests on collaborative and social features of library marketing methods.

Keywords: Intelligent Library System, Library Marketing, Collaborative Filtering, Intelligent Bookshelf (IBS), Data Analysis, Data Mining.

1 Introduction

Library is one of the important public service organizations in our society. It provides us with information in the form of books, magazines, CDs, DVDs, and so on. It also provides with other services such as advising us for finding appropriate information and helping us for training information literacy skills as well in the reference desk service.

As a service organization, it is a key concern for libraries to have better customer satisfaction, or patron satisfaction (PS). They have been putting much effort in order to improve PS. This was well described in the well-known "The Five Laws of Library Science" by the very famous Indian library scientist S. R. Ranganathan [9] nearly a half century ago. They are (1) Books are for use, (2) Every reader his book, (3) Every book his reader, (4) Save the time of the reader, and (5) The library is a growing organism. They are still applicable to current libraries by replacing "book" with "information", "material", or "service".

Due to the development and popularization of information and communications technology (ICT), our ways of life have been changed quite a lot and still is rapidly changing. More precicely, quite a lot of people carry mobile phones all the time and

N.T. Nguyen, M.T. Le, and J. Świątek (Eds.): ACIIDS 2010, Part II, LNAI 5991, pp. 237–246, 2010.

they can access to the Internet at anytime from anywhere. We now live in a ubiquitous society.

According to the American Marketing Association (AMA) [1], the concept of marketing is; "Marketing is the activity, set of institutions, and processes for creating, communicating, delivering, and exchanging offerings that have value for customers, clients, partners, and society at large." As we apply this concept to library we would be able to say that library marketing is the activity that provides values not only to the library's patrons but also to our society at large.

In this paper we put focus on the library marketing (LM) [5, 6, 8] that utilizes the subjective data that can be collected by libraries; such as circulation data, reference records, material usage data, patrons' profiles, etc. In this type of LM, firstly library collects data, then analyse them, and extract useful information, knowledge, and tips for developing patron services and for improving library management. Further, we put emphasis on collaborative and social natures of LM because we believe such natures are going to be more and more important for the patrons in our ubiquitous society.

The rest of this paper is organized as follows: In the following Section 2 we will recognize the importance and potential of the library marketing concept and methodology through some examples. Then in Section 3 we will propose a model of collaborative library marketing (CLM) system that aims to apply the analysis methods to collaborations between patron and patron, patron and library, and library and library. We will propose some supposedly useful application examples. In Section 4 we deal with LM and CLM by utilizing RFID (Radio Frequency Identification) technology [2], which is considered to be the representative one of AIDC (Automatic Identification and Data Capture) technologies. By using RFID we can strengthen CLM functions. In Section 5, we take mobile phone (MP) as terminal equipment for information access. Current mobile phone is rather PDA (Personal Digital Assistant) with communication function than just a "mobile phone." Further, quite a lot of MPs are equipped with cameras and they can recognize two-dimensional code such as QR (Quick Response) code; which is quite an important advantage of MPs. Finally in Section 6, we will conclude our discussions in this paper.

2   Library Marketing: Case Study

The librarians of Gwacheon Public Library of Korea had a series of experimental research on the usage of rooms and other areas in the library [3]. They put barcode readers at the entrance/exit of the rooms and asked the patrons to let their ID cards to scan as they moved from one area to another. They also collected the exit records as the patrons left the library building during the experiment period.

By analyzing the collected data, they got some results such as, for example, two reading rooms, one in the 3rd floor and another one in the 4th floor, were used by many patrons, and based on the result they proposed to rearrange the rooms so that if the library provided the patrons with a larger reading room, they would be more convenient in the library for reading books [3, 4].

In Ehime University Library in Japan, Yamada analyzed the circulation data of the library and extracted the result how often the books are borrowed along with the passage of year after publication [10]. He concluded that new books are more preferred, i.e. are borrowed more frequently than old books. He also found an interesting fact that some books are constantly borrowed even many years have passed after publication. He called such books as "ever-green books" and he proposed that these books should be considered as must-collect books for university libraries.

In Kyushu University Library in Japan, Kim and Minami reported the analysis results from the data that told how the rooms and areas were used [5, 8]. They found that the information salon, which is the area in the entrance lobby where patrons are able to use PCs, was popularly used, so it would be useful to extend such area for better patron satisfaction. They also found that some rooms located far from the entrance were not used very much around an hour after opening. From this fact they proposed to suspend the turning of the lights of these rooms for an hour after opening time in order to reduce the energy without giving much inconvenience to the patrons.

We hope that such researches on library marketing are becoming more popularly done so that we can share such tips and knowledge for improving patron satisfaction.

3   Collaborative Library Marketing

3.1   A Model of Collaborative Library Marketing System

A general model of library marketing system is described in Figure 1. In the center is a library, which has a library server for collecting, storing, and using the data for library marketing (LM). People and equipment in the library are connected with the server. Inside of the library, the patrons, librarians, and even some equipment like intelligent bookshelves (IBSs)[1], which are the bookshelves that have readers that detect the existence of books in the shelves, are connected to the server and provide data for LM. Sometimes they will get information or data from the server, oppositely.

The patrons outside of the library will also provide with and accept information from the server. Some patrons may be staying at home and they can access the library server with their computers and sometimes with their IBS(s). Some other patrons may want to access to the server as they walk in a street or in a supermarket as they shop.

In the right hand of Figure 1 illustrates the library network. Precisely the library server of each library forms a network and shares some data and information with other libraries so that the group of libraries is able to provide better services to their patrons based on the bigger data and the information obtained in different ways of data and analysis methods.

The data analysis LM in a library is carried out not only automatically by the server but also manually by the librarians in cooperation with the library system. The librarians' major jobs relating to (collaborative) LM are (1) to check the analysis results obtained in the server, (2) to provide the server with information and knowledge that are supposed to be useful for (collaborative) LM, and (3) to develop new analysis methods and patron services.

---

[1] We will deal with IBS more precisely in Section 4.2.

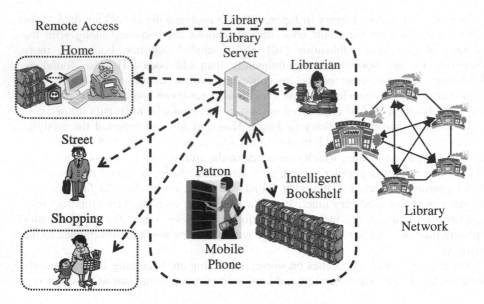

Fig. 1. A Model of Collaborative Library Marketing System

3.2 Collaboration between Patrons in Assistances from Library

The most important mission of library is to provide its patrons with good services so that it is considered to be a reliable organization by the patrons and it will get a good patron satisfaction as a result. From this point of view, it has essential importance to provide with collaboration services not only from the library but also from other patrons.

Firstly we would like to discriminate two types of collaboration in this paper; explicit and implicit [7]. Explicit collaboration is the ones that patrons do something together or teach each other. On the other hand in an implicit collaboration the patrons do something, learn for example, and it looks like doing alone. The data in the process of leaning or something are collected and stored in the service server. The data are analyzed and the results will be used in order to help the patrons as they learn. In such cases the collaboration is taken place in an implicit manner. Eventhough the patrons do not intend to help other patrons, they help each other as a result. We would like to emphasize in this paper that implicit collaboration is as important as explicit collaboration.

Suppose that the patron at home is studying a subject. He will borrow some text books and reference books. He will ask for help to librarians. He may have some bookshelves that are able to detect what books are shelved in real time, i.e. intelligent bookshelves (IBSs). We suppose that the libraries having collaborative library marketing system will provide their patrons with learning assistant service. Then the registered patrons will provide the data that are collected in their learning process to the library. The library will assist the patrons as they learn based on the data and the librarians' knowledge and skill, which is in other words, managing and helping the patrons in their learning.

By collecting, storing, and analyzing such data the library will be able to help their patrons based on the analysis results what would be the best materials for each patron by considering the learning records, learning style, such as to learn one by one or learning randomly etc., studying skills, such as how fast the patron learn in comparison with average learners, and so on. This is a good example of (implicit) collaborative learning and collaborative LM system. A method for recommending learning materials is proposed in [7].

For the patrons on the street or in a store for shopping, the similar methods will be applicable as well. By providing such services, future libraries will be able to play a social role not only providing books, magazines, and information but also providing assistant services for intelligent production processes.

4 Collaborative Library Marketing with RFID Technology

4.1 RFID Technology for Library

The principles of RFID tag system [2] is illustrated in Figure 2. The RFID tag system consists of two major components; tags and reader/writers (R/Ws). A passive tag has no battery for the IC chip. Instead of that it has an antenna with which it gets electric energy from the antenna of an R/W when they are close enough each other. Then they transmit data from the R/W to the tag (i.e. write) and vice versa (i.e. read). The R/W is controlled by some application program, probably a part of the library system, running on a computer.

RFID technology has the following advantages when it is applied to library. (1) Efficient Checkout and Checkin: Checkout and checkin processing with RFID are much faster than with barcode. Furthermore use of RFID self-checkout and self-checkin machines will reduce the burden of library staff a lot as well. (2) Easy to Use Checkout and Checkin: Furthermore, the self-checkout/checkin machines are much easier to use for patrons because the locationing restrictions for RFID tags are far less than barcode tags, and because of the multiple reading with anti-collision mechanism of RFID that is impossible with barcode. (3) Efficient Inventory: The inventory time reduces notably, probably from half to one tenth or even less.

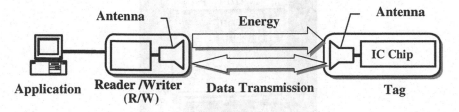

Fig. 2. Principle of RFID (Passive) Tag System

4.2 Intelligent Bookshelf (IBS) System with RFID Technology

An intelligent bookshelf (IBS) is a bookshelf which has shelf type RFID R/W, where its RFID antennas are installed in the bookshelves so that the R/W can detect what books are put in which shelf in real time [6]. Figure 3 is an example of IBS applied to a book truck, or a book trolley.

By using such equipments we can collect the usage data of books. There are two types of IBSs in Figure 1. One type is in the library and another one is at home. IBSs will detect when a book is taken out from the shelf and when it is returned. These data are stored in controlling computers and they are transmitted to the library server.

By analyzing the usage data we may extract useful knowledge for library marketing. For example, we can get the information how often a specific book is used and the differences of usage patterns according to the day of the week, time zone in a day, etc. These usage patterns will differ from IBS to IBS. The IBS at home will reflect the learning pattern of the patron and the IBS in the library will reflect the usage pattern of visiting patrons as a whole.

One possible method of analysis of these data that should be useful for collaborative service is to use the collaborative filtering (CF). We suppse that the library has the learning history data of several patrons, that are collected by the learning support system. By analysing the data and the circulation data of each of the patrons, the system creates profile data of the patron; his/her interest, profile, learning periods for various subjects, privacy data, and many others. When one patron starts learning a subject, the system will search the database for patrons who have similar profiles. By using the learning history of such similar patrons, the system will estimate how long will it take for the learning patron to finish the subject, what materials of the library might be useful for the patron, what kind of librarians' support will be helpful, and so on and so forth. This is a typical example of collaborative learning style of patrons under the support of the library. It is possible to privide with other collaborative learning services using circulation data, additional to the learning data or themselves, and also by using other kinds of data such as information from the Web documents, librarians' network, etc.

Fig. 3. An Intelligent Bookshelf (IBS) for Library Marketing

4.3 Collaborative Library Marketing with IBS

Intelligent bookshelf (IBS) is quite useful itself. It will be even more useful if we put other equipments such as display screen, LED lamp, shelf ID mark, etc. Figure 4 is an illustration of such an "extended IBS". Different from the ordinary IBS that is just input equipment for the library server an extended IBS becomes input/output equipment so that it can interact with patrons.

By using such IBS, the library is able to provide more advanced services to its patrons. One possible service is like as follows. If a patron is interested in a book of the IBS, he will take out the book and put it at the reader of the IBS. Then the information about the book will be displayed on the screen. The information displayed on the screen might include its evaluation information, probably provided by other patrons, librarians, or Web pages like blogs and others. From this action the system can get the data that some patron is interested in the book strong enough to read the related information together with timestamp data. It might be more useful if the screen is a touch panel so that the patron can go forward to find additional information of the book.

The IBS can be used for navigating the patrons to the appropriate location. In one example case we suppose that a patron wants to find a bookshelf where the books in the intended field are shelved. Then he or she put the information about the field, with decimal classification number for example, and then the floor plan of the library will be displayed on the shelf screen and tells which direction to go in order to find the target bookshelf. When the patron reaches to the target bookshelf, or tier, then the LED lamp will turn on and shows where the intended books will be found.

In this section we have supposed that such services are provided by using only RFID tags and other equipments. In the following section, we will deal with mobile phones with cameras that can recognize 2D-code. With these equipments library can provide even more convenient services.

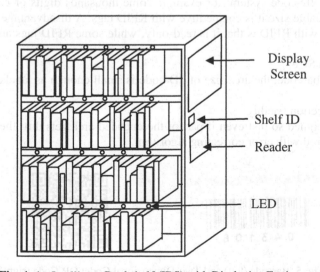

Fig. 4. An Intelligent Bookshelf (IBS) with Displaying Equipment

5 Collaborative Library Marketing with Mobile Phone

5.1 Advantages of Mobile Phones (MPs) for Marketing

Thanks to the development and popularization of the mobile phones (MPs) together with their accessibility to the Internet as well as they are equipped with display facility, our information environment has greatly changed. With MPs we can get virtually any information as we walk, as we are in a bus or train, and as we shop just like as we stay at home. Now we live in a ubiquitous society. As is shown in Figure 1 in Section 3.1 we assume such environment in our collaborative library marketing system.

Furthermore quite a lot of recent MPs are equipped with cameras and it is possible not only to take pictures but also to recognize two-dimensional code (hear after also called 2D-code for brevity) marked on various goods with them [4]. QR code (Figure 5 (right)) is most popularly used in Japan among several other kinds of 2D-codes. Other 2D-codes include PDF417, DataMatrix, and Maxi Code.

Because of such pupularity, QR code is applied to quite a lot of fields such as discount coupons for shopping and foods in restaurants, as entrance tickets for exhibitions, and naturally enough as to let the customer access to the companies' Web pages, and various others.

Barcode (Figure 5 (left)) has been used worldwide including libraries. 2D-code can be considered as the advanced version of barcode and thus it will replace the role of barcode in the near future. 2D-code has the following features.

(1) Low Cost:
Just like barcode, it is not expensive and we can print with ordinary printers.
(2) Large Recordable Data Size:
The recordable, or printable, size of data is about tens of digits in barcode, while it is much large in 2D-code system; for example some thousands digits or characters. In terms of recordable size it is competitive with RFID tags. A disadvantage of 2D-code in comparison with RFID is that it is read-only, while some RFID tags are read-write type.
(3) Small Area Size:
Comparing to barcode, the area size of 2D-code is small enough as labels for back of books.
(4) Error Correction Facility:
2D-code is designed so that even if part of the encoded area gets dirt, the whole data may be recovered with error correcting algorithm.

Fig. 5. Barcode (left) and Two-Dimensional Code: QR Code (right)

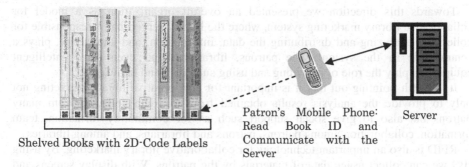

Shelved Books with 2D-Code Labels

Patron's Mobile Phone: Read Book ID and Communicate with the Server

Server

Fig. 6. Books that Two Dimensional Code Labels are Attached and the Service to Get Information about the Books with a Mobile Phone

5.2 Utilization of Mobile Phones for Collaborative Library Marketing

Two-dimensional code has some advantage not only over barcode as was discussed, but also over RFID tag in some points of view. Probably the most important one is that, also as was pointed out in the previous section, virtually all patrons have its reader, as an application of their mobile phones. It is technologically realizable to develop a mobile phone that has RFID reader. However it is quite difficult practically because such MP will be very expensive so that most people will not buy.

One of the good applications of 2D-code in book management in library is to put a 2D-code label on each of the books. A typical image of this application is shown in Figure 6. 2D-code can include a lot more of data compared with barcode, we can put not only the book ID but also other useful data such as the classification ID, book title, author name(s), the dates it has been borrowed, the average/shortest/longest days of borrowing, and so on. The shelves might look like the one shown in the left part of Figure 6. A patron can identify the ID of a book that he or she is interested in. By aiming the mobile phone and take a picture of the 2D-code, the book's ID data is easily taken into the mobile phone. Then the ID data is transmitted to the server for it, and the information relating to the book will be displayed on the phone. Starting from this point, the patron can go further. For example, he or she can get further information, put the data to his or her own blog, record the book to the personal virtual shelf, and even go to an e-commerce site and purchase the book.

6   Concluding Remarks

The aim of this paper is to demonstrate the importance and usefulness of collaborative nature of library marketing in order for libraries to have better patron satisfaction. Generally speaking the ordinary services such as providing patrons with appropriate books, magazines, and other information has the major importance for libraries. However in addition to such services, other types of services should gain growing importance for the libraries towards the future society where patrons ask for help from the libraries. In this paper we put special emphasis on the collaborative services as the representative ones among various such services.

Towards this direction we presented an overall organization as a model for collaborative library marketing system, where the library server that is responsible for collecting, analyzing and distributing the data, information, and knowledge, plays a central role in the system. The patrons, librarians, and even some intelligent equipments play the role of providing and using such data and knowledge.

It is worth pointing out that it is important for collaborative library marketing not only to provide the analysis results obtained from the data collected from many patrons but also to give social support such as for community formation, team formation, collaboration among patrons, patrons and librarians, and among libraries.

RFID is also an important technology for collaborative library marketing. By using IBS we can collect usage data of materials by the patrons. With display screens and other equipment, IBS will become even more useful for patrons and librarians as well.

Furthermore mobile phones are essential tools as terminal equipment, or PDAs, for library marketing system. Without them realization of ubiquitous society where people can access to any information they need on the fly so that they can solve various problems as soon as it come across to the patrons.

We are developing a prototype system based on the multi-agent system (MAS) so that we can evaluate our concept for collaborative library marketing system. We have to collect the practical data for library marketing in our experiments and apply our methods on these data and assess their usefulness and get tips for improving the algorithm for analysis for the practical applications to the libraries in the near future.

References

1. American Marketing Association (AMA), http://www.marketingpower.com/
2. Finkenzeller, K.: RFID Handbook, 2nd edn. John Wiley & Sons, Chichester (2003)
3. Kim, E.: A Study on the Space Organization by the User's Behavior in Public Library. Gyeonggi University (2008) (in Korean)
4. Minami, T.: A perspective to the Library in Network-Oriented Society – Ubiquitous Library Services through PDAs, Bulletin of Kyushu Institute of Information Sciences (2008) (in Japanese)
5. Kim, E., Minami, T.: Library's Space Organization Improvement Based on Patrons' Behavioral Research—An Attempt for Library Marketing at Gwacheon Public Library in Korea and Kyushu University Library, Research and Development Division Annual Report 2008/2009, Kyushu University Library (2009) (in Japanese)
6. Minami, T.: A Design for Library Marketing System and its Possible Applications. In: Richards, D., Kang, B.-H. (eds.) Knowledge Acquisition: Approaches, Algorithms and Applications. LNCS (LNAI), vol. 5465, pp. 183–197. Springer, Heidelberg (2009)
7. Minami, T.: Towards Library Supported Collaborative Learning. In: Proc. CSEDU 2009 (2009)
8. Minami, T., Kim, E.: Data Analysis Methods for Library Marketing. In: Lee, Y.-h., Kim, T.-h., Fang, W.-c., Ślęzak, D. (eds.) FGIT 2009. LNCS, vol. 5899, pp. 27–34. Springer, Heidelberg (2009)
9. Ranganathan, S.R.: The Five Laws of Library Science. Bombay Asia Publishing House (1963)
10. Yamada, S.: Analysis of Library Book Circulation Data: Turnover of Open-shelf Books. Journal of College and University Libraries 69, 27–33 (2003) (in Japanese)

DAG Scheduling on Heterogeneous Distributed Systems Using Learning Automata

Habib Moti Ghader[1], Davood KeyKhosravi[2], and Ali HosseinAliPour[3]

[1] Islamic Azad University- Tabriz Branch-Young Research Club, Tabriz, Iran
Habib_Moti@Yahoo.Com
[2] Islamic Azad University-Osku Branch- Departmant of Computer Engineering, Osku, Iran
kaikhosravi2003@yahoo.com
[3] Islamic Azad University- Tabriz Branch, Departmant of Computer Engineering, Tabriz, Iran
Ali.hosseinalipour@Yahoo.com

Abstract. DAG scheduling is of great importance to optimal distribution of tasks in parallel and distributed systems. In this paper a novel approach to DAG scheduling, utilizing learning automata across distributed systems, is proposed. The learning process begins with an initial population of randomly generated learning automata. Each automaton by itself represents a stochastic scheduling. The scheduling is optimized within a learning process. Compared with current genetic approaches to DAG scheduling better results are achieved. The main reason underlying this achievement is that an evolutionary approach such as genetics looks for the best chromosomes within genetic populations whilst in the approach presented in this paper learning automata is applied to find the most suitable position for the genes in addition to looking for the best chromosomes. The scheduling resulted from applying our scheduling algorithm to some benchmark task graphs are compared with the existing ones.

Keywords: DAG, Distributed Systems, Learning Automata, Scheduling.

1 Introduction

DAG scheduling on Distributed Systems is of great importance to distribution of programs across the networks. DAG scheduling is an NP-hard problem [8-9]. Therefore, to avoid the NP-hardness, heuristic approaches are conventionally applied to find near optimal scheduling [12-15]. Genetic approaches provide near-optimal scheduling for large task graphs in an almost acceptable amount of time [10]. These approaches start with an initial population of randomly selected solutions or chromosomes. Within an evolutionary process the population is improved and new generations are created. The process terminates mostly after a predefined number of generations or when the qualities of the chromosomes within a generation are almost similar. The time required for this evolutionary process depends on the size of the problem search space and the method applied to evolve individual chromosomes within a population. However, not only the quality of the chromosomes but also the

N.T. Nguyen, M.T. Le, and J. Świątek (Eds.): ACIIDS 2010, Part II, LNAI 5991, pp. 247–257, 2010.

position of genes within each chromosome affects the speed of the genetic algorithms. A major shortcoming of the genetic approaches is that they do not look for the best genes within chromosomes. Genetic approaches only focus on the chromosomes and do not consider genes. Learning automates could be applied to find the best position for the genes in each chromosome. Thereby, the evolutionary process could be speeded up.

In this paper a new evolutionary approach based on learning automates is presented. The approach applies Object Migration Automates (OMA) to represent different possible scheduling [16]. As described in Section 3 each action within an object migration automaton represents a single task within the task graph. Each action itself may have n different states, where each state defines the quality of each task or gene within a possible scheduling or automata. The evolutionary approach starts with an initial population of randomly generated object migration learning automata. Each automata itself represents a scheduling of the task graph. Within a learning process the quality of these automatons is improved. Our experimental results, demonstrates the quality and the speed of finding near optimal- scheduling for a given task graph. The remaining parts of this article are organized as follows: In Section 2, learning automata and in section 3 a novel algorithm for solving task graph scheduling problem will be introduced. In section 4, experimental results are presented and finally in part 5, conclusion will be discussed.

2 Learning Automata

Learning automata is an abstract model that chooses randomly an operation from a set of finite operations and then applies it on the environment to the selected operation environment is evaluated by learning automata and then informs the evaluated result by the help of a reinforcement signal to the learning automata. The learning automaton updates its interior situation by utilizing selected operation and reinforcement signal and selects the next operation afterwards [1-5, 9].

3 The Proposed Algorithm

For a task graph with n nodes, different n! Permutation exists in DAG scheduling and if learning automata is used for solving DAG scheduling problem, learning automata must be n! actions. Because much more actions lead to the reduction of converge speed, so due to this fact, Object Migration Automata proposed by Oommen and Ma is utilized, that one of them is OMA based on Tsetline automata. The represented algorithm utilizes Tsetline automata for solving task graph scheduling problem.

In this automaton $\underline{\alpha} = \{\alpha_1, \alpha_2, ..., \alpha_r\}$ is a set of suitable actions for learning automata. This automaton has K action (the member of actions is equal the number of graph task). $\underline{\phi} = \{\phi_1, \phi_2, ..., \phi_{KN}\}$ is a set of statuses and N is memory depth for automata. The set of those statuses of this automaton are partitioned to K sub-division $\{\phi_1, \phi_2, ..., \phi_N\}$, $\{\phi_{N+1}, \phi_{N+2}, ..., \phi_{2N}\}$, ..., $\{\phi_{(K-1)N+1}, \phi_{(K-1)N+2}, ..., \phi_{KN}\}$ and each task is classified according to its status. In K action status set $\phi_{(k-1)N+1}$ is called inter status and ϕ_{kN} status is called border status. Situated node in $\phi_{(k-1)N+1}$ is a node with maximum importance and ϕ_{kN} is a node with minimum importance.

3.1 Initial Population

At first, P (number of population) random generated automata and then all tasks are allocated to interior automata statuses by the use of Algorithm Fig.1.

Procedure 1: assign task's to state's	
1. Input: automata L, number of automata p, number of tasks k, memory depth n, automata state S.	4. For i=1 to k
	5. Assign t_i to L_j.S(1+[(i-1)*n])
2. Output: p automata.	6. end //end of for.
3. For j=1 to p	7. End //end of procedure.

Fig. 1. Generating initial population

Then a number is allocated randomly to all actions. The random allocated number to actions includes 2 concepts:

1. Task priority.
2. Processor's number, which executes the task.

After allocating random numbers to automata actions, the values of the actions are interchanged with the number of $N/2$ load. Fig.2 shows the manner of running this work.

Procedure 2: Assign Random Numbers to Actions	
1. input: number of processors m, number of tasks n	8. $j \leftarrow$ random[1, n];
2. output: Automata $v(j)$	9. $k \leftarrow$ random[1, n];
3. begin	10. if $j \neq k$ then
4. for $j = 1$ to n	11. swap {$v(j)$, vk(k)};
5. $v(j) \leftarrow m * j$ -random[0,m − 1];	12. end //end of for.
6. end //end of for.	13. output: the Automata $v(j)$;
7. for $i = 1$ to $\lfloor n/2 \rfloor$	14. end //end of procedure.

Fig. 2. Allocating tasks to processors

We are going to perform, the shown graph in Fig.3, in a system having 2 processing units. Referring to Fig.4, you can observe how tasks are allocated to interior status in order. Since there are two processors in the system, so odd numbers indicate P1 processor and even numbers indicate P2 processor. If the system includes more than two processors, processor's number will consist of the result of the allocated number to the number of all processors.

The manner of swapping actions values of an automaton is described in Fig. 5.

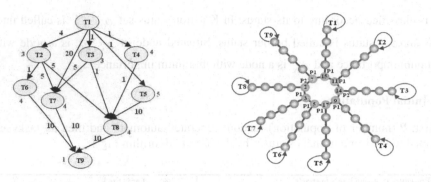

Fig. 3. Example of DAG with 9 tasks   **Fig. 4.** Example of automata for Fig. 3 task graph

a. automata action status after   b. Selection of two random actions   c. The output for automata after
allocating values randomly   swapping actions randomly

Fig. 5. Example of swapping 2 random actions

3.2 Task Execution on Processors

While a program is running over parallel processors, data dependency between tasks must be taken into consideration. In fact, a task is not able to be executed unless all its parent's tasks have been already executed. In this part, the manner of task execution on processors will be described. Each task is run according to Fig.4 automata in related processor; in a way that among all ready tasks, a task with high priority than others will be executed. Ready task means that its parent's entire task has been executed. Regarding Fig.4 automata and Fig.3 DAG, for example, in first step just T_1 task is ready. Thus T_1 will be run in first step. After T_1 is executed, other tasks T_4، T_3، T_2 and T_5 will become ready, in this step, T_5 task priority is more than other ready tasks; therefore, T5 task will be run. So all tasks will be executed on their related processors [10]. Fig. 6 specifies tasks execution on related processors in order. With regard to Fig.4 automata, tasks $T_1, T_5, T_2, T_4, T_6, T_7, T_9$ will be run on P1 processor and T_3, T_8 on P2 processor as well.

Procedure 3: One Task Sequence Growth		
1.	input: number of tasks n, $Automata\ v(j)$, the set of nodes $\overline{S}$	7. $j^* \leftarrow \arg\max\{v(j)\|j \in \overline{S}\ \};$
2.	output: task sequence TS	8. $\overline{S} \leftarrow \overline{S} \setminus j^*;$
3.	begin	9. $TS \leftarrow TS \cup j^*;$
4.	$\overline{S} \leftarrow$ empty, $TS \leftarrow$ empty;	10. $j \leftarrow j^*;$
5.	$n \leftarrow 0, j \leftarrow 0;$	11. end //end of while.
6.	while $(j \leq n)$ do	12. output: task sequence TS;
		13. End //end of procedure.

Fig. 6. Algorithm for specifying task execution order

After running algorithm Fig.6 for graph Fig.3, tasks execution order will be as Fig.7.

J	$\overline{S}$	$v(j)$	J^*	TS
0	{1}	v(1)=15	1	S={1}
2	{2, 3, 4, 5}	V(2)=11, v(3)=14, v(4)=9, v(5)=17	2	S={1, 5}
1	{2, 3, 4}	V(2)=11, v(3)=14, v(4)=9	3	S={1, 5, 3}
3	{2, 4}	V(2)=11, v(4)=9	4	S={1, 5, 3, 2}
5	{4, 6, 7}	V(4)=9, v(6)=5, v(7)=3	5	S={1, 5, 3, 2, 4}
4	{6, 7, 8}	V(6)=5, v(7)=3, v(8)=2	6	S={1, 5, 3, 2, 4, 6}
7	{7, 8}	V(7)=3, v(8)=2	7	S={1, 5, 3, 2, 4, 6,7}
9	{8}	V(8)=2	8	S={1, 5, 3, 2, 4, 6,7, 8}
6	{9}	V(9)=7	9	S={1, 5, 3, 2, 4, 6,7, 8,9}

Fig. 7. Graph Fig.3 task execution order

All tasks, now, will be executed according to Fig.7 table. The manner of tasks execution is described in Fig.8. As it is clear from Fig.8, acquired output from this algorithm is *makespan*.

Procedure 4: Assigning Tasks to Processors	
1. input: processing time *pk*, task sequence *TS*, Automata *v(j)*, the communication delay τ_{jk}	16. $tk \leftarrow ek + pk$;
	17. $S \leftarrow S \in Sk\{Pi\,; ek + pk\}$;
	18. end // end of if.
	19. else
2. output: makespan *f*, schedule *S*	20. begin
3. begin	21. $ek \leftarrow \max\{tj\,|j \in pre(k)\}$;
4. $Pi \leftarrow 0, i = 1, 2, \ldots, m$;	22. $tk \leftarrow ek + pk$;
5. $tk \leftarrow 0, k = 1, 2, \ldots, n$;	23. $S \leftarrow S \cup Sk\{Pi\,; ek + pk\}$;
6. $s \leftarrow 0$;	24. end //end of else.
7. for *j* = 1 to *n*	25. else
8. $s \leftarrow T\,Si$;	26. begin
9. $Pi \leftarrow v(s)\%m$;	27. $ek \leftarrow \max\{tj\,|j \in pre(k)\}$;
10. if *Pi* = 0 then $Pi \leftarrow m$;	28. $tk \leftarrow ek + pk$;
11. if assigned task *Tj* $\prec$ *Tk* then	29. $S \leftarrow S \cup Sk\{Pi\,; ek + pk\}$;
12. if $Pi \neq Pl$ then	30. end //end of if.
13. begin	31. end // end of for
14. $ek \leftarrow \max\{tj\,|j \in pre(k)\} + \tau_{jk}$;	32. output: schedule *S*
15. $tk \leftarrow ek + pk$;	33. makespan*f* $\leftarrow \max\{tk, k = 1, 2, \ldots, n\}$;
	34. end //enf of procedure.

Fig. 8. Algorithm-task execution

Fig.9 shows task execution order on processors in details.

J^*	S	p_i	$t_i = e_i + p_j$
1	P1={1}, p2={}	1	T1= 0+2=2
5	P1={1, 5}, p2={}	1	T5=2+5=7
3	P1={1, 5}, p2={3}	2	T3=3+1.5=4.5
2	P1={1, 5, 2}, p2={3}	1	T2=7+3=10
4	P1={1, 5, 2, 4}, p2={3}	1	T4=10+4=14
6	P1={1, 5, 2, 4, 6}, p2={3}	1	T6=14+4=18
7	P1={1, 5, 2, 4, 6, 7}, p2={3}	1	T7=18+4=22
8	P1={1, 5, 2, 4, 6, 7}, p2={3, 8}	2	T8=17+2=19
9	P1={1, 5, 2, 4, 6, 7, 9}, p2={3, 8}	1	T9=29+1=30

Fig. 9. Display of task execution order in details

After that algorithm Fig.8 is executed for displayed automata in Fig.4, task execution will be as given Gantt chart in Fig.10.

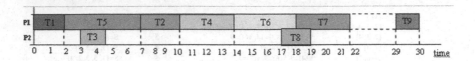

Fig. 10. Display of task graph 7.a execution by automata Fig.4

3.3   Reward and Penalty Operator

In every automaton, an action is selected randomly and then is rewarded or penalized as well. Regarding the action's reward or penalty, the action's status varies in the set of related action status. If an action is situated in a border status, its penalizing would cause the change of that action and as a result, a new *makespan* is established. Considering the learning automata type, reward and penalty operator will be different. Fig.11 shows the

---
Procedure 5 : Reward
1.   **Input: number of state s , number of tasks n, Automata L**
2.   **output: Automata L**
3.   **begin**
4.   **If ((L.s-1)%n<>0) then**
5.   **L.s--;**
6.   **End //end of procedure.**

Fig. 11. Performing an automata reward

Rewarding will be done in which the rate of task evaluation is smaller than Threshold rate.

Task evaluation rate consist of $\dfrac{x}{y}$.

x: A set of communication cost of all parents nodes and t_i child node as. $\left[\sum c(t_i,t_j) \quad if(p_{t_i} \neq p_{t_j})\right] P_{t_i} \neq P_{t_j}$.

y: A set of communication cost of all child nodes and t_i parent node. $\sum c(t_i,t_j)$

p_{t_i} : A processor that t_i task is performed on it.

p_{t_j} : A processor that t_j task is performed on it.

$c(t_i, t_j)$: Communication cost between t_j and t_i tasks.

Threshold rate is equal $\dfrac{T}{Ntasks}$.

T: Consist of a number of related tasks to t_i task that is executed on a processor which t_i task is run in it.

Ntasks: The number of all graph tasks.

The more t_i tasks evaluation rate is inclined towards zero, the more the communication cost between processors will have inclination towards zero, so if evaluation rate of t_i task is equal zero, the reason is that all related t_i tasks are run over a processor. As indicated in Fig.1, this automaton includes 9 tasks and Fig.10 shows the manner of task execution on processors. As rewarding a task, attributed number to the related action changes its status. If the chosen action is a t_3 task. Then the manner of rewarding will be according to Fig.12 automata, but if the attributed number to a task is located in an ultimate status(for example: in automata Fig.2 status), So by implementing rewarding nothing will occur.

If evaluation rate of a task is bigger or equal the threshold value, so the vertex of given task is penalized. Fig.13 displays the way a vertex is penalized. Upon penalizing a vertex, two kind of state may take place:

1. The given vertex is placed in a status except border one. Therefore, penalizing the vertex would minimize its importance. The manner of penalizing the related vertex, for example to t_7 task is indicated in Fig.14.

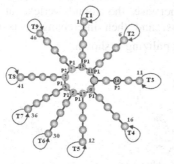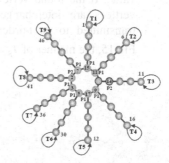

a. Automata status before rewarding t_3 .   **b.** Automata status after rewarding t_3

Fig. 12. Manner of rewarding t_3 task

Procedure 6: Penalize	
1. Input: State S, Number of state u, Number of tasks n, Automata L, Action A	11. Create scheduling *LA'* from *LA* by swapping u and U
2. output: Automata L	12. if Lenght(makespan (LA') <best makespan Length then
3. Begin	13. begin
4. repeat	14. best makespan Lenght = Lenght(makespan(LA');
5. for u = 1 to n do	15. bestNode = U;
6. if (L.s(U)) mod N <> 0 then	16. end //enf of if.
7. L.s(U)++;	17. end //end of for.
8. end //end of for.	18. L.S(bestNode) = L.A(bestNode)*N;
9. until at least one node appears in the boundary state Best sequence makespan;	19. L.S(u) = L.A(u)*N;
10. for U = 1 to n do	20. Swap(L.S(u),L.S(bestNode));
	end //end of procedure.

Fig. 13. Penalizing a vertex

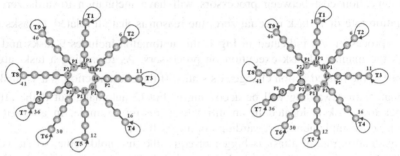

a. Automata status before penalizing t_7 task.   **b.** Automata status after penalizing t_7 task.

Fig. 14. The manner of t_7 task penalizing

2. The given vertex is situated in border status. In this case, a vertex from a graph is to be found in a way that processor's swapping (attributed number to vertices) will have the most reduction in Finish Time (FT) value. If the found vertex is placed in border status, the places of two vertices are interchanged, otherwise the found vertex, at first, is transmitted to its border status, and then interchanging is done. In Fig.15, the manner of t_8 task penalizing is shown.

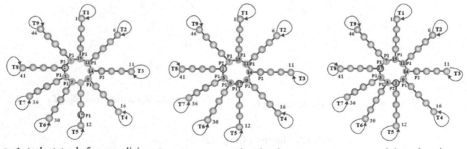

a. t_8 task status before penalizing. **b.** t_5 vertex transmit to border status. **c.** t_8, t_5 task interchanging.

Fig. 15. t_8 task penalizing

4 Experimental Results

In this part, the proposed algorithm is compared and evaluated with other algorithms. At first, with regard to Fig.16.a graph, the acquired results of LDCP, HEFT, DLS algorithm [11] and GA [10] and the proposed algorithm (PLA[1]) are displayed in Fig. 17. Fig. 18 shows the scheduling Finish time of the given algorithms utilizing 2 processors.

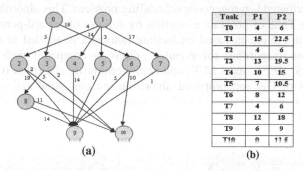

Task	P1	P2
T0	4	6
T1	15	22.5
T2	4	6
T3	13	19.5
T4	10	15
T5	7	10.5
T6	8	12
T7	4	6
T8	12	18
T9	6	9
T10	0	11.6

(a)   (b)

Fig. 16. a. An example of task graph. **b.** Computation cost matrix

The genetic algorithm and learning automata applied the following parameters throughout the simulations:

	Population Size	Generation	Memory Depth	Crossover Probability	Mutation Probability
PLA	50	100	5	-	-
GA	50	100	-	0.7	00.3

Algorithm	DLS	HEFT	LDCP	GA	PLA
No. Processor	2	2	2	2	2
Finish Time	65.5	65.5	64	58.5	56

Fig. 17. Comparison of the PLA with other algorithms (for displayed graph in Fig. 16.a)

Fig. 19 shows the learning process of the proposed algorithm.

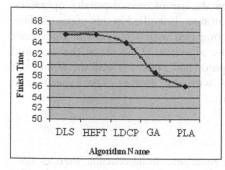

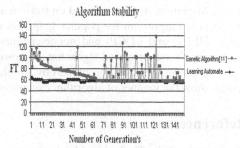

Fig. 18. Comparison of the PLA with other

Fig. 19. Learning process for the PLA with once execution by 150 frequent times for Fig.3 graph.

---

[1] Proposed Learning Automata.

Now we are going to run the proposed algorithm for multiple graphs and compare the result with the acquired results of GA. Fig. 20.a indicates the execution results on a distributed system with P1 and P2 processors. It should be mentioned that the throughput of P1 doubles the P2 processing power. Fig. 20.b shows the execution results on a distributed system with 3 processors (P1, P2, P3). It should be started that the throughput of P1 and p2 are equal but P3 processing power is half of P1.

In this paper, a nondeterministic algorithm based on learning automata is proposed to solve the distributed heterogeneous scheduling problem. This algorithm is trying to optimize Finish Time of program execution by the use of reward-penalty operators. Object Migration Automata based on tsetline automata is applied in this paper. At last, the proposed algorithm for popular graphs is compared with deterministic algorithms (DLS, HEFT, and LDCP) and nondeterministic algorithm (GA). Then the acquired results of GA and proposed algorithm are compared and examined with the number of multiple processors and for numerous graphs with one another. The experimental results show the manner of algorithm execution.

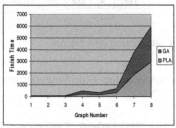
a. distributed system with 2 processors

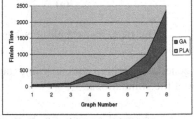
b. distributed system with 3 processors

Fig. 20. a. Comparison of the acquired results of the PLA and GA for random graphs in two different distributed systems

5 Conclusion

In this paper, a nondeterministic algorithm based on learning automata is proposed to solve the distributed heterogeneous scheduling problem. This algorithm is trying to optimize Finish Time of program execution by the use of reward-penalty operators. Object Migration Automata based on tsetline automata is applied in this paper. At last, the proposed algorithm for popular graphs is compared with deterministic algorithms (DLS, HEFT, and LDCP) and nondeterministic algorithm (GA). Then the acquired results of GA and proposed algorithm are compared and examined with the number of multiple processors and for numerous graphs with one another. The experimental results show the manner of algorithm execution.

References

1. Narendra, K.S., Thathachar, M.A.L.: Learning Automata: An introduction. Prentice Hall, Englewood Cliffs (1989)
2. Meybodi, M.R., Beigy, H.: New Class of Learning Automata Based Scheme for Adaptation of Back propagation Algorithm Parameters. In: Proc. of EUFIT 1998, Achen, Germany, September 7-10, pp. 339–344 (1998)

3. Oommen, B.J., Ma, D.C.Y.: Deterministic Learning Automata Solution to the Keyboard Optimization Problem. IEEE Trans. on Computers 37(1), 2–3 (1988)
4. Beigy, H., Meybodi, M.R.: Optimization of Topology of Neural Networks Using Learning Automata. In: Proc. of 3th Annual Int. Computer Society of Iran Computer Conf. CSICC 1998, Tehran, Iran, pp. 417–428 (1999)
5. Hashim, A.A., Amir, S., Mars, P.: Application of Learning Automata to Data Compression. In: Narendra, K.S. (ed.) Adaptive and Learning Systems, pp. 229–234. Plenum Press, New York (1986)
6. Daoud, M.I., Kharma, N.: Ahigh performance algorithm for static task scheduling inheterogeneous distributed computing systems. J. Parallel Distrib. Comput. 68, 399–409 (2008)
7. Wu, M., Dajski, D.: Hypertool: a programming aid for message passing systems. IEEE Trans. Parallel Distributed Systems 1(3), 330–343 (1990)
8. Sarkar, V.: Partitionning and Scheduling parallel Programs for Execution on Multiprocessors. MIT Press, Cambridge (1989)
9. Garey, M.R., Johnson, D.S.: Computers and Instractability: A Cuid to Theory of NP\Completeness. Freeman, New York (1979)
10. Hwang, R., Gen, M., Katayama, H.: A comparison of multiprocessor task scheduling algorithms with communication costs. Computers & Operations Research 35, 976–993 (2008)
11. Daoud, M.I., Kharma, N.: A high performance algorithm for static task scheduling in heterogeneous distributed computing systems. J. Parallel Distrib. Comput. 68, 399–409 (2008)
12. Wu, A.S., Yu, H., Jin, S., Lin, K.C., Schiavone, G.: An Incremental Genetic Algorithm Approach to Multiprocessor Scheduling. IEEE Transaction on Parallel and Distributed Systems 15(9), 824–834 (2004)
13. Salleh, S., Zomaya, A.Y.: Multiprocessor scheduling using mean-field annealing. Elsevier, Future Generation computer Systems 14, 393–408 (1998)
14. McCreary, C.L., Khan, A.A., Thompson, J.J., McArdle, M.E.: A Comparison of Heuristics for Scheduling DAGS on Multiprocessors. In: Proceedings of the 8th International Parallel Processing Symposium, April 1994, pp. 446–451 (1994)
15. Kwok, Y.K., Ahmad, I.: Static scheduling algorithms for allocating directed task graphs to multiprocessors. ACM Comput. Surv. 31(4), 406–471 (1999)
16. Zarei, B., Meybodi, M.R., Abbaszadeh, M.: A Hybrid Method for Solving Traveling Salesman Problem. In: 6th IEEE/ACIS International Conference on Computer and Information Science. ICIS 2007, July 11-13, pp. 394–399 (2007)
17. Oommen, B.J., Ma, D.C.Y.: Deterministic Learning Automata Solution to the Keyboard Optimization Problem. IEEE Trans. on Computers 37(1), 2–3 (1988)
18. Poznyak, A.S., Najim, K.: Learning Automata and Stochastic optimization. Springer-verlag London Limited, Heidelberg (1997)
19. Motie Ghader, H., Parsa, S., Nejad, M.H.: Application of Learning Automata for DAG Scheduling on Homogeneous Networks. In: 17th Iranian Conference on Electrical Engineering, Iran University of Science and Technology, ICEE 2009 (2009)

Visualization of the Similar Protein Structures Using SOM Neural Network and Graph Spectra

Do Phuc and Nguyen Thi Kim Phung

Department of Information Systems
University of Information Technology
phucdo@uit.edu.vn, phungntk@uit.edu.vn

Abstract. In this paper, we would like to present our research result to build a graph clustering system using the SOM neural network and graph spectra. We use this system to support the visualization of similar protein structures in graph database of protein structures. Graph spectra is a set of eigenvalues of the normalized Laplacian matrix representing the graph. These eigenvalues are sorted in descendant order. We create a feature vector of sorted eigenvalues in descendant order to represent graph. SOM neural network is used to cluster the graph spectra; graph distance is Euclidean distance between graph spectra. Using graph spectra, we can improve the speed of training phase of SOM neural network. After clustering, the 2D SOM output layer will create the clusters of similar protein structures. By putting 2D SOM output layer on the computer display, we can visualize the similar protein structures of database by moving around the computer display. Our proposed solution was tested with the protein structures downloaded from SCOP database which was created by manual inspection and automated methods for description of the structural and evolutionary relationships between all proteins known. Our results are compared with the SCOP.

Keywords: graph clustering, graph spectra, protein structure, SOM, visualization.

1 Introduction

Graph is an effective way to represent data objects. Currently, graphs are used to represent the 3D objects such as the structure of chemical elements, protein structure, XML [2, 7, 9, 11]. With the growth of data, the data volume increases rapidly requiring to organize databases to store data represented by graphs. In this article, graphs are represented by graph spectra. We use SOM neural network to cluster the graph spectra. Using graph spectra, we can speed up the process of training the SOM neural network. Finally, we use SOM output layer to visualize graphs representing the protein structure on the computer display. The paper is organized as follows: 1) Introduction 2) Organizing data in relational databases 3) Graph distance 4) Graph distance based on the graph spectra 5) Clustering graphs by using SOM neural network 6) Graph database of protein structure 7) Using SOM network and graph spectra for visualization and similarity query in graph database of protein structures 8) Experimental and discussion 9) Conclusions.

N.T. Nguyen, M.T. Le, and J. Świątek (Eds.): ACIIDS 2010, Part II, LNAI 5991, pp. 258–267, 2010.

2 Organizing Data in Relational Databases

The labeled graphs is a triple G (V, E, Lv) with set of vertices V, set of edges E and set of vertex labels Lv. The vertices may have same labels. The graphs were labeled and stored in three tables of relational database as follows:

a) Table 1: graphmaster (graphID, NumVerts, NumEdges)
This table stores graph ID, number of vertices and edges of each graph
b) Table 2: graphnode (graphID, NodeID, NodeLabel)
This table stores graph ID, each node is numbered separately by Node ID within the graph, the nodes may have same labels.
c) Table 3: graphlink (graphID, NodeIDStart, NodeIDEnd, Weight)
This table stores the edges of graphs including Node Id of the start node, Node ID of end node and weight of edge.

3 Graph Distance

To measure the distance between two graphs, we can use following ways:

- **Graph edit distance:** We use a series of edit operations to change the first graph to the second graph. Edit operations include deletion, insertion of vertices or edges. Each edit operation has own cost. A series of edit operations with minimal cost is the graph distance. The calculation of graph edit distance has high complexity and belongs to class of NP complete [4].
- **Graph distance based on the maximal common sub-graph**: H Bunke [4] proposed the formula to calculate the graph distance between two graphs. For two graphs G_1 and G_2, the distance between two graphs G_1, G_2, denoted by $d(G_1, G_2)$ is calculated as follows:

$$d(G_1,G_2) = 1 - \frac{|mcs(G_1,G_2)|}{\max(|G_1|,|G_2|)} \qquad (3.1)$$

Where mcs (G_1, G_2) is the maximal common sub-graph of two graphs G_1 and G_2. The problem of calculating the maximal common sub-graph of two graphs is a NP complete [12]

4 Graph Distance Based on Graph Spectra

Identifying protein clusters is an important task carried out in the field of bioinformatics. There are several methods to store protein structures. We use graph spectra to do this operation by using normalized Laplacian matrix to represent graph [7, 9]. Then set of eigenvalues of normalized Laplacian matrix are calculated to create graph spectra after sorting the eigenvalues in descendant order. Using graph spectra allows to speed up the SOM network training in comparison with the direct calculation on graphs [3, 10]. The normalized Laplacian matrix is defined as follows:

$$L_G(u,v) = \begin{cases} degree(v) : if\ u = v \\ -1 : if\ u \neq v \\ 0\ otherwise \end{cases}$$

Where degree (v) is the degree of vertex v. Given two graphs G1 and G2 (see Figure 1):

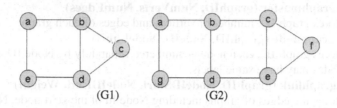

(G1)   (G2)

Fig. 1. Two labeled graphs

The normalized Laplacian matrices of graph G1 and graph G2 are shown in figure 2.

	A	B	C	D	E
A	2	-1	0	0	-1
B	-1	3	-1	-1	0
C	0	-1	2	-1	0
D	0	-1	-1	3	-1
E	-1	0	0	-1	2

a)   Normalized Laplacian Matrix of graph G1

	A	B	C	D	E	F	G
A	2	-1	0	0	0	0	-1
B	-1	3	-1	-1	0	0	0
C	0	-1	3	0	-1	-1	0
D	0	-1	0	3	-1	0	-1
E	0	0	-1	-1	3	-1	0
F	0	0	-1	0	-1	2	0
G	-1	0	0	-1	0	0	2

b) Normalized Laplacian Matrix of graph G2

Fig. 2. Normalized Laplacian matrices of graph G1 and graph G2

4.1 Graph Spectra

Graph spectra is a set of eigenvalues of the normalized Laplacian matrix representing the graphs. For graphs with n vertices, the graph spectra is as follows:

$$s = \{\lambda_1, \lambda_2, \ldots, \lambda_n\}$$

Where $\lambda_1 \geq \lambda_2 \geq \ldots \geq \lambda_n$

4.2 Graph Distance Based on Graph Spectra

Graph distance based on graph spectra between two graphs is Euclidean distance between graph spectra of two graphs. Let s and t be graph spectra of two graphs. Graph distance based on graph spectra of graph s and t is calculated as follows:

$$d_s(s,t) = \sqrt{\sum_i (s_i - t_i)^2} \tag{4.1}$$

Using graph distance based on graph spectra to measure the distance between two graphs is an effect way for faster comparison with the complexity of graphs having more than 400 vertices. It reduces the comparison time between graphs when performing. To calculate the distance between two graphs with different number of vertices, we use function Pad (v, x, k). This function will fill k values of x to the end of the vector v [2]. In the above example, the graph spectra of graph G1 and graph G2 are as follows:

λ (G1) = (4.62, 3.62, 2.38, 1.38, 0.00)
λ (G2) = (5.25, 3.80, 3.55, 2.45, 2.20, 0.75, 0.00)

Since graph G1 has number of vertices less than graph G2, we need to insert two zeros at the end of the graph spectra of graph G1. Graph spectra of graph G1 after filling value 0 is:

λ (G1) = (4.62, 3.62, 2.38, 1.38, 0.00, 0.00, 0.00)
λ (G2) = (5.25, 3.80, 3.55, 2.45, 2.20, 0.75, 0.00)

The distance between graph G1 and graph G2 is Euclidean distance between graph spectra λ(G1) and graph spectra λ(G2). In this case, the distance between graph G1 and G2 is 2.89.

5   Clustering Graphs by Using SOM Neural Network

SOM neural network is a kind of neural network. SOM neural network can reduce the size of high dimensional data objects by projecting them to the lower dimensional space. In our research, we use SOM neural networks with one input layer and one output layer. Each link between input and output of the SOM network corresponds to a weight. Total of inputs of each neuron in SOM layer is equal to the total of weights of the inputs of this neuron.

5.1   Clustering Graphs by Using SOM

The process of training the SOM neural network will adjust the weight gradually based on the input vectors. Results of training phase will create several clusters on the SOM output layer. The data objects with the proximity will be grouped into cluster that has the nearest distance from the input pattern to the neuron representing this cluster. According to the nature of the training algorithm of SOM neural network, the clusters that are near each other on the SOM output layer will contain data objects of a high similarity level (proteins with similar structure) [5, 6].

5.2 Training Algorithm of Traditional SOM Network

Basic function of the training algorithm of the traditional SOM neural network is to cluster the weight vector of neurons of the SOM output layers into separate clusters. The training algorithm of the traditional SOM neural network is as follows:

Step 1: Initializing randomly weight of neurons of the SOM output layer and assign value to $N_C(t)$ (the radius of the neighbor area). Assign value 1 to cycle t.

Step 2: Presenting an input vector (t) to SOM input layer and normalize v(t) then calculate the Euclidean distance from input vector v (t) to all the vectors of all neurons of the SOM output layer and choose neuron with the minimal Euclidean distance from input vector v (t) in to the corresponding neuron.

$$d_E(v, w_{ic\ jc}) = \min(d_E(v_i, w_{ij}))$$

Where i and j are the valid indices and to be set to the size of the SOM output layer.

Step 3: Updating the weight of the node to be located in the neighboring area of the node which contains winning neuron (i_c, j_c) by formula:

$$\underline{w}_{ij}(t+1) = \underline{w}_{ij}(t) + \gamma(\underline{v} - \underline{w}_{ij}(t)) \tag{5.1}$$

Where $i_c - N_c(t) \leq i \leq i_c + N_c(t)$ và $j_c - N_c(t) \leq j \leq j_c + N_c(t)$
Factor γ is in the [0,1], this number will decrease over time.

Step 4. Update t = t + 1, present the next input vector to the input of SOM and return to step 2 until achieving the convergence or exceed the number of iterations.

6 Graph Databases of Protein Structures

Protein is considered as polypeptide chains linked together. A polypeptide chain is a chain of amino acids (amino acid residues) linked together by peptide bonds. An amino acid consists of a central carbon atom (usually alpha Carbon C_α) and an amino group (NH2), a hydrogen atom (H), a Carboxy group (COOH) and a side chain (R) bound to the C_α. The backbone of the polypeptide is given by the repeated sequence of three atoms of each residue in the chain: the amide N, the alpha Carbon C_α and the Carbonyl C.

PDB is Protein Database Bank which contains the structure of the proteins. From protein data structure in the PDB, we determined coordinates of each atom C_{alpha} on the C chain polypeptide and use graph model to represent protein structure [9]. In this model, each vertex is an amino acid. We selected C_{alpha} atom to represent amino acid. Two vertices have an edge when their C_{alpha} atom are neighbors of each other, in other words, the distance between C_{alpha} atoms of two amino acids is smaller than a given threshold δ, the threshold is usually selected in range 6.5-8.5°A. Figure 3 shows the protein structure and the graph representing this protein structure.

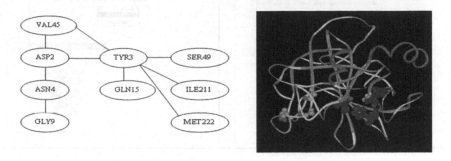

Fig. 3. The graph representing the protein structure

7 Using SOM Neural Network and Graph Spectra for Visualizing and Similarity Queries in Graph Databases of Protein Structures

We use the SOM neural network with one input and one output layer to cluster the graph spectra representing the protein structure.

7.1 Visualizing the Graph Database of Protein Structures

After clustering the graphs representing protein structure, the SOM output layer includes clusters of proteins with similar structure. The output layer is placed on the computer display. Through the output layer, we can easily survey the similar structure proteins. We can also click on each node of the output layer to show a list of proteins with similar structures and click on the name of a particular protein to see more about the structure of this protein.

7.2 Similarity Query

To process a similarity query, we use the following steps:

- Input is a graph query representing the structure of the protein you want to be searched
- Calculating the graph spectra of the graph query
- Determining node on the SOM output layer with smallest distance between weight vector of this node and graph spectra of query graph. Highlight the neighboring areas containing this node and display a window of proteins with high similarity

Figure 4 shows the 3D structure of protein with the PDB access code 1ssx and 2h5c in the same cluster of SOM 2D output layer.

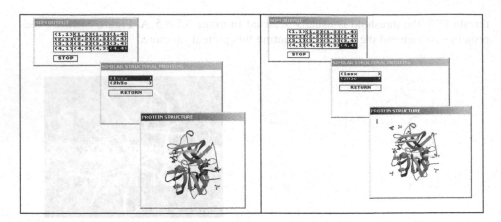

Fig. 4. Proteins with the similar structure on the SOM 2D output layer

8   Experiment and Discussion

8.1   Discussion of the Accuracy of SOM Clustering

Classification of protein domains based on their structure provides a valuable resource that can be used to understand protein function and evolutionary relationships. SCOP is the most widely used databases. This database is hierarchically organized and used to protein domains as a basic unit of classification. We recognize that the manual method to be used in SCOP produces high quality classification hierarchies. We use SCOP as a gold standard for measuring the quality of our proposed method. We also evaluated the quality of clusters using a measure called "cluster purity"[1]. It is 1 when all domains in the same cluster have perfect agreement in their class labels, and it is defined as:

$$ClusterPurity(M,S) = \frac{1}{N} \sum_{A \in M} \max_{B \in S} | \frac{A \cap B}{A} | \qquad (8.1)$$

In the above equation, A is a cluster in the set of clusters M created by SOM, B is a cluster in the set of S families of SCOP and N is the cardinality of M. The SCOP structure is very large, so we select only the proteins and the clusters belonging to the test set. It means that, we have:

$$\bigcup_{A \in M} A = \bigcup_{B \in S} B \qquad (8.2)$$

Cluster Purity is a value between 0 and 1. If the cluster value is 1 then the cluster A of M will be the subset of a cluster B of S. The higher value of cluster purity, the higher quality of cluster. We use the structural classification of proteins from the SCOP database [8] as gold standard to evaluate our method. Both prokaryotic and eukaryotic serine proteases in the super family 'Trypsin-like serine proteases is used to test SOM based graph clustering method. Based on SCOP, we select the proteins in the group ([50497], [50499], [50501],....), download data of protein structure from PDB and build databases containing graphs representing protein structures. Then, Jacobi algorithm is used to calculate the eigenvalues of normalized Laplacian matrix representing graphs.

Table 1. The cluster purity

	M1	M2	M3	M4	M5	M6
S1	0.25	0.0	0.0	0.0	0.0	0.0
S2	0.75	0.0	0.0	0.0	0.0	0.0
S3	0.0	1.0	0.0	0.0	0.0	0.0
S4	0.0	0.0	1.0	0.0	0.0	0.0
S5	0.0	0.0	0.0	1.0	0.7	0.0
S6	0.0	0.0	0.0	0.0	0.3	0.2
S7	0.0	0.0	0.0	0.0	0.0	0.8
Max	0.75	1.0	1.0	1.0	0.7	0.8

We also extended the protein with acid amid less than 200 by the insertion of value 0 to the right of universal graphs. Finally we have a set of graph spectra with 200 elements, each graph spectra represents for a specific structure of the protein. To experiment, we download the structure of 100 proteins from the PDB database and create 7 clusters from S1, S2, ..., S7 of SCOP containing this protein. After clustering by using SOM, we have clusters of M = (M1, M2, M3, ..., M6).

We use the cluster purity [1]. The value of each cell in this table is $|Sj| / |Mi|$. Last line of this table is the value of each cell with the largest value in each column.

The total cluster purity of SOM compared with SCOP is as follows:

$$\frac{1}{6}(0.75 + 1.0 + 1.0 + 1.0 + 1.0 + 0.9) = 0.87$$

This value shows the promise of our method of using SOM with graph spectra for clustering graphs.

8.2 Time Complexity

We perform 2 tasks: calculation of graph spectra and training time of SOM neural network. Jacobi algorithm is used to find the eigenvalues of the normalized Laplacian matrix representing graphs. The complexity of this algorithm is $O(n^3)$ with n is the number of graphs [2]. We use the traditional learning algorithm of SOM neural network with input vector as graph spectra.

Figure 5 shows the calculation time and the time for training SOM network computer Pentium IV, 3GHz speed, 1G RAM memory. The time of calculating graph spectra for graphs with number of vertices changing from 40 to 200 vertices, the calculation time of graph spectra changing from 2.928 seconds to 469.211 seconds. Time duration for training SOM network, we use SOM neural network with the size of output layer is 5x5, using graphs with 200 vertices (200 amino acids), training cycle is 5,000 and the number of graphs changing from 50 to 500 graphs. Time duration changes from 1,200 seconds to 9,000 seconds.

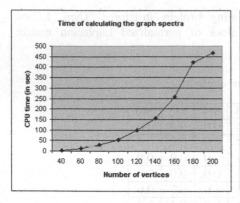
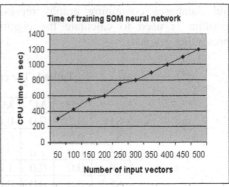

Fig. 5. Calculation time of eigenvalues and time duration of training SOM neural network

9 Conclusions

In this paper, we would like to present our research result to build a system for clustering graphs. Graph is represented by graph spectra. By using the SOM neural network and eigenvalues we cluster graph database. The system can support the visualization and similarity search in the graph database containing the protein structures. By using SOM output layer, we easily search the protein structure because the similar protein structures will be arranged in the neighboring area of the SOM output layer and this output layer is put on the computer display. In similarity search, we locate and highlight node on the 2D SOM output layer that is most similar to the query graph. We test and analyze the results on the graph database containing protein structures and compare with the SCOP. Accuracy of the method exceeds 85%, this result shows the promise of our method. Besides the calculation of eigenvalues and time duration for training SOM neural network are acceptable. We also continue to do research in the phenomena of co-spectrality of graphs to enhance the accuracy and improve the training algorithm of SOM neural network.

References

1. Brew, C.: Schulte im Walde Spectral Clustering for German Verbs. In: Proc. of the Conf. in Natural Language Processing, Philadelphia, PA, pp. 117–124 (2002)
2. Auffarth, B.: Spectral Graph Clustering. Universitat de Barcelona, course report for Technicas Avanzadas de Aprendizaj, at Universitat Politecnica de Catalunya (2007)
3. Phuc, D., Hung, M.X.: Using SOM based graph clustering for extracting main ideas from documents. In: Proc. of IEEE RIVF 2008, pp. 209–214 (2008)
4. Bunke, H., Shearer, K.: Graph distance metric based on the maximal common sub-graph. Pattern Recognition letter 19, 225–229 (1998)
5. Vesanto, J.: SOM based data visualization. Helsinki University of Technology, Finland (1999)
6. Kaski, S., Honkela, T., Lagus, K., Kohonen, T.: WEBSOM–self-organizing maps of document collections. Neuro computing 21 (1998)

7. Wilson, R.C., Zhu, P.: A Study of graph spectra for comparing graphs and trees. CS Department, University of York, UK (2008)
8. Murzin, A.G., Brenner, S., Hubbard, T., Chothia, C.: SCOP: a structural classification of proteins database for the investigation of sequences and structures. Journal of Molecular Biology 247, 536–540 (1995)
9. Vishveshwara, S., et al.: Protein structure insights from graph theory. Journal of Theoretical and Computational Chemistry 1(1) (2002)
10. Günter, S., Bunke, H.: Self-organizing map for clustering in the graph domain. Pattern Recognition Letters 23(4), 405–417 (2002)
11. Lang, S.: Protein domain decomposition using spectral graph partition. CS Department, University of York, UK (2008)
12. Henry Suters, W.: A new approach and faster exact methods for the maximum common sub-graph problem. In: Proceedings of 11th International Computing and Combinatorics (2005)

Real Time Traffic Sign Detection
Using Color and Shape-Based Features

Tam T. Le[1], Son T. Tran[2], Seichii Mita[2], and Thuc D. Nguyen[1]

[1] University of Science, VNU-HCMC, Vietnam
{lttam,ndthuc}@fit.hcmus.edu.vn
[2] Toyota Technological Institute of Nagoya, Japan
{ttson,smita}@toyota-ti.ac.jp

Abstract. This paper presents a new approach for color detection and segmentation based on Support Vector Machine (SVM) to retrieve candidate regions of traffic signs in real-time video processing. Instead of processing on each pixel, this approach utilizes a block of pixels as an input vector of SVM for color classification, where the dimension of each vector can be extended by a group of neighboring pixels. This helps to handle the diversification of data on both training and testing samples. After that, Hough transform and contour detection are applied to verify the candidate regions by detecting shapes of circle and triangle. The experimental results are highly accurate and robust for our testing database, where samples are recorded on various states of environment.

Keywords: traffic-sign detection, color detection and segmentation, SVM.

1 Introduction

Traffic signs on road tell drivers about traffic rules and states of road such as warning, prohibition, limitation of speed, and so on. With that useful information, our traffic is safer and more convenient. In vision-based methods, the traffic-sign detection and recognition have many difficulties due to changes of environment and speed of vehicles. For instance, traffic signs are usually faded when they are exposed a long time under the sunlight. Moreover, their colors are changeable for various lights influenced by weather conditions such as fog, cloud, or snow. It is not only affected by the illuminant color of the daylight but also by the weak nightlight. The different directions of view also make traffic signs difficult to detect and recognize from camera positioning inside of cars. In addition, traffic signs may be also occluded by trees, buildings, or pedestrians. At the present time, the evaluation of a new approach by comparing with the existing methods is not a simple task because there is not any standard traffic sign database.

Most reviewed literature has used color as one of the main features to detect traffic signs beside shape-based features. The first approach of color detection and segmentation has utilized thresholds in a suitable color space, where HSV color space is the first choice [1, 2] because it is more intuitive than the others and we can separate the color information from the brightness to generate a continuous space of color, where

N.T. Nguyen, M.T. Le, and J. Świątek (Eds.): ACIIDS 2010, Part II, LNAI 5991, pp. 268–278, 2010.

it is not supported in RGB color. This approach is simple, but it is difficult to define threshold of an interested color due to continuity of color space. A de la Escalcra [3], et. al., proposed two look up tables on hue and saturation channel in HSI color space to enhance the ability of interested-color separation. H. Fleyeh [4] presented algorithms of the dynamic threshold, fuzzy color segmentation, and shadow highlight invariance. These algorithms utilize one global value of all pixels in image as an additional constraint to detect and segment color more accurately. C. Fang [5], et. al., presented two-layer neural network on the hue channel in HSI color space to detect interested color. This method matched intensity values of testing pixels to those of interested color on hue channel by the neural network.

Support Vector Machine (SVM) is an efficient learning method, especially in traffic-sign detection problem. Satumino[6], et. al., and Kiran[7], et. al., applied this method for the shape classification phase as well as the recognition phase. And in this paper, we present a new approach for the color detection and segmentation based on SVM to retrieve candidate regions of traffic signs in real-time video processing. Instead of processing on each pixel, our approach utilizes a block of pixels as an input vector of SVM for color classification, where the dimension of each vector can be extended by a group of neighboring pixels. This helps to handle the diversification of data on both training and testing data. Our contribution is the first method of applying SVM in color classification for traffic sign detection.

2 Model of Traffic Sign Detection

We propose an algorithm of color detection and segmentation based on SVM to retrieve candidate regions. The algorithm assigns input of SVM as a block of pixels, a group of neighboring pixels located at each position in the image, to enhance the precision as well as the recall. Especially, the algorithm can reduce a part of noise when traffic signs are exposed in bad weather such as cloudy and snowy days. Figure 1 presents steps of our algorithm to detect traffic signs in real-time video processing. Our method is based on color detection and segmentation using SVM on blocks of pixels.

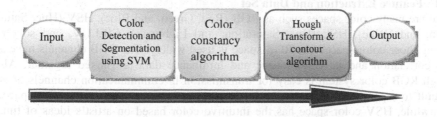

Fig. 1. The proposed model to detect traffic signs

To speed up the processing time, the size (640x480) of the input image is downscaled to (320 x 240). The color detection and segmentation using SVM is applied on the downscaled image. Then, the candidate regions are extracted from the original image by a projection from each position of the candidate region on downscaled image to that of the input image. The candidate regions are enhanced by color constancy algorithm [8]. Figure 2 presents one result of the color constancy algorithm. We can see in Fig. 2 that the edge and color features in the right image are enhanced from the left one.

Fig. 2. The result of using color constancy algorithm

Finally, Hough transform and contour algorithm are utilized to verify the candidate regions. The next sections present the color detection by using SVM and traffic-sign segmentation by using Hough transform and contour algorithm.

2.1 Color Detection and Segmentation by Using SVMs

Different from most of existing methods dealing with each pixel as input, our approach concerns blocks of pixels; therefore, the information of neighbor pixels can help to handle the diversification of both training and testing data. It means that SVMs can return a recall rate better than that of single-pixel-based algorithms. In this approach, instead of deciding whether a pixel has an interested color or not, our method chooses blocks of interested color through the results of SVMs. Using feature extraction of a pixel block presented in the section of 2.1.1 will help to reduce complexity of the calculation in SVMs because the dimensions of the input vectors, support vectors, and the hyper-plane are only equal to two times of the number of pixels in each block.

2.1.1 Feature Extraction and Data Set

There are many color spaces such as RGB (Red, Green, and Blue), HSV (Hue, Saturation, and Value), HSL (Hue, Saturation, and Lightness), CIELUV, and so on. Among those color spaces, the RGB space is used widely. The RGB channels have a high correlation and its chrominance and luminance data are mixed together. Although RGB color system is easy for transmission in communication channels, it is difficult to separate the interested color because it is not a continuous color space. Meanwhile, HSV color space has the intuitive color based on artist's ideas of tint, saturation, and tone. In the HSV system, the Hue channel contains dominant color and it is invariant with highlight in white light sources. The saturation channel represents the colorfulness of an area in proportion to its brightness and the value of the channel contains color luminance. CIELUV color space was proposed by G. Wyszecki and standardized by CIE (Commission International de L'Eclairage), it is a system of perceptually uniform color. Our approach chooses ratios of RGB channel in [9] because it can reduce the dimension and allow defining the specific features for the interested color.

Let I denote an image of $(n \times m)$ and blk denote a block of (2×2). Each block on the image is not overlapped with the other blocks.

$$I = \bigcup_{i=1}^{\frac{n}{2} \times \frac{m}{2}} blk_i \tag{1}$$

Each pixel p of block blk has a value of (r, g, b) in the RGB color space. We define the value of α and β as follows;

$$\alpha = \frac{r}{g} \qquad \beta = \frac{r}{b} \tag{2}$$

Then, each pixel p can be expressed by $p = (\alpha, \beta)$, and block blk is defined as the follow;

$$blk = \begin{pmatrix} P_{ij} & P_{i(j+1)} \\ P_{(i+1)j} & P_{(i+1)(j+1)} \end{pmatrix} \tag{3}$$

where $i = 1,..,n-1$ and $j = 1,..,m-1$.

Hence, we can express the block feature denoted by FoB as an 8-dimension vector

$$FoB = \begin{pmatrix} \alpha_{x_{ij}} & \beta_{x_{ij}} & \alpha_{x_{i(j+1)}} & \beta_{x_{i(j+1)}} & \alpha_{x_{(i+1)j}} & \beta_{x_{(i+1)j}} & \alpha_{x_{(i+1)(j+1)}} & \beta_{x_{(i+1)(j+1)}} \end{pmatrix} \tag{4}$$

and it is utilized as an input vector of SVM.

Fig. 3. Some examples of training data built by K-Means

In the training stage, the absence of standard traffic sign dataset is the main reason to build our dataset. The traffic-sign images are collected from the Internet with various states such as color distortion, blur, highlight as well as nightlight color, and so on. K-means method is applied to cluster block feature FoB in eq. (4) for building the database of interested color blocks. In this case, there are two groups (Red and Blue) of interested color blocks. In Fig. 3, the left group of images is used to learn the red color and the right one for the blue color. Each small image in Fig. 3 is a set of (2x2) blocks extracted from K-Mean algorithm. The three top lines of subgroup are sample sets of interested color and the other lines are set of non-interested color.

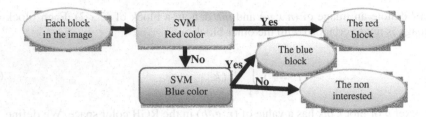

Fig. 4. The flow chart of using SVM to detect the interested blocks

2.1.2 SVMs for Color Classification

SVM is proposed by Vapnik and his group at AT&T Bell laboratory in 1992 [10]. From the database built in 2.1.1, let x denote a feature vector of a block - *FoB*, y denote a label of classification, $y \in \{1, -1\}$, b denote a bias value, and w denote the hyperplane of SVM. The primal formulation of SVM is to minimize $\|w\|_2$ such that the constraint of $y_k(w^T x_k + b) \geq 1$ is satisfied with all $k = 1, 2,.., N$. Applying Lagrange multipliers, we have the formula for the primal problem:

$$L(w,b,\alpha) = \frac{1}{2}w^T w - \sum_{k=1}^{N} \alpha_k (y_k(w^T x_k + b) - 1) \tag{5}$$

Therefore, the solution for this problem

$$\max_{\alpha} \min_{w,b} L(w,b,\alpha) \tag{6}$$

And the equivalent dual problem in the Lagrange multipliers α_k which can be solved in the quadratic programming is as follows:

$$\max_{\alpha} J_D(\alpha) = -\frac{1}{2}\sum_{k,l=1}^{N} y_k y_l x_k^T x_l \alpha_k \alpha_l + \sum_{k=1}^{N} \alpha_k \tag{7}$$

$$\text{s.t} \quad \sum_{k=1}^{N} \alpha_k y_k = 0 \tag{8}$$

The strong point of the dual problem is sparseness property. It means that most of $\alpha_k = 0$ except support vectors. Therefore, the complexity of the testing step is not large. Equation (10) presents the testing process of one block x.

$$y(x) = sign(\sum_{k=1}^{\#SV} \alpha_k y_k x_k^T x + b) = \begin{cases} 1, & \text{Red or Blue} \\ -1, & \text{Others,} \end{cases} \tag{9}$$

where the values of y_k, b, and α_k are the results of the learning step. The value of x_k is support vectors selected from the learning data. Note that we have two groups of (y_k, α_k, b, x_k), where one is correspondent to Red and the others for Blue. Figure 4 presents a process of retrieving Red and Blue blocks in an image by using eq. (9) of SVMs. Through experiments, we realize that the linear SVM returns a high accuracy with a low complexity to classify blocks of interested color - *FoB*. Although non-linear SVM can return a little better result, its calculation complexity is too large to apply on a real-time process. Therefore, we select the linear SVM instead of using non-linear model.

In practice, we propose a pre-processing step with the rough threshold γ on the HSL color space to enhance the performance of calculation. Its values are presented in eq. (10)-(11).

$$\gamma(red) = \begin{cases} \begin{cases} saturation \geq 51 \\ \begin{bmatrix} hue \leq 9 \\ hue \geq 170 \end{bmatrix} \rightarrow red \\ others \rightarrow non-red \end{cases} \end{cases} \tag{10}$$

$$\gamma(blue) = \begin{cases} \begin{cases} saturation > 51 \\ 99 \leq hue \leq 112 \end{cases} \rightarrow blue \\ others \rightarrow non-blue \end{cases} \tag{11}$$

The rough threshold γ is applied to the value of Red and Blue color defined on the HSL space with each pixel of (2x2) block. As a result, the SVM method only classifies blocks which are positive error or correct answer of the pre-processing. This step helps to reduce the number of blocks which will be checked by SVM, and it makes the complexity of detection lower.

2.1.3 Grouping Region

After color-interesting blocks are selected by the classification on SVM, we make an extension of breadth-first-search (BFS) algorithm with pre-defined distance to group controlled block and retrieve candidate regions. We define the region $r = (x_{ij}, w, h)$, where x_{ij} denotes the top left pixel of region, w and h are the width and height of the region, respectively. So, the distance of two regions $r = (x_{ij}, w, h)$ and $r' = (x'_{ij}, w', h')$ is defined as the following;

$$DoR(r, r') = \begin{cases} \infty & T_1 = \emptyset \wedge T_2 = \emptyset \\ 0 & T_1 \neq \emptyset \wedge T_2 \neq \emptyset \\ \min(|i-i'-w'|, |i+w-i'|) & T_1 = \emptyset \wedge T_2 \neq \emptyset \\ \min(|j-j'-h'|, |j+h-j'|) & T_1 \neq \emptyset \wedge T_2 = \emptyset \end{cases} \tag{12}$$

With $T_1 = [i, i+w] \cap [i', i'+w']$ and $T_2 = [j, j+h] \cap [j', j'+h']$. Using eq. (12) to calculate distance of two regions, we have a value of *DoR*. If *DoR* is equal or smaller than the pre-defined distance d, the two regions will have a connection. Then, applying the extension of BFS algorithm can reduce the number of candidate regions to only a few color-interested regions from a set of color-interested blocks. As a result, we only need to recognize a few color-interested regions instead of a large number of all color-interested blocks in the classification phase. Figure 5(a)-(h) present an instance of the BFS result. Each red block in Fig. 5 is a block of interested color which is classified by the linear SVMs. The BFS algorithm will group color-interested blocks in Fig. 5(a) to retrieve a color-interested region bounded by a rectangle in Fig. 5(f). In this instance, the pre-defined distance d is set to 2.

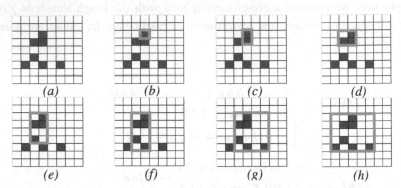

Fig. 5. Group region algorithm – extension of BFS with predefined distance (d=2)

Based on the position of car, we set a location of traffic signs positioning in the upper two-third part of images. Figure 6(a), (b), and (c) presents one result of our color detection by using SVM. This test is applied on database provided by the Research Center for Integration of Advanced Intelligent Systems and Devices, Toyota Technological Institute, Japan.

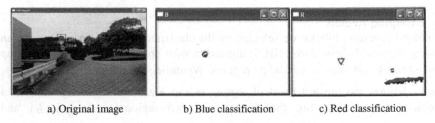

a) Original image   b) Blue classification   c) Red classification

Fig. 6. Example of color detection and segmentation using SVM on block of pixels

2.2 Using Hough Transform and Contour to Verify the Candidate Regions

After retrieving candidate regions, we apply Hough transform to detect circle traffic signs. In case of triangle traffic signs, we utilize Canny-edge detection with multi thresholds to get binary candidate regions and apply contour detection algorithm to extract geometry properties such as the number of edges and the size of angles between two lines for triangle verification. The complexity of classification phase depends on the number of candidate regions. Since the area of candidate regions is small, the complexity is small enough to process in real time. In our system, this classification phase averagely consumes less than 30 milliseconds per frame. Figure 7 presents the flow chart of traffic sign segmentation by using Hough transform and contour algorithm. It shows that one traffic sign is detected by both color and shape in our method.

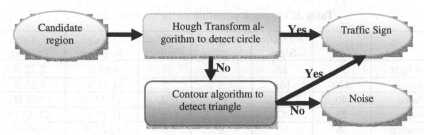

Fig. 7. Hough Transform and contour algorithm to verify candidate region

3 Experiment and Result

We utilize C language on a computer - 1.66 GHz core dual CPU and non-preprocessing images for our experiments. SVM-Light [11] is utilized to evaluate the proposed method.

Table 1. Results of linear-SVM training process

	Red color	Blue color
Positive samples	30000	30000
Negative samples	45000	60000
Recall	88,24%	98,03%
Precision	88,28%	98,03%

The learning data is defined in Table 1. After preprocessing with rough threshold γ in HSL, we set the value of γ such that there are only positive errors and color-interested blocks as outputs, so we have a bias in favor of negative samples towards positive ones. By using preprocessing with rough threshold γ in HSL, the algorithm is speeded up approximately 5 times. We can see the advantage of pre-processing step on the time consuming in Table 2. In addition, the average time consuming for all process in an image (640x480) is just about 50 milliseconds. It means that the proposed method can process about 20 fps.

Table 2. Time consuming of classifying color-interested blocks in an image (320x240)

	Linear SVM (Only)		Linear SVM + pre-processing γ	
	Red color	Blue color	Red color	Blue color
Worse case	0.14s	0.14s	0.032s	0.032s
Best case	0.09s	0.09s	0.015s	0.015s
Average case	0.11s	0.11s	0.023s	0.023s

To evaluate our method, we collect sampling images on the internet, some images in [4] as well as images of TTI lab to build a challenging database which contains the diversified images collected in many conditions of environment such as bad light, snow fall, or blurred, faded, occluded, and damaged traffic signs, and it is also published for free at http://www.fit.hcmus.edu.vn/~lttam/LTTam-TestingDatabase.rar

Table 3. The result of testing on the database

	# Images	# Traffic Signs	# False Negative	# False Positive	% False Negative	% False Positive
Bad light	30	45	3	1	6.67 %	2.22 %
Blurred	120	161	7	7	4.35 %	4.35 %
Faded	40	65	11	2	16.92 %	3.08 %
Noise	23	42	7	9	16.67 %	21.43 %
Occluded	36	66	8	6	12.12 %	9.09 %
Damaged	20	33	5	3	15.15 %	9.09 %
Snowfall	30	35	1	5	2.86 %	14.29 %
TTI campus	67	99	3	1	3.03%	1.01%
JP urban road	39	51	2	8	3.92%	15.69%
Total	**405**	**597**	**47**	**42**	**7.87%**	**7.04%**

Table 3 is the testing result of our proposed method on this database, the precision is 92.91% and the recall is 92.13%. Our result shows that the color classification using linear SVM gives a robust result in conditions of daylight, bad light, and snowfall.

We present some illustrating results of our test in Fig. 8 (a), (b), and (c) for the foggy, snowy weather, and Japanese urban road, respectively. Our algorithm utilized SVM for color detection combined with Hough transform for circle detection and contour algorithm for triangle detection. The result in Fig. 8 shows that our method can detect a far, small, and blurred traffic sign, where the camera is mounted inside of the car.

a) Foggy weather

b) Snowy weather

c) Japanese urban road

Fig. 8. Some examples from the testing database

Table 4 shows our comparison to previous works based on our database. The result shows that our method returns a high accuracy in a real time processing (average 20 fps) for autonomous-driving system.

Table 4. Results of our method and some related methods

	Detection Rate	False alarm Rate	Time consuming
Soetedjo [2]	85%	10%	2440 ms
Shneier [9]	88%	58%	50 ms
Our method	**92.91%**	**7.04%**	**50 ms**

4 Conclusions

In this paper, we presented a real-time processing method of traffic-sign detection to apply in autonomous driving system. Our proposed method utilized linear SVM to classify color by a low complexity (average 23 milliseconds per frame). After that shape matching has been applied to eliminate positive errors. We achieved 92.91 percent of detection accuracy and it has been applied on real-time autonomous driving system with the processing speed of 20fps, where the maximum speed of car is limited at 30 km per hour. In the near future, we will combine our processes of detection and recognition to generate a vision-based system of guidance and warning for the autonomous driving system.

References

1. Paclik, P., Novovicova, J., Pudil, P., Somol, P.: Road sign classification using Laplace kernel classifier. Pattern Recognition Letters 21, 1165–1173 (2000)
2. Soetedjo, A., Yamada, K.: An Efficient Algorithm for Traffic Sign Detection. Journal of Advanced Computational Intelligence and Intelligent Informatics 10(3), 409–418 (2006)
3. de la Escalcra, A., Moreno, L., Puente, E., Salichs, M.: Neural traffic sign recognition for autonomous vehicles. In: Proc. 20th Inter. Conf. on Industrial Electronics Control and Instrumentation, Bologna, Italy (1994)
4. Fleyeh, H.: Traffic and road sign recognition. PhD thesis at Dalarna University, Sweden (2008)
5. Fang, C., Fuh, C., Chen, S., Yen, P.: A road sign recognition system based on dynamic visual model. In: Proc. IEEE Computer Society Conf., Computer Vision and Pattern Recognition (CVPR), Madison, Wisconsin (2003)
6. Satumino, M.-B., Sergio, L.-A., Pedro, G.-J., Hilario, G.-M., Francisco, L.-F.: Road-sign detection and recognition based on support vector machines. IEEE Transactions on Intelligent Transportation Systems 8(2), 264–278 (2007)
7. Kiran, C.G., Lekhesh, V.: Prabhu, Abdu Rahiman V., Rajeev K.: Traffic Sign Detection and Pattern Recognition Using Support Vector Machine. In: Seventh International Conference on Advances in Pattern Recognition, pp. 87–90 (2009)

8. Ebner, M.: Color constancy using local color shifts. In: Pajdla, T., Matas, J(G.) (eds.) ECCV 2004. LNCS, vol. 3023, pp. 276–287. Springer, Heidelberg (2004)
9. Shneier, M.: Road Sign Detection and Recognition. In: Proceeding of SPIE, the International Society for Optical Engineering, Florida, USA (2006)
10. Boser, B.E., Guyon, I.M., Vapnik, V.N.: A training algorithm for optimal margin classifiers. In: Haussler, D. (ed.) 5th Annual ACM Workshop on COLT, Pittsburgh, PA, pp. 144–152 (1992)
11. Joachims, T.: Making large-Scale SVM Learning Practical. MIT Press, Cambridge (1999); Burges, C., Smola, A.: Advances in Kernel Methods – Support Vector Learning. MIT Press, Cambridge (1999)

Standard Additive Fuzzy System for Stock Price Forecasting

Sang Thanh Do[1], Thi Thanh Nguyen[1], Dong-Min Woo[2], and Dong-Chul Park[2]

[1] Faculty of Computer Science and Engineering,
Ho Chi Minh City University of Technology, Vietnam
{sang.dothanh,thi.nguyenthanh}@gmail.com
[2] Department of Information Engineering, Myongji University, Korea
{dmwoo,parkd}@mju.ac.kr

Abstract. Stock price forecasting has attracted tremendous attention of researchers over the past several decades. Many techniques thus have been proposed so far to deal with the problem. This paper presents an application of a computational intelligence technique - a fuzzy inference system, namely Standard Additive Model (SAM), for predicting stock price time series data. The modelling and learning power of the SAM have been benefited to build the model that is capable of prediction functionalities. Experimental results have demonstrated that the proposed approach outperforms the traditional Auto Regressive Moving Average (ARMA) model in terms of the forecasting performance.

Keywords: Standard Additive Fuzzy System, non-linear function approximation, stock price forecasting.

1 Introduction

Demands of the future prediction are critical not only to governments of countries but also to profit or nonprofit organizations and individuals. Although prediction techniques have been suggested from the 19th century, but they have just become emerging recently when information technology has developed rapidly. Thanks to this, prediction technologies are being deployed and/or implemented more and more easily and effectively. These newly popularized prediction techniques are referred to learning machine or artificial intelligence.

The most well-known learning machine method applied in prediction is Neural Network (NN) [4], [13], [15]; however it still shows shortcomings. The first is the determination of the network structure to suit best the problem being solved. The second is the initialization of parameters of network nodes. Generally, a neural network is considered as a black box in which it is not handy to keep track adjustments of parameters occurring in the system.

Due to the mentioned NN's drawbacks, it is the fuzzy inference system that is recommended instead [9], [11], [12]. Hence, this paper focuses on the application of the Standard Additive Model (SAM) fuzzy system [11] in time series prediction based on the function approximation theory. When prediction of the economic

N.T. Nguyen, M.T. Le, and J. Świątek (Eds.): ACIIDS 2010, Part II, LNAI 5991, pp. 279–288, 2010.

indicators of a market is crucial, it is this reason that motivates us to experiment the SAM approach in economic time series forecasting. The rest of the paper is organized as follows. The next section will present the basic configuration of SAM. Application of SAM for non-linear function approximation will be stated in the third section while the SAM learning process will be presented in the fourth section. The two last sections will be devoted to presentation of experiments, including data set and results, and conclusions.

2   Standard Additive Model

A fuzzy inference system consists of m fuzzy rules (Fig. 1). Each fuzzy rule is a conditional IF-THEN proposition of the form R_j: IF x = A_j THEN y = B_j; $j = \overline{1, m}$; where $x \in R_n$, $y \in R_p$, are n-dimensional vectors, A_j and B_j are fuzzy sets on the input space X and the output space Y respectively.

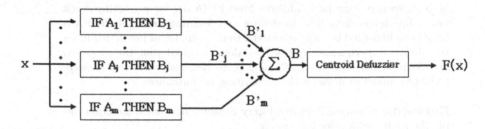

Fig. 1. SAM's Parallel Combination Structure

Because of SUM (summary) combination of fuzzy rules, this fuzzy system is named Standard Additive Model. In SAM, each input x fires j^{th} fuzzy rule and results in fuzzy set $B_j{'}$ determined by the PRODUCT operation between membership degree of if-part $a_j(x)$ in [0,1] and then-part fuzzy set $B_j : B_j' = a_j(x)B_j$. Then-part fuzzy set $B_j \subset R_p$ has set function $b_j : R_p \rightarrow [0,1]$ and volume V_j and centroid (center of gravity) c_j of fuzzy set. The defuzzification method of SAM is CoG (Centroid of Gravity):

$$F(x) = \text{Centroid}(B) = \text{Centroid}(\sum_{j=1}^{m} w_j a_j(x) B_j) = \frac{\sum_{j=1}^{m} w_j a_j(x) V_j c_j}{\sum_{j=1}^{m} w_j a_j(x) V_j} = \sum_{j=1}^{m} p_j(x) c_j$$

where w_j is the corresponding weight of the jth rule. The convex weight:

$$p_j(x) = \frac{w_j a_j(x) V_j}{\sum_{k=1}^{m} w_j a_k(x) V_k}$$

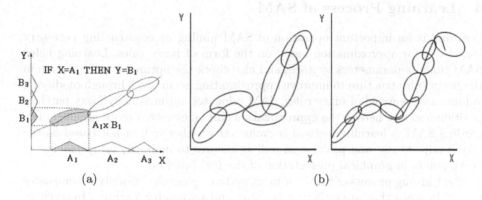

Fig. 2. (a) Mapping of Input Space into Output Space; (b) SAM Approximates a Non-linear Function f(x)

3 Applying SAM in Nonlinear Function Approximation

The SAM fuzzy system can uniformly approximate continuous and bounded measurable functions on compact domains. If y = f(x) is not analytically known, we cannot write an equation in explicit form. However, we can use the relationship between the input space X and the output space Y which given by Y = F(x), a relationship that links subsets of the input space X to subsets of the output space Y. Fig. 2a illustrates the fuzzy rule patches in the input-output space and how these patches cover the graph of f(x).

The approximation capability of SAM is outlined as follows:

– Each fuzzy rule R_j is formed by the combination of if-part fuzzy set A_j (its space defined by the X axis) and then-part fuzzy set B_j (space defined by the Y axis). The fuzzy patch $R_j = A_j \times B_j$ is determined in the product space $X \times Y$.
– By defuzzifying the combination of fuzzy patches $A_j \times B_j$, SAM can cover the graph of the unknown function f(x).

It is obvious from Fig. 2b that the more the fuzzy patches; the closer the SAM can approximate the function f(x). This means a fuzzy system having more fuzzy rules will result in more accurate outcomes in approximating process than systems having less fuzzy rules. This is actually a trade-off problem between accuracy and number of fuzzy rules.

The key is how to build fuzzy patches that are necessary for function approximation. This is crucial because the number of fuzzy rules and their parameters significantly influence the SAM's efficiency. In order to solve the problem, the power of SAM needs to be enabled by a learning process that is presented in detail in the following section.

4 Learning Process of SAM

Learning is an important operation of SAM aiming at constructing necessary knowledge for approximation based on the form of fuzzy rules. Learning helps SAM tune its parameters by itself, and also check the optimum configuration in the fuzzy rule structure to improve approximating accuracy. Through modifying volume and centroid of fuzzy rules, SAM relocates automatically fuzzy patches' position and size hereby the approximation can be expected more accurate. Regarding SAM, a learning method is evaluated whether well or not based on the way it adjusts size and position as well as ensures to maintain fuzzy patches at curve points in graphical presentation of the f(x) function.

The learning process of SAM (or fuzzy system, generally) usually encompasses two main steps that are structure learning and parameter learning. In order to improve SAM's efficiency, we integrate an optimal learning. Hence the process includes:

- Structure learning: This step is a self-study technique - unsupervised learning. By clustering input data, SAM will detect fuzzy rules needed for approximating specific nonlinear function [1-3], [8], [17].
- Parameter learning: After building up necessary fuzzy rules, the next task is to adjust parameters such as rule weights, parameters of if-part fuzzy sets, volumes, and centroids of then-part sets. This step employs the supervised learning. By computing errors between results of the system and expected values, SAM will tune parameters according to the gradient descent law.
- Optimal learning: This step enables SAM to eliminate unnecessary fuzzy rules using the genetic algorithm (GA). SAM will be more concise in order to enhance the processing speed and accuracy.

4.1 Unsupervised Learning

Unsupervised learning begins with categorizing data patterns into clusters. The purpose is that from a limited set of learning data patterns, we establish a set of clusters in which data patterns in a cluster are as same as possible and data patterns in different clusters are as different as possible. Clustering problem can be stated as follows:

- Set $X = \{x_1, x_2, ..., x_{ntd}\}$, $x_j \in R_n$, is a limited set of learning data, ntd is the number of training data or learning data.
- Denote V_{cntd} is matrix $c \times ntd$, $c \in Z^+$, $1 \leq c \leq ntd$.
- Determine a fuzzy cluster is presented by a centroid vector set: $V = \{v_1, v_2, ..., v_c\}$, $v_i \in R_n$.

Matrix $U = \{u_{ij}\} \in V_{cntd}, \forall i = \overline{1, c}, \forall j = \overline{1, ntd}$ where u_{ij} is real value in range [0,1] presenting for dependent level of x_j corresponding with centroid vector v_i and satisfying two following constrains: $u_{ij} \in [0,1], \sum_{i=1}^{c} u_{ij} = 1, \forall j, j = \overline{1, ntd}$ and

$$0 < \sum_{j=1}^{ntd} u_{ij} < ntd, \forall i, i = \overline{1, c}$$

The goal of fuzzy clustering issue is to minimize the target function J:

$$J(U,V) = \sum_{i=1}^{c} \sum_{x_j \in X} \sum_{k=1}^{c} g[w(x_i), u_{ij}]d(x_j, v_k)$$

where:

- $w(x_i)$ is initial weight of x_i.
- $d(x_j, v_k)$ represents difference between x_j and centroid vector v_k of the kth cluster. It must satisfy two following constraints: $d(x_j, v_k) > 0$ and $d(x_j, v_k) = d(v_k, x_j)$.

The Adaptive Fuzzy Leader Cluster (AFLC) algorithm has been applied in this research. This algorithm originates from Fuzzy C-Means (FCM) and integrates competitive learning method. The number of cluster is 0 at beginning, and then the algorithm will detect the number of clusters automatically. Upon each learning data pattern x_j, it looks for clusters which have centroid vector v_w as long as v_w is closest to x_j. The choice must satisfy the condition.

$$d(x_j, v_w) = \min\{d(x_j, v_i)\}, \; i = \overline{1, c}$$

where c is the number of existing clusters. The evaluation function d is as follows:

$$d(x_j, v_i) = \frac{\|x_j - v_i\|}{\frac{1}{ntd_i} \sum_{k=1}^{ntd_i} \|x_k - v_i\|} < \tau$$

where:

- τ: is the given threshold value.
- ntd_i: the number of checked data patterns which belong to the ith cluster.

If there is not any cluster that satisfy the above, create a new cluster c++ by which coordinates of its centroid v_{c++} equals x_j, otherwise x_j belongs to the wth cluster. In both cases, matrices U and V are both updated.

Case 1st: x_j belongs to wth cluster:

- wth row of matrix U is updated:

$$u_{wj} = \frac{\left(\frac{1}{\|x_j - v_w\|^2}\right)^{\frac{1}{1-m}}}{\sum_{k=1}^{c} \left(\frac{1}{\|x_j - v_w\|^2}\right)^{\frac{1}{1-m}}}$$

where $m > 1$ is the fuzzy level of matrix U.
- Vector v_w is updated:

$$v_w = \frac{\sum_{j=1}^{ntd} u_{wj}^m \cdot x_j}{\sum_{j=1}^{ntd} u_{wj}^m}$$

Case 2nd: x_j doesn't belong to wth cluster:

- c = c + 1.
- Add c^{th} row into matrix U: $u_{ck} = 0, \forall k \neq j; uc_j = 1$.
- Add vector $v_c = x_j$ into matrix V.

After clustering data patterns, the next task is to build up fuzzy rules from their centroid vectors using fuzzy sets. Indentifying/choosing membership functions of fuzzy sets is important when it affects how well fuzzy system approximates continuous functions.

In this paper we use trapezoid membership function (Fig. 3) [19] because of its simplification. This function includes four parameters: l, ml, mr, r where $ml < mr \in R$. $l > 0$ and $r > 0$ represent covered distance to left-side and right-side of ml and mr. We could put centroid m = (ml + mr).

$$\mu(x, ml, mr, l, r) = max\left(min\left(\tfrac{x-(ml-l)}{l}, 1, \tfrac{mr+r-x}{r}\right), 0\right)$$

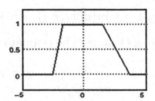

Fig. 3. Trapezoid function with m = 0, l = 1, ml = -2, mr = 2 and r = 2

4.2 Supervised Learning

The supervised learning helps decrease error between system's outcomes and expected results. The problem is stated as follows:

- Given input training data set $\{x_j\}$ and expected result set $\{y_j\}, j = \overline{1, ntd}$, where ntd is number of training data.
- Given SAM's fuzzy rules and parameters.
- Adjust parameters of if-part fuzzy set, then-part and weights as long as error between system's outcomes and expected results reach the stopping criteria.

There gradient descent algorithm [4], [16], [18] has been deployed in this step. The learning rule applies for variable ξ in SAM has following form:

$$\xi(t+1) = \xi(t) - \mu_t \tfrac{\partial E}{\partial \xi}$$

where μ_t is the learning rate. The aim of parameter adjustment learning phase is to minimize the square of error:

$$E(x) = \tfrac{1}{2}(f(x) - F(x))^2$$

$$\tfrac{\partial E}{\partial F} = -(f(x) - F(x)) = -\varepsilon(x)$$

The learning rule for each parameter is expanded in detail as follows.

$$c_j(t+1) = c_j(t) + \mu_t.\varepsilon(x).p_j(x)$$

$$V_j(t+1) = V_j(t) + \mu_t.\varepsilon(x).[c_j - F(x)].\frac{p_j(x)}{V_j}$$

$$w_j(t+1) = w_j(t) + \mu_t.\varepsilon(x).[c_j - F(x)].\frac{p_j(x)}{w_j}$$

Formulas for tuning parameters of trapezoid membership function are as follows:

$$ml_j(t+1) = \begin{cases} ml_j(t) - \mu_t\varepsilon(x)[c_j - F(x)]\frac{p_j(x)}{a_j(x)}\frac{1}{l_j}, & if \quad ml_j - l_j < x < ml_j \\ ml_j(t), & otherwise. \end{cases}$$

$$mr_j(t+1) = \begin{cases} mr_j(t) - \mu_t\varepsilon(x)[c_j - F(x)]\frac{p_j(x)}{a_j(x)}\frac{1}{r_j}, & if \quad mr_j < x < mr_j + r_j \\ mr_j(t), & otherwise. \end{cases}$$

$$l_j(t+1) = \begin{cases} l_j(t) - \mu_t\varepsilon(x)[c_j - F(x)]\frac{p_j(x)}{a_j(x)}\frac{ml_j - x}{l_j^2}, & if \quad ml_j - l_j < x < ml_j \\ l_j(t), & otherwise. \end{cases}$$

$$r_j(t+1) = \begin{cases} r_j(t) - \mu_t\varepsilon(x)[c_j - F(x)]\frac{p_j(x)}{a_j(x)}\frac{x - mr_j}{r_j^2}, & if \quad mr_j < x < mr_j + r_j \\ r_j(t), & otherwise. \end{cases}$$

4.3   Optimal Learning

Theoretically, regarding the fuzzy system in general or the SAM in particular, the more number of fuzzy rules the more accuracy in approximation process. Nevertheless, if a system has too many fuzzy rules, it would take a long time in learning process. An optimal system will only keep necessary fuzzy rules.

One of solutions for the above problem is using Genetic Algorithm (GA) [5], [7] which proposed by John Holland in 1975. The detailed GA for fuzzy system is as follows:

- **Step 1:** Initialize ten chromosomes. Each chromosome is a chain of binary values describing status of corresponding rules in SAM. Value "0" means the rule is omitted while value "1" means the rule is selected. Every generation only uses ten chromosomes. One of individuals in the first generation contains all rules (all gene values of chromosome are equal to "1").
- **Step 2:** Create new chromosomes by crossover (probability 0.5) and mutation (probability 0.01).
- **Step 3:** Use roulette wheel with adaptive function to select ten best chromosomes which have the minimal Fit(.) value.

$$Fit(m) = ln\left(\bar{\sigma}_\varepsilon^2\right) + \frac{log_n(m)}{n}$$

where:

$$\bar{\sigma}_\varepsilon = \frac{1}{n}\sum_{j=1}^{n}(y_j - F(x_j))^2$$

- • m: number of used rules.
  - • n: number of training data samples.
- **Step 4:** If the stopping condition (i.e. expected error) is not satisfied, return step 2.
- **Step 5:** Choose the best one in ten chromosomes at the final population.

The found binary chain of the best chromosome will be used for eliminating unnecessary rules.

5   Data Set and Experiments

In this paper, the forecasting results of SAM are compared to those of the Autoregressive Moving Average (ARMA) model [10], which is usually traditionally applied for time series prediction. The ARMA model composes of two parts, the Auto Regressive (AR) and the Moving Average (MA). Conventionally, the model is referred to ARMA (p, q) where p is the order of the AR part and q is the order of the MA part (as defined below).

$$Y_t = \varepsilon_t + \sum_{i=1}^{p} \varphi_i Y_{t-i} + \sum_{i=1}^{q} \theta_i \varepsilon_{t-i}$$

where ε_t is the white noise term at time t that is identically independently distributed (IID) with mean of 0 and variance of σ_ε^2 [i.e. IID(0, σ_ε^2)]; $\varphi_1, ..., \varphi_i$ are the parameters of the AR part; $\theta_1, ..., \theta_i$ are the parameters of the MA part. We use the ARMA Excel Add-In written by Kurt Annen [14] to determine parameters of the ARMA model and run the forecasts.

The data sets are used for experiments in this study are stock prices of the International Business Machines Corp. (IBM) and Microsoft Corporation (MSFT) downloaded from the Yahoo Finance website: http://finance.yahoo.com. The IBM dataset chosen spans from 26 August 2005 to 25 August 2009 with 1006 samples (Fig. 4a) whereas this number of the MSFT dataset is 502 (Fig. 4b), spanning from 29 August 2007 to 25 August 2009. The daily closing prices of the stocks are utilized in the experiments; other values are omitted such as open,

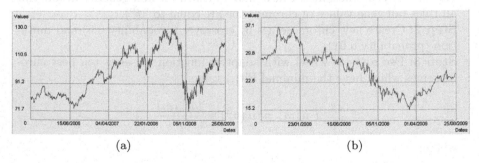

(a)   (b)

Fig. 4. (a) The graphical presentation of the IBM stock price; (b) The graphical presentation of the MSFT stock price

high, low and volume. Each dataset was split into two parts: training data set and testing data set which consists of 30 latest data samples.

The configuration of SAM model in both data set is identical. In detail, the AFLC algorithm is applied for unsupervised learning; gradient descent is applied for supervised learning; the learning rate is 0.001.

Clearly shown in Table 1, is the accuracy of the SAM prediction compared to the ARMA model. Although the number of training samples in the IBM stock forecasting is larger than that in the MSFT forecasting but the accuracy of IBM forecast is not higher than that of the MSFT forecast. It is due to the much non-linear variance of the IBM series occurring at the last samples (Fig. 4a). The less non-linear variance of the MSFT series leads to the more accurate prediction in both ARMA and SAM models. This also means that the accuracy of a forecasting model is highly sensitive to variance of the forecasting series itself. However, in both cases (much or less non-linear variances), the SAM always outperforms the ARMA model in terms of accuracy, as is found in this study.

Table 1. Mean Absolute Error (MAE) results of the ARMA and SAM models

	IBM	MSFT
Number of training samples	976	472
Parameters of ARMA (1,1)	$\varphi= 0.9997$ $\theta= -0.0316$ StdError $= 1.5822$	$\varphi= 1.0002$ $\theta= -0.1140$ StdError $= 0.6641$
MAE of ARMA	13.9078	5.9604
MAE of SAM	7.9026	0.4472

6 Conclusions

In stead of applying the traditional linear regression models for stock price fore-casting, this paper presented an application of a newly popularized technique: the SAM fuzzy system for forecasting problems. SAM offers an optimizing solu-tion for the stock price prediction and can be definitely an alternative approach for traditional models such as ARMA. A general drawback of the SAM model is time-consuming in processing. Nevertheless, when the information technology has been advanced rapidly recently, SAM in particular and machine learning techniques in general shows their absolute dominations compared to simple tra-ditional models.

References

1. Abraham, K.: Fuzzy Mathematical Techniques with Applications. Addison-Wesley Publishing Company, Reading (1986)
2. Antony, B.: Neural Network Analysis, Architectures and Application, Institute of Physics Publishing, pp. 143-149 (1997)

3. Baraldi, A., Blonda, P.: A Survey of Fuzzy Clustering Algorithms for Pattern Recognition, International Computer Science Institute, ICSI Technical Report TR-98-038 (1998)
4. Cauwenberghs, G.: A Fast Stochastic Error-Descent Algorithm for Supervised Learning and Optimization. Advance Neural Information Processing Systems 5, 244–251 (1993)
5. Chatterjee, S., Laudato, M., Lynch, L.A.: Genetic algorithms and their statistical applications: an introduction. The MIT Press, Cambridge (1996)
6. Chun, H.K., Jin, C.C.: Fuzzy Macromodel for Dynamic Simulation of Microelectromechanical Systems. IEEE Transactions on fuzzy systems, man, and cybernetics - part A: Systems and humans 36(4), 823–830 (2006)
7. David, G.: Genetic Algorithms in Search, Optimization and Machine Learning. Addison-Wesley Publishing Company, Reading (1989)
8. Frank, H., Frank, K., Rudolf, K., Thomas, R.: Fuzzy Cluster Analysis. John Wiley & Sons, Inc., Chichester (1999)
9. George, J.K., Bo, Y.: Fuzzy Sets and Fuzzy Logic: Theory and Applications. Prentice Hall PTR, Englewood Cliffs (1995)
10. Keith, W.H., Ian, A.M.: Time series modeling of water resources and environmental systems. Elsevier Science B.V., Amsterdam (1994)
11. Kosko, B.: Fuzzy Engineering. Prentice Hall PTR, Englewood Cliffs (1996)
12. Kosko, B.: Global Stability of Generalized Additive Fuzzy Systems, Systems, Man, and Cybernetics. IEEE Transactions on Part C: Applications and Reviews 28(3) (1998)
13. Kosko, B.: Neural Networks and Fuzzy Systems: A Dynamical Systems Approach to Machine Intelligence. Prentice Hall PTR, Englewood Cliffs (1991)
14. Kurt, A.: [web:reg] ARMA Excel Add-In, www.web-reg.de (accessed August 2009)
15. Madan, M.G., Liang, J., Noriyasu, H.: Static and Dynamic Neural Networks: From Fundamentals to Advanced Theory. Wiley-IEEE Press (2003)
16. Magoulas, G.D., Vrahatis, M.N., Androulakis, G.S.: Improving the Convergence of the Backpropagation Algorithm Using Learning Rate Adaptation Methods. Neural computaition 11(7), 1769–1796 (1999)
17. Pavel, B.: Grouping Multidimensional Data, pp. 25–71. Springer, Berlin (2006)
18. Robert, M.F.: The Steepest Descent Algorithm for Unconstrained Optimization and a Bisection Line-search Method. Massachusetts Institute of Technology (2004)
19. Sanya, M., Kosko, B.: The Shape of Fuzzy Sets in Adaptive Function Approximation. IEEE Transactions on fuzzy systems 9(4), 637–656 (2001)

Complex Neuro-Fuzzy Self-learning Approach to Function Approximation

Chunshien Li and Tai-Wei Chiang

Laboratory of Intelligent Systems and Applications
Department of Information Management
National Central University, Taiwan, ROC
jamesli@mgt.ncu.edu.tw

Abstract. A new complex neuro-fuzzy self-learning approach to the problem of function approximation is proposed, where complex fuzzy sets are used to design a complex neuro-fuzzy system as the function approximator. Particle swarm optimization (PSO) algorithm and recursive least square estimator (RLSE) algorithm are used in hybrid way to adjust the free parameters of the proposed complex neuro-fuzzy systems (CNFS). The hybrid PSO-RLSE learning method is used for the CNFS parameters to converge efficiently and quickly to optimal or near-optimal solution. From the experimental results, the proposed CNFS shows better performance than the traditional neuro-fuzzy system (NFS) that is designed with regular fuzzy sets. Moreover, the PSO-RLSE hybrid learning method for the CNFS improves the rate of learning convergence, and shows better performance in accuracy. Three benchmark functions are used. With the performance comparisons shown in the paper, excellent performance by the proposed approach has been observed.

Keywords: complex fuzzy set, complex neuro-fuzzy system (CNFS), PSO, RLSE, function approximation, machine learning.

1 Introduction

Theories of fuzzy logic and neural networks for system identification or system modeling have been widely investigated for applications [1]-[2]. By observing input-output data pairs for an unknown system of interest, a model can be set up for the system, using an intelligent modeling approach. With the model, the relationship of input-output behavior can be approximated. This process can be viewed as system identification, which is also known as system modeling or function approximation. Thus, system identification or modeling can be viewed as the problem of function approximation, for which an optimization process is usually involved to search for the optimal solution to the problem. However, due to the complexity and nonlinearity in real-world application problems, mathematical approaches for system identification are usually laborious and difficult. Since neural networks and fuzzy inference systems are universal approximator [3]-[4], neuro-fuzzy systems, which incorporate the advantages of fuzzy inference, neural structure and learning flexibility, have become popular and fundamental issues in modeling problems. Fuzzy sets can be used to

N.T. Nguyen, M.T. Le, and J. Świątek (Eds.): ACIIDS 2010, Part II, LNAI 5991, pp. 289–299, 2010.

reflect human concepts and thoughts, which tend to be imprecise, incomplete and vague. Complex fuzzy set (CFS) [5]-[7] is a new development in the theory of fuzzy systems. The concept of CFS is an extension of fuzzy set, by which the membership for each element of a complex fuzzy set is extended to complex-valued state. In a complex fuzzy set, membership values are complex numbers in the unit disc of the complex plane [5]-[6]. Although the introductory theory of the CFS has been presented [5], the research on complex fuzzy system designs and applications using the concept of CFS is found rarely. In this paper, a complex neuro-fuzzy system (CNFS) with a hybrid learning method is proposed to the problem of function approximation. The hybrid learning method includes the well-known particle swarm optimization (PSO) algorithm and the recursive least square estimator (RLSE) algorithm to train the proposed CNFS. The proposed approach shows better adaptability in approximating capability than a traditional neuro-fuzzy system, in terms of approximation accuracy and learning convergence rate.

In section 2 the proposed complex neuro-fuzzy approach is specified. In section 3 the PSO-RLSE hybrid learning method is given. In section 4 experimental results for function approximation are given. Finally, the paper is discussed and concluded.

2 Methodology for Complex Neuro-Fuzzy System

There are two frequently used fuzzy inference systems (FISs). The first is Mandani type FIS and the other is Takagi-Sugeno (T-S) type FIS. The difference between them lies in the consequents of fuzzy rules. For Mamdani fuzzy model, the consequents are specified with linguistic terms, which can be defined with fuzzy sets [8]. For T-S fuzzy model [9], the consequents are expressed as polynomial functions of the input variables. In this paper, the design of the proposed CNFS in extended from the concept of traditional neuro-fuzzy system (NFS), and the fuzzy T-S model is used in the proposed CNFS. The fuzzy theory can be used to represent uncertain or imprecise data, information and concept. The values of applying fuzzy sets for modeling uncertainty, for representing subjective human knowledge, and for emulating human reasoning processes, have been validated [5]. The concept of fuzzy sets is extended to complex fuzzy set (CFS) [5]-[7], which expands the range of membership from the unit interval [0, 1] to the unit disc in the complex plane. Assume there is a complex fuzzy set S whose membership function is given as follows.

$$
\begin{aligned}
\mu_s(h) &= r_s(h)\exp(j\omega_s(h)) \\
&= \mathrm{Re}(\mu_s(h)) + j\,\mathrm{Im}(\mu_s(h)) \\
&= r_s(h)\cos(\omega_s(h)) + jr_s(h)\sin(\omega_s(h))
\end{aligned}
\tag{1}
$$

where h is the base variable for the complex fuzzy set, $r_s(h)$ is the amplitude function of the complex membership, $\omega_s(h)$ is the phase function. The complex fuzzy set S is expressed as follows.

$$
S = \{(h, \mu_s(h)) \mid h \in U\}
\tag{2}
$$

In the case that $\omega_s(h)$ equals to 0, a traditional fuzzy set is regarded as a special case of a complex fuzzy set. Assume there is a fuzzy rule with the form of " If (x_A=A and

$x_B=B$) Then...", where A and B represent two different conditions in the rule. The two conditions can be described using two complex fuzzy sets, given as follows.

$$\mu_A(h_A) = r_A(h_A)\exp(j\omega_A(h_A)) \tag{3}$$

$$\mu_B(h_B) = r_B(h_B)\exp(j\omega_B(h_B)) \tag{4}$$

where h_A and h_B are the base variables; x_A and x_B are the linguistic variables for h_A and h_B, respectively. Intersection of the complex fuzzy sets A and B is expressed as follows.

$$\mu_{A\cap B} \equiv [r_A(h_A)*r_B(h_B)]\exp(j\omega_{A\cap B}) \tag{5}$$

where $*$ is for t-norm operator (intersection operator), $\omega_{A\cap B}= \wedge(\omega_A(h_A),\omega_B(h_B))$ is the intersection, and $\wedge(.,.)$ denotes the phase intersection operation. Union of the complex fuzzy sets A and B is expressed as follows.

$$\mu_{A\cup B} \equiv [r_A(h_A)\oplus r_B(h_B)]\exp(j\omega_{A\cup B}) \tag{6}$$

where $\oplus$ is for s-norm operator (union operator), $\omega_{A\cup B}=\vee(\omega_A(h_A),\omega_B(h_B))$ is the union of phase, and $\vee(.,.)$ is represented the union operator of phase. Assume we have a complex fuzzy system with K T-S fuzzy rules, given as follows.

Rule i: IF (x_1 is $A_1^i(h_1(t))$) and (x_2 is $A_2^i(h_2(t))$) ...

$$\text{and (} x_M \text{ is } A_M^i(h_M(t))) \text{ Then } z^i = a_0^i + \sum_{j=1}^{M} a_j^i h_j \tag{7}$$

$i =1,2,...,K$, where x_j is the j-th input linguistic variable, h_j is the j-th input of base variables, $A_j^i(h_j)$ is the complex fuzzy set for the j-th condition in the i-th rule, z^i is the output of the i-th rule, and a_j^i, $i=1,2...K$ and $j=0,1,...M$ are the consequent parameters . For the proposed CNFS, the complex fuzzy system is cast into the framework with six layered neuro-fuzzy network. The complex fuzzy reasoning for the CNFS from input to output is explained as follows.

Layer 0: The layer is called the input layer, which receives the inputs and transmits them to the next layer directly. The input vector is given as follows.

$$H(t)=[h_1(t),h_2(t),...,h_M(t)]^T \tag{8}$$

Layer 1: The layer is called the fuzzy-set layer. Nodes in the layer are used to represent the complex fuzzy sets for the premise part of the CNFS and to calculate the membership degrees.

Layer 2: This layer is for the firing-strengths. The firing strength of the i-th rule is calculated as follows.

$$\beta^i(t) = \mu_1^i(h_1(t))*\mu_2^i(h_2(t))*\cdots*\mu_M^i(h_M(t))$$
$$= \bigwedge_{j=1}^{M} r_j^i(h_j(t))\exp(j\omega_{A_1^i\cap...\cap A_M^i}) \tag{9}$$

$i = 1,2,...,K$, where min operator is used for the t-norm calculation of the firing strength. r^i_j is the amplitude of complex membership degree for the j-th fuzzy set of the i-th rule.

Layer 3: This layer is for the normalization of the firing strengths. The normalized firing strength for the i-th rule is represented as follows.

$$\lambda^i(t) = \frac{\beta^i(t)}{\sum\limits_{i=1}^{K}\beta^i(t)} = \frac{(\bigwedge\limits_{j=1}^{M} r^i_j(h_j(t)))\exp(j\omega_{A^i_1\cap..\cap A^i_M})}{\sum\limits_{i=1}^{K}(\bigwedge\limits_{j=1}^{M} r^i_j(h_j(t)))\exp(j\omega_{A^i_1\cap..\cap A^i_M})} \tag{10}$$

Layer 4: The layer is for normalized consequents. The normalized consequent of the i-th rule is represented as follows.

$$\xi^i(t) = \lambda^i(t) \times z^i(t)$$

$$= \lambda^i(t) \times \left(a^i_0 + \sum\limits_{j=1}^{M} a^i_j h_j(t) \right) \tag{11}$$

$$= \frac{(\bigwedge\limits_{j=1}^{M} r^i_j(h_j(t)))\exp(j\omega_{A^i_1\cap..\cap A^i_M})}{\sum\limits_{i=1}^{K}(\bigwedge\limits_{j=1}^{M} r^i_j(h_j(t)))\exp(j\omega_{A^i_1\cap..\cap A^i_M})} \times \left(a^i_0 + \sum\limits_{j=1}^{M} a^i_j h_j(t) \right)$$

Layer 5: This layer is called the output layer. The normalized consequents from Layer 4 are congregated in the layer to produce the CNFS output, given as follows.

$$\xi(t) = \sum\limits_{i=1}^{K}\xi^i(t) = \sum\limits_{i=1}^{K}\lambda^i(t) \times z^i(t)$$

$$= \sum\limits_{i=1}^{K}\frac{(\bigwedge\limits_{j=1}^{M} r^i_j(h_j(t)))\exp(j\omega_{A^i_1\cap..\cap A^i_M})}{\sum\limits_{i=1}^{K}(\bigwedge\limits_{j=1}^{M} r^i_j(h_j(t)))\exp(j\omega_{A^i_1\cap..\cap A^i_M})} \times \left(a^i_0 + \sum\limits_{j=1}^{M} a^i_j h_j(t) \right) \tag{12}$$

Generally the output of the CNFS is represented as follows.

$$\xi(t) = \xi_{\text{Re}}(t) + j\xi_{\text{Im}}(t)$$

$$= |\xi(t)| \times \exp(j\omega_\xi) \tag{13}$$

$$= |\xi(t)| \times \cos(\omega_\xi) + j|\xi(t)| \times \sin(\omega_\xi)$$

where $\xi_{\text{Re}}(t)$ is the real part of the output for the CNFS, $\xi_{\text{Im}}(t)$ is the imaginary part, the absolute value of the complex output is given in (14), and the phase of the complex output is expressed in (15).

$$\xi(t) = \sqrt{(\xi_{\text{Re}}(t))^2 + (\xi_{\text{Im}}(t))^2} \tag{14}$$

$$\omega_\xi = \tan^{-1}(\frac{\xi_{\text{Im}}}{\xi_{\text{Re}}}) \tag{15}$$

Based on (12), the complex inference system can be viewed as a complex function system, expressed as follows.

$$\xi(t) = F(H(t), W) = F_{\text{Re}}(H(t), W) + jF_{\text{Im}}(H(t), W) \tag{16}$$

where $F_{\text{Re}}(.)$ is the real part of the CNFS output, $F_{\text{Im}}(.)$ is the imaginary part of the output, $H(t)$ is the input vector to the CNFS, W denotes the parameter set of the CNFS. The parameter set W can be divided into two subsets, which are the premise-part subset and the consequent-part subset, denoted as W_{If} and W_{Then}, respectively.

3   Hybrid PSO-RLSE Learning for CNFS

Particle swarm optimization (PSO) [10] is a population-based optimization method, which is motivated by the food searching behaviour of bird flocking or fish schooling. Each bird in the swarm in viewed as a particle. Assume the location of food is viewed as the optimal solution in the problem space. Each particle is viewed as a potential solution in the search space. Each particle location can be mapped to a fitness (or called a cost) with some given fitness function (or called cost function). The particles in the swarm compare to each other to become the winner, which is known as **gbest**. The best location of a particle during the evolution process is called **pbest**. The location and the velocity of a particle in the swarm are updated using the information of **gbest** and its **pbest**. Assume the problem space is with Q dimensions. The method of PSO is expressed as follows.

$$V_i(k+1) = V_i(k) + c_1 \cdot \xi_1 \cdot (\textbf{\textit{pbest}}_i(k) - L_i(k))$$
$$+ c_2 \cdot \xi_2 \cdot (\textbf{\textit{gbest}}(k) - L_i(k)) \tag{17a}$$

$$L_i(k+1) = L_i(k) + V_i(k+1) \tag{17b}$$

$$V_i(k) = [v_{i,1}(k), v_{i,2}(k), \ldots, v_{i,Q}(k)]^{\text{T}} \tag{18}$$

$$L_i(k) = [l_{i,1}(k), l_{i,2}(k), \ldots, l_{i,Q}(k)]^{\text{T}} \tag{19}$$

where $V_i(k)$ is the velocity for the i-th particle on k-th iteration, $L_i(k)$ is the location for the i-th particle, $\{c_1, c_2\}$ are the parameters for PSO, and $\{\xi_1, \xi_2\}$ are random numbers in [0,1].

The RLSE [11] is used for the identification of the parameters of consequent part. For a general least-squares estimation problem, the output of a linear model, y, is specified by the linearly parameterized expression, given as follows.

$$y = \theta_1 f_1(u) + \theta_2 f_2(u) + \cdots + \theta_m f_m(u) \tag{20}$$

where u is the model's input, $f_i(.)$ is known function of u and θ_i, $i=1,2,\ldots,m$ represents unknown parameters to be estimated. Here θ_i can be viewed as the consequent parameters of the proposed T-S fuzzy approximator. To estimate the unknown parameters

$\{\theta_i, i=1,2,...,m\}$ for a unknown target system (or function), a set of input-output data pairs are used as training data, denoted as follows.

$$TD = \{(u_i, y_i), i = 1,2,...,N\} \tag{21}$$

Substituting data pairs into (17), a set of N linear equations are given as follows.

$$
\begin{aligned}
f_1(u_1)\theta_1 + f_2(u_1)\theta_2 + \cdots + f_m(u_1)\theta_m &= y_1 \\
f_1(u_2)\theta_1 + f_2(u_2)\theta_2 + \cdots + f_m(u_2)\theta_m &= y_2 \\
\vdots \qquad\qquad \vdots \qquad\qquad\qquad \vdots \qquad \vdots \\
f_1(u_N)\theta_1 + f_2(u_N)\theta_2 + \cdots + f_m(u_N)\theta_m &= y_N
\end{aligned}
\tag{22}
$$

The optimal estimation for θ can be calculated using the following RLSE equations.

$$P_{k+1} = P_k - \frac{P_k b_{k+1} b_{k+1}{}^{T} P_k}{1 + b_{k+1}{}^{T} P_k b_{k+1}}, \tag{23a}$$

$$\theta_{k+1} = \theta_k + P_{k+1} b_{k+1}(y_{k+1} - b_{k+1}{}^{T}\theta_k) \tag{23b}$$

$k=0,1,...,N\text{-}1$, where $[b^{T}_k, y_k]$ in the k-th row of $[A, y]$. To start the RLSE algorithm in (20), we need to select the initial values for θ_0 and P_0 is given as follows.

$$P_0 = \alpha I \tag{24}$$

where α is a large value and I is the identity matrix, and θ_0 is initially set to zeros.

For the training of the proposed CNFS, the hybrid PSO-RLSE learning method is applied to update the premise parameters and the consequent parameters. In hybrid way, the PSO is used with the RLSE for fast learning convergence. The premise parameters and the consequent parameters of the CNFS are updated by the PSO given in (17) and the RLSE given in (23), respectively. The PSO is used to update the premise parameters of the CNFS. The PSO is a heuristic method retaining characteristics of evolutionary search algorithms. Each location by **gbest** of the PSO provides a potential premise solution. The RLSE is used to update the consequent parameters, with the normalized firing strengths. With (20), to estimate the consequent parameters, the row vector b and the vector θ are arranged as follows.

$$b_{k+1} = [bb^1(k+1) \quad bb^2(k+1) \quad \cdots \quad bb^K(k+1)] \tag{25}$$

$$bb^i(k+1) = [\lambda^i \quad h_1(k+1)\lambda^i \quad \cdots \quad h_M(k+1)\lambda^i] \tag{26}$$

$$\theta = [\tau^1 \quad \tau^2 \quad \cdots \quad \tau^K] \tag{27}$$

$$\tau^i = [a_0^i \quad a_1^i \quad \cdots \quad a_M^i] \tag{28}$$

$i = 1,2,...,K$, and $k=0,1,...,N\text{-}1$, At each iteration for the hybrid PSO-RLSE learning, the output of the CNFS approximator can be obtained in (13). The error between output of the target and the CNFS is defined as follows.

$$\text{RMSE} = \left(\frac{1}{N} \sum_{t=1}^{N} (y(t) - \xi(t))^2 \right)^{\frac{1}{2}} = \left(\frac{1}{N} \sum_{t=1}^{N} (y(t) - F(H(t), W))^2 \right)^{\frac{1}{2}} \tag{29}$$

The error is used further to define the root mean square error (RMSE), which is used as the performance index in the study. The square of RMSE is called the mean square error (MSE).

4 Experiments for the Proposed Approach

Experiments for function approximation are conducted in this section to estimate and verify the performance of the proposed approach. Two subsections are given in the section. In the first subsection, the proposed approach using the CNFS approximator and the hybrid PSO-RLSE learning method is compared to two other compared approaches. The first compared approach uses a traditional NFS approximator and the PSO learning method, and the second compared approach uses the same CNFS approximator and the PSO method alone. The tooth function is used in the 1st subsection. In the 2nd subsection, the proposed approach is compared to the approach in [12]. Two benchmark functions are used in the 2nd subsection for performance comparison.

4.1 Comparison for the Proposed Approach to the PSO for CNFS, and the PSO for NFS

The "tooth" function is given as follows.

$$\begin{aligned} y = 0.08 \times \{&1.2 \times [(u-1) \times (\cos(3u))] \\ &+ [(u-(u-1) \times (\cos(3u))) \times \sin(u)]\}) \\ &3 \leq u \leq 7 \end{aligned} \tag{30}$$

The generated 100 data pairs from the tooth function are used to train the CNFS which is designed with complex fuzzy sets and the NFS which is designed with traditional fuzzy sets. Two inputs are used to the CNFS and the traditional NFS. Each input possesses three fuzzy sets. There are 9 rules in CNFS and the NFS, where 12 premise parameters and 27 consequent parameters are to be updated by the PSO algorithm. In the hybrid PSO-RLSE for the CNFS, PSO is used to adjust the 12 premise parameter of CNFS and the RLSE is to update the 27 consequent parameters. For the output of the CNFS, we select the real part to represent the approximator output. For the PSO settings for the proposed CNFS and the NFS, $\{c_1, c_2\} = \{2, 2\}$ and the population size = 635 are given. And, for the hybrid PSO-RLSE settings, $\{c_1, c_2\} = \{2, 2\}$, $\alpha = 10^4$, and $\theta_0 =$ zero-valued vector are given.

The learning curve and the result by the proposed PSO-RLSE for the CNFS are given in Figs. 1 to 2. The performance comparison for the proposed approach and the two compared approach is given in Table 1. The approximation errors by the three approaches are shown in Fig. 3.

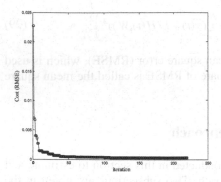

Fig. 1. Learning curve by the proposed hybrid PSO-RLSE for the CNFS for the tooth function

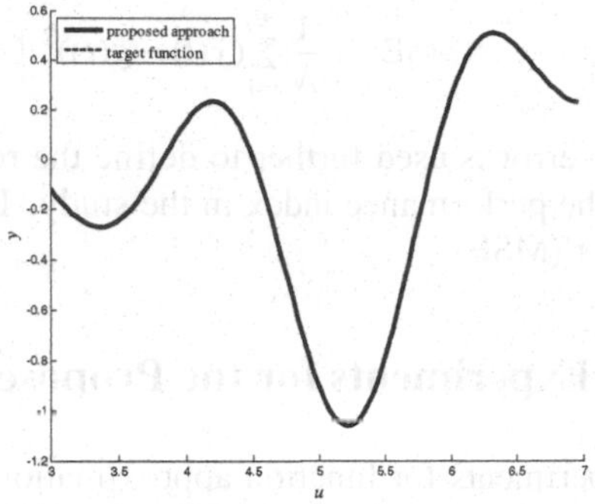

Fig. 2. Response by the proposed CNFS with PSO-RLSE for the tooth function

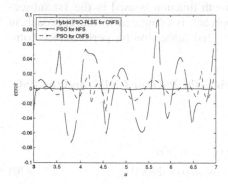

Fig. 3. Approximation errors by the PSO for NFS, the PSO for CNFS, and the hybrid PSO-RLSE for CNFS

Table 1. Performance Comparison

Methods	RLSE
PSO for NFS	7.86×10^{-3}
PSO for CNFS	1.56×10^{-3}
Hybrid PSO-RLSE for CNFS	2.73×10^{-4}

4.2   Comparison for the Proposed Approach to the Approach in [12]

The hybrid PSO-RLSE method for the proposed CNFS is employed to approximate the two benchmark functions which are the exponential function in [-4, 2] and the hyperbolic tangent function in [-5, 5]. For each of the benchmark functions, the error norm for function approximation is based on the mean square errors (MSE), as defined before. The training and testing data for each benchmark function are 400 and 200 sampled pairs. With the 2 benchmark function, Table 2 shows the performance comparisons in MSE for 20 experimental trials using the proposed approach and the compared approach [12]. The approximation responses and errors for the 2 functions by the proposed approach are shown in Figs 4 to 7.

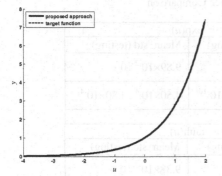

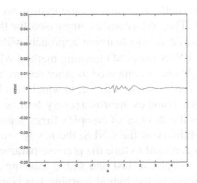

Fig. 4. Result by the proposed CNFS hybrid learning approach for the exponential function

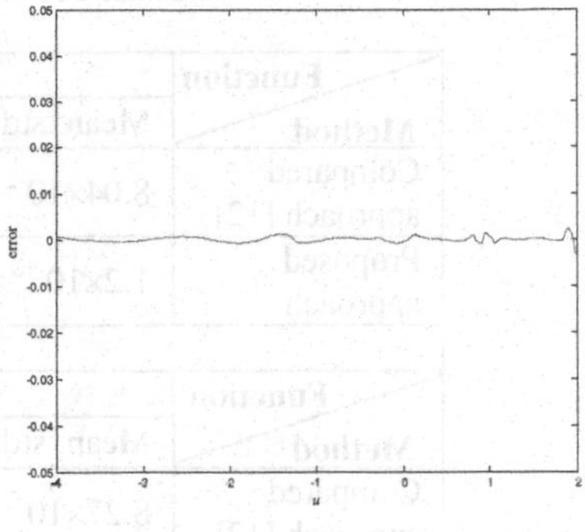

Fig. 5. Approximation error by the proposed CNFS hybrid learning approach for the exponential function

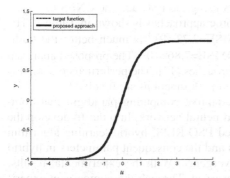

Fig. 6. Result by the proposed CNFS hybrid learning approach for the hyperbolic tangent function

Fig. 7. Approximation error by proposed CNFS hybrid learning approach for the hyperbolic tangent function

5 Discussion and Conclusion

The proposed complex neuro-fuzzy system (CNFS) has been presented to the problem of function approximation to verify the mapping performance of the proposed CNFS. The hybrid PSO-RLSE learning method has been applied to theproposed CNFS to adapt its system parameters. The system parameters are divided into two subsets to make easier the learning process for the optimal solution to application performance. The two subsets are the premise set of parameters and the consequent set of parameters. The former subset includes the parameters in defining the premise fuzzy sets for the CNFS, and the later subset collects the consequent parameters in defining the consequent parts of the rules in the CNFS. The well-known PSO is used to update the premise subset of parameters and the RLSE is for the consequent subset of parameters. This hybrid learning method is very efficient to find the optimal (or near optimal) solution for the CNFS in application performance.

Table 2. Performance Comparison

Function	$\exp(u)$	
Method	Mean±std (training)	Mean±std (testing)
Compared approach [12]	$8.04\times10^{-2}\pm0$	$9.89\times10^{-2}\pm0$
Proposed approach	$1.2\times10^{-10}\pm1.04\times10^{-10}$	$4.50\times10^{-4}\pm3.80\times10^{-4}$

Function	$\tanh(u)$	
Method	Mean±std (training)	Mean±std (testing)
Compared approach [12]	$8.27\times10^{-2}\pm0$	$9.38\times10^{-2}\pm0$
Proposed approach	$3.86\times10^{-8}\pm1.69\times10^{-8}$	$3.58\times10^{-5}\pm1.77\times10^{-5}$

It is found the proposed hybrid learning approach is superior to the two other compared methods, which are the traditional NFS using the PSO and the CNFS using the PSO. The performance comparison for the three approaches is shown in Table 1. The proposed hybrid learning approach with RMSE=2.73×10^{-4} is much better the traditional NFS with PSO learning method with RMSE=7.86×10^{-3}. The proposed approach has also been compared to other research approaches [12]. The performance comparisons are shown in Table 2, in which two benchmark functions are involved.

The complex neuro-fuzzy system is an adaptive computing paradigm that combines the theories of complex fuzzy logic and neural network. In order to develop the adaptability of the CNFS, the newly proposed PSO-RLSE hybrid learning algorithm has been used to tune the premise parameters and the consequent parameters in hybrid way to achieve fast and stable learning convergence. With the experimental results, the merit of the hybrid learning has been observed. Through the comparison experiments, the proposed approach has shown excellent performance.

References

1. Juang, C.F., Lin, C.T.: An online self-constructing neural fuzzy inference network and its applications. IEEE Transactions on Fuzzy Systems 6, 12–32 (1998)
2. Paul, S., Kumar, S.: Subsethood-product fuzzy neural inference system. IEEE Transactions on Neural Networks 13, 578–599 (2002)
3. Hornik, K., Stinchcombe, M., White, H.: Multilayer feed forward networks are universal approximators. Neural networks 2, 359–366 (1989)
4. Wang, L.X., Mendel, J.M.: Fuzzy basis functions, universal approximation, and orthogonalleast-squares learning. IEEE Transactions on Neural Networks 3, 807–814 (1992)
5. Kandel, A., Ramot, D., Milo, R., Friedman, M.: Complex Fuzzy Sets. IEEE Transactions on Fuzzy Systems 10, 171–186 (2002)
6. Dick, S.: Toward complex fuzzy logic. IEEE Transactions on Fuzzy Systems 13, 405–414 (2005)
7. Ramot, D., Friedman, M., Langholz, G., Kandel, A.: Complex fuzzy logic. IEEE Transactions on Fuzzy Systems 11, 450–461 (2003)

8. Farag, W.A., Quintana, V.H., Lambert-Torres, G.: A genetic-based neuro-fuzzy approach for modeling and control of dynamical systems. IEEE Transactions on Neural Networks 9, 756–767 (1998)
9. Takagi, T., Sugeno, M.: Fuzzy identification of systems and its applications to modeling and control. IEEE transactions on systems, man, and cybernetics 15, 116–132 (1985)
10. Kennedy, J., Eberhart, R.: Particle swarm optimization. In: IEEE International Conference on Neural Networks (1995)
11. Hsia, T.C.: System identification: Least-squares methods. D. C. Heath and Company (1977)
12. Wu, J.M., Lin, Z.H., Hus, P.H.: Function approximation using generalized adalines. IEEE Transactions on Neural Networks 17, 541–558 (2006)

On the Effectiveness of Gene Selection for Microarray Classification Methods

Zhongwei Zhang[1], Jiuyong Li[2], Hong Hu[3], and Hong Zhou[4]

[1] Department of Mathematics and Computing, University of Southern Queensland
QLD 4350, Australia
zhongwei@usq.edu.au
[2] School of Computer and Information Science
University of South Australia,
Mawson Lakes, Adelaide, SA 5001, Australia
Jiuyong.Li@unisa.edu.au
[3] Planning and Quality Office, University of Southern Queensland
QLD 4350, Australia
huhong@usq.edu.au
[4] Faculty of Engineering, University of Southern Queensland
QLD 4350, Australia
hzhou@usq.edu.au

Abstract. Microarray data usually contains a high level of noisy gene data, the noisy gene data include incorrect, noise and irrelevant genes. Before Microarray data classification takes place, it is desirable to eliminate as much noisy data as possible. An approach to improving the accuracy and efficiency of Microarray data classification is to make a small selection from the large volume of high dimensional gene expression dataset. An effective gene selection helps to clean up the existing Microarray data and therefore the quality of Microarray data has been improved. In this paper, we study the effectiveness of the gene selection technology for Microarray classification methods. We have conducted some experiments on the effectiveness of gene selection for Microarray classification methods such as two benchmark algorithms: SVMs and C4.5. We observed that although in general the performance of SVMs and C4.5 are improved by using the preprocessed datasets rather than the original data sets in terms of accuracy and efficiency, while an inappropriate choice of gene data can only be detrimental to the power of prediction. Our results also implied that with preprocessing, the number of genes selected affects the classification accuracy.

1 Introduction

Gene selection technology has been widely used by many researchers in the past decades to select the most effective genes from high dimensional Microarray data. The Microarray gene data acquired from Microarray technology is quite different than that from the normal relational databases. Normal relational databases contain a small number of attributes and a large number of samples. In contrast, gene expression Microarray data usually contains a very large number of

N.T. Nguyen, M.T. Le, and J. Świątek (Eds.): ACIIDS 2010, Part II, LNAI 5991, pp. 300–309, 2010.
© Springer-Verlag Berlin Heidelberg 2010

attributes but a small number of usable samples. With a large number of genes, it is absolutely desirable to have a large number of samples accordingly in order to build reliable Microarray classification models. However, the reality is that for most Microarray experiments, a limited number of samples are available due to the huge cost of producing such Microarray data and other factors, such as privacy and availability. As an example, for cancer Microarray data, the number of samples is usually less than 200.

In short, high dimensionality renders many classification methods not applicable for analyzing raw gene Microarray data. Furthermore, high dimension Microarray data with noisy attributes leads to unreliable and low accuracy analysis results. Consequently, reducing irrelevant and removing noise gene expression values from the original Microarray data are crucial for applying classification algorithms to analyze gene expression Microarray data.

Many researches have shown that gene selection can improve the performance of Microarray classification [1, 2, 3, 4, 5]. But these research haven't answered the question. *Can good gene selection methods enhance the prediction performance of all types of Microarray classification methods, wrapper classification methods in particular?*

This paper is organized as follows. In the preceding section, we identify problems in gene expression Microarray data classification and highlight the importance of gene selection for gene expression Microarray data. In Section 2, we review a number of gene selection methods. In Section 3, we present the design of methods for comparing the accuracy of SVMs and C4.5 using different gene selection methods. In Section 4, we test four different gene selection methods with six data sets. In Section 5, we present a discussion of the results. In Section 6, we conclude the paper.

2   Gene Selection Methods

To deal with the problems caused by high dimensionality and noisy Microarray gene data, a preprocessing phase should be introduced to reduce the noise and irrelevant genes before the Microarray data classifications are applied. As a preprocessing method of Microarray data classification, gene selection is a very effective way for eliminating the noisy genes. In essence, gene selection aims to select a relatively small set of genes from a high dimensional gene expression data set. Gene selection helps to clean up the existing Microarray data and therefore improve the quality of Microarray data. In other words, removing irrelevant and noisy genes is helpful for improving the accuracy of Microarray data classification. The resultant classification models of Microarray gene data would therefore better characterize the true relationships among genes and hence be easier to be interpreted by biologists. Arguably, a good gene selection method not only increases the accuracy of classification through the improvement of the Microarray data quality, but also speeds up the classification process through the cutdown of high dimensionality.

Based on the dependency on classification algorithms, gene selection methods can be roughly divided into wrapper and filter methods [6]. A filter method performs gene selection independently from a classification method. It preprocesses a Microarray data set before the data set is used for classification analysis. Some filter gene selection methods are: ranking gene selection methods [7], and information gain gene selection method [8], Markov blanket-embedded genetic algorithm for gene selection [9], and so on. One-gene-at-a-time filter methods, such as ranking [7], signal-to-noise [4], information gain [10], are fast and scalable but do not take the relationships between genes into account. Some genes among the selected genes may have similar expression levels among classes, and they are redundant since no additional information is gained for classification algorithms by keeping them all in the dataset. To this end, Koller and Sahami [11] developed an optimal gene selection method called *Markov blanket filtering* which models feature dependencies and can eliminate redundant genes. Further to this method, Yu and Liu [12] proposed the Redundancy Based Filter(RBF) method, which is able to deal with redundant problems. Favorable results have been achieved.

In contrast, a wrapper method embeds a gene selection method [13] within a classification algorithm. An example of a wrapper method is SVMs [14], which uses a recursive feature elimination(RFE) approach to eliminate the features iteratively in a greedy fashion until the largest margin of separation is reached. Wrapper methods are not as efficient as filter methods due to the fact that they usually run on the original high dimensional Microarray dataset. However, Kohavi and John [6] discovered that wrapper methods could significantly improve the accuracy of classification algorithms over filter methods. This discovery indicates that the performance of a classification algorithm is largely dependent on the chosen gene selection method. Nevertheless, no single gene selection method can universally improve the performance of classification algorithms in terms of efficiency and accuracy.

In Section 3, we design some experiments to investigate the dependency between gene selection methods and Microarray data classification methods.

3   Experimental Design and Methodology

Our approach is to use different existing gene selection methods to preprocess Microarray data for classification. We have carried out our experiments by comparing with benchmark algorithms SVMs and C4.5. Note that this choice is based on the following considerations.

Consideration of benchmark systems: For years, SVMs and C4.5 have been regarded as benchmark classification algorithms. SVMs was proposed by Cottes and Vapnik [15] in 1995. It has been one of the most influential classification algorithms. SVMs has been applied to many domains, for example, text categorization [16], image classification [17], cancer classification [18, 19]. SVMs can easily deal with high dimensional data sets with a wrapper gene selection method. SVMs also can achieve a higher performance compared to most existing classification algorithms.

Considering of wrapper methods: SVMs and C4.5 are not only benchmark classification systems, but each of them contains a wrapper gene selection method. SVMs uses a recursive feature elimination(RFE) approach to eliminate the features iteratively in a greedy fashion until the largest margin of separation is reached. Decision tree method can also be treated as a gene selection method. It selects the gene with the highest information gain at each step and all selected genes appear in the decision tree.

A ranking method identifies one gene at a time with differentially expressed levels among predefined classes and puts all genes in decreasing order. After a specified significance expressed level or number of genes is selected, the genes lower than the significance level or given number of genes are filtered out. The advantages of these methods is that they are intuitive, simple and easy to implement. In this study, we choose and implement four popular ranking methods collected by Cho and Won [20], namely Signal-to-Noise ratio (SNR), correlation coefficient (CC), Euclidean (EU) and Cosine (CO) ranking methods.

To evaluate the performance of different gene selection methods, six datasets from Kent Ridge Biological Data Set Repository [21] were selected. These data sets were collected from some influential journal papers, namely the breast cancer, lung cancer, Leukemia, lymphoma, colon and prostate data sets 3. Each Microarray dataset is described by the following parameters.

1. Genes: the number of genes or attributes
2. Class: the number of classes,
3. Record: the number of samples in the dataset

Table 1. Gene expression Microarray data sets

		Dataset name	Genes	Class	Sample
1		Breast Cancer	24481	2	97
2		Lung Cancer	12533	2	181
3		Lymphoma	4026	2	47
4		Leukemia	7129	2	72
5		Colon	2000	2	62
6		Prostate	12600	2	21

During the gene expression Microarray data preprocessing stage, we define the number of selected genes as 20, 50 and 100 and 200 for all filter gene selection methods. In our experiments, a tenfold cross-validation method is also carried out for each classification method to test its accuracy.

4   Experimental Results and Discussions

Figure 1 - 12 show the detailed results for SVMs and C4.5 tested on six different datasets preprocessed by four different filter gene selection methods.

From these experimental results, we make the following observations.

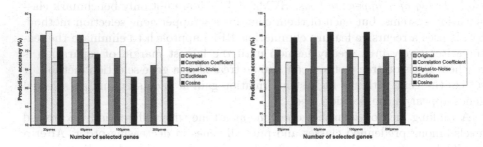

Fig. 1. C4.5 tested on Breast cancer dataset

Fig. 2. C4.5 tested on lung cancer dataset

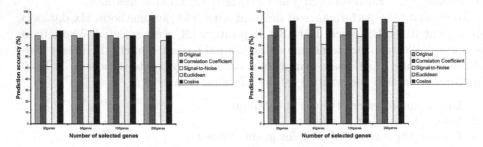

Fig. 3. C4.5 tested on Lymphoma dataset

Fig. 4. C4.5 tested on Leukemia dataset

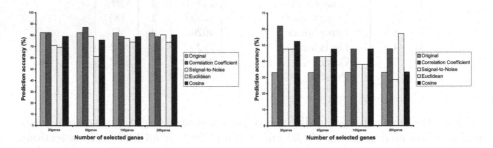

Fig. 5. C4.5 tested on Colon dataset

Fig. 6. C4.5 tested on Prostate dataset

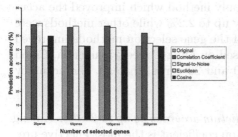

Fig. 7. SVM tested on Breast cancer dataset

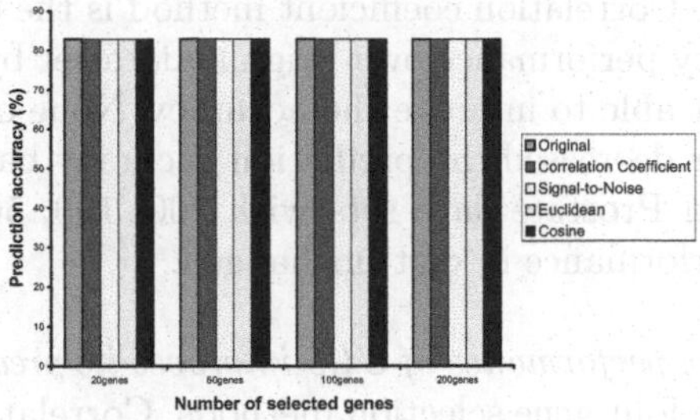

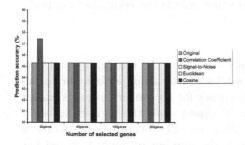

Fig. 8. SVM tested on lung cancer dataset

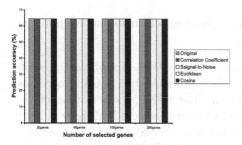

Fig. 9. SVM tested on Lymphoma data set

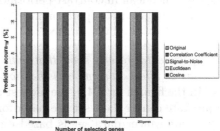

Fig. 10. SVM tested on Leukemia data set

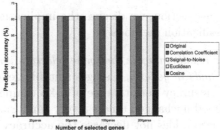

Fig. 11. SVM tested on Colon dataset

Fig. 12. SVM tested on Prostate dataset

When Microarray data sets are preprocessed, SVMs improves its prediction accuracy on Breast Cancer and Lymphoma data sets only. Signal-to-Noise and Correlation coefficient methods performed best and improved the accuracy up to 16.5% and 15.5% respectively on Cancer data. The Cosine method also improved the accuracy by up to 7.2% on Cancer data. On the Lymphoma data set, the Correlation coefficient method is the only method which improved the accuracy performance over original data set by up to 2.2% while other methods were not able to improve the accuracy. None of the gene selection methods improved nor decreased the prediction accuracy based on Lung Cancer, Leukemia, Colon and Prostate data sets with 200, 100, 50 and 20 genes. Instead, the accuracy performance is kept unchanged.

The performance of C4.5 improves its prediction accuracy by up to 28.6%. Among the four gene selection methods, Correlation coefficient is the most effective preprocessing method with an improvement of accuracy up to 7.6% on average, followed by Cosine 7.3%, and Signal-to-Noise 6.0%. The Signal-to-Noise gene selection method performed consistently better on Breast Cancer and Leukemia data sets with improved accuracy by up to 12.4%, but failed on the other cancer data sets. Euclidean in contrast performed worst among the compared methods, decreasing the accuracy performance on all provided cancer data sets except Breast and Prostate by up to 18.7%.

The experimental results show that with preprocessing, the number of genes selected has an affect on some classification methods in terms of performance accuracy. In the figures for C4.5, the highest accuracy for all cancer data sets except the prostate cancer data set are based on 50 genes; while the highest accuracy for prostate cancer data set is based on 20 genes. The overall performance is better when data sets contain 50. However, the number of genes selected has little impact on the performance of SVMs.

5   Discussion of Experimental Results

In this section, we discuss the implication of gene selection methods upon the classification methods.

The results indicate that gene selection improves the performance of classification methods in general. Using a suitable gene selection method with C4.5 increases the accuracy performance of C4.5 dramatically. For SVMs, its performance remained unchanged unless a very small size of genes was selected. Moreover, gene selection does not decrease the accuracy performance of SVMs. This result ensures that we can reduce the number of genes to a smaller size without hurting the accuracy performance of classification. This is very helpful for noisy Microarray data classification as most irrelevant genes in Microarray data classification would be reduced. It increases the performance of classification significantly in terms of speeding up the efficiency of Microarray data classification.

These results indicate that not all gene selection methods help the performance of Microarray classification methods in terms of improving the prediction accuracy of classification. Their performance depends on which Microarray classification method they are combined with. For C4.5, with the help of some gene selection, such as the Correlation coefficient method, the accuracy performance improved significantly. The Signal-to-Noise method generated mixed results combined with C4.5; while the Euclidean method is not a suitable gene selection method for C4.5 as it failed to improved the accuracy performance of C4.5 on most data sets. So to apply gene selection to C4.5, we have to seriously consider which gene selection algorithm to use to achieve maximum improvement. With SVMs, only the Correlation coefficient method managed to improve the accuracy performance on up to two data sets.

Gene selection may have little impact on some classification methods. The figures show that SVMs is insensitive to the gene selection methods used and hence data preprocessing does not increase its performance in most cases. This indicates that the SVMs classification method can initially handle noise data very well. Moreover, it would require little effort to select a gene selection method for SVMs.

The observations indicate that a data set with less genes or attributes does not necessarily guarantee the highest prediction accuracy. The number of genes selected by a preprocessing method should not be too small. At this stage, the objective of gene selection is just to eliminate irrelevant and noise genes. However, less informative genes can sometimes enhance the power of classification if they are co-related with the most informative genes. If the number of genes has been eliminated too harshly, it can also decrease the performance of the classification. So during the preprocessing, we need to make sure that a reasonable number of genes are left for classification.

Those results remind us that when selecting the gene selection method for data preprocessing, we must consider which classification method the gene selection is for. For example, if we select SVMs as a classification algorithm, then the Correlation coefficient or Signal-to-Noise gene selection methods are better for data preprocessing. An inappropriate choice can only harm the power of classification prediction.

6   Conclusions

In this paper, we have looked into the gene selection technique to improve the quality of Microarray data sets on Microarray data classification methods: SVMs and C4.5, which themselves contain a wrapped method. We observed that although in general the performance of SVMs and C4.5 are improved by using the preprocessed datasets rather than original data sets in terms of accuracy and efficiency, not all gene selection methods help improve the performance of classification. The rule-of-thumb is that some gene selection methods are suitable for some specific classification algorithms. For example, if we select SVMs as the classification algorithm, then a Correlation coefficient or Signal-to-Noise gene selection method is better for data preprocessing. On the contrary, an inappropriate choice can only harm the power of prediction. Our results also implied that with preprocessing, the number of genes selected affects the classification accuracy.

References

[1] Ding, C.H.Q.: Unsupervised feature selection via two-way ordering in gene expression analysis. Bioinformatics 19(10), 1259–1266 (2003)

[2] Li, S., Wu, X., Hu, X.: Gene selection using genetic algorithm and support vectors machines. Soft Comput. 12(7), 693–698 (2008)

[3] Song, M., Rajasekaran, S.: A greedy correlation-incorporated SVM-based algorithm for gene selection. In: AINA Workshops (1), pp. 657–661. IEEE Computer Society, Los Alamitos (2007)

[4] Golub, T.R., Slonim, D.K., Tamayo, P., et al.: Molecular classification of cancer: Class discovery and class prediction by gene expression monitoring. Science 286, 531–537 (1999)

[5] Veer, L.V., Dai, H., de Vijver, M.V., et al.: Gene expression profiling predicts clinical outcome of breast cancer. Nature 415, 530–536 (2002)

[6] Kohavi, R., John, G.H.: Wrappers for feature subset selection. Artificial Intelligence 97(1-2), 273–324 (1997)

[7] Mukkamala, S., Liu, Q., Veeraghattam, R., Sung, A.H.: Feature selection and ranking of key genes for tumor classification: Using Microarray gene expression data. In: Rutkowski, L., Tadeusiewicz, R., Zadeh, L.A., Żurada, J.M. (eds.) ICAISC 2006. LNCS (LNAI), vol. 4029, pp. 951–961. Springer, Heidelberg (2006)

[8] Liu, X., Krishnan, A., Mondry, A.: An entropy-based gene selection method for cancer classification using Microarray data. BMC Bioinformatics 6, 76 (2005)

[9] Zhu, Z., Ong, Y.S., Dash, M.: Markov blanket-embedded genetic algorithm for gene selection. Pattern Recognition 40(11), 3236–3248 (2007)

[10] Quinlan, J.R.: C4.5: Programs for Machine Learning. Morgan Kaufmann, San Mateo (1993)

[11] Koller, D., Sahami, M.: Toward optimal feature selection. In: International Conference on Machine Learning, pp. 284–292 (1996)

[12] Yu, L., Liu, H.: Redundancy based feature selection for Microarray data. In: Proceedings of the Tenth ACM SIGKDD International Conference on Knowledge Discovery and Data Mining, Seattle, Washington, USA, pp. 737–742 (2004)

[13] Blanco, R., Larrañaga, P., Inza, I., Sierra, B.: Gene selection for cancer classification using wrapper approaches. IJPRAI 18(8), 1373–1390 (2004)

[14] Guyon, I., Weston, J., Barnhill, S., Vapnik, V.: Gene selection for cancer classification using support vector machines. Machine Learning 46(1-3), 389–422 (2002)

[15] Cortes, C., Vapnik, V.: Support-Vector Networks. Machine Learning 20(3), 273–297 (1995)

[16] Joachims, T.: Text categorization with support vector machines: learning with many relevant features. In: Nédellec, C., Rouveirol, C. (eds.) ECML 1998. LNCS, vol. 1398, pp. 137–142. Springer, Heidelberg (1998)

[17] Osuna, E., Freund, R., Girosi, F.: Training support vector machines:an application to face detection. In: Proceedings of the IEEE Computer Society Conference on Computer Vision and Pattern Recognition (1997)

[18] Furey, T.S., Christianini, N., Duffy, N., Bednarski, D.W., Schummer, M., Hauessler, D.: Support vector machine classification and validation of cancer tissue samples using Microarray expression data. Bioinformatics 16(10), 906–914 (2000)

[19] Brown, M., Grundy, W., Lin, D., Cristianini, N., Sugnet, C., Furey, T., Jr., M., Haussler, D.: Knowledge-based analysis of Microarray gene expression data by using suport vector machines. Proc. Natl. Acad. Sci. 97, 262–267 (2000)
[20] Cho, S.B., Won, H.H.: Machine learning in DNA Microarray analysis for cancer classification. In: CRPITS'19: Proceedings of the First Asia-Pacific bioinformatics conference on Bioinformatics 2003, Darlinghurst, Australia, pp. 189–198. Australian Computer Society, Inc. (2003)
[21] Li, J., Liu, H.: Kent ridge bio-medical data set repository (2002)

A Multiple Combining Method for Optimizing Dissimilarity-Based Classification*

Sang-Woon Kim[1] and Seunghwan Kim[2]

[1] *Senior Member, IEEE.* Address: Dept. of Computer Science and Engineering, Myongji University, Yongin, 449-728 South Korea
kimsw@mju.ac.kr
[2] School of Electrical Engineering, University of Waterloo, Waterloo, ON N2L 3G1, Canada
s36kim@engmail.uwaterloo.ca

Abstract. This paper reports an experimental study on a multiple combining method for optimizing dissimilarity-based classifications (DBCs) by simultaneously using a dynamic time warping (DTW) and a multiple fusion strategy (MFS). DBCs are a way of defining classifiers among classes; they are not based on the feature measurements of individual samples, but rather on a suitable dissimilarity measure among the samples. In DTW, the dissimilarity is measured in two steps: first, we adjust the object samples by finding the best warping path with a correlation coefficient-based DTW technique. We then compute the dissimilarity distance between the adjusted objects with conventional measures. In MFS, fusion strategies are repeatedly used in generating dissimilarity matrices as well as in designing classifiers: we first combine the dissimilarity matrices obtained with the DTW technique to a new matrix. After training some base classifiers in the new matrix, we again combine the results of the base classifiers. Our experimental results for well-known benchmark databases demonstrate that the proposed mechanism works well and achieves further improved results in terms of the classification accuracy compared with the previous approaches.

Keywords: machine learning, pattern recognition, dissimilarity-based classifications, multiple combining strategies, dynamic time warping.

1 Introduction

One of the most recent and novel developments in statistical pattern recognition is the concept of dissimilarity-based classifications (DBCs) proposed by Duin and his co-authors [9], [10], [11]. DBCs are a way of defining classifiers between the classes, which are not based on the feature measurements of the individual patterns, but rather on a suitable *dissimilarity measure* between them [5], [9]. There are a few ways by which the classification efficiency of DBCs can be optimized. The major task of this study is to deal with how the dissimilarity measure can be effectively computed and

* This work was supported by the National Research Foundation of Korea funded by the Korean Government (NRF-2009-0071283). The second author is with Department of Computer Science and Engineering (Pattern Recognition Lab.), Myongji University, as a research assistant.

N.T. Nguyen, M.T. Le, and J. Świątek (Eds.): ACIIDS 2010, Part II, LNAI 5991, pp. 310–319, 2010.

how a classifier in the dissimilarity space can be designed. The reason we set this task as our goal comes from the necessity to measure the inter-pattern dissimilarities for all training samples such that there is no zero distance between objects of different classes. Consequently, the classes do not overlap, and therefore the lower error bound is zero. In image classification tasks, such as face recognition, one of the most intractable problems is the distortion and lack of information caused by the differences in face directions and sizes. Thus, by simply measuring the differences in facial images for each class, we are unable to obtain a good representation. If the representational capability is insufficient to cover the possible variations of data, it is difficult to improve the performance of DBCs in the dissimilarity space.

To overcome these limitations and thereby improve the performance of DBCs, we study a way of enriching the representational capability of dissimilarity measures as well as the classification capability of base classifiers built in dissimilarity spaces. In particular, this goal can be achieved by simultaneously employing a dynamic time warping (DTW) [12], [13], [14] and a multiple fusion strategy (MFS) [3]. In DTW, the dissimilarity is measured in two steps: first we adjust the faces by finding the best warping path, w, with a correlation coefficient-based DTW method; we then compute the dissimilarity matrix with the adjusted faces [4]. On the other hand, combination systems which fuse "pieces" of information have received considerable attention because of its potential to improve the performance of individual systems [6], [7]. From this perspective, to increase the classification accuracy of DBCs further, in MFS, we first combine dissimilarity matrices obtained with the correlation coefficient-based DTW method to a new representation matrix. Then, we again combine the results of the base classifiers trained in the new matrix to increase the classification accuracy.

The main contribution of this paper is to demonstrate that the classification accuracies of dissimilarity-based classifiers can be improved by simultaneously employing a dynamic time warping and a multiple fusion strategy. This has been achieved by incorporating an *adjust-measure* operation into the dissimilarity measuring process and by successively employing fusion strategies in generating dissimilarity matrices as well as in designing classifiers. Although the result presented is only for a case when the task is an appearance-based face recognition, the proposed approach can also be used in other high-dimensional classification tasks, such as multimedia retrieval, bioinformatics, computer vision, and text classification.

2   Schema for the Proposed Solution

2.1   Dissimilarity-Based Classifications

A dissimilarity representation of a set of samples, $T = \{x_i\}_{i=1}^n \in \mathbb{R}^d$, is based on pairwise comparisons and is expressed, for example, as an $n \times m$ dissimilarity matrix $D_{T,Y}[\cdot, \cdot]$, where $Y = \{y_1, \cdots, y_m\}$, a prototype set, is extracted from T, and the subscripts of D represent the set of elements, on which the dissimilarities are evaluated. Thus, each entry, $D_{T,Y}[i, j]$, corresponds to the dissimilarity between the pairs of objects, $\langle x_i, y_j \rangle$, where $x_i \in T$ and $y_j \in Y$. Consequently, an object, x_i, is represented as a row vector as follows:

$$[d(x_i, y_1), d(x_i, y_2), \cdots, d(x_i, y_m)]', 1 \leq i \leq n. \tag{1}$$

Here, the dissimilarity matrix, $D_{T,Y}[\cdot, \cdot]$, is defined as a *dissimilarity space*, on which the d-dimensional object, x, given in the feature space, is represented as an m-dimensional vector, $\delta_Y(x)$. In this paper, the dissimilarity matrix and its row vectors are simply denoted by $D(T, Y)$ and $\delta(x)$, respectively. The details of the DBCs are omitted here in the interest of compactness, but can be found in the literature, including [9].

2.2 Correlation Coefficient-Based DTW

With regard to measuring the dissimilarity of the sample points, we prefer not to directly measure the dissimilarity from the object points; rather, we utilize a way of using a correlation coefficient-based DTW (dynamic time warping) technique to *adjust* or *scale* the object samples. This measure of dissimilarity effectively serves as a new "feature" component in the dissimilarity space.

Consider the two sequences of $s = \{x_1, \cdots, x_n\}$, $x_i \in \mathbb{R}$, and $t = \{y_1, \cdots, y_m\}$, $y_i \in \mathbb{R}$, where x_i or y_i is the i-th element (gray value) of a column or row vector of an image sample. First, an alignment from s to t can be represented by a warping $w = \{w(1), w(2), \cdots, w(n)\}$, where $j = w(i), (j \in [1, m], i \in [1, n])$ means that the i-th element in s is aligned to the j-th element in t. Then, to find the best warping path w that minimizes the distance $d_w(s, t)$, we can use a correlation coefficient [8] estimated from the s and t, $\rho(s, t)$, which is defined as:

$$\rho(s, t) = \frac{cov(s, t)}{\sigma_s \sigma_t} = E((s - \mu_s)(t - \mu_t)')/\sigma_s \sigma_t, \tag{2}$$

where μ_s and σ_s are, respectively, the expected value, $\mu_s = \frac{1}{n} \sum_{i=1}^{n} x_i$, and the standard deviation, $\sigma_s = \sqrt{\frac{1}{n-1} \sum_{i=1}^{n} (x_i - \mu_s)^2}$, for the gray-level x_i of the vector s (μ_t and σ_t have the same expressions for the vector t). Also, E is an operator for the expected values. Here, since x_i and y_i are all one-dimensional random variables, the reader should note that the correlation coefficient estimated is:

$$\rho = \begin{bmatrix} \rho_{xx} & \rho_{xy} \\ \rho_{yx} & \rho_{yy} \end{bmatrix} = \begin{bmatrix} 1 & \rho_{xy} \\ \rho_{yx} & 1 \end{bmatrix}, \tag{3}$$

where ρ_{xy} ($=\rho_{yx}$) denotes the estimated correlation coefficient of the two random variables under any assumed distribution model.

For example, consider three column vectors of cardinality of n, which are artificially generated as follows: *s=sort(randn(n,1)); t1=sort(randn(n,1)); t2=randn(n,1)*, where *sort* and *randn* are the functions of *sorting* in ascending order and *generating* normally distributed random numbers, respectively. In this case, we can easily imagine that the correlation coefficient estimated between the two vectors of s and $t1$ is very *large*, while the correlation coefficient obtained from the vectors of s and $t2$ is relatively *small*. To make things clear, consider a graphical example. First, Fig. 1 shows a plot of the three column vectors, s, $t1$, and $t2$, generated with the cardinality of $100(= n)$. Here, the width of the column vectors is represented with three pixels to enhance the visibility. Next, Fig. 2(a) and (b) show distributions of the paired random variables $\{(x_i, y_i)\}_{i=1}^{n}$, which are constructed from $(s, t1)$ and $(s, t2)$, respectively.

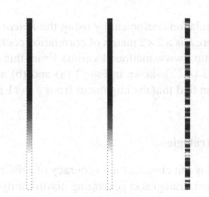

Fig. 1. A plot of three column vectors generated with the cardinality of $100(= n)$: (a) left; (b) middle; and (c) right. Here, (a), (b), and (c) are, respectively, of the s, $t1$, and $t2$ vectors and their widths are represented with three pixels to enhance the visibility.

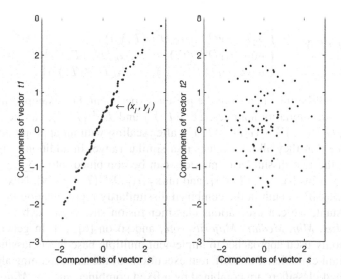

Fig. 2. Distributions of n paired random variables $\{(x_i, y_i)\}_{i=1}^{n}$: (a) left and (b) right. Here, the pictures of (a) and (b) are constructed with the sample pairs collected from $(s, t1)$ and $(s, t2)$, respectively. (a) exhibits positive correlation between x_i and y_i, while randomly scattered points in (b) indicate uncorrelated x_i and y_i.

Given the sample pairs $\{(x_i, y_i)\}_{i=1}^{n}$, there are several ways to estimate the correlation coefficient ρ (or more specifically, ρ_{xy}), such as *sample correlation coefficient* and *maximum likelihood estimates* [2]. Relative to alternative methods for estimating the correlation coefficient, the sample correlation method has the advantage of simplicity since it is an explicit function of the data samples. Furthermore, on the surface it appears not to require any specific model for the joint distribution function. The details of the maximum likelihood estimates are omitted here, but can be found in the literature [2]. Based on the sample correlation coefficient method, on the other hand, we can directly compute the correlation coefficient by using the $R=corrcoef(s,t)$ function of a package like Matlab, where R is a 2×2 matrix of correlation coefficients for two column vectors s and t (refer to http://www.mathworks.com). From this computation, the ρ_{xy}'s obtained from $(s,t1)$ and $(s,t2)$ shown in Fig. 2 (a) and (b) are 0.9833 and 0.1730, respectively. Thus, we can find that the alignment from s to $t1$ is better than that from s to $t2$.

2.3 Multiple Fusion Strategies

In the interest of increasing the classification accuracy of DBCs, we employ a method of successively using fusion strategies in generating dissimilarity spaces (representation step) as well as in designing classifiers (generalization step) [3]. First, to combine dissimilarity matrices obtained with different measuring systems to a new one, we make use of representation combining strategies, such as *Average*, *Product*, *Min*, or *Max* rules, defined as follows [9]:

$$D(T,Y) = \begin{cases} \sum_{i=1}^{k} \alpha_i D^{(i)}(T,Y), \\ \sum_{i=1}^{k} \log\left(1 + \alpha_i D^{(i)}(T,Y)\right), \\ \min\left\{\alpha_1 D^{(1)}(T,Y), \cdots, \alpha_k D^{(k)}(T,Y)\right\}, \\ \max\left\{\alpha_1 D^{(1)}(T,Y), \cdots, \alpha_k D^{(k)}(T,Y)\right\}. \end{cases} \tag{4}$$

Some of them will be investigated in the present experiment. For example, in the *Average* rule, two dissimilarity matrices, $D^{(1)}(T,Y)$ and $D^{(2)}(T,Y)$, can be averaged to: $D^{(3)} = \alpha_1 D^{(1)}(T,Y) + \alpha_2 D^{(2)}(T,Y)$, after scaling with an appropriate weight, α_i, to guarantee that they all take values in a similar range. In addition to this averaging method, the two dissimilarity matrices can be combined into : $\sum_{\tau=1}^{2} \log(1 + \alpha_\tau D^{(\tau)}(T,Y))$, $\min_\tau\{\alpha_\tau D^{(\tau)}(T,Y)\}$, and $\max_\tau\{\alpha_\tau D^{(\tau)}(T,Y)\}$ [9]. Next, to combine the base classifiers built in the combined dissimilarity representation matrix, i.e., $D^{(3)}$ as an instant, we can use various classifier fusion strategies, such as *Product*, *Mean*, *Sum*, *Max*, *Min*, *Median*, *Majority vote*, and so on [6], [7]. In general, there are two commonly used approaches to implement multiple base-level classifiers. One method is to train classifiers from different executions of the same learning algorithm. Then the resulted classifiers are combined by a fixed combiner, such as *Majority vote* combiner. The other method is to apply some different learning algorithms to the same data set. Then the derived classifiers are merged to a good ensemble classifier with a trainable combiner, such as *weighted Average* combiner. The fixed combiner has no extra parameter that needs to be trained, while the trainable combiner needs additional training.

2.4 The Algorithm

To solve the classification problem, we first combine dissimilarity matrices constructed with various measuring systems including the dynamic time warping for the entire training samples, and then again combine all of the results of the base classifiers (DBCs) designed in the combined dissimilarity space to reduce the classification error rates. The proposed algorithm for the combined DBCs is summarized in the following:

1. Select the entire training samples T as the representative set Y.
2. Using Eq. (1), compute dissimilarity matrices, $D^{(1)}(T,Y), \cdots, D^{(k)}(T,Y)$, by using the k different dissimilarity measures for all $x \in T$ and $y \in Y$.
3. For any *combined* dissimilarity representation matrix, $D^{(j)}, (j = 1, \cdots, l)$, perform classification of an input sample, z, with *combined* classifiers built in the combined dissimilarity matrix as follows:

   (a) Compute a dissimilarity row vector, $\delta^{(j)}(z)$, for the input sample z, with the same method as in generating the $D^{(j)}$.
   (b) Classify $\delta^{(j)}(z)$ by invoking a group of DBCs as the *base* classifiers designed with n m-dimensional vectors in the dissimilarity space. The classification results are labeled as $class_1, class_2, \cdots$, respectively.

4. Obtain the final result from the $class_1, class_2, \cdots$, by combining the base classifiers designed in the above step, where the base classifiers are combined to form the final decision in *fixed* or *trainable* fashion.

The rationale of this strategy is presented in a later section together with the experimental results.

3 Experimental Results

3.1 Experimental Data

The proposed method has been tested and compared with conventional methods. This was done by performing experiments on two well-known face benchmark databases, namely AT&T[1] and Yale[2] . The face database captioned "AT&T," formerly the ORL database of faces, consists of ten different images of 40 distinct objects, for a total of 400 images. The size of each image is 112×92 pixels, for a total dimensionality of 10304 pixels. The "Yale" face database contains 165 gray-scale images of 15 individuals. The size of each image is 243×320 pixels, for a total dimensionality of 77760 pixels. To reduce the computational complexity of this experiment, each image in the Yale database was down-sampled into 178×236 pixels and then represented by a centered vector of normalized intensity values. To obtain a different data set, a partial image, which is 64×78 pixels in size and contains only facial components without background, was extracted from each of the Yale images. This database is referred to as "Yale2" in the following sections.

3.2 Experimental Method

Since the construction of the dissimilarity matrix is a very time-consuming job, we constructed a 50×50 (or 55×55 for Yale) dissimilarity matrix, instead of 400×400 (or 165×165 for Yale) matrix for the five objects randomly selected from the AT&T or

---

[1] http://www.cl.cam.ac.uk/Research/DTG/ attarchive/ facedatabase.html
[2] http://www1.cs.columbia.edu/ belhumeur/ pub/ images/ yalefaces

Yale database[3]. We repeated this experiment 20 times and obtained a final result (mean values and standard deviations) by averaging them. In the experiments, all evaluations were performed using the "leave-one-out" strategy. That is, to classify an image of an object, that image was removed from the training set, and the dissimilarity matrix was computed with the remaining $n - 1$ images.

To construct the dissimilarity matrix, all training samples were selected as representatives and the dissimilarities between each sample and the representatives were measured with the Euclidean distance and the regional distance [1], [15]. Here the two distance measures are named "ED" and "RD", respectively[4]. The distance measure RD is defined as the average of the minimum difference between the gray value of a pixel and the gray value of each pixel in the 5×5 neighborhood of the corresponding pixel. In this case, the regional distance compensates for a displacement of up to three pixels of the images. For the interest of brevity, the details of the distance measure are omitted here, but can be found in the literature including [1].

The dissimilarity representation matrices were generated with WED, DTW, and MUL methods. In WED (weighted Euclidean distance), the dissimilarity between object images was measured after normalization, where the scaling factors were obtained from the *mean face*, which was achieved by simply averaging all the training images. Also, in DTW (DTW based distance), the dissimilarity between object images was measured after adjusting them with the DTW technique and the best warping path was found by using the correlation coefficient computed from the paired two images. Fig. 3 shows an example of computing the correlation coefficients between two column vectors, s and t, which are marked with a bold vertical bar in the facial images.

In Fig. 3(a), (b), and (c), three column vectors, which are, respectively, the 15th, 15th, and the 35th columns in each picture, are denoted as x_{15}, y_{15}, and y_{35}, respectively. Then, we can compute the correlation coefficients for these column vectors, $\rho^1 = \rho(x_{15}, y_{15})$ and $\rho^2 = \rho(x_{15}, y_{35})$, as follows:

$$\rho^1 = \begin{bmatrix} 1.0000 & 0.1765 \\ 0.1765 & 1.0000 \end{bmatrix}, \quad \rho^2 = \begin{bmatrix} 1.0000 & 0.7114 \\ 0.7114 & 1.0000 \end{bmatrix}.$$

From this computation, we know that x_{15} shown in (a) is preferable to be aligned to y_{35} of (c), rather than y_{15} of (b). After adjusting the two images in x-axis direction using the best warping path found for all the column vectors, we repeated the same process to adjust them in y-axis direction using the best warping path found for all the row vectors. Finally, in MUL, two dissimilarity matrices, $D^{(1)}(T, Y)$ and $D^{(2)}(T, Y)$, generated with WED and DTW methods were combined to a new dissimilarity matrix after

---

[3] We experimented with the simpler 50×50 matrix here. However, it is possible to construct the 400×400 dissimilarity matrix rapidly by employing a computational technique. We are currently investigating how the experiment with the full matrix can be performed at a higher speed.

[4] Here, we experimented with two simple measures, namely ED and RD measures. However, it should be mentioned that we can have numerous solutions, depending on dissimilarity measures, such as the Hamming distance, the modified Hausdorff distances, the blurred Euclidean distance, etc. From this perspective, the question "what is the best measure ?" is an interesting issue for further study.

Fig. 3. An example of computing the correlation coefficients between two column vectors marked with bold vertical bars of the faces which are different from each other in their pose(direction): (a) left; (b) middle; and (c) right. The warping of the two column vectors shown in (a) and (c) is preferable to that of (a) and (b).

performing normalization with 21 different scaling factors; $\alpha_1 = 0.0, 0.05, \cdots, 1.0$ and $\alpha_2 = 1 - \alpha_1$.

We then combined again all the results of the base classifiers trained in the new dissimilarity space in *fixed* or *trained* fashion. That is, after training three base classifiers, such as nearest mean classifier, linear Bayes normal classifier, and k-nearest neighbor classifier, we evaluated the classification accuracies of *Product* combining classifier (a fixed combiner) and *Mean* combining classifier (a trained combiner). All of the base classifiers and the combiners mentioned above were implemented with PRTools[5].

3.3 Experimental Results

Table 1 shows a comparison between the classification accuracy rates (%) of combined DBCs for the three databases.

From Table 1, first of all, we can see that there is an improvement in the achieved classification accuracies of MUL methods. An example of this is the classification accuracies of ED measure for AT&T data. For the *three* dissimilarity spaces constructed with WED, DTW, and MUL methods, the classification accuracies of the *fixed* or *trained* combiners are 95.10(%), 89.40(%), and 97.00(%), respectively. From this consideration, the reader can observe that the classification performances of DBCs trained in MUL are usually better than those of DBCs built in WED and DTW spaces, which leads to the conclusion that combining dissimilarity matrices is helpful. The other observations obtained from the experiment are the followings:

- The results of RD measures for MUL methods are better than those of ED measures for almost all of the data sets, which means that RD's representational capability of dissimilarity measures is better than ED's.

- The results of *fixed* or *trained* combiner are about equal to those of the best base classifier and it thereby operates as a selector. Consequently, the combining classifier makes the system more robust.

- The standard deviations presented in the brackets are significant because they have been obtained with the 20 iterations for the objects of randomly chosen classes. Consequently, the accuracy of the combined DBCs seems to be sensitive to the class characteristics.

---

[5] PRTools is a Matlab toolbox for pattern recognition (refer to http://prtools.org/).

Table 1. A comparison of classification accuracies (%) of combined DBCs trained with the Euclidean (WED), dynamic time warping (DTW), and multiple fusion (MUL) methods for the three databases. Here, the numbers represented in the brackets of each row are the standard deviations. Also, the values underlined are the *highest* ones in the two combiners per each measure.

experimental datasets	distance measures	combined matrices	combined classifiers (DBCs)	
			fixed combiner	*trained* combiner
AT&T	ED	WED	95.10(2.71)	95.10(2.71)
		DTW	89.40(5.62)	89.40(5.62)
		MUL	97.00(2.00)	97.00(2.00)
	RD	WED	98.60(1.96)	98.60(1.96)
		DTW	93.90(4.33)	93.90(4.33)
		MUL	98.70(1.87)	98.70(1.87)
Yale	ED	WED	92.56(3.70)	88.00(4.04)
		DTW	92.56(4.96)	87.00(4.96)
		MUL	93.56(3.13)	89.22(4.16)
	RD	WED	94.67(3.18)	86.56(3.57)
		DTW	93.89(3.67)	87.11(3.86)
		MUL	96.33(2.63)	87.00(3.62)
Yale2	ED	WED	80.67(4.73)	80.67(4.73)
		DTW	80.22(6.03)	80.11(5.97)
		MUL	81.33(5.32)	81.44(5.26)
	RD	WED	86.27(4.06)	86.36(4.07)
		DTW	84.36(5.35)	84.36(5.45)
		MUL	87.09(3.00)	87.18(2.86)

- Finally, it is interesting to point out that the scaling factors, $\alpha_1 : \alpha_2$, used for combining the representation matrices measured with ED methods for AT&T, Yale, and Yale2 databases are $0.75 : 0.25$, $0.55 : 0.45$, and $0.35 : 0.65$, respectively.

4   Conclusions

In this paper we studied a multiple combining method of simultaneously using the dynamic programming and multiple fusion strategies to optimize dissimilarity-based classifications (DBCs). The proposed combining scheme was analyzed by experiments and compared with the previous methods for the well-known benchmark face databases. Our experimental results demonstrated the possibility that the proposed method could be used efficiently for optimizing DBCs.

Despite this success, problems remain to be addressed. First of all, in this experiment, we combined only two dissimilarity matrices generated with the *same* measuring system and implemented three base classifiers in the combined matrix. The problems of investigating the fusion of different measuring systems, i.e., ED and RD, and training sufficient number of the base classifiers remain unchallenged. Then, the problem of automatically selecting an optimal scaling factor for different dissimilarity matrices for a given application remains open. Finally, the experimental results showed that the highest accuracies were achieved when the combined representation was classified with the fixed or trained combiner. The problem of theoretically analyzing this observation remains unresolved. Future research will address these concerns.

References

1. Adini, Y., Moses, Y., Ullman, S.: Face recognition: the problem of compensating for changes in illumination direction. IEEE Trans. Pattern Anal. and Machine Intell. 19(7), 721–732 (1997)
2. Kim, J., Fessler, J.A.: Intensity-based image registration using robust correlation coefficients. IEEE Trans. Medical Imaging 23(11), 1430–1444 (2004)
3. Kim, S.-W., Duin, R.P.W.: On optimizing dissimilarity-based classifier using multi-level fusion strategies. Journal of Institute of Electronics Engineers of Korea 45-CI(5), 15–24 (2008) (in Korean); A preliminary version of this paper was presented at the 20th Canadian Conference on Artificial Intelligence. LNCS (LNAI), vol. 4509, pp. 110–121. Springer, Heidelberg (2007)
4. Kim, S.-W., Gao, J.: A dynamic programming technique for optimizing dissimilarity based classifiers. In: da Vitoria Lobo, N., Kasparis, T., Roli, F., Kwok, J.T., Georgiopoulos, M., Anagnostopoulos, G.C., Loog, M. (eds.) S+SSPR 2008. LNCS, vol. 5342, pp. 654–663. Springer, Heidelberg (2008)
5. Kim, S.-W., Oommen, B.J.: On using prototype reduction schemes to optimize dissimilarity-based classification. Pattern Recognition 40, 2946–2957 (2007)
6. Kittler, J., Hatef, M., Duin, R.P.W., Matas, J.: On combining classifiers. IEEE Trans. Pattern Anal. and Machine Intell. 20(3), 226–239 (1998)
7. Kuncheva, L.I.: Combining Pattern Classifiers - Methods and Algorithms. John Wiley & Sons, Chichester (2004)
8. Milton, J.S., Arnold, J.C.: Introduction to Probability and Statistics - Principles and Applications for Engineering and the Computer Sciences, pp. 157–164. McGraw-Hill, New York (1990)
9. Pekalska, E., Duin, R.P.W.: The Dissimilarity Representation for Pattern Recognition: Foundations and Applications. World Scientific Publishing, Singapore (2005)
10. Pekalska, E., Duin, R.P.W., Paclik, P.: Prototype selection for dissimilarity-based classifiers. Pattern Recognition 39, 189–208 (2006)
11. Pekalska, E., Paclik, P., Duin, R.P.W.: A generalized kernel approach to dissimilarity-based classification. Journal of Machine Learning Research 2(2), 175–211 (2002)
12. Qiao, Y., Yasuhara, M.: Affine invariant dynamic time warping and its application to online rotated handwriting recognition. In: Proc. of International Conference on Pattern Recognition, ICPR 2006 (2006)
13. Ratan, A.L., Grimson, W.E.L., Wells, W.M.: Object detection and localization by dynamic template warping. International Journal of Computer Vision 36(2), 131–147 (2000)
14. Sahbi, H., Boujemaa, N.: Robust face recognition using dynamic space warping. In: Tistarelli, M., Bigun, J., Jain, A.K. (eds.) ECCV 2002. LNCS, vol. 2359, pp. 121–132. Springer, Heidelberg (2002)
15. Wang, L., Zhang, Y., Feng, J.: On the Euclidean distance of images. IEEE Trans. Pattern Anal. and Machine Intell. 27(8), 1334–1339 (2005)

A Comparative Study on the Performance of Several Ensemble Methods with Low Subsampling Ratio

Zaman Faisal* and Hideo Hirose

Kyushu Institute of Technology,
680-4 Kawazu, Iizuka-shi, Fukuoka, Japan
zaman@ume98.ces.kyutech.ac.jp,
hirose@ces.kyutech.ac.jp

Abstract. In ensemble methods each base learner is trained on a re-sampled version of the original training sample with the same size. In this paper we have used resampling without replacement or subsampling to train base classifiers with low subsample ratio i.e., the size of each subsample is smaller than the original training sample. The main objective of this paper is to check if the scalability performance of several well known ensemble methods with low subsample ratio are competent and compare them with their original counterpart. We have selected three ensemble methods: Bagging, Adaboost and Bundling. In all the ensemble methods a full decision tree is used as the base classifier. We have applied the subsampled version of the above ensembles in several well known benchmark datasets to check the error rate. We have also checked the time complexity of each ensemble method with low subsampling ratio. From the experiments, it is apparent that in the case of bagging and adaboost with low subsampling ratio for most of the cases the error rate is inversely related with subsample size, while for bundling it is opposite. Overall performance of the ensemble methods with low subsampling ratio from experiments showed that bundling is superior in accuracy with low subsampling ratio in almost all the datasets, while bagging is superior in reducing time complexity.

Keywords: Low Subsampling ratio, Bagging, Adaboost, Bundling, Scalability.

1 Introduction

An ensemble of classifiers is a set of classifiers whose individual decisions are combined in some way to classify new examples. Many popular meta-learning techniques in computer science can be conveniently described within this framework. An ensemble consists of a set of possibly different classifier types. The output of each classifier is combined in one of many different ways in order to reach a final classification. This definition is broad, however it encompasses many

---

* Corresponding author.

N.T. Nguyen, M.T. Le, and J. Świątek (Eds.): ACIIDS 2010, Part II, LNAI 5991, pp. 320–329, 2010.
© Springer-Verlag Berlin Heidelberg 2010

different popular techniques within the same framework. Numerous theoretical and empirical studies have been published to establish the advantages of combining learning machines [14], [11], for example using Boosting or Bagging methods, [2] [8] very often leads to improved generalization performance, and a number of theoretical explanations have been proposed [4], [8].

In standard bagging individual classifiers are trained on independent bootstrap samples that are generated with replacement from the set of labeled training samples, where as in adaboost by *resampling* the fixed training sample size and training examples resampled according to a probability distribution are used in each iteration the distribution of the training data depends on the performance of the classifier trained in the previous iteration. Unlike these two ensemble methods, Hothorn and Lausen proposed "Double Bagging"[12] and "Bundling" [13] to add the outcomes of arbitrary classifiers to the original feature set for bagging of classification trees. Double bagging is a combination of linear discriminant analysis and classification trees. As in the bootstrap method, approximately $\frac{1}{3}$ of the observations in the original training set are not part of a single bootstrap sample in bagging [3], and Breiman termed the set constituted by these observations as an out-of-bag sample. In double bagging, an out-of-bag sample (OOBS) is utilized to estimate the coefficients of a linear discriminant function and the corresponding linear discriminant variables computed for the bootstrap sample are used as additional features for training a classification tree. Bundling is a generalized form of double bagging, where instead of a single classifier model a set of additional classifier models is constructed on OOBS, in [13] authors used 5-NN(Nearest Neighbor), 10-NN, Logistic classifier model and stabilized linear discriminant classifier(sLDA) as the additional classifier models.

In the less explored variant of bagging also known as Subagging, instead of with replacement sampling a m-out-of-n without replacement sampling is used [5], A common choice for the sample size in subagging is $m_{wor} = n/2$ [10]. In fact, provided that higher-order terms can be neglected, half-sampling without replacement is expected to behave similarly as $m = n$ sampling with replacement [10]. As bagging is not a scalable approach to data mining since each individual bag is the size of the full training set, use of subsampling can be more advantageous. In [18] in each iteration of the adaboost algorithm a subsample is drawn from the original training set with replacement according to the probability distribution maintained over the training set, then a classifier is trained by applying the given weak learning algorithm to the subsample. The basic idea of that work is based on Friedman's gradient boosting [9]. In [16] authors used support vector machine as the additional classifier in double bagging and different subsampling fractions to generate the training set for the base classifier. They showed that with smaller subsampling ratio the performance of double bagging is superior to most of the contemporary ensemble methods. In [17] authors reported the effect of small subsampling ratio on different bagging type ensemble of stable (linear) classifiers.

This paper is organized as follows: in Section 2, we have explained in brief the insertion of small subsample ratios in three ensemble methods we used in our experiments and later we have discussed the effect of small subsampling ratio on these ensemble methods from a theoretical point of view. In Section 3, firstly we have discussed the goal and setup of the experiments and later presented the results of the experiments with discussion. It is followed by the Conclusion of the investigations.

2   Small Subsampling Ratio in Bagging, Adaboost and Bundling

In this section we briefly discuss how to construct the three ensemble methods with small subsamples and later illustrate how much effective will be the sub-sampled versions of these ensemble methods in classification. In all the ensemble methods the subsamples are used for generating the training sets for the base classifier. In bagging [2] which is a statistical re-sample and combining technique, the subsamples are used instead of bootstrap samples to generate multiple train-ing set, and construct a classifier on each set. This method is defined as *Subagging* [5], we implemented the algorithm as stated in this paper. The algorithm con-sist of two major steps, *subsampling* and *aggregating*. The Adaboost algorithm, introduced by Freund and Schapire [8], is adaptive in the sense that it adapts to the error rates of the individual *weak* classifier and boosts its performance. In this paper we have used the adaboost by resample and in each iteration a small subsample is drawn from the original training set. In our experiments we have used multi-class adaboost algorithm, Adaboost.M1 [7]. In the paper [18], authors used the subsampled adaboost for binary classification, but in our paper we implemented the multi-class version of the adaboost. "Bundling" method was proposed by Hothorn and Lausen [13] to construct ensemble of decision tree, considering that other general splits such as linear ones may produce more ac-curate trees. In bundling framework the out-of-bag sample is used to train a set of additional classifier models to integrate the outputs with the base learning model. In this paper we have inserted small subsamples to train the base classi-fier in bundling and defining this method as, "Subbundling". For bundling (and subbundling), a radial basis support vector machine (RSVM), a linear support vector machine (LSVM), a logistic linear classifier and a stabilized linear dis-criminant classifier(sLDA) are used as the additional bundle of classifiers to be trained on the out-of-bag samples.

In this paper we have used small subsamples instead of bootstrap samples to train base classifiers for bagging, adaboost and bundling. There is a rea-son to believe that the performance of the $m(= 0.5n)$-out-of-n (subsampling) bootstrapping to perform similar to n-out-of-n bootstrap. The effective size of resample in the n-out-of-n bootstrapping in terms of amount if information it contains is

$$\frac{(\sum N_i)^2}{\sum N_i^2} \approx \frac{1}{2}n$$

where N_i denotes how many times the ith data value is repeated in the sub-sample. In [10] it is shown that the bias and variance of the regression bagging ensemble can be reduced by smaller subsamples. Based on this theory we have used small subsampling ratio to create subagging ensemble for classification task. Also as error rate of an ensemble is inversely related to the sample size and on the other hand the variance of an ensemble increase with the decreasing of the subsample size, these two competing effect will not increase the error rate with decreasing sample size as usually expected in prediction. In Section 3, we shall observe the results supporting this theory. For adaboost, small subsample increase the variance of each base classifier and there is less assurance that the small subsampling fractions will be beneficial for classification task as universally so as with regression problems. In bundling, use of small subsamples means larger OOBS (smaller subsample) for the additional classifier models. In [17] authors showed that the performance of subagging of stable (linear) classifiers are conversely related with double bagging of stable classifiers regarding the size of the subsamples. The main reason for this is, with larger OOBS the additional classifier models will be trained better(the stable base classifier will not be trained well) and will produce more accurate estimates of class probability as the additional predictors for the base decision tree. But this has a disadvantage from the time complexity point of view; with large OOBS the additional classifier models will take more time to train than smaller OOBS. Results in Section 3, will provide the experimental evidence for these.

3   Experiments and Discussion of Results

In this section we have firstly stated the aim and setup of the experiments and then we have discussed the results obtained from the experiments.

3.1   Aim and Setup of the Experiments

The main aim of the experiments of this paper was to find an efficient scalable ensemble method for classification task. To achieve this goal we setup an experiment with three fold objectives. First, to find the most effective ensemble method with smaller subsamples in terms of accuracy and time complexity among bagging, adaboost and bundling, then find out the most effective subsample ratio in reducing misclassification error and time complexity and finally compare the most effective ensemble with small subsamples with the contemporary most accurate ensemble method Rotation Forest [15]. We have used 12 UCI [1] datasets for the first two experiments and 15 UCI datasets for the last experiment. The dataset descriptions are given Table 1. In all the experiments we have reported average error of 15 ten-fold cross-validation results i.e., each cell in each table consists of an average value of total 150 (15×10-CV) testing.

For bagging, adaboost and bundling we have used a full decision tree as the base classifier. For all the ensembles the size is fixed to 50. We have used five subsample ratios 0.1 − 0.5. We report the total training time (in seconds) for each ensemble method for these subsample sizes and also for their bootstrapped (original) version. For each ensemble method the average error rate and time

Table 1. Description of the 15 Data used in this paper

Dataset	Objects	Classes	Features
Diabetes	768	2	8
German-credit	1000	2	20
Glass	214	7	9
Cleveland-Heart	297	5	13
Heart-Statlog	270	2	13
Ionosphere	351	2	34
Iris	150	3	4
Liver-disorder	345	2	6
Lymphography	148	4	18
Sonar	208	2	60
Vehicle	846	4	18
Vote	435	2	16
Wisconsin-breast	699	2	9
Wine	178	3	13

with their corresponding average ranks are also reported in the each table. The classifier with the lowest average rank for errors are marked bold and underlined. In each table for each row (dataset) the best performing ensemble method with lowest error rate is marked in bold and underlined, while the ensemble with lowest training time is underlined for each dataset. For each table we performed the comparison of the methods as proposed by Demšar [6], which is described below:

- First perform the Friedman test, to check the null hypothesis that all methods have equal performance.
- If the null hypothesis is rejected, then perform the Nemenyi posthoc test to detect the methods which are not significantly different from each other.
- Perform the Wilcoxon Sign Rank test, where we compare the *best* performing classifier with all other classifiers. We select best classifier on the basis of lowest average rank for errors. If there are several classifiers with lower average rank for errors with very small difference, then we select the one with lowest average rank for training time among them.

In each table in the second last row we have indicated the classifiers which are not significantly different. We have notified the classifiers of the same group as, "group a". The best performing classifier is also marked with a (*) sign in each table. The Wilcoxon Sign Rank test will clarify the performance of the *best* ensembles in each table if there is no significant difference detected by the Nemenyi posthoc test. We have given the p-values of the Wilcoxon test in the last row of each table. A high p-value in a column indicates that there is not much difference between the ensemble method of that corresponding column and the best method regarding the generalization performance in that table and vice versa.

3.2 Discussion of the Results

We have presented results as: firstly the bagging results, then the adaboost results, then bundling results and finally we made an overall comparison of the best ensemble methods from these three tables with rotation forest in the final table. In Table 2 we reported the error and time (in seconds) of bagging and subagging with subsample ratios 0.1 − 0.5. From Table 2 we see that, the time complexity increases with the increase of subsample ratio, where the standard bagging taking highest time to train for each dataset. In terms of average accuracy for all the datasets, subagging with subsample ratio 0.2 (denoting this ensemble as subbag20 from now on) producing the lowest value 0.1842, while the subagging with 10% subsample ratio giving lowest average time 1.10 for all the datasets. In terms of average rank for accuracy, subbag20 has the lowest average rank 3.08, with average rank for training time 2. In this table, the subbag20 giving best average rank interms of accuracy and best average rank for lesser training time among the more accurate ensembles, so we will select this classifier for our final experiment for comparison with other ensemble methods. From the results in Table 2 we can conclude that subagging with very small subsample ratio produced better performance than standard bagging.

Table 2. Error rate and time (in seconds) of bagging and subagging in benchmark datasets

Datasets	Bagging Full		Subagging with small subsampling ratio 0.1		0.2*		0.3		0.4		0.5	
	Error	Time	Error	Time	Error	Time	Error	Time	Error	Time	Error	Time
Diabetes	0.2820	7.51	0.2404	**1.23**	**0.2362**	2.01	0.2401	2.95	0.2609	3.87	0.2742	5.09
German	0.2992	19.34	**0.2814**	**1.61**	0.3024	3.19	0.3088	5.23	0.3040	7.85	0.2998	10.66
Heart	0.3037	5.65	**0.1919**	**0.84**	0.2067	1.06	0.2431	1.34	0.2774	1.61	0.3037	2.03
Hearts	0.2874	5.48	**0.1844**	**0.82**	0.1941	0.99	0.2163	1.19	0.2489	1.41	0.2733	1.74
Ion	**0.0832**	3.61	0.1482	**0.85**	0.1214	0.98	0.1105	1.12	0.1048	1.22	0.0957	1.36
Liver	0.3177	3.51	**0.3044**	**0.93**	0.3107	1.27	0.3229	1.65	0.3444	2.08	0.3496	2.44
Lymph	0.1775	3.06	0.1789	**1.03**	0.1578	1.27	**0.1465**	1.40	0.1493	1.58	0.1634	1.77
Sonar	0.2289	3.52	0.2539	**0.77**	0.2308	0.89	0.2202	1.01	0.2183	1.10	**0.2087**	1.30
Vehicle	**0.2525**	17.95	0.2946	**2.40**	0.2794	3.42	0.2700	4.30	0.2700	5.25	0.2619	6.05
Vote	0.0515	3.63	**0.0455**	**0.87**	0.0483	1.05	0.0478	1.26	0.0487	1.55	0.0524	1.90
Wbc	**0.0299**	3.59	0.0360	**0.91**	0.0334	0.98	0.0316	1.15	0.0316	1.17	0.0302	1.29
Wine	**0.0315**	2.38	0.1315	**0.89**	0.0888	1.04	0.0697	1.09	0.0596	1.19	0.0506	1.23
Average Values	0.1954	4.60	0.1909	**1.10**	**0.1842**	1.51	0.1856	1.97	0.1932	2.49	0.1970	3.07
Average Rank	3.75	6	4.83	**1**	**3.08**	2	3.5	3	3.58	4	3.58	5
CD(Critical Difference) = 1.96 , DF(difference in average ranks between best and worst method)=1.75												
Groups	The test cannot detect significant difference between the methods as CD >DF											
p-values	0.7508		0.9310				0.9310		0.7950		0.7075	
*= The best performing ensemble considering best average rank in accuracy and time to train												

From Table 3 we see that the performance of the subsampled adaboost are not so satisfactory in terms of accuracy, where as the adaboost algorithm as usual performing far better than the subsampled adaboost, but compensating the time complexity (adaboost taking nearly 6 times more than subbag20 to train and nearly double than bagging). Also we see that the average error of adaboost 0.1779 is not remarkably less than that of subbag20 0.1842.

Table 3. Error rate and time (in seconds) of adaboost and adaboost with small subsamples in benchmark datasets

Datasets	Adaboost Full*		Adaboost with small subsampling ratio 0.1		0.2		0.3		0.4		0.5	
	Error	Time	Error	Time	Error	Time	Error	Time	Error	Time	Error	Time
Diabetes	0.2929	13.70	0.3249	<u>2.29</u>	0.2849	2.87	<u>0.2820</u>	3.89	0.2919	4.73	0.2953	5.81
German	0.3300	28.86	<u>0.3064</u>	<u>2.59</u>	0.3254	3.90	0.3520	5.77	0.3614	7.40	0.3590	10.57
Heart	0.2336	5.35	0.2340	<u>1.77</u>	<u>0.2330</u>	1.95	0.2555	2.29	0.2639	2.55	0.2747	2.84
Hearts	<u>0.2191</u>	5.13	0.2247	<u>1.86</u>	0.2274	1.95	0.2311	2.16	0.2437	2.48	0.2696	2.76
Ion	<u>0.0695</u>	3.42	0.1705	<u>1.88</u>	0.1145	1.94	0.0974	2.13	0.0900	2.25	0.0712	2.40
Liver	<u>0.3059</u>	6.91	0.3581	<u>1.91</u>	0.3466	2.20	0.3478	2.68	0.3710	2.95	0.3570	3.55
Lymph	0.1602	3.23	0.2623	<u>1.09</u>	0.2042	1.45	0.2041	1.74	<u>0.1585</u>	2.02	0.1592	2.49
Sonar	<u>0.1798</u>	2.95	0.2943	<u>1.73</u>	0.2769	1.94	0.2288	2.01	0.2317	2.13	0.2028	2.14
Vehicle	<u>0.2368</u>	15.38	0.3628	<u>3.80</u>	0.3080	5.44	0.2858	6.61	0.2706	7.97	0.2553	9.12
Vote	0.0508	5.99	0.0732	<u>1.81</u>	0.0652	1.99	0.0611	2.53	0.0611	2.26	<u>0.0501</u>	2.90
Wbc	<u>0.0331</u>	4.25	0.0489	<u>1.37</u>	0.0405	1.78	0.0401	2.61	0.0345	3.38	0.0345	3.97
Wine	<u>0.0331</u>	3.06	0.1206	<u>1.02</u>	0.1236	1.39	0.0793	1.54	0.0583	2.03	0.0449	2.59
Average Values	<u>0.1779</u>	8.19	0.2333	<u>1.93</u>	0.2125	2.40	0.2054	3.00	0.2031	3.51	0.1978	4.26
Average Rank	<u>1.92</u>	6	4.66	<u>1</u>	3.66	2	3.5	3	4	4	3.44	5
CD(Critical Difference) = 1.96 , DF(difference in average ranks between best and worst method)=2.74												
Groups	Group 1		Group 2		Group 1,2		Group 1,2		Group 2		Group 1,2	
p-values			0.1938		0.4704		0.5066		0.5443		0.6235	
*= The best performing ensemble considering best average rank in accuracy and time to train												

The performance of bundling and subbundling is interesting in Table 4. As we already mentioned in Section 2 that, with small subsample the time complexity of the bundling (subbundling) may increase; we see exactly that, in Table 4. The time complexity is conversely related with subsample size in most of the datasets. The bootstrapped bundling (bundling) and sbundle50 (denoting the bundling with 50% subsample ratio) taking nearly same time to train on average (for sbundle50 it is 1.66 seconds and for bundling it is 1.83 seconds) as the size of training samples in both the ensemble is 67.3% and 50% of the original training sample. But the difference in average accuracy is high, as sbundle50 have average rank 3.08 for accuracy, while bundling have average rank 4.08 for accuracy. The overall average accuracy for all the dataset is lowest for sbundle50, 0.1448, while bundling producing lowest average time 15.05 (in compare to 17.91 seconds of sbundle50). From the p-values of the Wilcoxon test, it is apparent that there is no significant difference between subbundling with 50% and subbundling with 40% subsamples. From results of Table 2 and Table 4 we can say that subsampling ratio definitely improve the performance of bagging and bundling, where as from Table 3 we can say that adaboost is not suitable for subsampled training sets. We shall select sbundle50 for our next experiment of overall comparison with rotation forest.

Table 4. Error rate and time (in seconds) of bundling and subbundling in benchmark datasets

Datasets	Bundling Full		Subbundling with small subsampling ratio									
	Full		0.1		0.2		0.3		0.4		0.5*	
	Error	Time	Error	Time	Error	Time	Error	Time	Error	Time	Error	Time
Diabetes	0.2477	_9.71_	**0.2369**	22.42	0.2447	19.84	0.2411	17.28	0.2381	15.07	0.2450	12.95
German	0.2245	_23.27_	0.2298	61.36	0.2316	61.15	_0.2324_	58.2	0.2222	48.81	0.2264	41.14
Heart	0.1571	18.33	0.1533	13.04	0.1307	12.43	0.1340	11.04	0.1302	11.20	_0.1433_	10.04
Hearts	0.1703	_8.82_	**0.1622**	10.95	0.1629	10.39	0.1629	9.92	0.1674	9.67	0.1696	8.99
Ion	0.0660	_16.74_	**0.0643**	21.75	0.0626	20.71	0.0643	19.41	0.0638	18.19	0.0649	17.20
Liver	0.2721	6.18	0.2759	8.516	0.2672	7.82	0.2695	7.12	0.2765	6.48	_0.2585_	_6.04_
Lymph	0.1621	_19.35_	0.1675	21.33	0.1648	22.14	0.1486	21.17	0.1513	20.74	_0.1486_	20.30
Sonar	**0.1808**	19.4	0.2413	21.89	0.2307	19.95	0.2057	18.59	0.1951	17.56	0.1903	_16.60_
Vehicle	0.1791	_21.36_	0.1924	61.08	0.1834	57.28	0.1851	50.32	_0.1758_	43.45	0.1829	37.07
Vote	0.0482	_11.24_	0.04557	17.11	0.0427	15.82	0.0432	14.85	_0.0418_	13.60	0.0422	12.51
Wbc	0.0299	23.15	**0.0286**	30.56	0.0307	27.15	0.0339	25.92	0.0322	24.98	0.0336	_23.14_
Wine	0.0382	_8.02_	0.0382	9.29	**0.0146**	9.15	0.0202	8.91	0.0247	8.59	0.0224	8.30
Average Values	0.1489	_15.05_	0.1539	24.94	0.1494	23.65	0.1467	21.96	0.1456	19.87	_0.1445_	17.91
Average Rank	4.08	1.83	4.08	5.66	3.16	5.08	3.75	4	_3_	2.58	3.08	_1.66_
CD(Critical Difference) = 1.96 , DF(difference in average ranks between best and worst method)=1.08												
Groups	The test cannot detect significant difference between the methods as CD >DF											
p-values	0.8852		0.7508		0.8852		0.9540		1.0000			
*= The best performing ensemble considering best average rank in accuracy and time to train												

For overall comparison we choose subbag20, bagging, adaboost, sbundle50, bundling for their superior performance over other ensemble methods in our experiments. In this experiment we have added three more datasets and another ensemble method, Rotation Forest. Our main aim of this experiment is to compare the subsampled version of bagging and bundling with rotation forest. We have reported all the results of this experiment in Table 5. From Table 5 we see that bagging is worst performing ensemble in terms of average accuracy 0.1845, where as the best one is sbundle50, producing 0.1359. In terms of average rank on accuracy, sbundle50 also have the lowest value 2.26 (meaning that sbundle50 is the more accurate ensemble on most of the datasets), where rotation forest and bundle scored a tie 2.42.

Table 5. Error rate of Subbag20, Bagging, Adaboost, Sbundle50, Bundling and Rotation Forest in benchmark datasets

Datasets	Subbag20	Bagging	Adaboost	Sbundle50*	Bundling	Rotation Forest
Diabetes	**0.2362**	0.2820	0.2929	0.2450	0.2477	0.2531
German	0.3024	0.2992	0.3300	0.2264	**0.2245**	0.2847
Glass	0.2495	0.2453	0.2456	0.2486	**0.2431**	0.2439
Heart	0.2067	0.3037	0.2336	**0.1493**	0.1674	0.1754
Hearts	0.1941	0.2874	0.2191	**0.1696**	0.1703	0.2109
Ion	0.1214	0.0832	0.0695	0.0649	0.0660	**0.0559**
Iris	0.1000	0.0747	0.0533	0.0453	0.0442	**0.0432**
Liver	0.3107	0.3177	0.3059	**0.2585**	0.2721	0.2968
Lymph	0.1578	0.1775	0.1602	**0.1486**	0.1621	0.1557
Sonar	0.2308	0.2289	0.1798	0.1903	0.1808	**0.1726**
Vehicle	0.2794	0.2525	0.2368	0.1829	**0.1791**	0.2182
Vote	0.0483	0.0515	0.0508	**0.0422**	0.0482	**0.0422**
Vowel	0.2073	0.0855	0.0517	**0.0121**	0.0202	0.0511
Wbc	0.0334	0.0299	0.0331	0.0336	0.0299	**0.0289**
Wine	0.0888	0.0315	0.0331	**0.0224**	0.0382	0.0307
Average Error	0.1845	0.1812	0.1664	**0.1359**	0.1396	0.1509
Average Rank	4.6	4.8	4.2	**2.26**	2.4	2.4

CD(Critical Difference) = 1.75 , DF(difference in average ranks between best and worst method)=2.54

Groups	Group 2	Group 2	Group 2	Group 1	Group 1	Group 1
p-values	0.1354	0.1249	0.3614		0.9339	0.6041

*= The best performing ensemble considering best average rank in accuracy and time to train

After performing the overall comparison test as proposed in [6], we can say that rotation forest and sbundle50 are not significantly different, where as subbag20 and bagging are also not significantly different. For an overall summary we can say that,

1. For a competitive ensemble method with faster training, subbag20 is suitable.
2. For superior accuracy, compensating the time complexity, sbundle50 is suitable. The main reason bundling took more time to train than adaboost is as it need to train four additional classifier models than adaboost and bagging. It should be noted that the time complexity of rotation forest is higher than bundling.

4   Conclusion

In this paper a comparative study on the effect of small subsampling ratio on different well known ensemble method has been done. From the experimental results it can be concluded that the performance of bagging and bundling can be improved, with small subsamples. The performance of bundling with small subsample is remarkably better than other classifiers, though the time for each ensemble to train take much higher time than adaboost and bagging in most of the cases. These findings are novel from the point of view that we are first to perform such an experiment to compare ensemble methods with different small subsamples in terms of accuracy and training time. Our main aim behind these experiments were to find a scalable algorithm suitable for data mining task. From the experiment we can say that the subbundling with 50% subsample ratio is most suitable among all the methods for this.

References

1. Blake, C.L., Merz, C.J.: UCI Repository of Machine Learning Databases, http://www.ics.uci.edu/mlearn/MLRepository.html
2. Breiman, L.: Bagging predictors. Machine Learning 24(2), 123–140 (1996a)
3. Breiman, L.: Out-of-bag estimation. Statistics Department, University of Berkeley CA 94708, Technical Report (1996b)
4. Breiman, L.: Heuristics of instability and stabilization in model selection. Annals of Statistics 24(6), 2350–2383 (1996c)
5. Bühlman, P.: Bagging, subagging and bragging for improving some prediction algorithms. In: Arkitas, M.G., Politis, D.N. (eds.) Recent Advances and Trends in Nonparametric Statistics, pp. 9–34. Elsevier, Amsterdam (2003)
6. Demšar, J.: Statistical comparisons of classifiers over multiple datasets. J. Mach. Learn. Research 7, 1–30 (2006)
7. Freund, Y., Schapire, R.: Experiments with a New boosting algorithm. In: Machine Learning: Proceedings to the Thirteenth International Conference, pp. 148–156. Morgan Kaufmann, San Francisco (1996)
8. Freund, Y., Schapire, R.: A decision-theoretic generalization of online learning and an application to boosting. J. Comput. System Sci. 55, 119–139 (1997)
9. Friedman, J.: Stochastic gradient boosting. Comput. Statist. Data Anal. 38, 367–378 (2002)
10. Friedman, J., Hall, P.: On Bagging and Non-linear Estimation. J. Statist. Planning and Infer. 137(3), 669–683 (2007)
11. Hastie, T., Tibshirani, R., Freidman, J.: The elements of statistical learning: data mining, inference and prediction. Springer, New York (2001)
12. Hothorn, T., Lausen, B.: Double-bagging: combining classifiers by bootstrap aggregation. Pattern Recognition 36(6), 1303–1309 (2003)
13. Hothorn, T., Lausen, B.: Bundling classifiers by bagging trees. Comput. Statist. Data Anal. 49, 1068–1078 (2005)
14. Kuncheva, L.I.: Combining Pattern Classifiers. Methods and Algorithms. John Wiley and Sons, Chichester (2004)
15. Rodríguez, J., Kuncheva, L., Alonso, C.: Rotation forest: A new classifier ensemble method. IEEE Trans. Patt. Analys. Mach. Intell. 28(10), 1619–1630 (2006)
16. Zaman, F., Hirose, H.: Double SVMbagging: A subsampling approach to SVM ensemble. To appear in Intelligent Automation and Computer Engineering. Springer, Heidelberg (2009)
17. Zaman, F., Hirose, H.: Effect of Subsampling Rate on Subbagging and Related Ensembles of Stable Classifiers. In: Chaudhury, S., Mitra, S., Murthy, C.A., Sastry, P.S., Pal, S.K. (eds.) PReMI 2009. LNCS, vol. 5909, pp. 44–49. Springer, Heidelberg (2009)
18. Zhang, C.X., Zhang, J.S., Zhang, G.Y.: An efficient modified boosting method for solving classification problems. J. Comput. Applied Mathemat. 214, 381–392 (2008)

Analysis of Bagging Ensembles of Fuzzy Models for Premises Valuation

Marek Krzystanek[1], Tadeusz Lasota[2], Zbigniew Telec[1], and Bogdan Trawiński[1]

[1] Wrocław University of Technology, Institute of Informatics,
Wybrzeże Wyspiańskiego 27, 50-370 Wrocław, Poland
[2] Wrocław University of Environmental and Life Sciences, Dept. of Spatial Management
Ul. Norwida 25/27, 50-375 Wrocław, Poland
marek.krzystanek@gmail.com, tadeusz.lasota@wp.pl,
zbigniew.telec@pwr.wroc.pl, bogdan.trawinski@pwr.wroc.pl

Abstract. The investigation of 16 fuzzy algorithms implemented in data mining system KEEL from the point of view of their usefulness to create bagging ensemble models to assist with real estate appraisal were presented in the paper. All the experiments were conducted with a real-world dataset derived from a cadastral system and registry of real estate transactions. The results showed there were significant differences in accuracy between individual algorithms. The analysis of measures of error diversity revealed that only the highest values of an average pairwise correlation of outputs were a profitable criterion for the selection of ensemble members.

Keywords: genetic fuzzy systems, bagging, real estate appraisal, KEEL.

1 Introduction

Ensemble methods have drawn considerable attention of many researchers for the last decade. Ensembles combining diverse machine learning models have been theoretically and empirically proved to ensure significantly better performance than their single original models. Although many multiple model creation techniques have been developed [28], according to [27] five commonly used groups of them can be distinguished, namely bagging [4], boosting [29], AdaBoost [13], stacked generalization [30], and mixture of experts [17].

Bagging, which originates from the term of bootstrap aggregating, belongs to the most intuitive and simplest ensemble algorithms providing a good performance. The diversity of regressors is obtained by using bootstrapped replicas of the training data. That is, different training data subsets are randomly drawn with replacement from the original training set. So obtained training data subsets, called also bags, are used then to train different regression models. Finally, individual regressors are combined through an algebraic expression, such as minimum, maximum, sum, mean, product, median, etc. [27]. Bagging has been employed to regression trees [3], Gaussian process [9], and neural networks [16]. There are also a few works on applying ensemble fuzzy systems to solve classification [6] and prediction problems [18].

N.T. Nguyen, M.T. Le, and J. Świątek (Eds.): ACIIDS 2010, Part II, LNAI 5991, pp. 330–339, 2010.

Many articles consider the bias-variance-covariance decomposition of error functions and the problem of diversity-accuracy trade off as the basis of the construction of hybrid ensembles [5], [7], [21].

Very often the whole generated ensemble turns out to be not an optimal solution, e.g. due to bias in the learners. Numerous approaches to select classifiers/regressors to compose ensembles were proposed i.e. genetic algorithms [10], [31], ensemble pruning [15], [26], instance selection [14], overproduce-and-choose strategy [12], negative correlation learning [25], and many others.

In our previous works [19], [20], [22] we tested different machine learning algorithms, among others genetic fuzzy systems trying to select the most appropriate ones to build data driven models for real estate appraisal using MATLAB and KEEL. We began also our study on the bagging approach, to investigate whether it could lead to the improvement of the accuracy machine learning regression models devoted to assist with real estate appraisals [23], [24]. In this paper we present the extension of our previous experiments with bagging ensembles to 16 fuzzy algorithms implemented in KEEL. We aimed to consider if this class of algorithms fits to design hybrid ensembles.

Actual data used to generate and learn appraisal models came from the cadastral system and the registry of real estate transactions referring to residential premises sold in one of big Polish cities at market prices within two years 2001 and 2002. They constituted an original dataset of 1098 instances of sales/purchase transactions. Four attributes were pointed out as price drivers: usable area of premises, floor on which premises were located, year of building construction, number of storeys in the building, in turn, price of premises was the output variable.

The concept of a data driven models for premises valuation, presented in the paper, was developed based on the sales comparison method. The appraiser accesses the system through the internet and input the values of the attributes of the premises being evaluated into the system, which calculates the output using a given model. The final result, that is a suggested value of the property, is sent back to the appraiser.

2 Plan of Experiments

The main goal of the investigations was to carry out the comparative analysis of 16 fuzzy algorithms employed to create ensemble models for premises property valuation. For this purpose KEEL, a non-commercial Java software tool designed to assess evolutionary algorithms for data mining problems, was used [2]. KEEL algorithms used in study are listed in Table 1, and details of the algorithms and references to source articles can be found on KEEL web site: www.keel.es. Most algorithms were evolutionary fuzzy ones. Two families of algorithms were distinguished during the analysis of results achieved: MOGUL - a methodology to obtain genetic fuzzy rule-based systems under the Iterative Rule Learning approach and the Wang-Mendel algorithm tuned by means of evolutionary post-processing algorithms.

Table 1. Fuzzy algorithms used in study

Alg.	KEEL name	Description
COR	Regr-COR_GA	Genetic fuzzy rule learning, COR algorithm inducing cooperation among rules
FRS	Regr-FRSBM	Fuzzy and random sets based modeling
FGP	Regr-Fuzzy-GAP	Fuzzy rule learning, grammar-based GP algorithm
PFC	Regr-Fuzzy-P_FCS1	Pittsburgh fuzzy classifier system #1
FSA	Regr-Fuzzy-SAP	Fuzzy rule learning, grammar GP based operators and simulated annealing based algorithm
SEF	Regr-Fuzzy-SEFC	Symbiotic evolution based fuzzy controller design method
THR	Regr-Thrift	Genetic fuzzy rule learning, Thrift algorithm
IRL	Regr-Fuzzy-MOGUL-IRL	Iterative rule learning of descriptive Mamdani rules
IHC	Regr-Fuzzy-MOGUL-IRLHC	Iterative rule learning of Mamdani rules - high constrained approach
ISC	Regr-Fuzzy-MOGUL-IRLSC	Iterative rule learning of Mamdani rules - small constrained approach
ITS	Regr-Fuzzy-MOGUL-TSK	Local evolutionary learning of TSK fuzzy rule based system
W-M	Regr-Fuzzy-WM	Fuzzy rule learning, Wang-Mendel algorithm
WAG	Regr-Fuzzy-WM & Post-A-G-Tuning-FRBSs	Wang-Mendel algorithm tuned using approxi-mative genetic tuning of FRBSs
WGG	Regr-Fuzzy-WM & Post-G-G-Tuning-FRBSs	Wang-Mendel algorithm tuned using global genetic tuning of the fuzzy partition of linguistic FRBSs
WGS	Regr-Fuzzy-WM & Post-G-S-Weight-RRBS	Wang-Mendel algorithm tuned using genetic selection of rules and rule weight tuning
WGT	Regr-Fuzzy-WM & Post-G-T-Weights-FRBSs	Wang-Mendel algorithm tuned using genetic tuning of FRBS weights

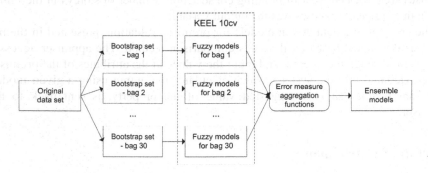

Fig. 1. Schema of bagging ensemble model development

Schema of the experiments is depicted in Fig. 1. On the basis of the original data set 30 bootstrap replicates (bags) of the cardinality equal to the original dataset were created. The bags were then used to generate models employing each of 16 above mentioned algorithms, and as the result 480 models for individual bags were obtained. For the reference 16 base models for individual algorithms using the original dataset were also produced. During the pre-processing phase normalization of data was performed using the min-max approach. As the accuracy measure the mean square error (MSE) was applied. All models were generated using 10-fold cross validation (10cv). As aggregation functions for combining the ensemble members' individual outputs simple averages were used.

3 Results of Experiments

At the first step the performance of single models built over the original data set was compared (see Fig. 2). It can be observed that MSE differentiate the algorithms substantially, ranging from 0.0025 for SEF and FRS to 0.060 for W-M and 0.0078 for PFC. The MSE values of individual base models were then used as reference levels when analyzing the bagging ensembles.

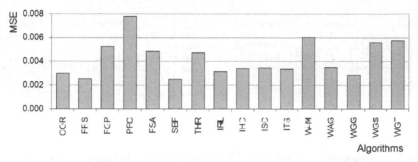

Fig. 2. Comparison of MSE for models built using the original data set

Having the values of MSE for individual algorithms over 30 bags, the Friedman and Iman-Davenport non-parametric statistical tests were performed in respect of average ranks, which use χ^2 and F statistics, respectively [11]. The calculated values of these statistics were 391.85 and 195.42, respectively, whereas the critical values at $\alpha=0.05$ are $\chi^2(15)=27.49$ and $F(15,435)=1.69$, so the null-hypothesis were rejected. Average ranks of individual algorithms are shown in Table 2, where the lower rank value the better algorithm.

Table 2. Average rank positions of individual algorithms over all 30 bags

Rank	Alg.	Rank	Alg.	Rank	Alg.	Rank	Alg.
2.30	ITS	4.40	SEF	7.93	WAG	12.80	FSA
4.03	IRL	5.37	IHC	10.10	THR	13.63	W-M
4.13	WGG	5.57	ISC	11.50	WGS	14.67	PFC
4.17	FRS	7.23	COR	12.37	WGT	15.80	FGA

Figures 3 a)-p) show the differences in accuracy of bagging ensembles built by means of individual evolutionary fuzzy algorithms over from 2 to 30 successive bags. The greater distance between the horizontal lines reflecting the MSE of original models and the tops of bars, which represent the MSE of a given ensemble, the bigger error reduction provided by that ensemble. Moreover, due to the same scale of all charts it is possible to assess visually the differences between respective models. So you can notice that PFC, THR, and all algorithms belonging to the Iterative Rule Learning and Wang-Mendel families ensure the best error reduction. Of course, the relative values of MSE should be taken into account, because in fact for PFC, THR, W-M, WGS, and WGT ensembles lead to the error reduction, but their accuracy remains still at a rather low level. Maximal percentage MSE reduction of bagged ensembles compared to MSE provided by the original models is shown in Table 3.

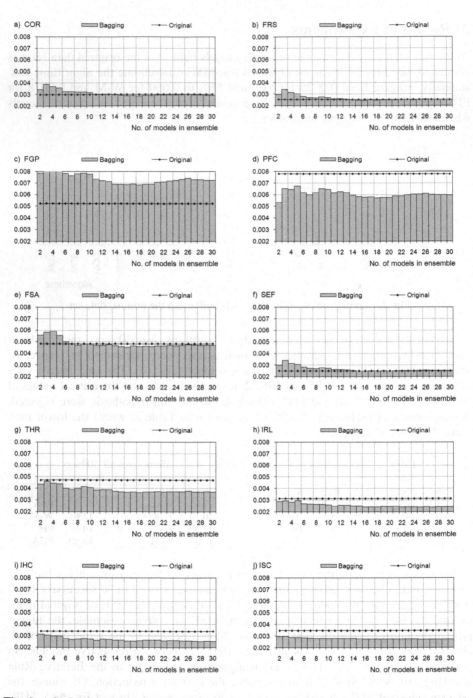

Fig. 3 a)-j). Performance of bagged ensembles compared with original model for individual algorithms in terms of MSE

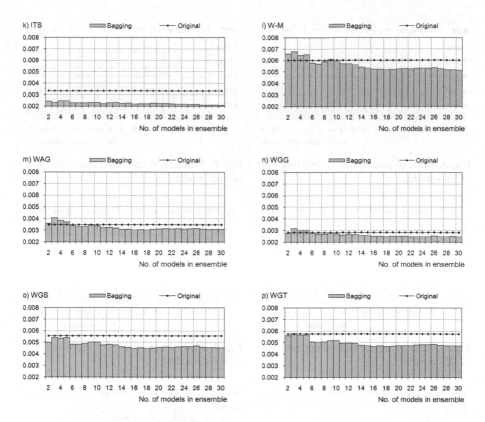

Fig. 3 k)-p). Performance of bagged ensembles compared with original model for individual algorithms in terms of MSE

Table 3. Maximal error reduction of bagged ensembles compared to single models for individual algorithms

Alg.	No. of bags	MSE	% red	Alg.	No. of bags	MSE	% red
COR	18	0.00286	4.39%	IRL	27	0.00240	23.56%
FRS	16	0.00243	3.89%	IHC	30	0.00252	25.75%
FGP	18	0.00685	-30.81%	ISC	25	0.00272	21.70%
PFC	2	0.00534	31.45%	ITS	30	0.00209	37.66%
FSA	16	0.00454	6.30%	W-M	30	0.00514	14.66%
SEF	16	0.00247	0.62%	WAG	18	0.00302	13.21%
THR	18	0.00364	22.63%	WGG	22	0.00245	13.18%

These observations encourage us to seek the optimal subset of bags composing an ensemble with the best performance. The simplest method is to take a number of models which reveal the lowest accuracy error. Figures 4, 5, 6 and 7 depict the ensembles comprising from 10 to 25 best models generated with the best representatives of FRS, SEF, ITS, and WGG. The gain in accuracy can be easily noticed. In Table 4 three best algorithms which provide the lowest MSE for each bag, and two worst ones were placed in five leftmost columns. So, still better results might provide hybrid ensembles comprising the best models created by different algorithms.

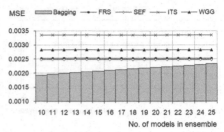

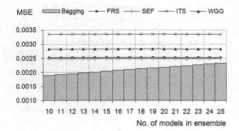

Fig. 4. MSE of ensembles comprising best of FRS models

Fig. 5. MSE of ensembles comprising best of SEF models

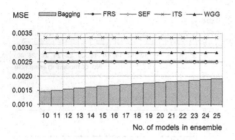

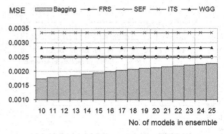

Fig. 6. MSE of ensembles comprising best of ITS models

Fig. 7. MSE of ensembles comprising best of WGG models

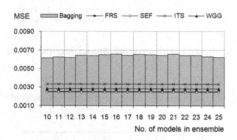

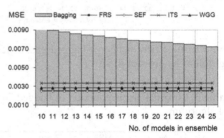

Fig. 8. MSE of ensembles comprising models with the lowest CV

Fig. 9. MSE of ensembles comprising models with the highest CV

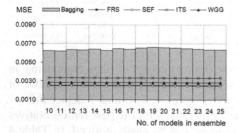

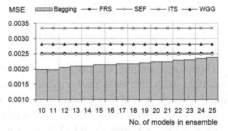

Fig. 10. MSE of ensembles comprising models with the lowest APC

Fig. 11. MSE of ensembles comprising models with the highest APC

Table 4. Fuzzy models holding highest and lowest rank positions according to accuracy and diversity criteria for individual bags

Bag	Best MSE			Worst MSE		CV		APC	
	1st	2nd	3rd	15th	16th	Lwst	Hgst	Lwst	Hgst
1	ITS	IHC	FRS	PFC	FGP	PFC	W-M	PFC	SEF
2	ITS	IRL	FRS	FWM	FGP	PFC	FGA	PFC	WGG
3	ITS	ISC	IHC	PFC	FGP	PFC	FGA	PFC	IRL
4	FRS	SEF	IRL	PFC	FGP	PFC	W-M	PFC	WGG
5	ITS	FRS	SEF	PFC	FGP	PFC	FGA	PFC	WGG
6	ITS	IRL	WGG	FSA	PFC	PFC	ITS	PFC	WGG
7	FRS	WGG	SEF	FWM	FGP	PFC	W-M	PFC	**WGG**
8	FRS	ITS	SEF	PFC	FGP	PFC	FGA	PFC	WGG
9	ITS	IRL	IHC	FGP	PFC	PFC	W-M	PFC	IRL
10	ITS	IRL	FRS	PFC	FGP	PFC	W-M	PFC	WGG
11	ITS	WGG	IRL	FGP	FSA	PFC	WGT	PFC	WGG
12	SEF	FRS	COR	FWM	PFC	PFC	FGA	PFC	WGG
13	FRS	SEF	WGG	FSA	FGP	PFC	W-M	PFC	WGG
14	ITS	IRL	WGG	FSA	FGP	PFC	WGT	FGA	WGG
15	FRS	SEF	COR	WGT	FGP	PFC	FGA	PFC	WGG
16	ITS	IRL	IHC	PFC	FGP	PFC	FGA	PFC	WGG
17	IIIC	IRL	ITS	PFC	FGP	PFC	FGA	PFC	IRL
18	SEF	FRS	WGG	PFC	FGP	PFC	W-M	PFC	WGG
19	SEF	FRS	WGG	PFC	FGP	PFC	FGA	PFC	WGG
20	ITS	WGG	FRS	FWM	FGP	PFC	W-M	ITS	WGG
21	ITS	WGG	SEF	PFC	FGP	PFC	FGA	PFC	WGG
22	WGG	IRL	ITS	PFC	FGP	PFC	FGA	PFC	WGG
23	ITS	IRL	ISC	PFC	FGP	PFC	W-M	PFC	WGG
24	SEF	FRS	ITS	PFC	FGP	PFC	W-M	PFC	IRL
25	ITS	IRL	ISC	PFC	FGP	PFC	FGA	PFC	WGG
26	ITS	IHC	ISC	PFC	FGP	PFC	FGA	PFC	WGG
27	ITS	WGG	IRL	PFC	FGP	PFC	WGT	PFC	WGG
28	IHC	ITS	WAG	PFC	FGP	PFC	WGS	PFC	WGS
29	ITS	FRS	SEF	PFC	FGP	PFC	W-M	PFC	ITS
30	ITS	ISC	IHC	PFC	FGP	PFC	FGA	PFC	WGG

Inspired by some works on the application of diversity of errors to construct ensembles [1], we tried to check the usability of two methods which measure the diversity of regressors' outputs regardless of whether they are correct or incorrect. These were the coefficient of variability (CV) measured for each instance for individual algorithms, hoping the higher CV the bigger error reduction could be obtained. But it was not the case. In Table 4 the algorithms with the lowest and the highest CV for each bag are enumerated, but Figures 8 and 9 indicate there is no gain in performance in both cases. Another diversity measure we tested was the average pairwise correlation (APC) of the results produced by algorithms over individual bags. In Table 4 the algorithms with the lowest and the highest APC for each bag are listed, and the performance was depicted in Figures 10 and 11. Ensembles comprising models of the lowest APC also did not lead to the improvement of accuracy. Only the highest APC turned to be beneficial and ensembles composed of those models provided noticeable error reduction.

4 Conclusions and Future Work

Sixteen fuzzy algorithms implemented in data mining system KEEL were tested from the point of view of their usefulness to create bagging ensemble models to assist with real estate appraisal. All the experiments were conducted with a real-world data set derived from cadastral system and the registry of real estate transactions. 30 bootstrap replicates (bags) were created by random drawing with replacement the elements contained in the original dataset of the cardinality equal to 1098. The bags were then used to generate models employing each of 16 above mentioned algorithms, and as the result 480 models for individual bags were obtained.

The results in terms of MSE as the accuracy measure showed there were significant differences between individual algorithms. The best performance in terms of both accuracy and error reduction rate revealed algorithms FRS, WGG, and whole family of Iterative Rule Learning. Further considerations led to the conclusion it was possible to construct hybrid bagging ensembles which would provide still better output. The analysis of two measures of error diversity showed that only the highest values of an average pairwise correlation of outputs were a valuable criterion for the selection of ensemble members.

Further research is planned to construct optimal hybrid ensembles of real estate appraisal models with the use new sets of actual data and with the consideration of time impact on the prices of land and premises. Moreover, an open problem remains the stability of such ensembles, therefore further investigations are needed.

Acknowledgments. This paper was supported by Ministry of Science and Higher Education of Poland under grant no. N N519 407437.

References

1. Aksela, M., Laaksonen, J.: Using diversity of errors for selecting members of a committee classifier. Pattern Recognition 39, 608–623 (2006)
2. Alcalá-Fdez, J., et al.: KEEL: A Software Tool to Assess Evolutionary Algorithms for Data Mining Problems. Soft Computing 13(3), 307–318 (2009)
3. Banfield, R.E., et al.: A Comparison of Decision Tree Ensemble Creation Techniques. IEEE Trans. on Pattern Analysis and Machine Intelligence 29(1), 173–180 (2007)
4. Breiman, L.: Bagging Predictors. Machine Learning 24(2), 123–140 (1996)
5. Brown, G., Wyatt, J., Harris, R., Yao, X.: Diversity Creation Methods: A Survey and Categorisation. Journal of Information Fusion 6(1), 5–20 (2005)
6. Canul-Reich, J., Shoemaker, L., Hall, L.O.: Ensembles of Fuzzy Classifiers. In: Proc. IEEE International Conference on Fuzzy Systems, FUZZ-IEEE 2007, pp. 1–6 (2007)
7. Chandra, A., Yao, X.: Evolving hybrid ensembles of learning machines for better generalisation. Neurocomputing 69, 686–700 (2006)
8. Chawla, N.V., Hall, L.O., Bowyer, K.W., Kegelmeyer, W.P.: Learning Ensembles From Bites: A Scalable and Accurate Approach. Journal of Machine Learning Research 5, 421–451 (2004)
9. Chen, T., Ren, J.: Bagging for Gaussian Process Regression. Neurocomputing 72(7-9), 1605–1610 (2009)
10. Cordón, O., Quirin, A.: Comparing Two Genetic Overproduce-and-choose Strategies for Fuzzy Rule-based Multiclassification Systems Generated by Bagging and Mutual Information-based Feature Selection. Int. J. Hybrid Intelligent Systems (2009) (in press)
11. Demšar, J.: Statistical comparisons of classifiers over multiple data sets. Journal of Machine Learning Research 7, 1–30 (2006)

12. Dos Santos, E.M., Sabourin, R., Maupin, P.: A dynamic overproduce-and-choose strategy for the selection of classifier ensembles. Pattern Recognition 41(10), 2993–3009 (2008)
13. Freund, Y., Schapire, R.E.: Decision-theoretic generalization of on-line learning and an application to boosting. J. Computer and System Sciences 55(1), 119–139 (1997)
14. García-Pedrajas, N.: Constructing Ensembles of Classifiers by Means of Weighted Instance Selection. IEEE Transactions on Neural Networks 20(2), 258–277 (2009)
15. Hernandez-Lobato, D., Martinez-Munoz, G., Suarez, A.: Pruning in ordered regression bagging ensembles. In: Yen, G.G. (ed.) Proceedings of the IEEE World Congress on Computational Intelligence, pp. 1266–1273 (2006)
16. Islam, M.M., et al.: Bagging and Boosting Negatively Correlated Neural Networks. IEEE Trans. on Systems, Man, and Cybernetics, Part B: Cyb. 38(3), 771–784 (2008)
17. Jacobs, R.A., Jordan, M.I., Nowlan, S.J., Hinton, G.E.: Adaptive mixtures of local experts. Neural Computation 3, 79–87 (1991)
18. Kim, D.: Improving the Fuzzy System Performance by Fuzzy System Ensemble. Fuzzy Sets and Systems 98(1), 43–56 (1998)
19. Król, D., Lasota, T., Trawiński, B., Trawiński, K.: Investigation of Evolutionary Optimization Methods of TSK Fuzzy Model for Real Estate Appraisal. International Journal of Hybrid Intelligent Systems 5(3), 111–128 (2008)
20. Krzystanek, M., Lasota, T., Trawiński, B.: Comparative Analysis of Evolutionary Fuzzy Models for Premises Valuation Using KEEL. In: Nguyen, N.T., Kowalczyk, R., Chen, S.-M. (eds.) ICCCI 2009. LNCS, vol. 5796, pp. 838–849. Springer, Heidelberg (2009)
21. Kuncheva, L.I., Whitaker, C.J.: Measures of Diversity in Classifier Ensembles and Their Relationship with the Ensemble Accuracy. Machine Learning 51, 181–207 (2003)
22. Lasota, T., Mazurkiewicz, J., Trawiński, B., Trawiński, K.: Comparison of Data Driven Models for the Validation of Residential Premises Using KEEL. International Journal of Hybrid Intelligent Systems (2009) (in press)
23. Lasota, T., Telec, Z., Trawiński, B., Trawiński, K.: A Multi-agent System to Assist with Real Estate Appraisals using Bagging Ensembles. In: Nguyen, N.T., Kowalczyk, R., Chen, S.-M. (eds.) ICCCI 2009. LNCS, vol. 5796, pp. 813–824. Springer, Heidelberg (2009)
24. Lasota, T., Telec, Z., Trawiński, B., Trawiński, K.: Exploration of Bagging Ensembles Comprising Genetic Fuzzy Models to Assist with Real Estate Appraisals. In: Corchado, E., Yin, H. (eds.) IDEAL 2009. LNCS, vol. 5788, pp. 554–561. Springer, Heidelberg (2009)
25. Liu, Y., Yao, X.: Ensemble learning via negative correlation. Neural Networks 12(10), 1399–1404 (1999)
26. Margineantu, D.D., Dietterich, T.G.: Pruning Adaptive Boosting. In: Proc. 14th Int. Conf. Machine Learning, pp. 211–218 (1997)
27. Polikar, R.: Ensemble Learning. Scholarpedia 4(1), 2776 (2009)
28. Rokach, L.: Taxonomy for characterizing ensemble methods in classification tasks: A review and annotated bibliography. Comp. Stat. and Data Analysis 53, 4046–4072 (2009)
29. Schapire, R.E.: The Strength of Weak Learnability. Machine Learning 5(2), 197–227 (1990)
30. Wolpert, D.H.: Stacked Generalization. Neural Networks 5(2), 241–259 (1992)
31. Zhou, Z.H., Wu, J., Tang, W.: Ensembling Neural Networks: Many Could Be Better Than All. Artificial Intelligence 137, 239–263 (2002)

Comparison of Bagging, Boosting and Stacking Ensembles Applied to Real Estate Appraisal

Magdalena Graczyk[1], Tadeusz Lasota[2], Bogdan Trawiński[1], and Krzysztof Trawiński[3]

[1] Wrocław University of Technology, Institute of Informatics,
Wybrzeże Wyspiańskiego 27, 50-370 Wrocław, Poland
[2] Wrocław University of Environmental and Life Sciences, Dept. of Spatial Management
Ul. Norwida 25/27, 50-375 Wroclaw, Poland
[3] European Centre for Soft Computing, Edificio Científico-Tecnológico, 3ª Planta,
C. Gonzalo Gutiérrez Quirós S/N, 33600 Mieres, Asturias, Spain
mag.graczyk@gmail.com, tadeusz.lasota@wp.pl,
bogdan.trawinski@pwr.wroc.pl,
krzysztof.trawinski@softcomputing.es

Abstract. The experiments, aimed to compare three methods to create ensemble models implemented in a popular data mining system called WEKA, were carried out. Six common algorithms comprising two neural network algorithms, two decision trees for regression, linear regression, and support vector machine were used to construct ensemble models. All algorithms were employed to real-world datasets derived from the cadastral system and the registry of real estate transactions. Nonparametric Wilcoxon signed-rank tests to evaluate the differences between ensembles and original models were conducted. The results obtained show there is no single algorithm which produces the best ensembles and it is worth to seek an optimal hybrid multi-model solution.

Keywords: ensemble models, bagging, stacking, boosting, property valuation.

1 Introduction

During the last decade the ensemble learning systems attracted attention of many researchers. This collection of methods combines the output of the machine learning systems, in literature called "weak learners" in due to its performance, from the group of learners in order to get smaller prediction errors (in regression) or lower error rates (in classification). The individual estimator must provide different patterns of generalization, thus the diversity plays a crucial role in the training process. Otherwise, the ensemble, called also committee, would end up having the same predictor and provide as good accuracy as the single one. It was proved [11], [12]; [15], [21] that the ensemble performs better when each individual machine learning system is accurate and makes error on the different instances at the same time.

There are many taxonomies for ensemble techniques [4],[17],[22],[24], and there is a recognized group so called data resampling, which generates different training sets to obtain unique learner. In this group it is included bagging [2],[5], boosting [9],[25], and stacking [3],[27]. In boostrap aggregation, i.e. bagging, each machine learning

N.T. Nguyen, M.T. Le, and J. Świątek (Eds.): ACIIDS 2010, Part II, LNAI 5991, pp. 340–350, 2010.

system is independently learned on resampled training set, which is randomly picked from the original samples of the training set. Hence, bagging is devoted to the unstable algorithms, where the small changes in the training set, result in large changes in the output of that system. Training of each predictor could be in parallel, in due to independence of training of each learning algorithm.

Boosting provides sequential learning of the predictors. The first one is learned on the whole data set, while the following are learned on the training set based on the performance of the previous one. In other words, the examples which were predicted improperly are marked. Then, such examples get more probability to appear in the training set of the next predictor. It results with different machines being specialized in predicting some areas of the dataset.

Stacking is composed of two phases. Firstly, usually different models are learned based on a dataset. Then, the output of each of the model are collected to create a new dataset. In the new dataset each instance is related to the real value that it is suppose to predict. Secondly, that dataset is used with a learning algorithm, the so-called meta-learning algorithm, in order to provide the final output.

However, above mentioned techniques are the most recognized ones, we may obtain diversity through applying randomness to our algorithm, e.g., random subspace [1], [14]. It is a feature selection algorithm, which generates different learners from different subset of attributes taken from the feature space.

Very often it is the case that the whole generated ensemble is not an optimal option, i.e. due to bias in the learners. It has been proven that small ensembles can perform better than large ensembles [14], [28], [29]. Although ensemble selection was mostly developed in the classifier environment like [7], there is some bunch of publications dedicated to ensemble selection for regression. The simplest idea consists in making a performance ranking of learners and selecting a number of the best ones [6]. Probably the most common way is to use a genetic algorithm [7],[29]. Other techniques comprise heuristic approaches, like Kappa pruning [20], greedy ones based on complementaries (biases among regressors) [13], defined diversity measure [23], or negative correlation learing [19].

So far we have investigated several methods to construct regression models to assist with real estate appraisal: evolutionary fuzzy systems, neural networks, decision trees, and statistical algorithms using MATLAB, KEEL, RapidMiner, and WEKA data mining systems [10], [16], [18]. In this paper we present results of several experiments with actual data of residential premises sales/purchase transactions aimed at the comparison of bagging, boosting and stacking ensemble techniques using WEKA [26]. Six machine learning algorithms implemented in WEKA were employed including neural networks, regression trees, support vector machine, and statistical linear regression.

2   Algorithms Used and Plan of Experiments

The main goal of our study was to compare three approaches to ensemble learning i.e. bagging, stacking and additive regression to examine how they improve the performance of models to assist with real estate appraisal. All experiments were conducted using *WEKA (Waikato Environment for Knowledge Analysis)*, a non-commercial and open-source data mining system [8],[26]. WEKA contains tools for data pre-processing, classification, regression, clustering, association rules, and visualization. It is also well-suited for developing new machine learning schemes.

Following WEKA algorithms for building, learning and optimizing models were employed to carry out the experiments.

MLP – MultiLayerPerceptron. Algorithm is performed on networks consisting of multiple layers, usually interconnected in a feed-forward way, where each neuron on layer has directed connections to the neurons of the subsequent layer.

RBF – Radial Basis Function Neural Network for Regression Problems. The algorithm is based on feed-forward neural networks with radial activation function on every hidden layer. The output layer represents a weighted sum of hidden neurons signals.

M5P – Pruned Model Tree. The algorithm is based on decision trees, however, instead of having values at tree's nodes, it contains a multivariate linear regression model at each node. The input space is divided into cells using training data and their outcomes, then a regression model is built in each cell as a leaf of the tree.

M5R – M5Rules. The algorithm divides the parameter space into areas (subspaces) and builds in each of them a linear regression model. It is based on M5 algorithm. In each iteration a M5 Tree is generated and its best rule is extracted according to a given heuristic. The algorithm terminates when all the examples are covered.

LRM - Linear Regression Model. Algorithm is a standard statistical approach to build a linear model predicting a value of the variable while knowing the values of the other variables. It uses the least mean square method in order to adjust the parameters of the linear model/function.

SVM – NU-Support Vector Machine. Algorithm constructs support vectors in high-dimensional feature space. Then, hyperplane with the maximal margin is constructed. Kernel function is used to transform the data, which augments the dimensionality of the data. This augmentation provokes that the data can be separated with an hyperplane with much higher probability, and establish a minimal prediction probability error measure.

Three metalearning methods were employed to create ensembles:

Additive regression – is an implementation of boosting in WEKA. The algorithm starts with an empty ensemble and incorporates new members sequentially. At each stage the model that maximizes the performance of the ensemble as a whole is added, without altering those already in the ensemble.

Bagging – consists in aggregating results of *n* models which were created on the basis of *N* bootstrap sets. The bootstrap sets are created out of the original dataset through feature selection or random drawing with replacement. The final result is calculated by averaging the outputs of individual models built over each bootstrap set.

Stacking – in stacking, the result of a set of different base learners at the level-0 is combined by a metalearner at the level-1. The role of the metalearner is to discover how best to combine the output of the base learners. In each run LRM, M5P, M5R, MLP, RBF, and SVM were the base learners and one of them was the metalearner.

Actual data used to generate and learn appraisal models came from the cadastral system and the registry of real estate transactions referring to residential premises sold in one of big Polish cities at market prices within two years 2001 and 2002. They constituted original dataset of 1098 instances of sales/purchase transactions. Four attributes were pointed out by an expert as price drivers: 1 – usable area of premises, 2 – number of storeys in the building where premises were located, 3 – floor on which premises were located, 4 – year of building construction, whereas price of premises was the output variable. In order to assure diversity five data sets were used in the experiments to build appraisal models: first one denoted by 1234 contained all four attributes of premises, and the other included all combinations of three attributes and were denoted as 123, 124, 134, and 234. However in this paper we present only the results obtained with 1234 and 124 data sets which revealed the best performance.

All runs of experiments were repeated for 1 to 30 steps of each ensemble method, 10-fold cross validation and root mean square error as accuracy measure were applied. Schema of the experiments with WEKA was depicted in Fig. 1. In the case of bagging *i-th* step means that the ensemble model was built with *i* bootstrap sets. As an output of each run WEKA provided two vectors of actual and predicted prices of premises for all input instances. Having such output we were able to calculate the final predictive accuracy of each committee in terms of MAPE (mean absolute percentage error). We also conducted nonparametric Wilcoxon signed-rank test to evaluate the differences between ensembles and original models.

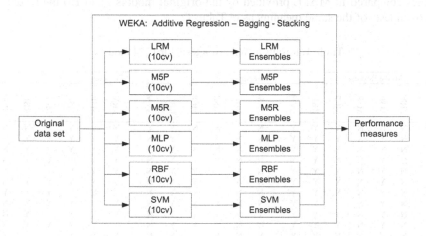

Fig. 1. Schema of the experiments with WEKA

3 Results of Experiments

The performance of the ensemble models built by additive regression, bagging and stacking for 2-30 steps was presented in Fig. 2,3 and 4 respectively. The outcome of 1234 models was placed in left columns whereas the right ones contain the results provided by 124 models. The charts allow you to observe how the accuracy of sequential ensembles changed when the number of steps was increased, so the scales of the axes were retained the same in each figure. The horizontal lines reflect the MAPE of original models. If the error reduction achieved by ensembles compared to an original model takes place, then the horizontal line is located above tops of bars representing the prediction accuracy of individual committees. In the case of *Additive*

Regression the predictive accuracy of ensembles remained at the same level as the number of was increased except for the M5P models. For *Bagging* it could be observed that the values of MAPE tended to decrease when the number of bags was getting bigger. *Stacking* was characterized by varying performance of models. In general there were no substantial differences between 1234 and 124 models.

In Table 1 the results of nonparametric Wilcoxon signed-rank test to evaluate the outcome of individual ensemble models were presented. The zero hypothesis H0 stated there were not significant differences in accuracy, in terms of MAPE, between a given ensemble model and the model built using an original (base) dataset. N denotes that there was no evidence against the H0 hypothesis, whereas Y means that there was evidence against the H0 hypothesis. In most cases the differences were significant except for LRM and M5R models for *Additive Regression*, LRM, M5P, and SVM models for *Bagging*, and SVM ones for *Stacking*.

In order to consider the possibility to apply a hybrid multi-model approach in Tables 2 and 3 the ensembles with the lowest values of MAPE for individual algorithms were placed for 1234 and 124 models respectively. The columns denoted by No contain the numbers of steps at which the best result was achieved, by Gain – the benefit of using committees i.e. the percentage reduction of MAPE values of respective ensembles compared to MAPE provided by the original models , and H0 the results of Wilcoxon tests of the same meaning as in Table 1.

Table 1. Results of Wilcoxon tests for 1234 models

No	Additive Regression						Bagging						Stacking					
	LRM	M5P	M5R	MLP	RBF	SVM	LRM	M5P	M5R	MLP	RBF	SVM	LRM	M5P	M5R	MLP	RBF	SVM
2	N	N	N	Y	Y	Y	N	Y	N	N	Y	Y	Y	Y	Y	Y	N	N
3	N	Y	N	Y	Y	Y	N	N	N	N	Y	N	Y	Y	Y	Y	Y	N
4	N	Y	N	Y	Y	Y	N	N	N	N	Y	N	Y	Y	Y	Y	Y	N
5	N	Y	N	Y	Y	Y	N	N	N	Y	Y	N	Y	Y	Y	Y	Y	N
6	N	Y	N	Y	Y	Y	N	N	N	Y	Y	N	Y	Y	Y	Y	Y	N
7	N	Y	N	Y	Y	Y	N	N	Y	Y	Y	N	Y	Y	Y	Y	Y	N
8	N	Y	N	Y	Y	Y	N	N	Y	Y	Y	N	Y	Y	Y	Y	Y	Y
9	N	Y	N	Y	Y	Y	N	N	Y	Y	Y	N	Y	Y	Y	Y	Y	N
10	N	Y	N	Y	Y	Y	Y	N	Y	Y	Y	N	Y	Y	Y	Y	Y	N
11	N	Y	N	Y	Y	Y	N	N	Y	Y	Y	N	Y	Y	Y	Y	Y	N
12	N	Y	N	Y	Y	Y	Y	N	Y	Y	Y	N	Y	Y	Y	Y	Y	Y
13	N	Y	N	Y	Y	Y	Y	N	Y	Y	Y	N	Y	Y	Y	Y	Y	Y
14	N	Y	N	Y	Y	Y	Y	N	Y	Y	Y	N	Y	Y	Y	Y	Y	N
15	N	Y	N	Y	Y	Y	Y	N	Y	Y	Y	N	Y	Y	Y	Y	Y	N
16	N	Y	N	Y	Y	Y	Y	N	Y	Y	Y	N	Y	Y	Y	Y	Y	Y
17	N	Y	N	Y	Y	Y	Y	N	Y	Y	Y	N	Y	Y	Y	Y	Y	N
18	N	Y	N	Y	Y	Y	Y	N	Y	Y	Y	N	Y	Y	Y	Y	Y	Y
19	N	Y	N	Y	Y	Y	Y	N	Y	Y	Y	N	Y	Y	Y	Y	Y	N
20	N	Y	N	Y	Y	Y	Y	N	Y	Y	Y	N	Y	Y	Y	Y	Y	N
21	N	Y	N	Y	Y	Y	Y	N	Y	Y	Y	N	Y	Y	Y	Y	Y	N
22	N	Y	N	Y	Y	Y	Y	N	Y	Y	Y	N	Y	Y	Y	Y	Y	N
23	N	N	N	Y	Y	Y	Y	N	Y	Y	Y	N	Y	Y	Y	Y	Y	N
24	N	Y	N	Y	Y	Y	Y	N	Y	Y	Y	N	Y	Y	Y	Y	Y	N
25	N	Y	N	Y	Y	Y	Y	N	Y	Y	Y	N	Y	Y	Y	Y	Y	N
26	N	Y	N	Y	Y	Y	N	N	Y	Y	Y	N	Y	Y	Y	Y	Y	Y
27	N	Y	N	Y	Y	Y	N	N	Y	Y	Y	N	Y	Y	Y	Y	Y	N
28	N	Y	N	Y	Y	Y	N	N	Y	Y	Y	N	Y	Y	Y	Y	Y	N
29	N	Y	N	Y	Y	Y	Y	N	Y	Y	Y	N	Y	Y	Y	Y	Y	Y
30	N	Y	N	Y	Y	Y	Y	N	Y	Y	Y	N	Y	Y	Y	Y	Y	N

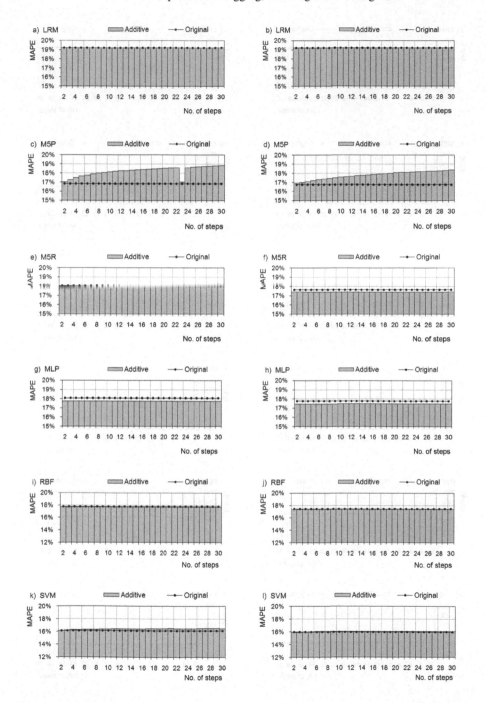

Fig. 2 a)-l). Performance of *Additive Regression* ensembles, in terms of MAPE, compared with original models for individual algorithms (1234 – left column, 124 – right column)

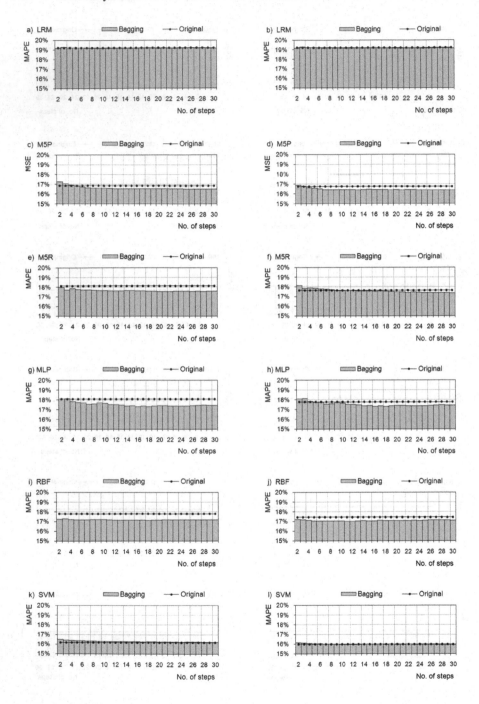

Fig. 3 a)-l). Performance of *Bagging* ensembles, in terms of MAPE, compared with original models for individual algorithms (1234 – left column, 124 – right column)

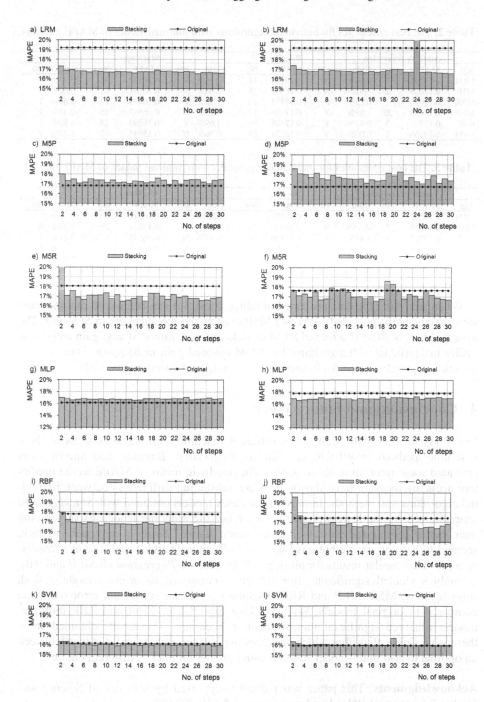

Fig. 4 a)-l). Performance of *Stacking* ensembles, in terms of MAPE, compared with original models for individual algorithms (1234 – left column, 124 – right column)

Table 2. The best ensembles for individual algorithms with minimal values of MAPE for 1234

Alg.	Additive regression				Bagging				Stacking			
	MAPE	No	Gain	H0	MAPE	No	Gain	H0	MAPE	No	Gain	H0
LRM	0.19202	2	0.00%	N	0.19133	14	0.36%	Y	0.16533	26	13.90%	Y
M5P	0.17025	2	-1.08%	N	0.16500	28	2.04%	N	0.16994	21	-0.60%	Y
M5R	0.18038	7	0.12%	N	0.17537	22	3.18%	Y	0.16498	13	8.65%	Y
MLP	0.17735	29	1.89%	Y	0.17306	18	4.27%	Y	0.16567	15	-2.48%	Y
RBF	0.17707	3	0.42%	Y	0.17126	18	3.69%	Y	0.16590	26	6.60%	Y
SVM	0.16166	2	0.00%	Y	0.16224	24	-0.36%	N	0.15935	12	1.43%	Y

Table 3. The best ensembles for individual algorithms with minimal values of MAPE for 124

Alg.	Additive regression				Bagging				Stacking			
	MAPE	No	Gain	H0	MAPE	No	Gain	H0	MAPE	No	Gain	H0
LRM	0.19202	2	0.00%	N	0.19121	16	0.42%	Y	0.16572	30	13.70%	Y
M5P	0.16859	2	-0.65%	N	0.16379	24	2.24%	Y	0.17048	25	-1.78%	N
M5R	0.17392	4	1.41%	Y	0.17370	30	1.54%	N	0.16545	17	6.21%	Y
MLP	0.17462	2	1.71%	Y	0.17306	18	2.59%	Y	0.16607	3	4.71%	Y
RBF	0.17374	3	0.34%	N	0.17020	11	2.38%	Y	0.16404	28	5.91%	Y
SVM	0.15899	2	0.46%	N	0.15962	26	0.07%	N	0.15931	28	0.26%	N

When analysing the data presented in tables 2 and 3 it can be noticed that the most substantial benefit provided LRM and M5P used as the metalearners in *Stacking*. The lowest values of MAPE achieved SVM models but with minor, if any, gain and statistically insignificant. All algorithms but SVM ensured gain in *Bagging*. These results are encouraging to undertake further research on hybrid ensemble methods.

4   Conclusions and Future Work

Our study on the application of six different machine learning algorithms to three ensemble methods in WEKA, i.e. *Additive Regression*, *Bagging*, and *Stacking* has provided some interesting observations. The results in terms of MAPE as the predictive accuracy measure have shown there are substantial differences between individual algorithms and methods. In most of the cases models obtained with *Stacking* were characterized by the lowest prediction error but the outcome tended to vary giving once better results and the other times much worse. *Bagging* approach, in turn, seemed to be more stable but gave worse results than *Stacking* and *Additive Regression* provided similar results for all steps. With *Additive Regression* all MLP and RBF ensembles yielded significant error reduction compared to original models. With *Bagging* most M5R, MLP and RBF ensembles produced significant error reduction compared to original models. With *Stacking* most LRM, M5R and RBF multiple-models achieved significant improvement in accuracy. The results obtained show there is no single algorithm which produces the best ensembles and it is worth to seek an optimal hybrid multi-model solution using greater number of different datasets.

Acknowledgments. This paper was partially supported by Ministry of Science and Higher Education of Poland under grant no. N N519 407437.

References

1. Banfield, R.E., et al.: A Comparison of Decision Tree Ensemble Creation Techniques. IEEE Trans. on Pattern Analysis and Machine Intelligence 29(1), 173–180 (2007)
2. Breiman, L.: Bagging Predictors. Machine Learning 24(2), 123–140 (1996)
3. Breiman, L.: Stacked Regressions. Machine Learning 24(1), 49–64 (1996)
4. Brown, G., Wyatt, J., Harris, R., Yao, X.: Diversity Creation Methods: A Survey and Categorisation. Journal of Information Fusion 6(1), 5–20 (2005)
5. Büchlmann, P., Yu, B.: Analyzing bagging. Annals of Statistics 30, 927–961 (2002)
6. Chawla, N.V., Hall, L.O., Bowyer, K.W., Kegelmeyer, W.P.: Learning Ensembles From Bites: A Scalable and Accurate Approach. J. of Mach. Learn. Res. 5, 421–451 (2004)
7. Cordón, O., Quirin, A.: Comparing Two Genetic Overproduce-and-choose Strategies for Fuzzy Rule-based Multiclassification Systems Generated by Bagging and Mutual Information-based Feature Selection. Int. J. Hybrid Intelligent Systems (2009) (in press)
8. Cunningham, S.J., Frank, E., Hall, M., Holmes, G., Trigg, L., Witten, I.H.: WEKA: Practical Machine Learning Tools and Techniques with Java Implementations. Morgan Kaufmann, New Zealand (2005)
9. Freund, Y., Schapire, R.E.: Decision-theoretic generalization of on-line learning and an application to boosting. J. Computer and System Sciences 55(1), 119–139 (1997)
10. Graczyk, M., Lasota, T., Trawiński, B.: Comparative Analysis of Premises Valuation Models Using KEEL, RapidMiner, and WEKA. In: Nguyen, N.T., Kowalczyk, R., Chen, S.-M. (eds.) ICCCI 2009. LNCS, vol. 5796, pp. 800–812. Springer, Heidelberg (2009)
11. Hansen, L., Salamon, P.: Neural network ensembles. IEEE Transactions on Pattern Analysis and Machine Intelligence 12(10), 993–1001 (1990)
12. Hashem, S.: Optimal linear combinations of neural networks. Neural Net. 10(4), 599–614 (1997)
13. Hernandez-Lobato, D., Martinez-Munoz, G., Suarez, A.: Pruning in ordered regression bagging ensembles. In: Yen, G.G. (ed.) Proceedings of the IEEE World Congress on Computational Intelligence, pp. 1266–1273 (2006)
14. Ho, K.T.: The random subspace method for constructing decision forests. IEEE Transactions on Pattern Analysis and Machine Intelligence 20(8), 832–844 (1998)
15. Krogh, A., Vedelsby, J.: Neural network ensembles, cross validation, and active learning. In: Advances in Neural Inf. Proc. Systems, pp. 231–238. MIT Press, Cambridge (1995)
16. Król, D., Lasota, T., Trawiński, B., Trawiński, K.: Investigation of Evolutionary Optimization Methods of TSK Fuzzy Model for Real Estate Appraisal. International Journal of Hybrid Intelligent Systems 5(3), 111–128 (2008)
17. Kuncheva, L.I.: Combining Pattern Classifiers: Methods and Algorithms. Wiley, Chichester (2004)
18. Lasota, T., Mazurkiewicz, J., Trawiński, B., Trawiński, K.: Comparison of Data Driven Models for the Validation of Residential Premises Using KEEL. International Journal of Hybrid Intelligent Systems (2009) (in press)
19. Liu, Y., Yao, X.: Ensemble learning via negative correlation. Neural Net. 12, 1399–1404 (1999)
20. Margineantu, D.D., Dietterich, T.G.: Pruning Adaptive Boosting. In: Proc. 14th Int. Conf. Machine Learning, pp. 211–218 (1997)
21. Opitz, D., Shavlik, J.W.: Actively searching for an effective neural network ensemble. Connection Science 8(3-4), 337–353 (1996)
22. Polikar, R.: Ensemble Learning. Scholarpedia 4(1), 2776 (2009)

23. Prodromidis, A.L., Chan, P.K., Stolfo, S.J.: Meta-Learning in a Distributed Data Mining System: Issues and Approaches. In: Kargupta, H., Chan, P.K. (eds.) Advances of Distributed Data Mining. AAAI Press, Menlo Park (2000)
24. Rokach, L.: Taxonomy for characterizing ensemble methods in classification tasks: A review and annotated bibliography. Comp. Stat. and Data Anal. 53, 4046–4072 (2009)
25. Schapire, R.E.: The Strength of Weak Learnability. Mach. Learning 5(2), 197–227 (1990)
26. Witten, I.H., Frank, E.: Data Mining: Practical machine learning tools and techniques, 2nd edn. Morgan Kaufmann, San Francisco (2005)
27. Wolpert, D.H.: Stacked Generalization. Neural Networks 5(2), 241–259 (1992)
28. Yao, X., Liu, Y.: Making Use of Population Information in Evolutionary Artificial Neural Networks. IEEE Trans. Systems, Man, and Cybernetics, Part B 28(3), 417–425 (1998)
29. Zhou, Z.H., Wu, J., Tang, W.: Ensembling Neural Networks: Many Could Be Better Than All. Artificial Intelligence 137, 239–263 (2002)

A Three-Scan Algorithm to Mine High On-Shelf Utility Itemsets

Guo-Cheng Lan[1], Tzung-Pei Hong[2,3], and Vincent S. Tseng[1,4]

[1] Department of Computer Science and Information Engineering,
National Cheng Kung University, Tainan, Taiwan
[2] Department of Computer Science and Information Engineering,
National University of Kaohsiung, Kaohsiung, Taiwan
[3] Department of Computer Science and Engineering,
National Sun Yat-sen University, Kaohsiung, Taiwan
[4] Institute of Medical Informatics,
National Cheng-Kung University, Tainan, Taiwan
rrfoheiay@idb.csie.ncku.edu.tw, tphong@nuk.edu.tw,
tsengsm@mail.ncku.edu.tw

Abstract. In this paper, we handle a new kind of patterns named high on-shelf utility itemsets, which considers not only individual profit and quantity of each item in a transaction but also common on-shelf time periods of a product combination. We propose a three-scan mining approach to effectively and efficiently discover high on-shelf utility itemsets. The proposed approach adopts an itemset-generation mechanism to prune redundant candidates early and to systematically check the itemsets from transactions. The experimental results on synthetic datasets also show the proposed approach has a good performance.

Keywords: data mining, utility mining, high utility itemsets, on-shelf data.

1 Introduction

Mining association rules [1] is an important issue in the field of data mining due to its wide applications. Agrawal *et al.* first proposed the most well-known algorithm, namely Apriori, for mining association rules from a transaction database [1]. Traditional association rules are, however, derived from frequent itemsets, which only consider the occurrence of items but do not reflect any other factors, such as prices or profits. The actual significance of an itemset cannot be easily recognized in this way since all the items in a database are assumed to have the same significance. Chan *et al.* thus proposed the utility mining to solve the problem [3]. They considered both individual profits and quantities of products (items) in transactions, and used them to find out actual utility values of itemsets. The high utility itemsets, which had their utility values larger than or equal to a predefined threshold, were then found as the desired. Several other researches about utility mining were proposed in these years, most of which emphasized on how to efficiently find out the high utility itemsets from databases [5][6][10][12][13][14].

N.T. Nguyen, M.T. Le, and J. Świątek (Eds.): ACIIDS 2010, Part II, LNAI 5991, pp. 351–358, 2010.

Temporal data mining has emerged and attracted much attention in these years because of its practicality [2][4][8][9][11]. For example, assume there is an association rule like "In the winter, customers usually purchase overcoats and stockings together". The itemset {overcoats, stockings} may be not frequent throughout the entire database, but may be with a high frequency in the winter. Mining time-related knowledge is thus interesting and useful.

In this paper, we thus handle the mining of a new kind of patterns, called high on-shelf utility itemsets, which consider not only individual profits and quantities of products in transactions, but also actual on-shelf time periods of products. The whole time interval to be analyzed is split into several time periods, and any itemset with a high utility value within the union of all its on-shelf time periods is thought of as a high on-shelf utility itemset. But utility mining is usually much more complicated than traditional association-rule mining. We thus propose a three-scan mining algorithm for effectively and efficiently finding high on-shelf utility itemsets from a database.

The remaining parts of this paper are organized as follows. The related works to utility mining are reviewed in Section 2. The table required in the proposed approach is described in Section 3. The proposed three-scan mining algorithm for finding high on-shelf utility itemsets from a database is stated in Section 4. The experimental evaluation is shown in Section 5. Conclusions and future works are finally given in Section 6.

2   Review of Utility Mining

In this section, some related researches of utility mining are briefly reviewed. In association-rule mining, only binary itemsets are considered. In real-world applications, however, products bought in transactions may contain profits and quantities. Especially, some high-profit products may occur with low frequencies in a database. For example, jewel and diamond are high utility items but may not be frequent when compared to food or drink in a database. The high-profit but low-frequency itemsets may not be found in traditional association-rule mining approaches. Chan *et al.* thus proposed utility mining to discover high utility itemsets [3]. Utility mining considered not only the quantities of the items in a product combination but also their profits. Formally, local transaction utility and external utility are used to measure the utility of an item. The local transaction utility of an item is directly obtained from the information stored in a transaction dataset, like the quantity of the item sold in a transaction. The external utility of an item is given by users, like its profit. External utility often reflects user preference and can be represented by a utility table or a utility function. By using a transaction dataset and a utility table together, the discovered itemset will better match a user's expectation than by only considering the transaction dataset itself.

Traditional association-rule mining keeps the downward-closure property, but utility mining does not. Therefore, utility mining is much harder than traditional association-rule mining. Liu et al. proposed a two-phase algorithm to discover high utility itemsets from a database by adopting the downward-closure property [10]. They named their downward-closure property as the transaction-weighted-utilization (TWU) model. It used the whole utility of a transaction as the upper bound of an itemset in that transaction to keep the downward-closure property. It consisted of two phases. In phase 1, the model found out possible candidate itemsets from a database. In phase 2, the database was rescanned again to update the actual utility of the possible candidate itemsets and found the ones with their actual utility values larger than or equal to a predefined threshold (called the minimum utility threshold). Several other algorithms for utility mining were also proposed [6][13][14], and some related studies are still in progress [5][12].

There has been little study for discovering high on-shelf utility itemsets in databases. This motivates our exploration of the issue of efficiently mining high on-shelf utility itemsets in databases.

3 The Table Required

In this section, the table used in the proposed algorithm is introduced. It is called On-Shelf (OS) table, used for increasing the execution efficiency.

In this paper, we assume the information about whether an item is on shelf of a store within a time period is known. The OS table is used to keep this information for all items. For example, assume there are three time periods and four products for sale in a store. Table 1 shows the on-shelf information of the four products A, B, C and D in the periods, where '1' represents "on shelf" and '0' represents "off shelf".

Table 1. An example of the OS table

Item \ Period	t_1	t_2	t_3
A	1	1	1
B	1	1	0
C	1	1	1
D	0	1	1

It is very easy to extend the OS table from an item to an itemset. The AND operation can be used to achieve the purpose. For example, the on-shelf and the off-shelf time periods of products A, B and D are represented by the bit strings as (1, 1, 1), (1, 1, 0) and (0, 1, 1), respectively. By the AND operation on them, the common sale periods for the itemset {ABD} on the shelf at the store is (0, 1, 0). It represents the time period for all the three products on the shelf is only t_2.

4 The Proposed Mining Algorithm

In this section, the proposed three-scan mining algorithm for discovering high on-shelf utility itemsets in a database is described. A high on-shelf utility itemset is the one with its sum of utilities in all on-shelf periods larger than or equal to a threshold. Besides, the PTTU table and the filtration mechanism are used to help the execution of the algorithm. The PTTU table is first described below.

4.1 The PTTU Table

The PTTU (Periodical Total Transaction Utility) table is designed in the algorithm to increase the execution efficiency. Let the transaction utility of a transaction be the sum of the profits of the items contained in it multiplied by their quantities. An entry in the table records the periodical total transaction utility of all the transactions occurring within a time period. For example, assume there are four transactions which all occur at the time period t_3. Also assume the transaction utility values of the transactions are 15, 18, 16 and 16, respectively. The sum of the transaction utilities for the four transactions within the time period t_3 is thus $15 + 18 + 16 + 16$, which is 65. It is then filled into the corresponding entry ($pttu_3$) of the PTTU table. An example of the PTTU table is shown in Table 2.

Table 2. An example of the PTTU table

Period	Periodical Total Transaction Utility
t_1	104
t_2	48
t_3	65

4.2 The Filtration Mechanism

To reduce the number of unnecessary itemsets during generating candidate itemsets, a filtration mechanism is designed to avoid generating unnecessary utility itemsets based on the high-transaction-weighted-utilization (*HTWU*) 2-itemsets. The mechanism is explained by the following example.

Example 1. Assume that a transaction T in a certain time period includes four items, $3A$, $2B$, $25C$ and $3D$, where the numbers represent the quantities of the items. Also assume their profit values are 3, 10, 1 and 6, respectively. Besides, suppose {AB}, {BC} and {CD} are three high-transaction-weighted-utilization 2-itemsets, which have been found. Fig. 1 shows the process of generating itemsets by using the filtration mechanism.

In Fig. 1, the proposed algorithm first fetches the first item A in the transaction T and allocates it to the first row of the two-dimensional vector array. The algorithm then fetches the second item B in T and allocates it to the second row of the vector array. Since only the item A lies in front of the fetched item B, the algorithm then checks whether items A and B have high-transaction-weighted-utilization relationship. In this example, {AB} is a high-transaction-weighted-utilization 2-itemset. It is thus generated and put into the back of {A} in the first row because the first item in {AB} is A. The algorithm then continues to fetch the third item C and performs the same process. It first puts {C} in the third row of the vector array. It then forms {AC} and checks whether {AC} is a high-transaction-weighted-utilization 2-itemset. In the example, {AC} is not, such that no combination of a subset in the first row with {C} is necessary. The algorithm then forms {BC} from the second row and finds it is a high-transaction-weighted-utilization 2-itemset. {BC} is thus put in the back of {B} in the second row. Two new

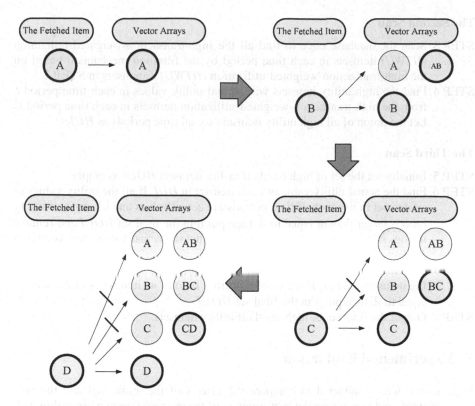

Fig. 1. The whole process of generating itemsets by using the filtration mechanism

subsets {BC} and {C} are generated for the third item. Similarly, the fourth item D is fetched and the above process is repeated again. The two new subsets {CD} and {D} are obtained and put into the vector array.

4.3 The Three-Scan Algorithm for Finding High On-Shelf Utility Itemsets

Based on the above mechanism and data structure, the proposed mining algorithm is stated as follows.

INPUT: A transaction database D with n transactions, each of which consists of transaction identification, transaction occurring time and items purchased, m
items in D, each with a profit value, an on-shelf (OS) table of items with k
desired time periods, and a minimum utility threshold λ.

OUTPUT: The set of high on-shelf utility itemsets ($HOUI$).

The First Scan

STEP 1: Initialize the $PTTU$ (Periodical Total Transaction Utility) table as the zero
table.

STEP 2: Scan the database once to construct the $PTTU$ table and to find the high-
transaction-weighted-utilization 2-itemsets in each time period.

The Second Scan

STEP 3: Scan the database once to find all the high-transaction-weighted-utilization (*HTWU*) itemsets in each time period by the filtration mechanism based on the high-transaction-weighted-utilization *(HTWU)* 2-itemsets in STEP 2.

STEP 4: Find the high utility itemsets with actual utility values in each time period t_j from the high-transaction-weighted-utilization itemsets in each time period t_j. Let the union of all high utility itemsets for all time periods as *HUI*.

The Third Scan

STEP 5: Initially set the set of high on-shelf utility itemsets *HOUI* as empty.

STEP 6: Find the actual utility value of each itemset in *HUI*. If all the utility values of an itemset in its on-shelf time periods are actual values and its on-shelf utility ratio is larger than or equal to λ, then put it in the final set *HOUI* and remove it from *HUI*; If one of its utility values is not the actual value, then keep it in *HUI*.

STEP 7: Scan the database again to find the actual utility value of each of the remaining itemsets in *HUI*. If the on-shelf utility ratio of an itemset is larger than or equal to λ, then put it in the final set *HOUI*.

STEP 8: Output the final set of high on-shelf utility itemsets.

5 Experimental Evaluation

Experiments were conducted to compare the effects of the traditional and the proposed patterns and to evaluate the performance of the proposed mining algorithm with different parameter values. The experiments were implemented in J2SDK 1.5.0 and executed on a PC with 3.0 GHz CPU and 1GB memory. The datasets in the experiments were generated by the IBM data generator [7]. Since our objective is to discover high on-shelf utility itemsets, we also develop a simulation model which is similar to the model used in [10].

Experiments were first made on the synthetic T10I4N4KD200K dataset to evaluate the difference between the high utility itemsets with and without considering common on-shelf time periods. The results were shown in Fig. 2, where the proposed high on-shelf utility itemsets were represented as *HOUI* and the traditional high utility itemsets were represented as *HUI*. Besides, *HOUI_P5* represented the proposed approach was executed with 5 time periods.

It could be observed from the figure that the number of high utility itemsets was always less that of high on-shelf utility itemsets. It could be easily explained as follows. A high on-shelf utility itemset always had the same utility value as its corresponding traditional high utility itemset, but the former considered a less number of transactions than the latter. A high on-shelf utility itemset was thus easier to satisfy the minimum utility threshold than its corresponding traditional high utility itemset. The traditional high utility itemsets were thus included in the set of the proposed high on-shelf utility itemsets under the same parameter settings.

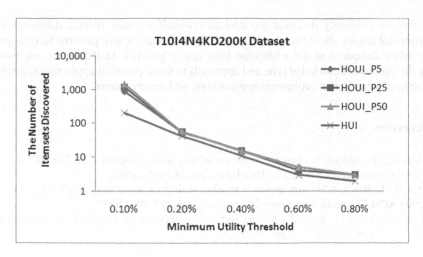

Fig. 2. The numbers of high on-shelf utility itemsets under different utility thresholds

Experiment was then made to evaluate the efficiency of the proposed mining algorithm. Fig. 3 showed the execution time on the T10I4N4KD200K dataset for different thresholds varying from 0.10% to 0.80%.

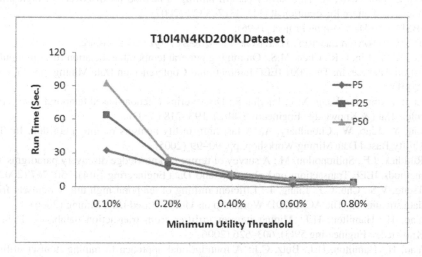

Fig. 3. The execution time of the proposed algorithm under different thresholds

It could be easily observed that the execution time of the proposed mining algorithm increased nearly linearly with the value of the threshold in this figure.

6 Conclusion

In this paper, we have proposed the high on-shelf utility itemsets, which consider the common on-shelf time periods for items. We have also proposed the three-scan mining algorithm to efficiently discover the desired on-shelf itemsets from a database. The experimental results show that the proposed high on-shelf utility patterns have a good effect when compared to the traditional high utility patterns. In the future, we would apply the proposed knowledge type and approach to some practical applications, such as data stream, supermarket promotion applications, and among others.

References

1. Agrawal, R., Srikant, R.: Fast algorithm for mining association rules. In: The 20th International Conference on Very Large Data Bases, pp. 487–499 (1994)
2. Ale, J.M., Rossi, G.H.: An approach to discovering temporal association rules. In: The 2000 ACM Symposium on Applied Computing, pp. 294–300 (2000)
3. Chan, R., Yang, Q., Shen, Y.: Mining high utility itemsets. In: The 3rd IEEE International Conference on Data Mining, pp. 19–26 (2003)
4. Chang, C.Y., Chen, M.S., Lee, C.H.: Mining general temporal association rules for items with different exhibition periods. In: The 3rd IEEE International Conference on Data Mining, pp. 59–66 (2002)
5. Chu, C.J., Tseng, V.S., Liang, T.: Mining temporal rare utility itemsets in large databases using relative utility thresholds. International Journal of Innovative Computing, Information and Control 4(11), 2775–2792 (2008)
6. Hu, J., Mojsilovic, A.: High-utility pattern mining: a method for discovery of high-utility item sets. Pattern Recognition 40(11), 3317–3324 (2007)
7. IBM Quest Data Mining Project (1996),
   http://www.almaden.ibm.com/cs/quest/syndata.html
8. Lee, C.H., Lin, C.R., Chen, M.S.: On mining general temporal association rules in a publication database. In: The 2001 IEEE International Conference on Data Mining, pp. 337–344 (2001)
9. Li, Y., Ning, P., Wang, X.S., Jajodia, S.: Discovering calendar-based temporal association rules. Data & Knowledge Engineering 44(2), 193–218 (2003)
10. Liu, Y., Liao, W., Choudhary, A.: A fast high utility itemsets mining algorithm. In: The Utility-Based Data Mining Workshop, pp. 90–99 (2005)
11. Roddick, J.F., Spiliopoilou, M.: A survey of temporal knowledge discovery paradigms and methods. IEEE Transactions on Knowledge and Data Engineering 14(4), 750–767 (2002)
12. Tseng, V.S., Chu, C.J., Liang, T.: Efficient mining of temporal high utility itemsets from data streams. In: The ACM KDD Workshop on Utility-Based Data Mining (2006)
13. Yao, H., Hamilton, H.J.: Mining itemset utilities from transaction databases. Data & Knowledge Engineering 59(3), 603–626 (2006)
14. Yao, H., Hamilton, H.J., Butz, C.J.: A foundational approach to mining itemset utilities from databases. In: The 4th SIAM International Conference on Data Mining, pp. 482–486 (2004)

Incremental Prediction for Sequential Data

Tomasz Kajdanowicz and Przemysław Kazienko

Wrocław University of Technology, Wyb. Wyspiańskiego 27, 50-370 Wrocław, Poland
{tomasz.kajdanowicz,kazienko}@pwr.wroc.pl

Abstract. Ensemble methods of incremental prediction for sequences refer to learning from new reference data that become available after the model has already been created from a previously available data set. The main obstacle in the prediction of sequential values in the real environment with huge amount of data is the integration of knowledge stored in the previously obtained models and the new knowledge derived from the incrementally acquired new increases of the data. In the paper, the new approach of the ensemble incremental learning for prediction of sequences was proposed as well as examined using real debt recovery data

Keywords: incremental prediction, ensemble prediction, prediction, sequential data, claim appraisal.

1 Introduction

The major trend of machine learning research has been focused on algorithms that can be trained based on the training data, in which cases are characterized by the fixed, unknown distribution. Assumption of the steady distribution was even applied in online and incremental learning, i.e. various forms of neural networks and other statistical approaches while incrementing are based on the idea that data distribution is constant. The problem of learning in changeable, non-stationary environment, where distribution of data changes over time, has received less attention. As more and more practical problems required sophisticated learning techniques such as spam detection, card fraud detection, etc., the non-stationary environment based learning methods have received increasing attention. It is getting even more complicated while the output of the process should reveal in prediction in sequence of values. Hence, the sequence prediction is closely related with the field of game theory, information theory and recently machine learning e.g. [1, 3, 7].

One of the research directions of incremental learning in non-stationary environment are techniques based on ensemble of classifiers. As stated in [13], multiple classifier systems usually use more than one classifier to reproduce changes in distribution of data. While dealing with many classifiers, especially in non-stationary environment, it is important to provide intelligent techniques for their fusion. Among others, there exist several approaches for this purpose such as incremental boosting [9, 12], selection of the most informative training sample [5] or incremental construction of support vector machines [14, 15].

In general, in the literature the taxonomy of ensemble based approaches for changing environments has been proposed in three general categories: stable ensemble with

N.T. Nguyen, M.T. Le, and J. Świątek (Eds.): ACIIDS 2010, Part II, LNAI 5991, pp. 359–367, 2010.

changing combination rules, updating online ensemble with stable combination technique and ensemble based on new members after new data became available [10].

In the paper a new method for incremental learning is presented. It consists of hybrid combination of two different approaches: new data (increases) is used to derive additional ensemble's member and according to the incremental steps combination method is aligned. The method does not refer to the previously learned data to ensure that the algorithm is truly incremental. Moreover, the algorithm does not discard any of previously generated models.

2   Problem Description

Systems presented in the following section solve three specific problems in prediction: incremental profile of learning data, prediction of sequential values and adjustment of references to the input, predicted (test) data.

We assume that the learning, reference data set is not given in advance and some new data portions (increases) are periodically provided. Since the newer data contain more valid and more reliable information, they should have greater influence on the output values. This idea is hardly implementable within the typical predictors.

The prediction of sequences means that for the single learning or test case we obtain a sequence of output values. It happens for example in debt recovery appraisal: for each valuated debt, we want to have the sequence of possible repayments in the following periods (months) [6, 7]. Furthermore, for the first items in the sequence we have more information than for the last ones – repayment history of many learning cases is not long enough.

The third problem is the wide heterogeneity of the learning data – its domain is often wider than the area of the test data. It implies the discovery and application of competence regions, i.e. separation of more specific subareas within the entire reference domain [2, 11].

3   Concept of Incremental Ensemble Prediction of Sequences

The general idea of incremental learning is to build separate predictors for each new data increase without need to access to the data from the previous increases, Fig. 1. In is of great importance in case of large amounts of data incoming in separate batches. Besides, the incremental prediction is performed for the sequence of values, or more precise, separately for each item in the sequence. In addition, the output of prediction for the previous items is utilized while processing the following items in the sequence.

Let us consider the process for the single learning data set k. First, the prepared historical data k is clustered into groups of similar cases – competence regions. Next, a set of models is created (learnt) separately for each cluster jk (jth for the kth increase) using the fixed set of common, predefined models, e.g. logistic regression, decision tree, bagging, etc. Subsequently, the best model is selected for each cluster jk and becomes the cluster's predictive model P_{ijk} for item i in the sequence, upon data set k. This assignment is done based on minimization of the standard deviation error. This

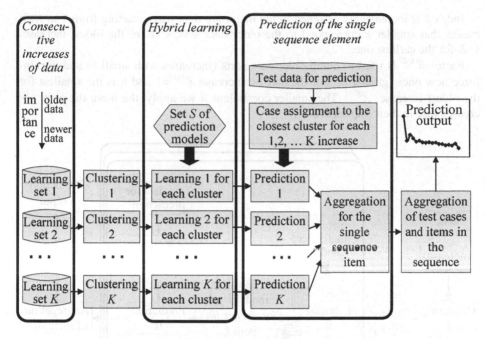

Fig. 1. The concept of incremental prediction of sequences

process is repeated independently for each new incoming learning data set (increase of data). Finally, we have K separate cluster sets and related with them best predictive models, Fig. 1.

This main learning phase is followed by the prediction stage, which in turn is performed separately for each item in the sequence. For each input predictive case, the closest cluster of historical data is determined and the prediction for this case is accomplished based on the model trained and assigned to that cluster. Since we have K sets of clusters related to incremental learning, we need to find K closest clusters and perform K predictions, one for each of K data increases. To aggregate these K prediction values the following formula is used:

$$p_i(x) = \frac{\sum_{k=1}^{K} p_{ik}(x)\lambda^{K-k}}{\sum_{k=1}^{K} \lambda^{K-k}}, \tag{1}$$

where

$p_i(x)$ – aggregated predicted value for test case x related to item i in the sequence;
$p_{ik}(x)$ – value predicted by predictor ik for test case x related to increase k and item i;
$\lambda \in [0;1]$ – the forgetting factor, the forgetting is stronger for lower values of λ;
K – total number of increment data parts (increases).

Index k is increasingly assigned to the incoming data parts, starting from 1 to K. It means that smaller k correspond to the older data sets; $k=1$ for the oldest increase, $k=K$ for the earliest one.

Factor $\lambda^{(K-k)}$ is used to dump older data parts (increases with small k) and to reinforce new ones (greater k). For the earliest increase $\lambda^{(K-k)}=1$ and it is the smallest for the oldest increase: $\lambda^{(K-1)}$. The smaller coefficient λ we apply, the more the older increases lose in their importance.

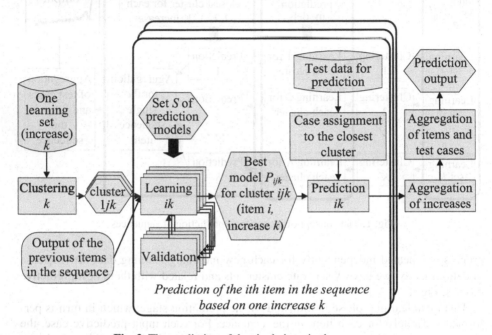

Fig. 2. The prediction of the single item in the sequence

Note that the above description concerns prediction for only one item i in the sequence, Fig. 2. To achieve the result for the entire sequence (M items), we need to maintain separate predictive models P_{ijk} for each such item i. The important characteristic of the proposed method is that the predicted value for item i is taken as the input variable to predict values for all the following items in the sequence, i.e. item $i+1$, $i+2$, ..., K, see Fig. 3.

Historical reference cases from increase k are clustered into N_k^G groups using partitioning method. For that reason, we achieve the following clusters: G_{11k}, G_{12k}, ..., $G_{1N_k^G k}$ for increase k, see also Fig. 2. Features directly available within the input data set are the only used in the clustering process. Besides, clustering is performed for the whole reference increase k, i.e. for learning cases having at least one item in the sequence of predictive variable. For the following items, e.g. for period i, cases with too short history (with the history shorter than i items), are just removed from their clusters without re-clustering. As a result, the quantity of each cluster G_{ijk} may vary

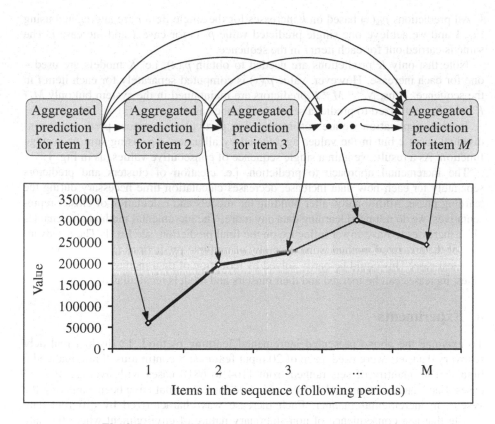

Fig. 3. Input variable dependency in prediction of sequences

depending on item i and it is smaller for greater i. For the ith item and the jth group G_{ijk}, we have: $card(G_{ijk}) \geq card(G_{(i+1)jk})$. In consequence, there are the same reference groups for all items but their content usually decreases for the following items. This happens in debt recovery valuation, because the debt collection company possesses many pending recovery cases, which can be used as references in prediction only for the beginning periods. If the quantity of one cluster for the greater period is too small then this cluster is merged with another, closest one for all following periods.

Next, the best prediction model P_{ijk} is separately assigned to each group G_{ijk} for item i. It means the more models if there are more clusters (N_k^G) and more items in the sequence (M); for the single increase we maintain $N_k^G * M$ models. Each of them is selected from a set S of considered models, Fig. 2. Hence, $N_k^G * M$ models are selected from total $N_k^G * M * card(S)$ models generated for a single increase.

Each group G_{ijk} possesses its own, separate representation and the common similarity function is used to evaluate closeness between group G_{ijk} and each input case x just being predicted. Next, the single closest group, or more precise the assigned model, is applied to the input case x. As a result, we obtain predicted value $p_{ik}(x)$ for increase k and item i, see Eq. 1. This process of clustering, modeling, best model selection, cluster assignment for case x is repeated separately for each learning data part – increase *k*. All predictions $p_{ik}(x)$ based on *k* increases for the single item *i* are aggregated using Eq. 1 and we achieve one single predicted value $p_i(x)$ for case *x* and increase *i*. The same is carried out for each item *i* in the sequence.

Note that only *K* predictions are needed to obtain $p_i(x)$, i.e. *K* models are used – one for each increase. However, value $p_i(x)$ is computed separately for each item *i* in the sequence. Thus, $N_k^G * M * K$ predictors are maintained in the system but only $M * K$ models are utilized to predict a sequence for one case *x*.

In some application we are not interested in particular sequences predicted for individual cases *x* but in the values aggregated by all cases, e.g. using sum or average function. As a result, we gain a single sequence of consecutive values, as in Fig. 1.

The incremental approach to prediction, i.e. creation of clusters and predictors separately for each new data increase, decreases calculation time necessary during the learning phase. Additionally, after building the models and calculation of cluster representatives, we do not need learning data any more. The incremental method also enables older increases to have lower influence on the final prediction, see Eq. 1. The disadvantage of the presented method is the necessity of multiple predictions (K predictions) for each test case. This could be partly solved by reduction of data increases, especially the oldest increases can be merged and their clusters and models recalculated.

4 Experiments

To examine the above presented incremental learning method, 12 distinct real debt recovery data sets were used, each of 20 input features: 5 continuous, 9 nominal and 6 binary. The quantity of sets ranged from 1104 to 6818 cases with average of 3723 cases. Each data set consisted of 6 equal in size increases that have been supplying the system in incremental manner. Each increase was characterized by different data profile that is a consequence of non-stationary nature of environment where the data is produced. The following increases have greater importance in the prediction process, in accordance with Eq. 1. The data for each increase was separately clustered and the average number of clusters was 15.2.

According to the process presented in Sec. 3 (Fig. 1), after each increase has become available, the next incremental learning step is performed. Each case from all data sets required to be predicted in the whole sequence. Besides, each predictor of the ensemble was learnt and evaluated using 5 cross-fold validations [4]. The following 6 predictors were used (set S, Fig. 2): two M5P trees (with two different parameters), logistic regression, Decision Stump, and two baggings (with two different parameters).

The main goal of the experiments was to evaluate and compare the proposed incremental method against the standard learning method that realizes prediction every time based on the whole data set. The main results of the experiments are presented in Tab. 1. The computation times have also been monitored; they are shown in Fig. 5. Some other studies on the influence of clustering of the learning set onto prediction accuracy can be found in [6, 7].

The research was implemented and conducted within the R statistical computing environment with some extended and customized algorithms based on RWeka, rJava and tree plug-ins. In the experiment, the debt recovery value predictions have been conducted for sequences of 10 items – 10 consecutive periods (months). As a result, for all data sets, 380 160 predictors were created and exploited.

Table 1. Error results of the standard and incremental learning method comparison

Data set	Standard method	Incremental method			
		λ			
		0.2	0.5	0.7	0.9
1	27.90%	36.73%	36.78%	36.52%	36.08%
2	33.87%	43.04%	42.03%	40.55%	38.73%
3	2.02%	2.64%	2.59%	2.54%	2.47%
4	7.27%	5.28%	4.70%	4.04%	3.43%
5	34.47%	44.02%	44.62%	45.08%	45.51%
6	11.35%	17.22%	16.23%	15.41%	14.62%
7	19.98%	20.36%	18.64%	17.22%	15.90%
8	18.98%	15.71%	14.41%	13.57%	12.96%
9	14.11%	11.74%	10.28%	9.28%	8.42%
10	18.67%	14.78%	16.67%	18.14%	19.74%
11	16.10%	13.12%	12.92%	12.59%	12.04%
12	0.81%	4.18%	3.87%	3.40%	2.50%
Average method error	17.13%	19.07%	18.65%	18.19%	17.70%

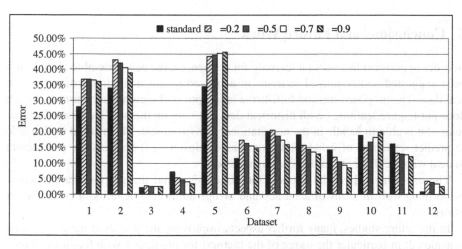

Fig. 4. Prediction error for the standard and incremental learning method with distinct forgetting factor λ values

As presented in Tab. 1 and Fig. 4, the proposed incremental method performing the prediction based only on the knowledge already stored in models from the past data increases, provides reasonably good results in comparison with the standard method. By adjustment of forgetting factor λ in the way that the current learning data is taking most responsibility in the prediction process, the average error of the method is reduced to the level of standard approach.

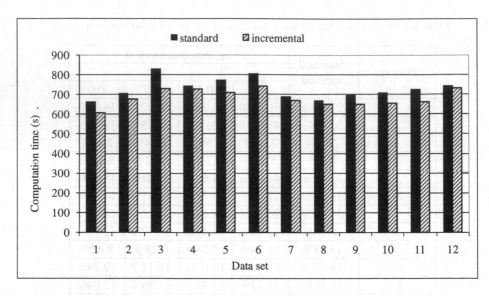

Fig. 5. Average computation time for the standard and incremental learning method

The incremental learning method is able to save from 3% up to 10% of computation time compared to the standard method, Fig. 5.

5 Conclusions and Future Work

In order to perform incremental learning for sequences, the new ensemble method has been suggested and compared in terms of achieved results and computation time on real data. The proposed method fulfills the assumption of incremental learning, which does not require learning on all historical data but only on the new increases. For that reason, it is able to handle in the non-stationary environment. The method is also able to predict sequences of values and by means of clustering it contains the mechanism to deal with the diversity in the learning data sets.

The experimental results support the conclusion that the proposed approach is performing well in terms of accuracy and brings additional benefits in computation effort.

In the future studies, many further aspects improving the proposed method will be considered; in particular the usage of the method for prediction with feedback spread in time in order to tune the existing ensemble.

The application of the similar concept is also considered to be applied for prediction in social-based systems [8].

Acknowledgments. This work was supported by The Polish Ministry of Science and Higher Education, the development project, 2009-11.

References

1. Aburto, L., Weber, R.: A Sequential Hybrid Forecasting System for Demand Prediction. In: Perner, P. (ed.) MLDM 2007. LNCS (LNAI), vol. 4571, pp. 518–532. Springer, Heidelberg (2007)
2. Bishop, C.M.: Pattern Recognition and Machine Learning. Springer Science+Business Media, New York (2006)
3. Dekel, O., Shalev-Shwartz, S., Singer, Y.: Individual Sequence Prediction using Memory-efficient Context Trees. IEEE Transactions on Information Theory 55(11), 5251–5262 (2009)
4. Dietterich, T.G.: Approximate statistical tests for comparing supervised classification learning algorithms. Neural Computation 10(7), 1895–1923 (1998)
5. Engelbrecht, A.P., Brits, R.: A Clustering Approach to Incremental Learning for Feedforward Neural Networks. In: Proceeding of the International Joint Conference on Neural Networks IJCNN 2001, Washington DC, USA, pp. 2019–2024 (2001)
6. Kajdanowicz, T., Kazienko, P.: Hybrid Repayment Prediction for Debt Portfolio. In: Nguyen, N.T., Kowalczyk, R., Chen, S.-M. (eds.) ICCCI 2009. LNCS, vol. 5796, pp. 850–857. Springer, Heidelberg (2009)
7. Kajdanowicz, T., Kazienko, P.: Prediction of Sequential Values for Debt Recovery. In: Bayro-Corrochan, E. (ed.) CIARP 2009. LNCS, vol. 5856, pp. 337–344. Springer, Heidelberg (2009)
8. Kazienko, P., Musiał, K.: Multidimensional Social Network and Its Application to the Social Recommender System. IEEE Transactions on Systems, Man, and Cybernetics - Part A: Systems and Humans (2009) (in press)
9. Kidera, T., Ozawa, S., Abe, S.: An Incremental Learning Algorithm of Ensemble Classifier Systems. In: Proceeding of the International Joint Conference on Neural Networks, IJCNN 2006, Vancouver, Canada, pp. 3421–3427 (2006)
10. Kuncheva, L.I.: Classifier ensembles for changing environments. In: Roli, F., Kittler, J., Windeatt, T. (eds.) MCS 2004. LNCS, vol. 3077, pp. 1–15. Springer, Heidelberg (2004)
11. Kuncheva, L.: Combining Pattern Classifiers. Methods and Algorithms. John Wiley & Sons, Inc., Chichester (2004)
12. Mohammed, H.S., Leander, J., Marbach, M., Polikar, R.: Comparison of Ensemble Techniques for Incremental Learning of New Concept Classes under Hostile Non-stationary Environments. In: IEEE International Conference on Systems, Man and Cybernetics, Taipei, Taiwan, pp. 4838–4844. IEEE Computer Society (2006)
13. Muhlbaier, M., Polikar, R.: Multiple Classifiers Based Incremental Learning Algorithm for Learning in Nonstationary Environments. In: Proceeding of the Sixth International Conference on Machine Learning and Cybernetics, Hong Kong, pp. 3618–3623. IEEE Computer Society, Los Alamitos (2007)
14. Shilton, A., Palaniswami, M., Ralph, D., Tsoi, C.: Incremental training of support vector machine. In: Proceeding of the International Joint Conference on Neural Networks, IJCNN 2001, Washington DC, USA, pp. 1197–1202 (2001)
15. Zhao, Y., He, Q.: An Incremental Learning Algorithm Based on Support Vector Domain Classifier. In: Proceeding of the International Conference on Cognitive Informatics, Lake Tahoe, USA, pp. 805–809 (2006)

Predictive Maintenance with Multi-target Classification Models

Mark Last[1], Alla Sinaiski[1], and Halasya Siva Subramania[2]

[1] Department of Information Systems Engineering,
Ben-Gurion University of the Negev, Beer-Sheva 84105, Israel
mlast@bgu.ac.il, sinaiski@gmail.com
[2] India Science Lab, General Motors Global Research and Development
GM Technical Centre India Pvt Ltd, Creator Building, International Tech Park Ltd.
Whitefield Road, Bangalore - 560 066, India
halasyasiva.subramania@gm.com

Abstract. Unexpected failures occurring in new cars during the warranty period increase the warranty costs of car manufacturers along with harming their brand reputation. A predictive maintenance strategy can reduce the amount of such costly incidents by suggesting the driver to schedule a visit to the dealer once the failure probability within certain time period exceeds a pre-defined threshold. The condition of each subsystem in a car can be monitored onboard vehicle telematics systems, which become increasingly available in modern cars. In this paper, we apply a multi-target probability estimation algorithm (M-IFN) to an integrated database of sensor measurements and warranty claims with the purpose of predicting the probability and the timing of a failure in a given subsystem. The multi-target algorithm performance is compared to a single-target probability estimation algorithm (IFN) and reliability modeling based on Weibull analysis.

Keywords: Predictive Maintenance, Telematics, Fault Prognostics, Vehicle Health Management, Reliability, Multi-Target Classification, Info-Fuzzy Networks.

1 Introduction

Three major maintenance policies include corrective, preventive, and predictive maintenance [12]. *Corrective maintenance* implies that equipment, such as a vehicle, is repaired after a failure has occurred. As long as a vehicle is under a warranty agreement, the vehicle owner does not usually pay for the repair though he or she can suffer a severe inconvenience as a result of an unexpected malfunction and the time wasted on a visit to an authorized car dealer. Most vehicles are also subject to a *preventive maintenance* policy, which requires performing certain inspections and other operations at a schedule predetermined by the car manufacturer, mostly on the basis of mileage and/or time in service. Since these actions are scheduled in advance, the customer has the flexibility to perform them at the time of his or her convenience. This policy is also expected to reduce warranty costs by avoiding claims related to

N.T. Nguyen, M.T. Le, and J. Świątek (Eds.): ACIIDS 2010, Part II, LNAI 5991, pp. 368–377, 2010.

unexpected failures that may occur if the periodic maintenance is not performed on time.

However, the preventive maintenance policy does not take into account the actual condition of the car subsystems as periodic maintenance activities are scheduled at a fixed mileage or time interval. In contrast, *predictive maintenance* (also called Condition Based Maintenance [5]) can schedule a visit to the dealer based on some sensory information representing the current condition of the car and its subsystems. This approach should, on one hand, minimize the risk of unexpected failures, which may occur before the next periodic maintenance operation, and on the other hand, reduce the amount of unnecessary preventive maintenance activities.

Extensive amounts of telematics data are collected nowadays by several car manufacturers. To utilize this sensory information for predictive maintenance we need to identify the features relevant to predicting each type of failure and to find accurate models that will predict the probability of failure as a function of elapsed time (one week, two weeks, etc.). Such time-based probability distribution will assist the customer to determine the urgency of his/her next visit to the dealer. Obviously, if the failure probability until the time of the next periodic maintenance activity is too low, an extra visit to the dealer would be redundant. Thus, the predictive maintenance policy should provide the maximum "peace of mind" for the customer.

Several data mining methods have been used in the past to support the predictive maintenance policy. Thus, the authors of [10] use a hybrid, dynamic Bayesian network to model the temporal behavior of the faults and determine fault probabilities in the Mercedes-Benz E320 sedan handling system. The continuous input to the diagnostic system comes from six sensors located on the vehicle: four wheel speed sensors, a yaw rate gyro, and a steering angle sensor. In the case study of [10], ten faults are considered: a failure of any of the four tires as well as faults of any of the six sensors. In [12], an artificial neural network model is developed to estimate the life percentile and failure times of roller bearings based on the available sensory information. Given the conditional probability distribution provided by the neural network model, the optimal replacement strategies are computed by a cost matrix method. The authors of [1] propose a two-step approach to predictive maintenance. First, a decision-tree algorithm is applied to sensory data in order to identify the relevant features for the condition of the equipment and the associated thresholds (split values of the features selected by the algorithm). Second, a reliability model, such as Weibull distribution, is fitted to the live data of the equipment for predicting the time interval to the next maintenance action. The method is demonstrated on a historical maintenance database of more than 2000 copiers.

In this paper, we apply a multi-target probability estimation algorithm (Multi-Target Info-Fuzzy Network) to an integrated synthetic dataset of sensory data and warranty failure data. The goal is to issue an early warning of a failure expected in an individual car as well as to estimate the approximate timing of the expected failure (e.g., within one week, within two weeks, etc.). Multi-target classification and probability estimation methods are presented in the next section followed by a brief description of Weibull reliability analysis. The results of a case study based on the synthetic data are described next. We conclude the paper with some directions for future applications of multi-target and multi-model data mining methods to the maintenance data.

2 Multi-target Classification

As indicated in [2], the common assumption of most data mining algorithms (decision trees, Naïve Bayes, artificial neural networks, etc.) that a learning task has only one objective is very restrictive. Data objects in many real-world databases may be simultaneously assigned multiple class labels related to multiple tasks (targets). These targets (dimensions) may be strongly related to each other, completely unrelated, or just weakly related. Thus, in case of predictive maintenance of a car, we may be interested to predict the occurrence and the timing of one or several failures in the same subsystem or even in completely different subsystems (e.g., batteries and tires).

The most straightforward approach to the problem of multi-target (sometimes called multi-objective) classification is to induce a separate model for each target using any single-target classification algorithm. Though the resulting models may be the best (e.g., the most accurate) ones for every individual target, the user may find a multi-target model much more comprehensible than a collection of single-target models. Moreover, as demonstrated by Caruana [2] and Last [6], the combination of several classification tasks in a single model may even increase the overall predictive accuracy.

In order to provide a unified framework for single-target and multi-target prediction tasks, Last [6] has defined an *extended classification task* using the following notation:

- $R = (A_1,..., A_k)$ - a set of k attributes in the dataset ($k \geq 2$).
- C - a non-empty subset of n *candidate input features* $(C \subset R,\ |C| = n \geq 1)$. The values of these features are usually known and they can be used to predict the values of target *attributes*. In case of predictive maintenance, these features may represent the sensory data.
- O - a non-empty subset of m target ("output") attributes $(O \subset R,\ |O| = m \geq 1)$. This is a subset of attributes representing the variables to predict, such as failure occurrence, time to failure, *etc.* The extended classification task is to build an accurate model (or models) for predicting the values of *all* target attributes, based on the corresponding *dependency subset* (or subsets) $I \subseteq C$ of *input features*.

The Multi-Target Classification task is different from *Multitask Learning* described by Caruana in [2]. The explicit goal of Multitask Learning is to improve the accuracy of predicting the values of a *single-dimensional class* (defined as the *main* learning task) by training the classification model, such as a neural network or a decision tree, on several *related* tasks (additional class dimensions). This is called *inductive transfer* between learning tasks. As emphasized by [2], the only concern of Multitask Learning is the generalization accuracy of the model, not its intelligibility. In contrast, this paper focuses on multi-target classification rather than on multi-task learning, since predictive maintenance involves two *equally important* prediction tasks: estimating the probability of failure and estimating the time to failure. Multi-target classification is also different from the *multi-objective approach to classification* [4], which is usually aimed at finding the best trade-off between accuracy and interpretability of a single-target classification model.

As shown in [6], an m-target classification function can be represented by a *multi-target info-fuzzy network* (M-IFN), where each terminal node is associated with the probability distributions of *all* target attributes (unlike the *bloomy decision tree* model [11] where the leaf nodes predict only a subset of class dimensions).   The M-IFN model is an extension of an Oblivious Read-Once Decision Graph (OODG) called *information network* (*IN*) [9].   Unlike the standard decision-tree models, the information network uses the same input attribute across all nodes of a given layer (level). The input attributes are selected incrementally by the IN induction algorithm to maximize a global decrease in the conditional entropy of the target attribute.   The IN induction algorithm is using the pre-pruning approach: when no attribute causes a statistically significant decrease in the entropy, the network construction is stopped. In [9], the IN induction algorithm is shown empirically to produce much more compact models than other methods of decision-tree learning, while preserving nearly the same level of classification accuracy.

The Multi-Target Info-Fuzzy Network (M-IFN) induction procedure starts with defining the target layer, which has a node for each category, or value, of every class dimension and the "root" node representing an empty set of input attributes. The M-IFN construction algorithm has only the growing (top-down) phase. The top-down construction is terminated (pre-pruned) by a statistical significance test and, consequently, there is no need in bottom-up post-pruning of the network branches. M-IFN construction is an *iterative* rather than a *recursive* process. At every iteration, the algorithm utilizes the entire set of training instances to choose an input (predicting) feature (from the set of unused "candidate input" features), which maximizes the decrease in the total conditional entropy of *all* class dimensions.

The conditional mutual information of a class dimension Y_i and an input feature X_n given the features $X_1, ..., X_{n-1}$ is calculated by:

$$MI(Y_i; X_n / X_1,...,X_{n-1}) = H(Y_i / X_1,...,X_{n-1}) - H(Y_i / X_1,...,X_n) =$$

$$\sum_{x_1 \in X_1,...,x_n \in X_n, y_i \in Y_i} p(x_1,...,x_n,y_i) \log \frac{p(y_i,x_n / x_1,...,x_{n-1})}{p(y_i / x_1,...,x_{n-1})p(x_n / x_1,...,x_{n-1})} \tag{1}$$

where $H(A/B)$ is the conditional entropy of the variable A given variable(s) B.

At *p-th* iteration, the M-IFN construction algorithm chooses the input feature X_{j*}, which maximizes the sum of information gains over all m class dimensions by finding

$$j* = \arg \max_{j \notin I} \sum_{i=1}^{m} MI(Y_i; X_j / X_1,...,X_{p-1}) \tag{2}$$

where I is the set of input features selected by the algorithm at $p-1$ iterations.

Further details on the M-IFN induction algorithm are provided in [6]. It is also shown in [6] that the resulting M-IFN model has the following information-theoretic properties:

- The *average accuracy* of a multi-target model in predicting the values of m target attributes will usually not be worse, or even will be better, than the average accuracy of m single-target models using the same set of input features.
- If all target attributes are either mutually independent or completely dependent on each other, the M-IFN induction algorithm is expected to produce the optimal (most accurate) model.

The M-IFN induction algorithm has been successfully applied to several real-world multi-target tasks such as prediction of grape and wine quality dimensions [8] and quality control of a complex manufacturing process [7].

3 Weibull Analysis

According to the Weibull reliability analysis, the probability of a component failure (such as a battery failure) up to some point in time can be calculated from the car age (in months in service) or, alternatively, from its mileage using the following formula:

$$F(x) = 1 - e^{-\left[\frac{x}{\alpha}\right]^{\beta}}$$

(3)

where β is called the shape parameter and α is the scale parameter. The x variable can represent either mileage or months in service.

Weibull distribution is capable of modeling Decreasing Failure Rate ($\beta < 1$), Constant Failure Rate ($\beta = 1$) and Increasing Failure Rate ($\beta > 1$) behavior. For new cars, we would expect the shape parameter to be smaller than one, indicating the "early-life" behavior, which models the first part of the well-known "bathtub curve".

Once the parameters of the Weibull distribution are known, the probability of failure over any period of time starting at the point x_1 and ending at the point x_2 can be calculated as $F(x_2) - F(x_1)$.

4 Case Study: Battery Failure Prediction

4.1 Data Description

We have applied the single-target and the multi-target Info-Fuzzy Network algorithms (IFN and M-IFN) to an integrated synthetic dataset of sensory and warranty data simulated for some high-end car model (with high electrical/electronic content) over a period of several months. In this case study, the focus is on battery (Lead Acid SLI batteries) failure prediction in each individual vehicle.

We have used the following definitions of the target attributes:

- *Battery Failure* – 1 if there was a failure after measurement (Repair Operation = Battery Recharge or Battery Replace), 0 - otherwise.
- *Elapsed Time* – the number of days elapsed between the data collection date (CREATED_TIMESTAMP) and the claim date (REPAIR_DATE). The value of this attribute (in months) was discretized to the following intervals: 0 (no failure), 0.001-0.5, 0.5-1, 1-2, 2+.

The generated dataset includes 46,418 records representing periodical battery sensory readings for 21,814 distinct vehicles. We have simulated the failure distribution based on battery warranty trends in real vehicles, resulting in 394 simulated battery failures, which is only 0.8% of the total number of readings. Thus, we are facing here a prediction problem characterized by extremely imbalanced classes.

The use of each data attribute in the prediction process is presented in Table 1 below.

Table 1. List of Attributes

Name	Use
OCV (open circuit voltage)	Candidate Input
Temp (temperature)	Candidate Input
SOC (state-of-charge)	Candidate Input
Load (amp-hours during ignition OFF)	Candidate Input
OASAH (off asleep amp hours)	Candidate Input
RSOC (run state-of-charge)	Candidate Input
Battery Failure	Target
SLOPE_VBAT (rate of battery voltage drop)	Candidate Input
Odometer	Candidate Input
Age (diff bet. sale date and read date)	Candidate Input
MPD (Average mileage per day)	Candidate Input
Elapsed Time (diff. bet. read date and repair date)	Target

4.2 Performance Measures

Each prediction rule produced by the IFN or the M-IFN construction algorithm provides an estimation of the failure probability given certain values of the input (predictive) features. We assume that the system issues an alarm if the Missed Alarm cost exceeds the False Alarm cost, i.e. *FN * CR > FP,* where *FP* (1 − *Probability (Failure))* is the False Alarm probability when an alarm is raised, *FN* (*Probability (Failure))* is the Missed Alarm probability when an alarm is not raised, and *CR* is the Cost Ratio between the cost of a missed alarm and the cost of a false alarm. Consequently, the expected misclassification cost of a failure prediction rule can be calculated as *min (FP * 1, FN * CR)*.

As in our case study there was no accurate information about the acceptable TP / FP rates and/or misclassification costs, prediction models were evaluated based on the entire area under its ROC (Receiver Operating Characteristics) curve (also known as Area under Curve, or AUC). ROC curves [3] are two-dimensional graphs in which the TP (True Positive) rate is plotted on the Y axis and the FP (False Positive) rate is plotted on the X axis. In the case of failure prediction models, an ROC curve depicts the relative tradeoff between true positives (actual failures predicted in advance) and false positives (unnecessary warnings issued to the drivers). In any ROC curve, the diagonal line $y = x$ represents the strategy of randomly guessing a class. A useful classifier should have an ROC curve above the diagonal line implying that its AUC is higher than 0.5. An ideal classifier, which is never wrong in its prediction, would have the maximum AUC of 1.0.

4.3 Single Target and Multi-target Models

The area under the ROC curve was estimated using 5-fold cross-validation. The IFN and M-IFN ROC Curves for the target attribute "Battery Failure" are shown in Figure 1 below. Both ROC curves are nearly identical resulting in the AUC of 0.6238 and 0.6165, respectively. The number of single-target prediction rules for "Battery Failure" and 'Elapsed Time" obtained with the IFN algorithm was 8 and 13, respectively, whereas the multi-target model included 14 rules estimating the probabilities of both target attributes. Thus, the multi-target approach has a clear advantage in terms of comprehensibility as it reduces the total number of prediction rules from 21 to 14, a decrease of 33%.

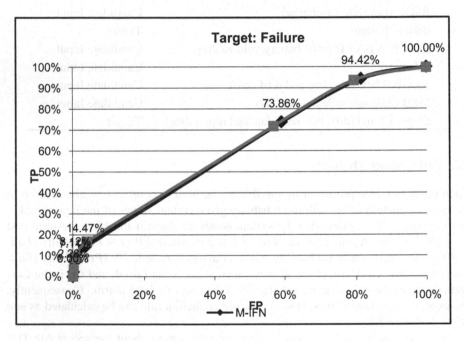

Fig. 1. ROC Curves - Single-Target vs. Multi-Target IFN

According to the single-target failure prediction model, the probability of failure, disregarding the elapsed time, may vary from 0.24% to 28.57% depending on the value of the three selected input features: SOC, Age, and OASAH. The second single-target prediction model reveals some minor differences (up to 1.5%) between the probabilities of various time-to-failure intervals as a function of the four selected input features: RSOC, Age, OASAH, and Odometer. A similar picture arises from the multi-target model except for one rule (If SOC is between 46.2745 and 99.2157 and AGE is between 4.14879 and 4.14942), which indicates a relatively high probability of failure (0.286) within 0.5 up to 2 months from the date of sensory reading. The following six features were selected by the M-IFN algorithm for simultaneous prediction of "Battery Failure" and 'Elapsed Time": SOC, Age, OASAH, Temp,

Load, and Odometer. The induced M-IFN model is shown in Figure 2, where the target nodes 0 and 1 represent the values of Battery Failure, whereas the target nodes 3-6 stand for the intervals of the Elapsed Time. The network has 14 terminal nodes associated with 14 probability estimation rules, but due to space limitations, we only show the complete connections of Node 20 to the nodes of the target layer. The numbers on the terminal-target connections represent the conditional probability of each class. These probabilities sum-up to one for each target attribute (Battery Failure and Elapsed Time). In case of the rule represented by Node 20, the probability of failure equals to 0.60 and it is distributed uniformly over three time-to-failure intervals (0.001-0.5, 0.5-1, and 1-2 months).

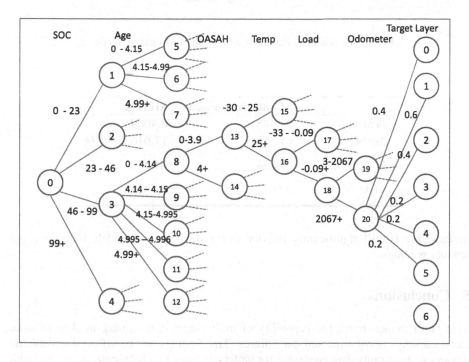

Fig. 2. Multi-Target Information-Fuzzy Network Model

4.4 Weibull Analysis

We also used the Weibull analysis to predict the probability of battery failure within one month of sensory reading. Figure 3 shows this probability as a function of the car age (Months in Service). As expected for the new cars, this chart reveals a decreasing failure rate. The Weibull distribution parameters and the Area under ROC curve for two optional predictive variables (Months in Service and Mileage) are shown in Table 2. In the case of mileage-based prediction, the mileage over the next month was extrapolated for each car using its Mileage per Day (MPD) value. Though it is clear from the AUC results that the car age is a better predictor of the battery failure probability, its

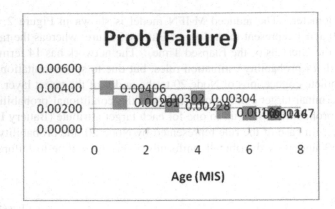

Fig. 3. Probability of Failure as a function of Months in Service

Table 2. Weibull Analysis

	Months in Service	Mileage
Beta	0.6443	0.4069
Alpha	61,898	17,616,308,874
R^2	0.9854	0.9559
AUC	0.5674	0.5437

performance is still significantly inferior to the data mining models induced using sensory readings.

5 Conclusions

This paper demonstrates the capability of multi-target data mining models to issue early warnings about expected car failures. The findings are based on the synthetic sensory and warranty data generated for battery failures in a high-end car model (with high electrical/electronic content) over a period of several months. The M-IFN construction algorithm has produced a compact and interpretable model of 14 rules estimating the probability distributions of two target attributes (Battery Failure and Elapsed Time). This model has outperformed the Weibull-based reliability analysis. Further improvement in predictive performance may be reached by combining the multi-target learning approach with ensemble methods, like bagging. Analyzing sensory and claims data of each vehicle over the entire warranty period should also bring more accurate results.

Acknowledgments. This work was supported in part by the General Motors Global Research & Development – India Science Lab.

References

1. Bey-Temsamani, A., Engels, M., Motten, A., Vandenplas, S., Ompusunggu, A.P.: A Practical Approach to Combine Data Mining and Prognostics for Improved Predictive Maintenance. In: The Data Mining Case Studies Workshop (DMCS), 15th ACM SIGKDD Conference on Knowledge Discovery and Data Mining (KDD 2009), Paris, pp. 37–44 (2009)
2. Caruana, R.: Multitask Learning. Machine Learning 28, 41–75 (1997)
3. Fawcett, T.: An introduction to ROC analysis. Pattern Recogn. Lett. 27(8), 861–874 (2006)
4. Ducange, P., Lazzerini, B., Marcelloni, F.: Multi-objective genetic fuzzy classifiers for imbalanced and cost-sensitive datasets. Soft Computing - A Fusion of Foundations, Methodologies and Applications (in press), doi:10.1007/s00500-009-0460-y
5. Gusikhin, O., Rychtyckyj, N., Filev, D.: Intelligent systems in the automotive industry: applications and trends. Knowl. Inf. Syst. 12(2), 147–168 (2007)
6. Last, M.: Multi-Objective Classification with Info-Fuzzy Networks. In: Boulicaut, J.-F., Esposito, F., Giannotti, F., Pedreschi, D. (eds.) ECML 2004. LNCS (LNAI), vol. 3201, pp. 239–249. Springer, Heidelberg (2004)
7. Last, M., Danon, G., Biderman, S., Miron, E.: Optimizing a Batch Manufacturing Process through Interpretable Data Mining Models. Journal of Intelligent Manufacturing 20(5), 523–534 (2009)
8. Last, M., Elnekave, S., Naor, A., Shonfeld, V.: Predicting Wine Quality from Agricultural Data with Single-Objective and Multi-Objective Data Mining Algorithms. In: Warren Liao, T., Triantaphyllou, E. (eds.) Recent Advances on Mining of Enterprise Data: Algorithms and Applications. Series on Computers and Operations Research, vol. 6, pp. 323–365. World Scientific, Singapore (2007)
9. Last, M., Maimon, O.: A Compact and Accurate Model for Classification. IEEE Transactions on Knowledge and Data Engineering 16(2), 203–215 (2004)
10. Schwall, M., Gerdes, J.C., Bäker, B., Forchert, T.: A probabilistic vehicle diagnostic system using multiple models. In: 15th Innovative Applications of Artificial Intelligence Conference (IAAI 2003), Acapulco, Mexico, pp. 123–128 (2003)
11. Suzuki, E., Gotoh, M., Choki, Y.: Bloomy Decision Tree for Multi-objective Classification. In: Siebes, A., De Raedt, L. (eds.) PKDD 2001. LNCS (LNAI), vol. 2168, pp. 436–447. Springer, Heidelberg (2001)
12. Wu, S., Gebraeel, N., Lawley, M., Yih, Y.: A Neural Network Integrated Decision Support System for Condition-Based Optimal Predictive Maintenance Policy. IEEE Transactions on Systems, Man and Cybernetics-Part A 37(2), 226–236 (2007)

Multiresolution Models and Algorithms of Movement Planning and Their Application for Multiresolution Battlefield Simulation

Zbigniew Tarapata

Military University of Technology, Cybernetics Faculty,
Gen. S. Kaliskiego Str. 2, 00-908 Warsaw, Poland
zbigniew.tarapata@wat.edu.pl

Abstract. In the paper a branch-and-bound algorithm for solving shortest paths problem in a large-scale terrain-based grid network, especially designed for multiresolution movement planning and simulation is discussed. The new approach deals with a specific method for merging the geographically adjacent nodes (squares) and the planning path to a "merged" graph. The merging is done by using geographically adjacent squares of primary graph (thus, we obtain nodes of a "merged" graph) and calculating costs in the "merged" graph as longest (or shortest) of the shortest paths between some subsets of nodes belonging to "merged" nodes. The properties of the algorithm are discussed and proved. Moreover, some remarks on how to parallelize and apply the presented algorithm for finding multiresolution shortest paths are proposed.

1 Introduction

Multiresolution paths are very interesting from many applications point of view (mobile robots [1], [9], [11], battlefield simulation [14], computer generated forces (CGF) [13], transportation or navigation [3], [12]). These are fields which describe either size of the environment or environment complexity (3D terrain). For example, in a battlefield simulation and planning models of movement a multiresolution environment is used. This is a nature of hierarchical structure of military units and methods of their behaviors on a simulated battlefield. For a company level of units greater precision of terrain (environment) model is required than e.g. for the brigade level. Battlefield movement planning and simulation is one of the elements of simulated battlefield automation. This is a domain of Computer Generated Forces (CGF) systems or semi-automated forces (SAF or SAFOR) [13]. CGF or SAF (SAFOR) are the technique which provides the simulated opponent using a computer system that generates and controls multiple simulation entities using software and possibly a human operator. The multiresolution paths problem is strongly connected with the problem of finding the shortest paths in large-scale networks. There are two main approaches to the shortest paths problem in large-scale networks: (a) to decompose a problem or environment (network, graph) in which we plan into smaller problems and then solve subproblems [1], [9], [12]; (b) to apply on-line algorithms which find and "merge" path cell-by-cell [4], [10], [14]. The first group of approaches is called multiresolution

N.T. Nguyen, M.T. Le, and J. Świątek (Eds.): ACIIDS 2010, Part II, LNAI 5991, pp. 378–389, 2010.

methods. As local algorithms inside all of these methods are used: modified Dijkstra's algorithm with priority queue represented by d-ary heap ($O(A \log_d V)$, where V – number of nodes of graph, A – number of edges (or arcs) of graph, $d = \max\{2, \lceil A/V \rceil\}$) proposed by Tarjan [16], with priority queue represented by Fibonacci heap ($O(E+V \log V)$) proposed by Fredman and Tarjan [5], A* algorithm (average time proportional to $O(\sqrt{V} \cdot V)$) [10]. Moreover, we can apply Bellmann-Ford's algorithm ($O(VE)$), Gabow-Tarjan's algorithm [6] ($O(\sqrt{V}E \log(VW))$) where W is the largest absolute weight of edges) or algorithm presented by Ahuja et.al. in [1] ($O(E+V\sqrt{\log W})$). For finding all-pairs shortest paths we can apply V times (for each node) the modified Dijkstra's algorithm ($O(VE\log_d V)$), Johnson's algorithm in sparse networks [8] ($O(V^2 log V+VE)$) or algorithms in DAGs (directed acyclic graphs) e.g. the Bellman algorithm ($O(V+E)$). The goal of all multiresolution methods is to reduce computational complexity by the reducing size of the problem (network size). For example, authors of [9] present a method based on cell decomposition and partitioning space into a quadtree and then use a staged search (similar to A* algorithm) to exploit the hierarchy. The goal of the approaches presented in [12] is to navigate a robot without violating terrain dependent constraints decomposing the terrain with wavelet decomposition. The paper [14] contains a review of various approaches for terrain representation (the Voronoi diagram, a straight-line dual of the Voronoi diagram (the Delaunay triangulation), visibility diagram, edge-dual graph, line-thinned skeleton, regular grid of squares) and path planning in the simulation systems.

In this paper presented a decomposition method (DSP – decomposition shortest paths) and its properties which decreases computational time of path searching in multiresolution graphs. The method significantly extends the concept described in [15]. The goal of the method is not only computation time reduction but, most of all, using it for multiresolution path planning. Presented in the last section is the method of how to use multiresolution battlefield modeling.

2   Definitions and Notations

Let graph $G = \langle V, A \rangle$ be given (see Fig.1b) as representation of e.g. terrain squares (see Fig.1a), where V describes a set of nodes (squares of terrain), $V=|V|$, A describes a set of arcs, $A = \{\langle x, y \rangle \subset V \times V : \text{square } x \text{ is adjacent to square } y\}$, $A=|A|$. For each arc $\langle x, y \rangle \in A$ we have cost $c(x,y)$ as crossing time ($c(x,x)=0$, $c(x,y)=+\infty$ when $\langle x, y \rangle \notin A$). The problem is to find the shortest path from node s to node t in G with the assumption that G is large in size and, simultaneously, to prepare data structure (some graph) for multiresolution path planning. The idea of the approach is to merge geographically adjacent small squares (nodes belonging to V) into bigger squares (called b-nodes, see Fig.1c) and to build b-graph G^* (graph based on the b-nodes, see Fig.1d) using specific transformation. This transformation is based on the assumption that we set an arc (b-arc) between two b-nodes $x^* \subset V, y^* \subset V$ when two such nodes as $x \in x^*, y \in y^*$ exist and that $\langle x, y \rangle \in A$ (x and y are called "border" nodes).

380   Z. Tarapata

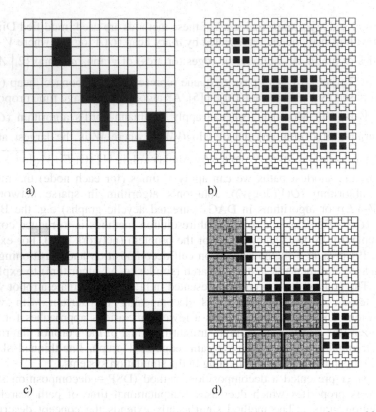

a)   b)

c)   d)

Fig. 1. Principles of G^* creation. Regions (squares) with black colour are impassable. a) Terrain space divided into a regular-size grid; b) Grid graph as a representation of terrain squares from a), only north-east-south-west arcs are permitted; c) Merging geographically adjacent small squares from b) into $n=16$ b-nodes (big squares); d) b-graph G^* for squares merging from c) with marked shortest s-t path in G^*.

Formal definition of the graph G^* is as follows: $G^* = \langle V^*, A^* \rangle$, $V^* = \{x_1^*, x_2^*, ..., x_n^*\}$ - set of b-nodes, $|V^*| = n$, $x_i^* = \{x_{i1}, x_{i2}, ..., x_{im_i}\} \subset V$ and each x_i^*, $i = \overline{1,n}$ generates subgraph of G,

$$A^* = \{\langle x^*, y^* \rangle \subset V^* \times V^* : \underset{x \in x^*, y \in y^*}{\exists} \langle x, y \rangle \in A\} \qquad (1)$$

Cost of the b-arc $\langle x^*, y^* \rangle \in A^*$ is set as $c^{*\min}(x^*, y^*)$ and $c^{*\max}(x^*, y^*)$: $c^{*\min}(x^*, y^*)$ is represented by cost vector of shortest of shortest paths from any node belonging to x^* to any node belonging to y^* for each predecessor z^* of x^*. This is a vector, because the cost from x^* to y^* depends on the node, from which we achieve x^* (therefore, for each predecessor of x^* we have a cost value, see Fig.2). This cost is calculated inside the subgraph built on the nodes belonging to x^*, y^* and z^*. Cost $c^{*\max}(x^*, y^*)$ is represented by the cost vector of longest of shortest paths from any node belonging to x^* to any node belonging to y^* for each predecessor z^* of x^*.

For further discussion we will use following notations: $W(x^*, y^*)$ - subset of nodes belonging to x^* which are adjacent ("border") to any node of y^*, $W(x^*, y^*) = \{x \in x^* : \underset{y \in y^*}{\exists} \langle x, y \rangle \in A\}$, $D(x,y)$ – set of paths between nodes x and y in graph G; $d(x, y) = (x_0 = x, x_1, ..., x_{l(d(x,y))} = y)$ - element of $D(x,y)$,

$\underset{i=0, l(d(x,y))-1}{\forall} \langle x_i, x_{i+1} \rangle \in A$, $L(d(x, y)) = \displaystyle\sum_{i=0}^{l(d(x,y))-1} c(x_i, x_{i+1})$ - cost of path $d(x,y)$

from x to y; $D^{\min}(W(x^*, z^*), W(y^*, v^*))$ - set of shortest paths in G between nodes belonging to $W(x^*, z^*)$ and $W(y^*, v^*)$:

$$D^{\min}\left(W(x^*, z^*), W(y^*, v^*)\right) = \left\{ \begin{array}{l} d^{\min}(x, y) \in D(x, y) : x \in W(x^*, z^*), y \in W(y^*, v^*), \\ L(d^{\min}(x, y)) = \underset{d(x,y) \in D(x,y)}{\min} L(d(x, y)) \end{array} \right\}$$

$c^{*\min}(x^*, y^*)$ - minimal of minimal cost vector for arc $\langle x^*, y^* \rangle \in A^*$ from x^* to y^*,

$c^{*\min}(x^*, y^*) = \left\langle c_{z^*}^{*\min}(x^*, y^*) \right\rangle_{z^* \in \{v^* \in V^* : \langle v^*, x^* \rangle \in A^*\}}$, $c_{z^*}^{*\min}(x^*, y^*)$ - minimal of minimal cost from x^* to y^* when the predecessor of x^* is z^*,

$$c_{z^*}^{*\min}(x^*, y^*) = \underset{d(\cdot,\cdot) \in D^{\min}(W(x^*, z^*), W(x^*, y^*))}{\min} L(d(\cdot, \cdot)) + \underset{d(\cdot,\cdot) \in D^{\min}(W(x^*, y^*), W(y^*, x^*))}{\min} L(d(\cdot, \cdot)) \quad (2)$$

$c^{*\max}(x^*, y^*)$ - maximal of minimal cost vector for arc $\langle x^*, y^* \rangle \in A^*$ from x^* to y^*,

$c^{*\max}(x^*, y^*) = \left\langle c_{z^*}^{*\max}(x^*, y^*) \right\rangle_{z^* \in \{v^* \in V^* : \langle v^*, x^* \rangle \in A^*\}}$, $c_{z^*}^{*\max}(x^*, y^*)$ - maximal of minimal cost for arc $\langle x^*, y^* \rangle \in A^*$ when the predecessor of x^* is z^*,

$$c_{z^*}^{*\max}(x^*, y^*) = \underset{d(\cdot,\cdot) \in D^{\min}(W(x^*, z^*), W(x^*, y^*))}{\max} L(d(\cdot, \cdot)) + \underset{d(\cdot,\cdot) \in D^{\min}(W(x^*, y^*), W(y^*, x^*))}{\max} L(d(\cdot, \cdot)) \quad (3)$$

$D^*(x^*, y^*)$ – set of paths between nodes x^* and y^* in graph G^*, $d^*(x^*, y^*) = \left(x_0^* = x^*, x_1^*, x_2^*, ..., x_{l^*(d^*(x^*, y^*))}^* = y^* \right)$ - element of $D^*(x^*, y^*)$,

$\underset{i=0, l^*(d^*(x^*, y^*))-1}{\forall} \langle x_i^*, x_{i+1}^* \rangle \in A$, $L^{*\min}(d^*(x^*, y^*))$ - cost of path $d^*(x^*, y^*)$ from x^* to y^* which is based on $c^{*\min}(x^*, y^*)$,

$$L^{*\min}(d^*(x^*, y^*)) = c_{p(x_0^*)}^{\min}(x_0^*, x_1^*) + \displaystyle\sum_{i=1}^{l^*(d^*(x^*, y^*))-1} c_{x_{i-1}^*}^{\min}(x_i^*, x_{i+1}^*) \quad (4)$$

where $p(x_0^*)$ denotes the predecessor of x_0^* in G^* representing the "direction", from which we start path planning in x_0^* (we use this interpretation, for example when $\langle x^*, y^* \rangle \in A^*$ and x^*, y^* represent internal nodes of some path $d^*(v^*, z^*)$; then $L^{*\min}(d^*(x^*, y^*)) = c_{p(x^*)}^{\min}(x^*, y^*)$ and $p(x^*)$ denotes the predecessor x^* on path $d^*(v^*, z^*)$). If the information about $p(x_0^*)$ is unimportant then $p(x_0^*) = x_1^*$. Let us note that the interpretation of $p(x_0^*)$ allows us to write (4) as the sum of length of parts of path $d^*(x^*, y^*)$ as follows:

$$L^{*min}(d^*(x^*, y^*)) = \sum_{i=0}^{l^*(d^*(x^*, y^*))-1} L^{*min}(d^*(x_i^*, x_{i+1}^*)) = \sum_{i=0}^{l^*(d^*(x^*, y^*))-1} c_{p(x_i^*)}^{min}(x_i^*, x_{i+1}^*)$$

$$p(x_i^*) = \begin{cases} x_{i-1}^*, & i > 0 \\ x_1^*, & i = 0 \end{cases}$$

(5)

Without the presented interpretation of $p(x_0^*)$ the calculation of the length of $d^*(x^*, y^*)$ as the sum of the length of its parts like in (5) would be impossible. We can define $L^{*max}(d^*(x^*, y^*))$ as the cost of path $d^*(x^*, y^*)$ from x^* to y^*, which is based on $c^{*max}(x^*, y^*)$, by analogy to (4)

$$L^{*max}(d^*(x^*, y^*)) = c_{p(x_0^*)}^{max}(x_0^*, x_1^*) + \sum_{i=1}^{l^*(d^*(x^*, y^*))-1} c_{x_{i-1}^*}^{max}(x_i^*, x_{i+1}^*)$$

(6)

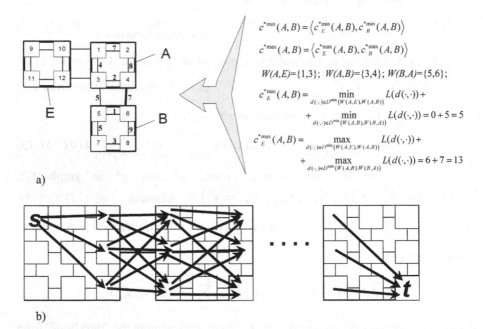

a)

b)

Fig. 2. (a) The interpretation and calculation method of $c_E^{*min}(A, B)$ and $c_E^{*max}(A, B)$ as components of $c^{*min}(A, B)$ and $c^{*max}(A, B)$; calculation of $c_B^{*min}(A, B)$ and $c_B^{*max}(A, B)$ in accordance. As "border" nodes of A to B we have W(A,B)={3,4}. (b) Constructing DAG for the last step of the DSP algorithm. Firstly, arcs link s with nodes inside x_s^* bordering on x_1^*, then link previous nodes with nodes of x_1^* bordering on x_s^*, etc.

Finally, we denote by $d^{*\max}(x^*, y^*)$ the shortest path in G^* from x^* to y^* with $c^{*\max}(x^*, y^*)$ cost function and by $d^{*\min}(x^*, y^*)$ the shortest path in G^* from x^* to y^* with $c^{*\min}(x^*, y^*)$ cost function. For $d^{*\max}(x^*, y^*)$ and $d^{*\min}(x^*, y^*)$ following conditions are satisfied:

$$L^{*\max}(d^{*\max}(\cdot,\cdot)) = \min_{d^*(\cdot,\cdot)\in D^*(\cdot,\cdot)} L^{*\max}(d^*(\cdot,\cdot)), \quad L^{*\min}(d^{*\min}(\cdot,\cdot)) = \min_{d^*(\cdot,\cdot)\in D^*(\cdot,\cdot)} L^{*\min}(d^*(\cdot,\cdot)),$$

where $D^*(\cdot,\cdot)$ describes set of paths in G^* between pairs of b-nodes.

3 The Branch-and-Bound (DSP) Algorithm for Shortest Paths

The branch-and-bound (decomposition) algorithm for shortest paths finding (DSP algorithm) consists of two main phases: (1) constructing graph G^* (steps 1-3); (2) finding the path from source s to destination t (steps 4-5). It uses Dijkstra's algorithm with k-ary heaps ($k - \max\{2, \lceil A/V \rceil\}$) (because the graph G is sparse and k-ary heap is very effective [16]) and may be presented in 5 steps.

1. to merge nodes from graph G (Fig.1b) into n big nodes (b-nodes) as subgraphs of G (Fig.1c) (n is the parameter of the algorithm);
2. to set each of subgraphs obtained in the step 1 as b-nodes and set b-arcs in this graph as described by (1) obtaining graph G^* (Fig.1d);
3. (a) for each $x^* \in V^*$ and for each $z^* \in V^*$ such that $\langle x^*, z^* \rangle \in A^*$ to determine the shortest paths trees (SPT) inside x^* for each $x \in W(x^*, z^*)$ as a source node;

    (b) to calculate costs $c^{*\min}(\cdot,\cdot)$ and $c^{*\max}(\cdot,\cdot)$ for each arc of G^* using (2)-(3);
4. to find shortest path $d^{*\min}(x_s^*, x_t^*)$ and $d^{*\max}(x_s^*, x_t^*)$ in G^* with cost functions $c^{*\min}(\cdot,\cdot)$ and $c^{*\max}(\cdot,\cdot)$ (lower and upper restriction on length of the path from s to t) between such pairs x_s^*, x_t^* of b-nodes that $s \in x_s^*$, $t \in x_t^*$ (see Fig.1d);
5. to find shortest path from s to t (s-t path) inside subgraph generated by nodes of G belonging to b-nodes of $d^{*\min}(x_s^*, x_t^*)$ ($(d^{*\max}(x_s^*, x_t^*))$):

    a) if $x_s^* = x_t^*$ then to find shortest s-t path inside the subgraph of G generated by nodes belonging to $x_s^* = x_t^*$ (use paths calculated in step 3a);

    b) otherwise if $x_s^* \neq x_t^*$ then s-t path may be found constructing the DAG with arcs directed from s to subset $W(x_0^* = x_s^*, x_1^*)$, then from $W(x_0^* = x_s^*, x_1^*)$ to $W(x_1^*, x_0^*)$, then from $W(x_1^*, x_0^*)$ to $W(x_1^*, x_2^*)$ etc. and lastly - from $W(x_{l^*(d^*(x_s^*,x_t^*))}^*, x_{l^*(d^*(x_s^*,x_t^*))-1}^*)$ to t (Fig.2b). The arc cost in DAG, is between nodes x and y, and the length of the shortest path is between x and y calculated in step 3a.

The DSP algorithm has some interesting properties. Theorem 1 shows lower and upper restriction on the length of the shortest path in G using the DSP algorithm. Theorem 2 shows the time and space complexity of the DSP algorithm.

Theorem 1. Let $L'(s, W(x_s^*, x_1^*))$ denote the length of the longest of the shortest paths from s to any node of $W(x_s^*, x_1^*)$ and x_1^* denote the direct successor of x_s^* on the path from x_s^* to x_t^*. For each s, $t \in V$ and $x_s^*, x_t^* \in V^*$, $x_s^* \neq x_t^*$ such that $s \in x_s^*$, $t \in x_t^*$ and the following formula is satisfied:

$$L^{*\max}(d^{*\max}(x_s^*, x_t^*)) + L'(s, W(x_s^*, x_1^*)) \geq L(d^{\min}(s,t)) \geq L^{*\min}(d^{*\min}(x_s^*, x_t^*)) \quad (7)$$

Proof is omitted. Conclusions resulting from Theorem 1:

- if in the G path from s to t exists then in the G^* path from x_s^* to x_t^* exists and the DSP algorithm will find it;
- if $G=G^*$ then the lower restriction equals the upper restriction (the DSP algorithm gives an optimal solution); otherwise, length $L(d^{\min}(s,t))$ of the shortest s-t path is restricted by the left and the right side of the inequality (7);

Theorem 2. Let digraph $G=(V, A)$, $|V|=V$, $|A|=A$, $s,t \in V$, $c : A \to R^+$ and cardinal n representing the number of b-nodes in G^* be given. Then the total time of the DSP algorithm (for preparing G^* and finding shortest s-t path) is equal $O\left(\sqrt{V^3/n}\log_k(V/n) + n\log_k n\right)$ and the space $O\left(\sqrt{V^3/n} + A + V\right)$, where $k = \max\{2, \lceil A/V \rceil\}$.

Proof is omitted.
Shown here the advantages of DSP algorithm for finding the all-pairs shortest paths in network G. We can formulate some acceleration functions $F_{Dijk}(V, n)$ and $F_{John}(V, n)$ as follows:

$$F_{Dijk}(V,n) = \frac{T_{Dijk}(V)}{T_{DSP}(V,n)} \qquad F_{John}(V,n) = \frac{T_{John}(V)}{T_{DSP}(V,n)} \quad (8)$$

where $T_{Dijk}(V)$, $T_{John}(V)$, $T_{DSP}(V,n)$ denotes, respectively, experimental average times of finding the all-pairs shortest paths in G with V nodes using: V times Dijkstra's algorithm with 4-ary heaps, Johnson's algorithm for sparse networks [8], the DSP algorithm with n b-nodes.

Let the grid network with V squares (nodes) be given. We can formulate the following optimization problem: to find such a cardinal n^*, for which

$$F(V,n^*) = \max_{n \in \{1,...,V\}} F(V,n) \quad (9)$$

In the Table 1 experimental impact of V on n^* and $F(V, n^*)$ is shown. The value of n^* may be approximated by function $n^* \approx 1{,}87 \cdot V^{0.34}$ and acceleration functions: $F_{Dijk}(V,n^*) \approx 0{,}39V^{0.67}$, $F_{John}(V,n^*) \approx 0{,}23V^{0.62}$, thus the average acceleration of the DSP algorithm with relation to the Dijkstra's and Johnson's algorithm is $\cong O(V^{0.65})$.

Table 1. Experimental impact of V on n^* and $F_{Dijk}(V, n^*)$, $F_{John}(V, n^*)$ for the all-pairs shortest paths problem for various numbers V of nodes from G

V	100	500	1000	5000	10000	100000	200000	1000000
n^*	9	16	21	36	46	100	130	220
$F_{Dijk}(V,n^*)$	9	25	40	118	187	865	1380	4000
$F_{John}(V,n^*)$	5	12	18	49	75	320	495	1400

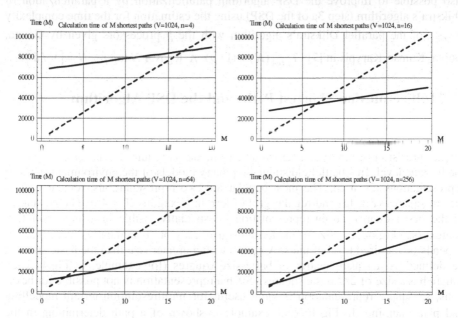

Fig. 3. Graphs of calculation time (represented by number of dominating operations) for finding M shortest paths using DSP algorithm (continuous line) and Dijkstra's algorithm (dashed line) between random pairs of nodes ($V=1024$, $n=4$, $n=16$, $n=64$, $n=256$)

However, the DSP algorithm gives a good result not only for all-pairs shortest paths problem (Table 1). Because the most complex steps of the algorithm (steps 1-3, "bottleneck") are done only one time (we build b-graph only one time – initial preprocessing) then if we compute a one-pair shortest path many times it allows us to amortize time of the "bottleneck". In the Fig.3 we present graphs of calculation time (represented by number of dominating operations) for finding M shortest paths using the DSP algorithm and Dijkstra's algorithm between random pairs of nodes. It easy to observe that the greater value of n (with the same value of V) the smaller number of shortest paths calculation to obtain shorter time for the DSP algorithm than for Dijkstra's one. For example (see Fig.3), to obtain the same calculation time for the DSP and Dijkstra's algorithm for $V=1024$, $n=4$ we must find $M^*=17$ shortest paths (for $M<17$ Dijsktra's algorithm is faster than the DSP one, otherwise the DSP algorithm is faster) and for $V=1024$, $n=64$ we must find only $M^*=3$ shortest paths (for $M<3$ Dijsktra's algorithm is faster than the DSP one, otherwise the DSP algorithm is faster).

The presented approach can be very easily computed in parallel. A very important problem from the point of view of parallelization effectiveness is to assign processors to the nodes (b-nodes) skillfully. We can assign each processor to subsets of nodes belonging to different b-nodes of the parallel DSP algorithm (PDSP). For example, by having n processors we can compute step 3 about n times faster assigning each of the processors to the single b-node (no communication between processors is required in this step because it can be computed for each of the n b-nodes independently). It is also possible to improve the DSP algorithm parallelization by a parallelization of Dijkstra's algorithm (step 3a of the DSP) using the estimation for the time complexity $T_{p,Dijk}$ of the parallel Dijkstra's algorithm with the p processors given by Grama, Gupta, Kumar, Karypis in [7]: $T_{p,Dijk}(N) = (1/p) \cdot A \cdot \log N + N \log p$, $N = \lceil V / n \rceil$.

4 Multiresolution Shortest Paths and the DSP Algorithm

Multiresolution environment is a nature of hierarchical structure of military units and methods of their behaviours on a simulated battlefield. For a company level of units greater precision of the terrain (environment) model is required than for example for the brigade level. In a battlefield simulation many models of the environment (terrain) representations are used. The most popular are two representations: regular grid of terrain squares (Fig.4a) and regular grid of terrain hexagons (Fig.4b). The advantage of the first (square) terrain representation is especially visible in a multiresolution context (see Fig.4c÷e). Size of the terrain square may be dynamically changed and it depends on required level of units. A square with a greater size than the basic size can be defined as a square matrix of basic-size squares (for example, in Fig.4d each square has a size of 2x2 basic squares). Such a representation is not possible for hexagons, so square representation is more useful for multiresolution terrain modelling and path planning. In Fig.4c÷e an example is shown of a path determining in the three-level graph: (c) the first level is the most detailed; (d) the second level is two times less detailed than the first; (e) the third level is four times less detailed than the first. These models may describe for example platoon, company and battalion levels on the battlefield. Let us note that it is easy to obtain a multiresolution model of terrain by defining graph G^* recurrently. If we establish that graph G defines a terrain model of the first level (e.g. company level) than G^* defines a model of the second (or higher) level (e.g. battalion level). This reasoning may be used to increase or to decrease each required level of model resolution. Parameter n of the DSP algorithm ($n \in \{1,...,V\}$) can be used to decide on the dimension of G^*. Then, the DSP algorithm may be used for finding multiresolution paths in such a multiresolution model of environment. For example, in Fig.4c $G^*=G$ and contains $n=256$ b-nodes (for the platoon level), in Fig.4d G^* contains $n=64$ b-nodes (for the company level) and in Fig.4e G^* contains $n=16$ b-nodes (e.g. for battalion level).

It is important to say that the presented method differs from very effective representations of terrain using quadtree [9] because of two main reasons: (1) elements of quadtree, which represent a terrain have non-regular size, (2) in majority applications (for example in [9]) quadtree represents only binary terrain with two types of regions: open (passable) and closed (impassable). The approach [9] is very effective for mobile robots, but it is not adequate to represent battlefield [14].

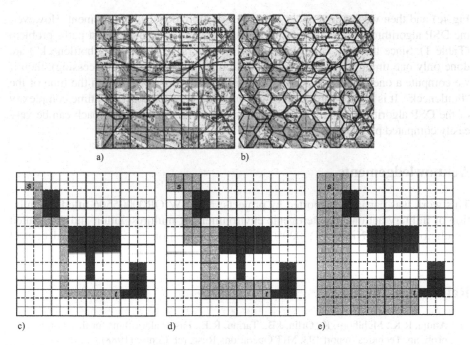

Fig. 4. Examples of terrain representation in a simulated battlefield: a) regular grid of terrain squares; b) regular grid of terrain hexagons. Multiresolution shortest path from s to t using the DSP algorithm in G^*: c) $G=G^*$ thus G^* contains 16×16 nodes; d) G^* contains 8×8 nodes; e) G^* contains 4×4 nodes.

Let us note that multiresolution approach for path planning represented by finding shortest paths in recurrently defined G^* can be also used for multistage path planning: first we can find a "rough" path $d^{*\min}(x_s^*, x_t^*)$ (or $d^{*\max}(x_s^*, x_t^*)$) - in a "rough" terrain represented by G^* (for example in Fig.4e) and then we can find an accurate path in a more detailed environment (represented by G with small squares, Fig.4c; more precisely: we find the shortest path from s to t (s-t path) inside the subgraph generated by nodes of G belonging to b-nodes of $d^{*\min}(x_s^*, x_t^*)$ (or ($d^{*\max}(x_s^*, x_t^*)$, see step 5 of the DSP algorithm). This is an example of top-down modelling.

5 Summary

The approach presented in this paper is dedicated especially for multiresolution path planning in grid graph-based route planning when the grid represents, for example a terrain environment as a regular grid of terrain squares. This is an element of battlefield simulation, which is strongly connected with military action planning and simulation. It can be shown that multiresolution approach for path planning represented by finding shortest paths in recurrently defined G^* can be also used for multistage path planning: first we can find a "rough" path in a "rough" terrain represented by G^* (for example in

Fig.4e) and then we can find an accurate path in more detailed environment. However, the DSP algorithm gives a good result not only for the all-pairs shortest paths problem (Table 1). Since the most complex steps of the algorithm (steps 1-3, "bottleneck") are done only one time (the b-graph is built only one time – initial preprocessing), then if we compute a one-pair shortest path many times it allows us to shorten the time of the "bottleneck". It is also possible to set a compromise between space and time complexity of the DSP algorithm. Moreover, it was shown that the presented approach can be very easily computed parallelly.

Acknowledgements

This work was partially supported by projects: MNiSW OR00005506 titled „Simulation of military actions in heterogenic environment of multiresolution and distributed simulation" and MNiSW OR00005006 titled „Integration of command and control systems".

References

1. Ahuja, R.K., Mehlhorn, K., Orlin, J.B., Tarjan, R.E.: Faster algorithms for the shortest path problem. Technical report 193, MIT Operations Research Center (1988)
2. Behnke, S.: Local Multiresolution Path Planning. In: Polani, D., Browning, B., Bonarini, A., Yoshida, K. (eds.) RoboCup 2003. LNCS (LNAI), vol. 3020, pp. 332–343. Springer, Heidelberg (2004)
3. Yu-Li, C., Romeijn, E., Smith, R.: Approximating Shortest Paths in Large-scale Networks with an Application to Intelligent Transportation Systems. Journal on Computing 10(2), 163–179 (1998)
4. Djidjev, H., Pantziou, G., Zaroliagis, C.D.: On-line and dynamic algorithms for shortest path problems. In: Mayr, E.W., Puech, C. (eds.) STACS 1995. LNCS, vol. 900, pp. 193–204. Springer, Heidelberg (1995)
5. Fredman, M.L., Tarjan, R.E.: Fibonacci heaps and their uses in improved network optimization algorithms. Journal of the Association for Computing Machinery 34, 596–615 (1987)
6. Gabow, H.N., Tarjan, R.E.: Faster scaling algorithms for network problems. SIAM Journal on Computing 18, 1013–1036 (1989)
7. Grama, A., Gupta, A., Karypis, G., Kumar, V.: Introduction to Parallel Computing, 2nd edn. Addison-Wesley, Reading (2003)
8. Johnson, D.B.: Efficient algorithm for shortest paths in sparse networks. Journal of the ACM 24, 1–13 (1977)
9. Kambhampati, S., Davis, L.S.: Multiresolution Path Planning for Mobile Robots. IEEE Journal of Robotics and Automation RA-2(3), 135–145 (1986)
10. Korf, R.E.: Artificial intelligence search algorithms. In: Algorithms Theory Computation Handbook. CRC Press, Boca Raton (1999)
11. Lavalle, S.: Planning algorithms. Cambridge University Press, Cambridge (2006)
12. Pai, D.K., Reissell, L.M.: Multiresolution Rough Terrain Motion Planning. IEEE Transactions on Robotics and Automation I, 19–33 (1998)

13. Petty, M.D.: Computer generated forces in Distributed Interactive Simulation. In: Proceedings of the Conference on Distributed Interactive Simulation Systems for Simulation and Training in the Aerospace Environment, The International Society for Optical Engineering, Orlando, USA, April 19-20, pp. 251–280 (1995)
14. Tarapata, Z.: Military route planning in battlefield simulation: effectiveness problems and potential solutions. Journal of Telecommunications and Information Technology 4, 47–56 (2003)
15. Tarapata, Z.: Decomposition algorithm for finding shortest paths in grid networks of large size. In: Proceedings of the 15th International Conference on Systems Science, Wroclaw, Poland, September 7-10, vol. III, pp. 209–216 (2004)
16. Tarjan, R.E.: Data Structures and Network Algorithms. Society for Industrial and Applied Mathematics, Philadelphia, Pennsylvania (1983)

An Algorithm for Generating Efficient Outcome Points for Convex Multiobjective Programming Problem*

Nguyen Thi Bach Kim[1] and Le Quang Thuy[2]

[1] Faculty of Applied Mathematics and Informatics
Hanoi University of Technology, Vietnam
`kimntb-fami@mail.hut.edu.vn`
[2] Faculty of Applied Mathematics and Informatics
Hanoi University of Technology, Vietnam
`thuylq-fami@mail.hut.edu.vn`

Abstract. In this article we propose a quite easy algorithm for generating efficient outcome points for convex multiobjective programming problems. To illustrate the performance of the new algorithm, we use it to generate efficient points for a sample problem.

Keywords: Convex multiobjective programming; efficient outcome point, Reverse Polyblock.

1 Introduction

Let Q be a nonempty set in the Euclidean space $\mathbb{R}^p$. We denote by Q_E and Q_{WE} the sets of all *efficient points* and *weakly efficient points* of Q respectively, that is

$$Q_E := \{q^* \in Q \mid \nexists q \in Q \text{ such that } q^* \geq q \text{ and } q^* \neq q\},$$

$$Q_{WE} := \{q^* \in Q \mid \nexists q \in Q \text{ such that } q^* \gg q\}.$$

Here for any two vectors $a, b \in \mathbb{R}^p$, the notation $a \geq b$ (respectively, $a \gg b$) means $a - b \in \mathbb{R}^p_+$ (respectively, $a - b \in \text{int}\mathbb{R}^p_+$), where $\mathbb{R}^p_+$ is the nonnegative orthant of $\mathbb{R}^p$. By the definition,

$$Q_E \subseteq Q_{WE}.$$

The following result (Theorem 2.10, Chapter 4 [8]) will be used.

Proposition 1. *Let $Q \subset \mathbb{R}^p$ be a nonempty convex set. A point $y^* \in Q$ is a weakly efficient of Q if and only if there is a nonzero vector $v \in \mathbb{R}^p$ and $v \geq 0$ such that y^* is an optimal solution to the convex programming problem*

$$\min\{\langle v, q \rangle \mid q \in Q\}.$$

---

* This paper is partially supported by the National Foundation for Science and Technology Development, Vietnam.

N.T. Nguyen, M.T. Le, and J. Świątek (Eds.): ACIIDS 2010, Part II, LNAI 5991, pp. 390–399, 2010.
© Springer-Verlag Berlin Heidelberg 2010

We consider the following convex multiobjective programming problem

$$\text{Min } f(x), \text{ s.t. } x \in X, \qquad\qquad (CMP)$$

where

$$f(x) = [f_1(x), f_2(x), ..., f_p(x)],$$

$p \geq 2$, $X \subset \mathbb{R}^n$ is a nonempty compact convex set, for each $j \in \{1, 2, ..., p\}$, $f_j : \mathbb{R}^n \to \mathbb{R}$ is convex on $\mathbb{R}^n$ and positive on X. In particular, if all the f_j, $j = 1, 2, ..., p$, are linear functions and X is a nonempty polyhedral convex set then problem (CMP) is said to be a linear multiobjective programming problem (LMP).

Let Y denote the set $f(X) = \{f(x)|x \in X\}$. The set $Y = f(X)$ is called the *outcome set* (or *image*) *of X under f*.

A point $x^0 \in X$ is said to be an *efficient solution* for problem (CMP) if $f(x^0) \in Y_E$. For simplicity of notation, let X_E denote the set of all efficient solutions for problem (CMP). When $f(x^0) \in Y_{WE}$, x^0 is called *weakly efficient solution* for problem (CMP) and the set of all weakly efficient solutions is denoted by X_{WE}.

It is clear that X_E (respectively, X_{WE}) is the pre-image under f of Y_E (respectively, Y_{WE}). We will refer to Y_E and Y_{WE} as the *efficient outcome set* and *weakly efficient outcome set* for problem (CMP), respectively.

It is well known that X_E, X_{WE}, Y_E and Y_{WE} are, in general, nonconvex sets, even in the case of linear multiobjective programming problem (LMP). Therefore solving the problem (CMP) is a hard task.

Problem (CMP) arises in a wide variety of applications in engineering, economics, network planning, production planning, operations research..., especially in multicriteria design and in multicriteria decision making (see, for instance, [4], [13], [15]).

In the case of the linear multiobjective programming problem (LMP), many algorithms have been proposed for generating either the set X_E, X_{WE}, Y_E and Y_{WE} or a representative portion thereof, without any input from decision maker; see, e.g., [1] , [2], [3], [5], [6], [7],... and references therein.

Several works have proposed to study the problem (CMP) and to generate an efficient solution or the weakly efficient solution set or the efficient solutions, see Z.H. Lin, D.L. Zhu and Z.P. Sheng ([10], 2003), D.T. Luc, T. Q. Phong and Michel V. ([9], 2005), W. Song and G. M. Yao ([12], 2008) and references therein.

The goal of this paper is to develop a very simple algorithm for generating efficient outcome points for convex multiobjective programming problem (CMP). It has been shown that, in practice, the decision maker prefers basing his or her choice of a most preferred solution primarily on Y_E rather than X_E. Arguments to this effect are given in [2].

The bases of the algorithm are presented in Section 2. The detail algorithm and an illustrative example is given in Section 3.

2   Theoretical Background

Assume henceforth that $X \subset \mathbb{R}^n$ is a nonempty, compact convex set given by

$$X := \{x \in \mathbb{R}^n | g_i(x) \le 0, \ i = 1, ..., m\},$$

where all the $g_1, g_2, ..., g_m$ are convex functions on $\mathbb{R}^n$.

For each $k = 1, ..., p$, let y_k^{opt} (respectively, $(x_1^k, \cdots, x_n^k, y_1^k, \cdots, y_p^k)$) be the optimal value (respectively, an optimal solution) for the following convex programming problem with linear objective function

$$\min \quad y_k \qquad\qquad (CP_k)$$

$$\text{s.t.} \begin{cases} f_j(x) - y_j \le 0, & j = 1, \cdots, p, \\ g_i(x) & \le 0, & i = 1, \cdots, m, \\ y_j & \ge 0, & j = 1, \cdots, p. \end{cases}$$

Let $y^M = (y_1^M, y_2^M, ..., y_p^M) \in \mathbb{R}^p$, where $y_j^M = \alpha$ for all $j = 1, 2, ..., p$ and

$$\alpha > \max \{ \ y_k^j, \ k = 1, \cdots, p, \ j = 1, \cdots, p \}.$$

Consider the set G given by

$$G = (Y + \mathbb{R}_+^p) \cap (y^M - \mathbb{R}_+^p) \qquad\qquad (1)$$

$$= \{y \in \mathbb{R}^p \mid f(x) \le y \le y^M \text{ for some } x \in X\}.$$

This set G is instrumental in the algorithm to be presented in Section 3 for generating efficient outcome points for problem (CMP).

Proposition 2. *The set G is a nonempty, full-dimension compact convex in $\mathbb{R}^p$.*

Proof. By definition, G is a nonempty, full-dimension compact set. Since $(y^M - \mathbb{R}_+^p)$ is a convex set, we have only to show that the $(Y + \mathbb{R}_+^p)$ is convex set. Indeed, the set

$$T = \{(x, y) \in \mathbb{R}^n \times \mathbb{R}^p | f_j(x) - y_j \le 0, j = 1, 2, ..., p; \ g_i(x) \le 0, i = 1, 2, ..., m\}$$

is a convex set because $f_j, \ j = 1, ..., p$ and $g_i, \ i = 1, ..., m$ are convex functions. It can easily be seen that $(Y + \mathbb{R}_+^p) = \{y \in \mathbb{R}^p \mid y \ge f(x), \ x \in X\}$ is the projection of T on $\mathbb{R}^p$. Therefore, $(Y + \mathbb{R}_+^p)$ is also convex set (see [11]). The proof is straight-forward.

Proposition 3. $Y_E = G_E$.

Proof. ($\Rightarrow$) Suppose that $y^1 \in Y_E$. By definition, we have $y^1 \in Y \subset (Y + \mathbb{R}_+^p)$ and $y^1 < y^M$, so that $y^1 \in G$.

Assume that $y^1 \notin G_E$. Then we may choose a point $y^2 \in G$ such that $y^2 \le y^1$ and $y^2 \ne y^1$. Since $G \subset (Y + \mathbb{R}_+^p)$, we have $y^2 = y^0 + u$ where $y^0 \in Y$ and $u \ge 0$.

Therefore, $y^1 \geq y^0$ and $y^1 \neq y^0$. Since $y^0 \in Y$, this contradicts that $y^1 \in Y_E$ and the assumption that $y^1 \notin G_E$ must be false, so that $Y_E \subseteq G_E$.

($\Leftarrow$) Suppose that $y^1 \in G_E$. Since $y^1 \in G$, then from (1) we have

$$y^1 = y^0 + u = y^M - v, \text{ with } y^0 \in Y, \ u \geq 0, \ v \geq 0$$

$$\Rightarrow y^0 = y^M - (v + u) \in (y^M - \mathbb{R}_+^p).$$

Therefore, $y^0 \in G$. If $u > 0$ then $y^1 \geq y^0$ and $y^1 \neq y^0$ which contradicts the fact that $y^1 \in G_E$. Hence $y^1 = y^0 \in Y$ (i.e., $u = 0$).

Assume that $y^1 \notin Y_E$. By definition, there is $y^2 \in Y$, $y^2 \neq y^1$ and $y^2 \leq y^1$. It means that $y^2 = y^1 - v$ with $v \geq 0$ and $v \neq 0$. Since $y^1 \in G$,

$$y^1 = y^M - t \text{ with } t \geq 0 \Rightarrow y^2 = y^1 - v = y^M - (t + v), \text{ with } (t + v) > 0.$$

Therefore $y^2 \in G$ and $y^2 \geq y^1$ and $y^2 \neq y^1$. This contradiction implies that $y^1 \in G_E$. Therefore $G_E \subseteq Y_E$ and the proof is completed.

Remark 1. From Proposition 2, we will refer to G as an *efficiency equivalent set* for Y. Notice that even though $Y_E = G_E$, the dimension of Y may be strictly less than $\dim G = p$.

For any $\hat{y} \in \mathbb{R}_+^p$ such that $\hat{y} \leq y^M$, we denote by

$$[\hat{y}, y^M] = \{y | \hat{y} \leq y \leq y^M\}$$

the *box* (or the *hyper-rectangle*) corresponding to vertex $\hat{y}$.

Let $y^m = (y_1^m, y_2^m, ..., y_p^m)$, where $y_j^m = y_j^{opt}$ for all $j = 1, 2, ..., p$, and

$$B^0 = (y^m + \mathbb{R}_+^p) \cap (y^M - \mathbb{R}_+^p) = [y^m, y^M].$$

It is clear that $G \subset B^0$ and $B_E^0 = \{y^m\}$.

The following fact will play an important role in establishing the validity of our algorithm.

Proposition 4. *For each point $\bar{y} \in B^0 \setminus G$, let y^w denote the unique point on the boundary of G that belongs to the line segment connecting $\bar{y}$ and y^M. Then $y^w \in G_E$.*

Proof. Let $D = G - y^w$. Then D is a nonempty compact set containing the origin 0 of the outcome space $\mathbb{R}^p$ and 0 belongs to the boundary of D because G is a nonempty compact set and y^w belongs to the boundary of G. According to the Separation Theorem [11], there is a nonzero vector $q \in \mathbb{R}^p$ such that

$$\langle q, u \rangle \geq 0 \text{ for all } u \in D. \tag{2}$$

Let $\bar{u} = y^M - y^w$. From the definition of D and (2), we have $\bar{u} \in D$, $\bar{u} \gg 0$ and

$$t \langle q, \bar{u} \rangle \geq 0 \text{ for all } t \geq 0. \tag{3}$$

The expression (2) can be written by

$$\langle q, y - y^w \rangle \geq 0 \text{ for all } y \in G,$$

i.e.,

$$\langle q, y \rangle \geq \langle q, y^w \rangle \text{ for all } y \in G. \tag{4}$$

Furthermore, it is obvious that (3) is only true when

$$q \geq 0. \tag{5}$$

Combining Proposition 1, (4) and (5) gives $y^w \in G_{WE}$.

To complete the proof it remains to show that $y^w \in G_E$. Assume the contrary that

$$y^w \in G_{WE} \setminus G_E.$$

This implies that there is $i_0 \in \{1, 2, ..., p\}$ such that $y_{i_0}^w = y_{i_0}^m$. Since $\bar{y} \in B^0 \setminus G$, we always have $\bar{y} = t y^w$ with $t < 1$. Therefore,

$$\bar{y}_{i_0} = t y_{i_0}^w = t y_{i_0}^m < y_{i_0}^m.$$

This is impossible because $\bar{y} \geq y^m$. The proof is straight forward.

Remark 2. According to Proposition 4, for each point $\bar{y} \in B^0 \setminus G$ chosen, we can find an efficient point $y^w \in G_E$. By Proposition 3, we have $y^w \in Y_E$.

To determine the efficient outcome point y^w, we solve the following convex programming problem with linear objective function

$$\min \lambda \tag{$T(\bar{y})$}$$

$$\text{s.t.} \begin{cases} f(x) - \lambda(y^M - \bar{y}) - \bar{y} \leq 0, \\ g_i(x) \leq 0, \quad i = 1, \cdots, m, \\ 0 \leq \lambda \leq 1. \end{cases}$$

Notice that the feasible region of problem $(T(\bar{y}))$ is a nonempty, compact convex set. Let (x^*, λ^*) be an optimal solution and let λ^* be the optimal value for this problem. Then we have

$$y^w = \bar{y} + \lambda^*(y^M - \bar{y}).$$

For the sake of convenience, y^w is said to be an efficient outcome point generated by $\bar{y}$.

In this way, by varying the choices of points $\bar{y}$ belonging to $B^0 \setminus G$, the decision maker can generate multiple points in Y_E.

3   The Algorithm for Generating Efficient Outcome Points

3.1   The Algorithm

The algorithm includes the two procedures. The goal of the first procedure is to construct the box $B^0 := [y^m, y^M]$ and the second procedure allows to generate an efficient outcome point for problem (CMP) if a point $\bar{y} \in B^0 \setminus G$ is known.

For the convenience of the reader, let us describe in detail these procedures.

Procedure (CB).(*Constructing the box* $[y^m, y^M]$)

- For each $k = 1, ..., p$, find an optimal solution and $(x_1^k, \cdots, x_n^k, y_1^k, \cdots, y_p^k)$ and the optimal value y_k^{opt} for the convex programming problem (CP_k).
- Let $y^m = (y_1^m, y_2^m, ..., y_p^m)$, where $y_j^m = y_j^{opt}$ for all $j = 1, 2, ..., p$.
- Let $y^M = (y_1^M, y_2^M, ..., y_p^M)$, where $y_j^m = \alpha$ for all $j = 1, 2, ..., p$ and

$$\alpha > \max \{ \ y_k^j, \ k = 1, \cdots, p, \ j = 1, \cdots, p \}.$$

Procedure (DE). (*Determining an efficient outcome point*)

- For a given vector $\bar{y} \in B^0 \setminus G$, solve the convex programming problem $(T(\bar{y}))$ to obtain an optimal solution (x^*, λ^*) and the optimal value λ^*;
- Let $y^w = \bar{y} + \lambda^*(y^M - \bar{y})$. The vector y^w is an efficient outcome point for problem (CMP).

The algorithm can be described below.

Algorithm

Step 1.(Initialization)

i) Construct B^0 by Procedure (CB);
ii) Let $OutputY_E := \emptyset$ (*the set of efficient outcome points*) and $S := \{y^m\}$ ($S \subset B^0 \setminus G$).

Step 2.(Iteration)

i) For each $\bar{y} \in S$, find an efficient outcome point y^w by Procedure (DE) and let $OutputY_E := OutputY_E \cup \{y^w\}$;
ii) Pick a finite set $\bar{S} \subset B^0 \setminus G$ such that $\bar{S} \cap S = \emptyset$. Set $S := \bar{S}$ and return the iteration step.

After the initialization step, the algorithm can execute the iteration step many times to generate the number of efficient outcome points for problem (CMP) depending on the user's requirement.

Let us conclude this section with some remarks on the implementation of the computational modules in the above algorithm.

Remark 3. i) Note that the constructing the box B^0 in Procedure (CB) involves solving p convex programming problems, each of which has a simple linear objective function and the same feasible region.

ii) For each efficient outcome point $y^w \in Y_E$ generated by $\bar{y}$, the choosing the finite set $\bar{S}$ belong to $B^0 \setminus G$ in Step 2(ii) can be accomplished via a technique cutting reverse polyblock, as shown in the next Proposition 5.

A set of the form $B = \bigcup_{y \in V} [y, y^M]$ with $V \subset [\hat{y}, y^M]$ and $|V| < +\infty$ is called a *reverse polyblock* in hyper-rectangle $[\hat{y}, y^M]$ with *vertex set* V. A vertex $y \in V$ is said to be *proper* if there is no $y' \in V \setminus \{y\}$ such that $[y, y^M] \subset [y', y^M]$. It is clear that a reverse polyblock is completely determined by its proper vertices.

Proposition 5. (see, e.g. [14]) *Let* $G = (Y + \mathbb{R}^p_+) \cap (y^M - \mathbb{R}^p_+)$ *be a nonempty compact convex set contained in a reverse polyblock* B *with vertex set* V. *Let* $v \in V \backslash G$ *and* y^w *is the unique point on the boundary of* G *that belongs to the line segment connecting* v *and* y^M. *Then* $\bar{B} = B \backslash [v, y^w]$ *is a reverse polyblock with vertex set* $\bar{V} = (V \backslash \{v\}) \cup \{v^1, v^2, \cdots, v^p\}$, *where*

$$v^i = y^w - (y^w_i - v_i)e^i, \quad i = 1, 2, \cdots, p,$$

where, as usual, e^i *denotes the* i-*th unit vector of* $\mathbb{R}^p_+$, *and*

$$G \subset \bar{B} \sqsubset B, \quad v \in B \backslash \bar{B}.$$

Remark 4. In view of Proposition 5, for each efficient outcome point y^w generated by $\bar{y}$, we can determine p new points belong to $B^0 \backslash G$ and differ $\bar{y}$. This can be applied to update the above algorithm in the choosing the finite set $\bar{S}$ in Step 2 (ii).

3.2   Example

To illustrate our algorithm, consider the convex multiobjective programming problem with $p = 2, n = 3$, where

$$f_1(x) = (x_1 - 2)^2 + x_2^2 + 1$$

$$f_2(x) = (x_2 - 4)^2 + x_3^2 + 1$$

and X is defined by the constraints

$$g_1(x) = x_1^2 + x_2^2 - x_3$$

$$g_2(x) = x_1^2 + x_2^2 + x_3 - 2$$

$$g_3(x) = x_1 + x_2 + x_3 - 2.$$

A brief summary of the results of executing this algorithm for generating seven different efficient outcome points for this problem can be described as follows

Step 1. (*Initialization*) (*Constructing the box* $B^0 = [y^m, y^M]$)
    We find the optimal value $y_1^{opt} = 2.000001$ (respectively, $y_2^{opt} = 10.999996$) and the optimal solution $(1, -0.000962, 1, 2.000001, 19.707308)$ (respectively, $(-0.000563, 1, 1, 8.128818, 10.999996)$) for the convex programming (CP_1) (respectively, (CP_2)).

We set $y^m = (y_1^{opt}, y_2^{opt}) = (2.000001, 10.999996); \quad y^M = (20, 20);$

$$B^0 = [y^m, y^M] = \{y \in \mathbb{R}^2 : 2.000001 \leq y_1 \leq 20; \ 10.999996 \leq y_2 \leq 20\};$$

We also set $OutputY_E := \emptyset; \ v^1 = y^m$ and $S := \{v^1 = (2.000001, 10.999996)\}$.

Step 2. (*The first time*)

Step 2(i) (**Generating a first efficient outcome point**)

We choose vector $\bar{y} := v^1 \in S$;

Solving the convex programming problem $(T(\bar{y}))$ to obtain an optimal solution $(x^*, \lambda^*) = (0.435109,\ 0.774977,\ 0.789911,\ 0.113860)$, the optimal value $\lambda^* = 0.113860$ and we have generated the first efficient outcome point y^w for problem (CMP), where $y^w = \bar{y} + \lambda^*(y^M - \bar{y}) = (4.049482, 12.024736)$. We set $y^1 := y^w = (4.049482, 12.024736)$ and

$$OutputY_E := OutputY_E \cup \{y^1\} = \{y^1\}.$$

Step 2(ii) (*Generating the finite* $\bar{S}$)

To derive $\bar{S}$, we set $B := B^0$ and $\bar{B} = B \setminus [v, y^w]$, where $v := v^1$ and $y^w = y^1$. By Proposition 5, we have $\bar{B}$ is a reverse polyblock with vertex set $\bar{V} = \{v^2 = (2.000001, 12.024736);\ v^3 = (4.049482, 10.999996)\}$. We set $\bar{S} := \bar{V},\ S := \bar{S} = \{v^2, v^3\}$ and return the iteration step.

Step 2. (*The second time*)

Step 2(i) (**Generating two another efficient outcome points**)

• We choose vector $\bar{y} := v^2 = (2.000001, 12.024736) \in S$.

Solving the convex programming problem $(T(\bar{y}))$ to obtain an optimal solution $(x^*, \lambda^*) = (0.588636,\ 0.646676,\ 0.764683,\ 0.078341)$, the optimal value $\lambda^* = 0.078341$ and we have generated the second efficient outcome point y^w for problem (CMP), where $y^w = \bar{y} + \lambda^*(y^M - \bar{y}) = (3.410139, 12.649527)$. We set $y^2 = y^w = (3.410139, 12.649527)$ and

$$OutputY_E := OutputY_E \cup \{y^2\} = \{y^1, y^2\}.$$

• Now we choose vector $\bar{y} := v^3 = (4.049482, 10.999996) \in S$.

Solving the convex programming problem $(T(\bar{y}))$ to obtain an optimal solution $(x^*, \lambda^*) = (0.261701,\ 0.885570,\ 0.852723,\ 0.047424)$, the optimal value $\lambda^* = 0.047424$ and the third efficient outcome point y^w for problem (CMP), where $y^w = \bar{y} + \lambda^*(y^M - \bar{y}) = (4.805918,\ 11.426812)$. We set $y^3 := y^w = (4.805918,\ 11.426812)$ and

$$OutputY_E := OutputY_E \cup \{y^3\} = \{y^1, y^2, y^3\}.$$

Step 2(ii) (*Generating the finite* $\bar{S}$)

We set $B := \bar{B}$ and $\bar{B} = B \setminus [v, y^w]$, where $v := v^2$ and $y^w = y^2$. The reverse polyblock $\bar{B}$ has the vertex set $\bar{V} = \{v^4 = (2.000001, 12.649527);\ v^5 = (3.410139, 12.024736), v^3 = (4.049482, 10.999996)\}$.

Now, we $B := \bar{B}$ and $\bar{B} = B \setminus [v, y^w]$, where $v := v^3$ and $y^w = y^3$. The reverse polyblock $\bar{B}$ has the vertex set $\bar{V} = \{v^4 = (2.000001, 12.649527);\ v^5 = (3.410139, 12.024736);\ v^6 = (4.049481, 11.426812);\ v^7 = (4.805918, 10.999996)\}$. We set $\bar{S} := \bar{V}$ and $S := \bar{S} = \{v^4, v^5, v^6, v^7\}$ and return the iteration step.

Step 2. (*The third time*)

Step 2(i) (**Generating four another efficient outcome points**)

- We choose vector $\bar{y} := v^4 = (2.000001, 12.649527) \in S$.

Solving the convex programming problem $(T(\bar{y}))$ to obtain an optimal solution $(x^*, \lambda^*) = (0.636254, 0.599507, 0.764231, 0.067734)$, the optimal value $\lambda^* = 0.067734$ and the fourth efficient outcome point y^w for problem (CMP), where $y^w = \bar{y} + \lambda^*(y^M - \bar{y}) = (3.219213, 13.147404))$. We set $y^4 := y^w = (3.219213, 13.147404)$ and

$$OutputY_E := OutputY_E \cup \{y^4\} = \{y^1, y^2, y^3, y^4\}.$$

- We choose vector $\bar{y} := v^5 = (3.410139, 12.024736) \in S$.

Solving the convex programming problem $(T(\bar{y}))$ to obtain an optimal solution $(x^*, \lambda^*) = (0.481621, 0.739522, 0.778852, 0.02665744)$, the optimal value $\lambda^* = 0.02665744$ and the fifth efficient outcome point y^w for problem (CMP), where $y^w = \bar{y} + \lambda^*(y^M - \bar{y}) = (3.852382, 12.237337)$. We set $y^5 := y^w = (3.852382, 12.237337)$ and

$$OutputY_E := OutputY_E \cup \{y^5\} = \{y^1, y^2, y^3, y^4, y^5\}.$$

- We choose vector $\bar{y} := v^6 = (4.049481, 11.426812) \in S$.

Solving the convex programming problem $(T(\bar{y}))$ to obtain an optimal solution $(x^*, \lambda^*) = (0.337559, 0.841074, 0.821356, 0.0264344)$, the optimal value $\lambda^* = 0.0264344$ and the sixth efficient outcome point y^w for problem (CMP), where $y^w = \bar{y} + \lambda^*(y^M - \bar{y}) = (4.471123, 11.653439)$. We set $y^6 := y^w = (4.471123, 11.653439)$ and

$$OutputY_E := OutputY_E \cup \{y^6\} = \{y^1, y^2, y^3, y^4, y^5, y^6\}.$$

- We choose vector $\bar{y} := v^7 = (4.805918, 10.999996) \in S$.

Solving the convex programming problem $(T(\bar{y}))$ to obtain an optimal solution $(x^*, \lambda^*) = (0.175685, 0.929489, 0.894817, 0.025416)$, the optimal value $\lambda^* = 0.025416$ and the seventh efficient outcome point y^w for problem (CMP), where $y^w = \bar{y} + \lambda^*(y^M - \bar{y}) = (5.190291, 11.228740)$. We set $y^7 := y^w = (5.190291, 11.228740)$ and

$$OutputY_E := OutputY_E \cup \{y^7\} = \{y^1, y^2, y^3, y^4, y^5, y^6, y^7\}.$$

References

1. Armand, P.: Finding all Maximal Efficient Faces in Multiobjective Linear Programming. Mathematical Programming 61, 357–375 (1993)
2. Benson, H.P.: An Outer Approximation Algorithm for Generating All Efficient Extreme Points in the Outcome Set of a Multiple Objective Linear Programming Problem. Journal of Global Optimization 13, 1–24 (1998)
3. Benson, H.P., Sun, E.: Outcome Space Partition of the Weight Set in Multiobjective Linear Programming 105(1), 17–36 (2000)
4. Jahn, J.: Vector Optimization: Theory, Applications and Extensions. Springer, Berlin (2004)

5. Kim, N.T.B., Luc, D.T.: Normal Cone to a Polyhedral Convex Set and Generating Efficient Faces in Linear Multiobjective Programming. Acta Mathematica Vietnamica 25(1), 101–124 (2000)
6. Kim, N.T.B., Luc, D.T.: Normal Cone Method in Solving Linear Multiobjective Problem. Journal of Statistics & Management System 5(1-3), 341–358 (2002)
7. Kim, N.T.B., Thien, N.T., Thuy, L.Q.: Generating All Efficient Extreme Solutions in Multiple Objective Linear Programming Problem and Its Application to Multiplicative Programming. East-West Journal of Mathematics 10(1), 1–14 (2008)
8. Luc, D.T.: Theory of Vector Optimization. Springer, Berlin (1989)
9. Luc, D.T., Phong, T.Q., Volle, M.: Scalarizing Functions for Generating the Weakly Efficient Solution Set in Convex Multiobjective Problems. SIAM Journal on Optimization 15(4), 987–1001 (2005)
10. Lin, Z.H., Zhu, D.L., Sheng, Z.P.: Finding a Minimal Efficient Solution of a Convex Multiobjective Program. Journal of Optimization Theory and Applications 118(3), 587–600 (2003)
11. Rockafellar, R.T.: Convex Analysis. Princeton University Press, Princeton (1970)
12. Song, W., Yao, G.M.: Homotopy Method for a General Multiobjective Programming Problem. Journal of Optimization Theory and Applications 138(1), 139–153 (2008)
13. Stewart, T.J., Van den Honert, R.C. (eds.): Trends in Multicriteria Decision Making. LNEMS, vol. 465. Springer, Berlin (1997)
14. Tuy, H., Nghia, N.D.: Reverse Polyblock approximation for generalized multiplicative/fractional programming. Vietnam Journal of Mathematics 31(4), 391–402 (2003)
15. Tabucanon, M.T.: Multiojective Programming for Industrial Engineers. In: Avriel, M., et al. (eds.) Mathematical Programming for Industrial Engineers, pp. 487–542. Marcel Dekker, New York (1996)

Resources Utilization in Distributed Environment for Complex Services

Adam Grzech

Institute of Computer Science, Wroclaw University of Technology,
Wybrzeze Wyspianskiego 27, 50-370 Wroclaw
adam.grzech@pwr.wroc.pl

Abstract. Service oriented architectures and approaches influenced significantly design and applications of services offered and delivered in decision support systems. The main idea of the service oriented approaches is based on natural and obvious assumption that the requested service may be assured by flexibly assembled from a variety of components, services and systems into multi-tier applications. Service oriented approach is an answer for required elasticity and flexibility according which the amount, functionality and quality of required services changes and the architecture should assure environment in which the services are defined on-line manner and allows to take into accounts knowledge and experiences of users. The aim of this paper is to present a model of complex services composed of atomic services and its utilization to estimate resources utilization in two compared attempts to provide complex service. The presented and discussed approaches are based on assumption that the same set of atomic services are performed sequentially and some parallelism of the set of atomic services may be applied to speed-up the requested complex service execution time.

Keywords: Service Oriented Architecture, services, resources utilization.

1 Introduction

Systems based on SOA (*Service Oriented Architecture*) paradigm offer services (complex services) which are delivered as composition of atomic services [8,9]. The main feature of such an attempt is that the required complex services may be efficiently and flexibly composed of available atomic services providing certain, well defined and required functionality. Requested complex services are characterized by set of parameters specifying both functional and nonfunctional requirements; the former define data processing procedures, while the latter describe various aspects of service quality. The set of parameters describing requested complex service form SLA (*Service Level Agreement*) [1,10].

Functionality of the requested complex service is available as a sum of atomic services functionalities. In order to deliver complex service with requested functional and non-functional properties appropriate atomic services must be chosen in the process of complex service composition [7]. Required functionality, uniquely defined in the SLA, determines set of required atomic services as well as a plan according which

N.T. Nguyen, M.T. Le, and J. Świątek (Eds.): ACIIDS 2010, Part II, LNAI 5991, pp. 400–409, 2010.

the atomic services are performed. Non-functionality of the requested complex service, which is mainly related to QoS (*Quality of Service*), in most cases, i.e., in distributed environment, may be assured or obtained by proper resources (processing and communication) and tasks (atomic services) allocation [4-6,12].

Discussed complex services delivery approach is available only at distributed environment; possible parallel execution of distinguishable atomic services requires allocation of proper amount of processing and communication resources in parallel manner. The distributed environment may be obtained both by allocation of separated or virtualized resources.

In order to obtain various required QoS in distributed environment well-known QoS strategies, i.e., best-effort, integrated services and differentiated services concepts may be applied. Usefulness of the mentioned concepts strongly depends on formulation of the non-functional SLA. Application of the best-effort concept, based on common resources sharing, leads to solution, where the same, higher enough, average quality of service is delivered to all performed services. The next two previously mentioned concepts offer differentiated quality of service for requested services (also guarantees) and are mainly based on resources reservation for individual requests (integrated services concept) or for classes of requests (differentiated services concept) [4,11].

One of the most important non-functional properties of the requested complex services, measuring the service QoS, is service response time. Value of the latter is mainly influenced by three factors: execution time of atomic services, load of the system (number of requests performed in parallel in the system) and communication delays introduced by communication channels among service units executing the atomic services. In order to guarantee stated in SLA service response time all these factors must be taken into account in the process of service composition [2,3].

The paper is devoted to discuss some selected issues concerning resources utilization in two case. The first case relates to service system, where the set of atomic services, specified by SLA of the requested complex service, are executed sequentially at distributed service units, while the second concerns distributed environment, where the same set of atomic services may be in some parts executed in parallel. The distributed character of the considered service system is mainly determined by the assumption that execution of particular atomic services requires some communication resources required to deliver both input and output data among the services. The second section is devoted to express resources utilization in the case when all atomic services are performed sequentially at distributed service units (processors) interconnected by some communication infrastructure which produces some communication delays. The next, third section contains expressions that allows to compute resources utilization when several service units – interconnected by some communication infrastructure - are applied in parallel to perform the requested set of atomic services. In the fourth section the resources utilization factors for the two considered cases are compared and discussed.

2 Complex Service as a Sequence of Particular Atomic Services

It is assumed that the considered system delivers complex services composed of atomic services; the latter is defined as a service with an indivisible functionality offered by known and well-specified place or places in the system. Moreover, it is also assumed that each atomic service may be performed at all processing units (processors) that are available at the services processing systems; various processing units offer for each considered atomic service possibility to obtain the required values of SLA's functional and non-functional parameters.

Let as assume that in order to obtain the requested k-th complex service s_k it is necessary to execute some well-defined and ordered set of atomic services $\{as_{ki}\}$ ($i = 1,2,...,I$). It is also assumed that the set of predefined (based on analysis of the required complex service's functionality) atomic services $\{as_{ki}\}$ are executed in the multi-processors distributed environment (service units network).

At the first discussed case, the atomic services from the set $\{as_{ki}\}$ are performed one by one according predefined order determined by required complex service functionality (described by the complex service SLA) (Figure 1).

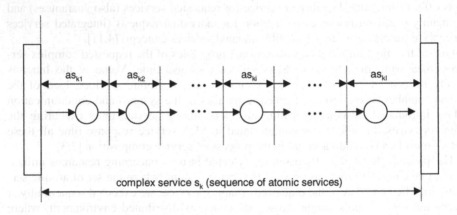

Fig. 1. Complex services performed as a sequence of atomic services

The total execution time of the k-th complex service s_k, uniquely defined by ordered set of atomic services $\{as_{ki}\}$, at the available, selected j-th service unit is equal to $T_{sj}(s_k)$, being a sum of execution times of all atomic services $T_j(as_{ki})$, is given by the following expression:

$$T_{sj}(s_k) = \sum_{i=1}^{I} T_j(as_{ki}),$$ (1)

where $T_j(as_{ki})$ is a service time of as_{ki} atomic service at the j-th ($j = 1,2,....,J$) service unit (processor).

If the system, where the atomic services are composed of identical services units (processors) than the times, required to complete the atomic services, are the same, i.e., $T_j(as_{ki}) = T_l(as_{ki})$ for $j \neq l$, where $j,l \in \{1,2,...,J\}$.

It is further assumed that the as_{ki} atomic service execution time at the j-th service unit ($T_j(as_{ki})$) is composed of two parts: atomic service as_{ki} completing delay $tc_j(as_{ki})$ (time required to complete data necessary to process the atomic service as_{ki} and to complete the results of the already executed service) and atomic service a_{ki} processing time $tp_j(as_{ki})$ (time required to process data according the procedures uniquely defined for the atomic service), i.e.,:

$$T_j(as_{ki}) = tc_j(as_{ki}) + tp_j(as_{ki}) \tag{2}$$

In order to obtain higher granularity of the service a_{ki} completing time $tc_j(as_{ki})$, the latter may be further decomposed into two times: before processing and after processing completing delays.

When the set of atomic services $\{as_{ki}\}$ is performed in sequence at the j-th ($j = 1,2,.....,J$) service unit and when the succeeded atomic services are served immediately one after another, than the total execution time for complex service s_k is equal to:

$$T_{sj}(s_k) = \sum_{i=1}^{I} T_j(as_{ki}) = \sum_{i=1}^{I} \left(tc_j(as_{ki}) + tp_j(as_{ki}) \right). \tag{3}$$

The above expression may be applied to find an optimal j^*-th ($j^* = 1,2,....,J$) service unit, i.e., a service unit guaranteeing the shortest completing time for given complex service s_k:

$$j^* \leftarrow \min_{j \in \{1,2,...,I\}} T_{sj}(s_k) = \min_{j \in \{1,2,...,I\}} \sum_{i=1}^{I} \left(tc_j(as_{ki}) + tp_j(as_{ki}) \right). \tag{4}$$

The utilization of the j-th ($j = 1,2,....,J$) processor, engaged at the system for servicing the sequence of atomic services $\{as_{ki}\}$, may be represented by sum $w_j(s_k)$ of the processor waiting times $tc_j(as_{ki})$; the sum is equal to:

$$w_j(s_k) = tc_j(s_k) = tc_j(\{as_{k1}, as_{k2},..., as_{ki},...., as_{kI}\}) = \sum_{i=1}^{I} tc_j(as_{ki}). \tag{5}$$

In such a case, the j-th ($j = 1,2,....,J$) service unit efficiency may be defined as a ratio of j-th service unit occupancy $tp_j(s_k) = tp_j\{as_{k1}, as_{k2},..., as_{kI}\}$ during processing the sequence of distinguished atomic services $\{as_{ki}\}$ and the total service time of the considered s_k complex service ($tc_j(s_k) + tp_j(s_k)$) is given by:

$$e_{sj}(s_k) = e_{sj}\{as_{ki}\} = \frac{tp_j(s_k)}{tc_j(s_k) + tp_j(s_k)}. \tag{6}$$

The efficiency $e_{sj}(s_k)$ may be increased by reducing processor waiting times, i.e., communication delays or by allocation of the required processing resources.

The discussed issue – performing the set $\{as_{ki}\}$ of ordered atomic services at the particular j-th ($j = 1,2,\ldots, J$) service unit (processor) - may be extended to the case when the s_k complex service is executed at already loaded service units which are characterized – for example – by service units availabilities.

3   Complex Service as a Sequence of Subsets of Atomic Services Executed in Parallel

Let as assume that given set of atomic services $\{as_{ki}\}$, required to assure functionality of the given complex service s_k, may be reordered in sequence of atomic services subsets such that atomic services, belonging to a particular subset, can be performed in parallel. Moreover, it is assumed that the resources granularity (number of processors) assures that all distinguished atomic services in any particular distinguished subset of atomic services may be performed independently (Figure 2).

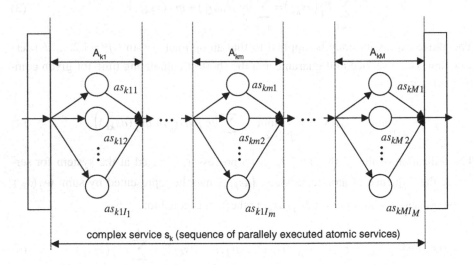

complex service s_k (sequence of parallely executed atomic services)

Fig. 2. Complex service performed utilizing possible atomic services parallelism.

Reordering of atomic services set $\{as_{ki}\}$ (required to assure s_k complex service) in gain to utilize possible service parallelism means, that the original set of atomic services $\{as_{ki}\}$ may be replaced by set of M distinguished subsets of atomic services $\{A_{k1}, A_{k2},\ldots, A_{km},\ldots, A_{kM}\}$, where $A_{km} = \{as_{km1}, as_{km2},\ldots, as_{kmI_m}\}$ is the m-th ($m = 1,2,\ldots, M$, $M \leq I$) subset of atomic services from the set $\{as_{ki}\}$ that can be performed in parallel.

In gain to assure generality of the discussed task, it is also assumed that the following conditions are satisfied: $1 \le I_m \le I$ for $m = 1,2,...,M$ and $I_1 + I_2 + ... + I_m + ... + I_M = I$ (if exist such an m that $I_m = I$, it means that $m = 1$ and all atomic services from the atomic services $\{as_{ki}\}$ may be performed in parallel; if $m = I$ it means that it is impossible to distinguish any parallelism in the atomic services set $\{as_{ki}\}$ and that all the services should be performed in sequence).

In such a case, assuming that all selected service units available at the service system are the same, the total time required to complete s_k complex service ($T_p(s_k)$) is equal to:

$$T_p(s_k) = \sum_{m=1}^{M} T_m(A_{km}) = \sum_{m=1}^{M} \max\left\{T(as_{km1}), T(as_{km2}),...,T(as_{kmI_m})\right\}. \tag{7}$$

Taking into account that execution of I_m atomic services in parallel needs some extra efforts (proportional to the number of atomic services being at the m-th subset) both before the processing ($\alpha_m I_m$) and after the processing ($\beta_m I_m$), the total execution time for the m-th subset of atomic services is given by the following:

$$\overline{T}_p(s_k) = \sum_{m=1}^{M} \left(\overline{T}_m(A_{km})\right) =$$

$$= \sum_{m=1}^{M} \left(\alpha_m I_m + \max\left\{T(as_{km1}), (T(as_{km2}),...,T(as_{kmI_m})\right\} + \beta_m I_m\right), \tag{8}$$

where α_m and β_m ($m = 1,2,...,M$) are unit times required to prepare data and to complete results before and after processing the I_m atomic services from A_{km}.

Assuming that the environment contains $p \ge I_{\max} = \max\{I_1, I_2,...,I_M\}$ service units (processors) (condition $p \ge I_{\max} = \max\{I_1, I_2,...,I_M\}$ guarantees that all atomic services in distinguished subsets A_{km} ($m = 1,2,...,M$) can be executed in parallel), the p service units (processors) utilization is defined as follows:

$$e_p(s_k) = \frac{\displaystyle\sum_{m=1}^{M} \sum_{i=1}^{I_m} tp(as_{kmi})}{p\overline{T}_p(s_k)} =$$

$$= \frac{\displaystyle\sum_{m=1}^{m} \sum_{i=1}^{I_m} tp(as_{kmi})}{p\displaystyle\sum_{m=1}^{M} \left(\alpha_m I_m + \max\left\{T(as_{km1}), T(as_{km2}),...,T(as_{kmI_m})\right\} + \beta_m I_m\right)}. \tag{9}$$

Unsatisfied $p \ge I_{max} = \max\{I_1, I_2, ..., I_M\}$ condition means that the parallelism of services - defined by grouping the set of atomic services $\{as_{ki}\}$ into the set of subsets of atomic services $\{A_{km}\}$ - cannot be realized. In such a case, the process of grouping of set of atomic services $\{as_{ki}\}$ should be repeated taking into account that the number of atomic services in A_{km} ($m = 1,2,..., M$) subsets should not be larger than given number of available service units (processors).

The utilization of available p processors, devoted at the system for serving the sequence of subsets of atomic services $\{A_{km}\}$, may be represented by sum of processors waiting times. The sum of waiting times when A_{km} ($m = 1,2,..., M$) during performing m-th subset of atomic services is given by the following expression:

$$w_p(A_{km}) = (p - I_m) \max_{l=1,2,...,I_m} T(a_{kml}) + \sum_{j=1}^{I_m} \left(\max_{l=1,2,...,I_m} T(as_{kml}) - tp(as_{kmj}) \right), \quad (10)$$

where the first component is a sum of waiting time of all processors (from the group of p processors) which are not engaged into performing A_{km} subset of atomic services ($p - I_m$), and where the second is a sum of waiting times of processors (from the group of I_m processors) for the most time-consuming atomic service.

Taking into account that all p processors wait during $\alpha_m I_m$ and $\beta_m I_m$ ($m = 1,2,..., M$) periods of time before and after performing every m-th subset A_{km} ($m = 1,2,..., M$), the total waiting time may be summarized and is equal to:

$$w_p(s_k) = \sum_{m=1}^{M} w_p(A_{km}) = \sum_{m=1}^{M} \left((\alpha_m + \beta_m) I_m + \left((p - I_m) \max_{l=1,2,...,I_m} T(as_{kml}) \right) \right) +$$

$$+ \sum_{m=1}^{M} \sum_{j=1}^{I_m} \left(\max_{l=1,2,...,I_m} \left(T(as_{kml}) - tp(as_{kmj}) \right) \right). \quad (11)$$

The latter component of the above equation may be expressed as follows:

$$\sum_{m=1}^{M} \sum_{j=1}^{I_m} \left(\max_{l=1,2,...,I_m} \left(T(as_{kml}) - tp(as_{kmj}) \right) \right) = \sum_{m=1}^{M} \sum_{j=1}^{I_m} \left(\max_{l=1,2,...,I_m} tc(as_{kmj}) \right). \quad (12)$$

If the number of available processor p, engaged in servicing the distinguished subsets of atomic services performing in parallel, is equal to the maximum size of parallelism in all distinguished A_{km} ($m = 1,2,..., M$) subsets, i.e., if $p = \max\{I_1, I_2, ..., I_M\}$, than the number of processing resources assure to perform the s_k complex service and minimize loss of processors utilization (processors waiting times).

The summarized waiting time of p processors may be considered as a price which has to be paid for speed-up the s_k complex service execution time $(T_p(s_k))$ performed by set of p parallel processors when compared with s_k complex service execution time performed on single processor $(T_s(s_k))$.

4   Resources Utilization Comparison

The above discussion, concerning resources utilization for the required complex service s_k is required, may be summarized as follows:

- the speed-up of s_k complex service composed of $\{a_{k1}, a_{k2}, ..., a_{ki}, ..., a_{kI}\}$ set of atomic services (performed sequentially at one processor) and regrouped into $\{A_{k1}, A_{k2}, ..., A_{km}, ..., A_{kM}\}$ subsets of atomic services (performed at $p = \max\{I_1, I_2, ..., I_M\}$ parallel processors) is:

$$s(1, p) = \frac{\overline{T}_p(s_k)}{T_s(s_k)} =$$

$$= \frac{\sum_{m=1}^{M} \left((\alpha_m + \beta_m)I_m + \max_{j=1,2,...,I_m} \left(T(as_{km1}), T(as_{km2}), ..., T(as_{kmI_m}) \right) \right)}{\sum_{i=1}^{T} T(as_{ki})}, \qquad (13)$$

- decreases resources utilization measured by ratio of summarized $p = \max\{I_1, I_2, ..., I_M\}$ processors and one processor waiting times:

$$u(1, p) = \frac{w_p(s_k)}{\sum_{i=1}^{I} tc(as_{ik})}. \qquad (14)$$

The latter is always greater than one due to the fact that parallelism of performing distinguished atomic services a_{ki}, grouped and performed within subsets A_{km}, always increase summarized processors waiting times according the following:

$$\sum_{i=1}^{I} tc(as_{ki}) = \sum_{m=1}^{M} \sum_{j=1}^{I_m} tc(as_{kml}) \leq \sum_{m=1}^{M} \sum_{j=1}^{I_m} \max_{l=1,2,...,I_m} tc(as_{kml} \in A_{km}), \qquad (15)$$

where the latter component (as well as the difference among the third and the second components) of the above expression reflect the fact that some processors, performing in parallel atomic services belonging to the groups A_{km} ($m = 1, 2, ..., M$) complete servicing earlier than another and than they have to wait for completing servicing by the processor which is most occupied.

Taking into account the above mentioned inequality, the expression for relative resources utilization in distributed environment composed of p processors acting in parallel (compared to utilization of one processor performing s_k service) may be rewritten as below and the following is satisfied:

$$u(1, p) \geq \frac{\sum_{m=1}^{M}\left((\alpha_m + \beta_m)p + (p - I_m) \max_{l=1,2,\ldots,I_m} T(as_{kml})\right) + \sum_{i-1}^{I} tc(as_{ki})}{\sum_{i=1}^{I} tc(as_{ik})} \geq 1, \quad (16)$$

due to the fact that the first component of the numerator is always nonnegative.

Comparison of efficiency factors characterizing the two discussed scenario allows to validate them (as well as all scenario combining elements of the two considered) and to name parameters of the distributed environment that may influence efficiency changes.

The discussed issues may be easily extended to the more general case i.e., when the service units (processors), available in the distributed environment and devoted to perform atomic services, are different as well as where communication cost among processor delivering services are not omitted.

5 Conclusions

In order to obtain expected features (flexibility, elasticity, personalization, etc.) of service supported by service oriented architectures, the latter are usually composed as a set of well-defined atomic services assuring required complex service's functionality.

The set of atomic services together with defined atomic services performance plan assures functionality described in the complex service's SLA. SLA defines also non-functionality of the complex service; the latter are defines expectations or requirements addressing quality of service issues. In most cases the quality of service issues concern complete service completion time. The latter may be extended to cases, where some other requirements are formulated (for example in area of data and/or service security). Assurance of the required quality of service in most cases means that the set of atomic services are performed in parallel if such kind of parallelism is possible (possible parallelism of atomic services is determined by the required functionality). Possible and potential parallelism in the set of atomic services is a source of possible speed-up of complex service delivery. The possible speed-up of the complex service completing time requires existing of resources (services units) available to perform atomic services in parallel manner.

The presented model and discussion is valid under assumption the atomic services are executed in the multi-processors distributed environment (service units network). In such an environment the systems performance is influenced by communication delays introduced by amount of data required to start particular atomic service (input data) and transfer the obtained data (output data). The case, when the distinguished set of atomic services are executed at one-service unit (processor) system is not considered; assumption about one-service units leads to eliminate communication delays.

In the paper issues concerning resources utilization in the above mentioned approach were discussed under some specified assumptions.

Acknowledgements

The research presented in this paper has been partially supported by the European Union within the European Regional Development Fund program no. POIG.01.03.01-00-008/08.

References

[1] Anderson, Š., Grau, A., Hughes, C.: Specification and satisfaction of SLAs in service oriented architectures. In: 5th Annual DIRC Research Conference, pp. 141–150 (2005)
[2] Garey, M., Johnson, D., Sethi, R.: The complexity of flowshop and jobshop scheduling. Mathematics of Operations Research 1, 117–129 (1976)
[3] Graham, R.L., Lawler, E.L., Lenstra, J.K., Rinnooy Kan, A.H.G.: Optimization and approximation in deterministic sequencing and scheduling: a survey. Annals of Discrete Mathematics 3, 287–326 (1979)
[4] Grzech, A.: Teletraffic control in the computer communication networks. Wroclaw University of Technology Publishing House (2002) (in Polish)
[5] Grzech, A., Świątek, P.: Parallel processing of connection streams in nodes of packet-switched computer communication networks. Cybernetics and Systems 39(2), 155–170 (2008)
[6] Grzech, A., Świątek, P.: The influence of load prediction methods on the quality of service of connections in the multiprocessor environment. Systems Science 35(3) (2009) (in press)
[7] Jaeger, M.C., Rojec-Goldmann, G., Muhl, G.: QoS aggregation in web service compositions. In: IEEE Int. Conf. on e-Technology, e-Commerce and e-Service, pp. 181–185 (2005)
[8] Johnson, R., Gamma, E., Helm, R., Vlisides, J.: Design patterns; elements of reusable object-oriented software. Addison-Wesley, Reading (1995)
[9] Milanovic, N., Malek, M.: Current Solutions for Web Service Composition. IEEE Internet Computing 8(6), 51–59 (2004)
[10] Narayanan, S., McIlraith, S.: Analysis and simulation of web services. Computer Networks: The International Journal of Computer and Telecommunications Networking 42(5), 675–693 (2003)
[11] Rajan, R., Verma, D., Kamat, S., Felstaine, E., Herzog, S.: A policy framework for Integrated and Differentiated Services in the Internet. IEEE Network, 34–41 (September 1999)
[12] Wang, Z.: Internet QoS: architecture and mechanisms for Quality of Service. Academic Press, London (2001)

DCA for Minimizing the Cost and Tardiness of Preventive Maintenance Tasks under Real-Time Allocation Constraint

Tran Duc Quynh[1], Le Thi Hoai An[1], and Kondo Hloindo Adjallah[2]

[1] Laboratory of Theoretical and Applied Computer Science LITA EA 3097,
University of Paul Verlaine - Metz, France
[2] LGIPM, INRIA Costeam, Ecole National d'Ingénieurs de Metz, France

Abstract. In this paper, we introduce a new approach based on DC (Difference of Convex functions) Programming and DCA (DC Algorithm) for minimizing the maintenance cost involving flow-time and tardiness penalties. The main idea is to divide the horizon considered into H intervals and the problem is first formulated as a mixed integer linear problem (MILP). It is afterward reformulated in the form of a DC program by an exact penalty technique. Solution method based on DCA is investigated to solve the resulting problem. The efficiency of DCA is compared with the algorithm based on the new flow-time and tardiness rule (FTR) given in [8]. The computational results on several test problems show that the solutions provided by DCA are better.

Keyword: DC Programming, DCA, flow-time, tardiness, release date.

1 Introduction

This paper addresses a job-shop scheduling problem for families of regenerative or "pseudo-periodic" jobs with unequal release dates, under real-time and identical resources availability constraints. Jobs are grouped in independent families, and jobs in a family are tied by precedence constraints. Jobs are processed parallel by independent resources. Except for the initial job of each family, the release dates of the other jobs need to be computed at the end of the precedent one, which induces real time decision. A job is processed only when a resource becomes available. One aims to maximize the number of jobs and to minimizing the make-span and tardiness of jobs on a given time window. This is a dynamic job-shop scheduling problem with assignment of $n(t)$ jobs without pre-emption to m processors and with constraints of precedence, real-time and resource availability on a time window where the number of tasks is function of the decision date. The problem may be modeled using mathematical programming. The non-dynamic problem has been shown to be an NP-hard in [14].

Under real-time decision conditions, exact methods for optimal solutions are time consuming, which explains the development of heuristic approaches. In [12], authors propose an algorithm with complexity $O(n)$ for solving the real-time decision making problem by turning it into a real-time insertion problem

N.T. Nguyen, M.T. Le, and J. Świątek (Eds.): ACIIDS 2010, Part II, LNAI 5991, pp. 410–419, 2010.

of new incoming unplanned jobs. Their algorithm may be easily turned into an heuristic. They considered a single machine but problem may apply to corrective maintenance job insertion into preventive maintenance jobs plan. The literature contains few works based on determinist methods for solving this problem to provide exact.

One will keep in mind that [8] has considered the optimization problem over a continuous time horizon. The approach introduced tries to bring a near optimal solution to realistic maintenance management of a system comprising N facilities, by using DCA model and algorithm. In this paper we consider a discrete time horizon by dividing the horizon considered into H intervals. The problem is first formulated as a mixed integer linear problem (MILP). It is afterward reformulated in the form of a DC program by the exact penalty techniques. Solution method based on DCA, an innovative approach in nonconvex programming framework ([1], [2], [5], [6]), is investigated to solve the resulting problem. The proposed DCA has interesting convergence properties: our algorithm converges to a local solution after a finitely many iterations and it consists of solving a linear program at each iteration. Moreover although the DCA is a continuous approach that works on a continuous domain, it provides an integer solution. This is unusual in continuous approaches and is original property of the proposed DCA. The efficiency of DCA is compared with the algorithm based on FTR rule [8]. The computational results on several test problems show that the solutions provided by DCA are better.

The rest of the paper is organized as follows. The problem statement is presented in Section 2 and the mathematical formulation of the problem is described in Section 3. The solution method via DC programming and DCA is developed in Section 4 while the computational results are reported in Section 5. Finally, we conclude the paper by Section 6.

2   Problem Statement

In this paper, we consider the discrete times that could be measured in hours or days, and each preventive maintenance action puts systematically the equipment to shutdown for a number of time periods. In the same time, we are interested in the preventive maintenance cost minimization of productive facilities, without distinction between the natures of actions (simple inspection, preventive replacement, adjustments to restore the capabilities, palliative repair for slowing down the degradation speed, etc. of the equipment) as often in most works.

We consider a system composed of N facilities that may request renewable maintenance jobs by q parallel processors. In production paradigm, facilities stand for the product that must be processed and the machines are maintenance operators or processors. The number of processors is limited compared to the number of facilities in the system $q << N$. No set-up time is required for the jobs. Tardiness and makespan of jobs pertaining to each equipment must be minimized while maximizing number of processed jobs on the finite horizon H.

Let us consider a finite time horizon divided into H intervals. Except for the initial job of each family, the release date of the job in k_i^{th} position on the ith facility is obtained by the relation, $r_{i,k_i} = c_{i,k_i-1} + \rho_i$. The value ρ_i is given for each facility and corresponds to the mean time between two consecutive maintenance actions.

Starting from the release date r_{i,k_i}, it is mandatory to complete the k_i^{th} job undertaken on the ith facility before the deadline d_{i,k_i} to not expand the tardiness penalties., i.e. for the reasons mentioned sooner, the k_i^{th} job must be completed within the interval $[r_{i,k_i}, d_{i,k_i}]$: $d_{i,k_i} = c_{i,k_i-1} + \delta_i$, with $\rho_i < \delta_i$. The value δ_i is also given for each facility.

As defined sooner, the makespan and the total tardiness of jobs on the time horizon H is expressed by:

$$C_H = \sum_{i=1}^{N} \left(\sum_{k_i/r_{i,k_i} \in H} \left(W_0(c_{i,k_i} - r_{i,k_i}) + W_1 \max(0, c_{i,k_i} - d_{i,k_i}) \right) \right)$$

where r_{i,k_i} denotes the job release date, c_{i,k_i} the completion date and d_{i,k_i} the due date of the k_i^{th} job on the ith equipment. W_0 and W_1 represent respectively the costs per time period of the jobs without and with tardiness.

We assume also that the jobs processing time p_i is specific and is given for each facility.

3   Mathematical Formulation

First, let us introduce the decision variables $x_{i,t}$ defined as: $x_{i,t} = 1$ if i^{th} equipment maintenance takes place at the time t, 0 otherwise. The required constraints on the number q of available processors is then expressed as

$$\sum_{i=1}^{N} x_{i,t} \le q \quad \forall t = 1, 2, .., H. \tag{1}$$

The conditions ensuring that the duration between two consecutive maintenance jobs on the ith equipment is greater than ρ_i can be written as

$$x_{i,t} + x_{i,t+1} + ... + x_{i,t+\rho_i+p_i-1} \le p_i \quad \forall t = 1, 2, ..., H - \rho_i - p_i + 1. \tag{2}$$

Moreover, since all the equipment is new at the beginning of the considered horizon, we know that

$$x_{i,t} = 0 \quad \forall i, \forall t < \rho_i. \tag{3}$$

Furthermore the following constraints assure that the job processing time of the ith equipment is p_i.

$$x_{i,H} + \sum_{t=1}^{H-1} |x_{i,t+1} - x_{i,t}| = \frac{2}{p_i} \sum_{t=1}^{H} x_{i,t}. \tag{4}$$

Indeed, if there is one job whose processing time of the ith equipment is not equal to p_i then from (2), we see that this job processing time is strictly smaller than p_i. Therefore, $\frac{2}{p_i} \sum_{t=1}^{H} x_{i,t} < 2T_i$, where T_i is the maintenance number of ith equipment in all horizon $[0, H]$. In other hand, $x_{i,H} + \sum_{t=1}^{H-1} |x_{i,t+1} - x_{i,t}| = 2.T_i$.

We can replace (4) with the next linear constraints, by using the auxiliary variables $z_{i,t}$:

$$0 \leq z_{i,t} \leq 1 \quad \forall i, \forall t = 1, 2, .., H; \tag{5}$$

$$\sum_{t=1}^{H} z_{i,t} = \frac{2}{p_i} \sum_{t=1}^{H} x_{i,t} \quad \forall i = 1, 2, ..., N; \tag{6}$$

$$x_{i,t+1} - x_{i,t} \leq z_{i,t} \quad \forall i = 1, 2, .., N \ \forall t = 1, 2, .., H - 1; \tag{7}$$

$$x_{i,t} - x_{i,t+1} \leq z_{i,t} \quad \forall i - 1, 2, .., N \ \forall t = 1, 2, .., H - 1; \tag{8}$$

$$z_{i,H} = x_{i,H} \quad \forall i = 1, 2, .., N. \tag{9}$$

We now define the objective function of our optimization problem. Since $\frac{1}{p_i} \sum_{i=1}^{H} x_{ik}^{t}$ is the maintenance number of ith equipment in horizon $[0, H]$, the flow-time of ith equipment in horizon $[0, H]$ is computed as follows:

$$\sum_{k_i/r_{i,k_i} \in H} (c_{i,k_i} - r_{i,k_i}) = H - (1 + \frac{1}{p_i} \sum_{i=1}^{H} x_{i,t}).\rho_i.$$

For computing the tardiness, we consider two consecutive maintenances (see Figure1). In this figure, t represents the completion date of the previous job. $t + \delta_i$, t', $t' - p_i$ are, respectively, the due date, the completion date, the beginning date of the next job.

Let us set $y_{i,t} = x_{i,t} + x_{i,t+1} + ... + x_{i,t+\delta_i - p_i - 1}$.

We can see that if $t' > t + \delta_i$, then the tardiness is $t' - t - \delta_i$. It is the number of variables $y_{i,t}$ which are equal to zero. In the case where $t' \leq t + \delta_i$ there is no variable $y_{i,t} = 0$. The tardiness cost is so expressed as

$$\sum_{k_i/r_{i,k_i} \in H} W_1 \max(0, c_{i,k_i} - d_{i,k_i}) = W_1. \sum_{t=1}^{H-\delta_i+p_i+1} [1 - min(y_{i,t}, 1)].$$

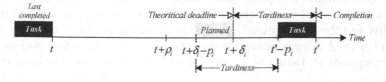

Fig. 1.

By considering the variables $w_{i,t}$ satisfying $0 \leq w_{i,t} \leq 1$ and $1 - y_{i,t} \leq$
$w_{i,t}$, we have $W_1 . \sum\limits_{t=1}^{H-\delta_i+p_i+1} [1 - \min(y_{i,t}, 1)] = W_1 . \sum\limits_{t=1}^{H-\delta_i+p_i+1} w_{i,t}$, and the cost
on the ith equipment is therefore expressed as $W_0 . H - W_0 (1 + \frac{1}{p_i} \sum\limits_{i=1}^{H} x_{i,t}) . \rho_i +$
$W_1 . \sum\limits_{t=1}^{H-\delta_i+p_i+1} w_{i,t}$. Finally the optimization model of our problem can be written
in the form

$$(P) \qquad \min \sum_{i=1}^{N} \left\{ W_0 . (H - \rho_i) - W_0 . \frac{\rho_i}{p_i} \sum_{t=1}^{H} x_{i,t} + W_1 . \sum_{t=1}^{H-\delta_i+p_i+1} w_{i,t} \right\},$$

subject to the constraints (1-3), (5-9) and

$$x_{i,t} + x_{i,t+1} + ... + x_{i,t+\delta_i-p_i-1} + w_{i,t} \geq 1 \quad \forall i, \forall t = 1, 2, .., H - \delta_i + p_i + 1. \quad (10)$$

$$0 \leq w_{i,t} \leq 1 \quad \forall i, \forall t = 1, 2, ..., H - \delta_i + p_i + 1. \quad (11)$$

$$x_{i,t} \in \{0, 1\} \quad \forall i, \forall t = 1, 2, .., H. \quad (12)$$

The above problem is a mixed integer linear program. In the next section we
will introduce a solution method via DC Programming and DCA.

4   DCA for Solving the Problem (P)

4.1   DC Programming and DCA for Mixed 0-1 Linear Program (M01LP)

DC programming and DCA introduced by Pham Dinh Tao in 1984 and extensively developed by Le Thi Hoai An and Pham Dinh Tao since 1994 (see [2],
[5], [6] and references therein) is an efficient approach for nonconvex continuous
optimization. They address a general DC program of the form

$$\alpha = \inf \{ \mathcal{F}(x) := g(x) - h(x) \; : \; x \in \mathbb{R}^n \} \quad (P_{dc})$$

where g and h are lower semi-continuous proper convex functions on $\mathbb{R}^n$. Such
a function $\mathcal{F}$ is called DC function, and $g - h$, DC decomposition of $\mathcal{F}$ while the
convex functions g and h are DC components of $\mathcal{F}$. It should be noted that a constrained DC program whose feasible set C is convex can always be transformed
into an unconstrained DC program by adding the indicator function χ_C of C
($\chi_C (x) = 0$ if $x \in C, +\infty$ otherwise) to the first DC component g. Based on the
DC duality and the local optimality conditions, the idea of DCA is simple : each
iteration l of DCA approximates the concave part $-h$ by its affine majorization
(that corresponds to taking $y^l \in \partial h(x^l)$) and minimizes the resulting convex
function

$$\min \{ g(x) - h(x^l) - \langle x - x^l, y^l \rangle : x \in \mathbb{R}^d \} \quad (P_l)$$

to obtain x^{l+1}. In the past years DCA has been successfully applied in several areas such as Machine Learning, Image processing, Network communication, Management science whose Suplly chain and Transport-Logistic, Finance, ect (see e.g. [1], [2], [5], [6] and the list of references in http://lita.sciences.univ-metz.fr/~lethi/DCA.html).

In [1] a continuous approach based on DCA has been developed for the well-known linearly constrained quadratic zero-one programming

$$\alpha = \min \left\{ f(x) := \frac{1}{2} x^T C x + c^T x : Ax \le b, x \in \{0,1\}^n \right\}, \quad \text{(01QP)}$$

where C is an $(n \times n)$ symmetric matrix, c, $x \in \mathbb{R}^n$, A is an $(m \times n)$ matrix and $b \in \mathbb{R}^m$. The idea is based on the following theorem:

Theorem 1. *[3] Let K be a nonempty bounded polyhedral convex set, f be a finite concave function on K. Then there exists $t_0 \ge 0$ such that for all $t > t_0$ the following problems have the same optimal value and the same solution set:*

$$(P_t) \qquad \alpha(t) = \inf\{f(x) + tp(x) : x \in K\}$$
$$(P) \qquad \alpha = \inf\{f(x) : x \in K, p(x) \le 0\}.$$

More precisely if the vertex set of K, denoted by $V(K)$, is contained in $\{x \in K, p(x) \le 0\}$, then $t_0 = 0$, otherwise $t_0 = \min\{\frac{f(x) - \alpha(0)}{S} : x \in K, p(x) \le 0\}$, where $S := \min\{p(x) : x \in V(K), p(x) > 0\} > 0$.

For applying the above theorem, the problem (01QP) is first rewritten in the equivalent strictly concave quadratic minimization problem

$$\text{(01CQP)} \quad \alpha = \min \left\{ \frac{1}{2} x^T (C - \bar\lambda I) x + (c + \frac{\bar\lambda}{2} e)^T x : Ax \le b, x \in \{0,1\}^n \right\},$$

where e is the vector of ones and $\bar\lambda > 0$ is chosen such that the matrix $C - \bar\lambda I$ is semi definite negative. By considering the penalty concave function p with nonnegative values on K defined as $p(x) = (1/2)[e^T x - x^T x]$, the problem (01CQP) can be written in the form

$$\text{(01QP)} \Longleftrightarrow (P) \quad \min\{f(x) := \frac{1}{2} x^T (C - \bar\lambda I) x + (c + \frac{\bar\lambda}{2} e)^T x : x \in K, p(x) \le 0\}.$$

From Theorem 1 above it follows that there exist $t_o \ge 0$ such that for all $t > t_o$ (P) is equivalent to the following problem

$$\min\{f(x) + tp(x) : x \in K\} = \min \left\{ \frac{1}{2} x^T (C - \bar\lambda I) x + (c + \frac{\bar\lambda}{2} e)^T x + tp(x) : x \in K \right\}.$$

Finally, the last problem can be formulated as the (strictly) concave quadratic minimization program:

$$\text{(CCQP)} \qquad \min \left\{ \frac{1}{2} x^T (C - \lambda I) x + (c + \frac{\lambda}{2} e)^T x : x \in K \right\},$$

with $\lambda = \bar{\lambda} + t$. Its solution set is so contained in the vertex set of K. A fast and efficient DCA scheme has been developed in [1] for the (CCQP) problem. The proposed DCA has interesting convergence properties: it converges to a local solution after a finitely many iterations and it consists of solving one linear program at each iteration. Moreover although the DCA is a continuous approach that works on a continuous domain, it provides an integer solution. This advantage of the DCA makes our method efficient in finding an integer ε-solution of (CCQP) in the large scale setting.

In the next subsection we will show how to adapt this approach for solving the Mixed 0-1 Linear Program (P).

4.2   DCA Applied to the Problem (P)

Let L be the number of variables of the problem (P), say $L = 2NH + \sum_{i=1}^{N}(H - \delta_i + p_i + 1)$. Denote by S the feasible set of (P) and let be the linear relaxation domain of S, namely $K := \{(x, z, w) \in \mathbb{R}^L$ satisfy (1-3), (5-11) and $x_{i,t} \in [0, 1]\}$.

We consider the function $p : \mathbb{R}^L \to \mathbb{R}$ defined by $p(x, z, w) = \sum_{i=1}^{N}\sum_{t=1}^{H} x_{i,t}(1 - x_{i,t})$. It is clear that $p(x, z, w)$ is concave and finite on K, $p(x, z, w) \geq 0 \; \forall (x, z, w) \in K$ and $S = \{(x, z, w) : p(x, z, w) \in K, p(x, z, w) \leq 0\}$.

Hence the problem (P) can be rewritten as

$$\min\left\{\sum_{i=1}^{N}\left\{W_0(H-\rho_i)-W_0\frac{\rho_i}{p_i}\sum_{t=1}^{H} x_{ik}^t+W_1.\sum_{t=1}^{H-\delta_i+p_i+1} w_{i,t}\right\} : (x, z, w) \in K, p(x, z, w) \leq 0\right\}. \tag{13}$$

By using theorem 1, we obtain, for the sufficiently large number η, the equivalent concave minimization problem to (13)

$$\min\left\{\sum_{i=1}^{N}\left\{W_0(H-\rho_i)-W_0\frac{\rho_i}{p_i}\sum_{t=1}^{H} x_{i,t}+W_1.\sum_{t=1}^{H-\delta_i+p_i+1} w_{i,t}\right\} - \eta.p(x, z, w) : (x, z, w) \in K\right\}. \tag{14}$$

(14) is a DC problem whose feasible set is K and the objective function is

$$f_\eta(x, z, w) = g(x, z, w) - h(x, z, w) \;, \text{ where} \tag{15}$$

$$g(x, z, w) := \chi_K(x, z, w);$$

$$h(x, z, w) := -\sum_{i=1}^{N}\left\{W_0(H-\rho_i)-W_0.\frac{\rho_i}{p_i}\sum_{t=1}^{H} x_{i,t}+W_1.\sum_{t=1}^{H-\delta_i+p_i+1} w_{i,t}\right\}+\eta.\sum_{i=1}^{N}\sum_{t=1}^{H} x_{i,t}(x_{i,t}-1).$$

DCA applied to the DC program (15) consists of computing, at each iteration l, the two sequences $\{u^l\}$ and $\{\varsigma^l = (x^l, z^l, w^l)\}$ such that $u^l \in \partial h(x^l, z^l, w^l)$ and $(x^{l+1}, z^{l+1}, w^{l+1})$ solves the next linear program

$$\min\{-\langle(x, z, w), u^l\rangle : (x, z, w) \in K\}.\tag{16}$$

From the definition of h, $u^l \in \partial h(x^l, z^l, w^l)$ is computed as

$$\begin{cases} u^l_{H.(i-1)+t} = 2.\delta x^l_{i,t} + W_0.\frac{\rho_i}{p_i} & \forall i = 1, 2, .., N, \forall t = 1, 2, ..., H, \\ u^l_j = 0 & \forall j = NH, NH + 1, ..., 2NH. \\ u^l_j = -W_1 & \text{otherwise.} \end{cases}\tag{17}$$

The DCA applied to (15) can be described as follows.
Initialization
Let ϵ be a sufficiently small positive number. Set $l = 0$ and $(x^0, z^0, w^0) \in \mathbb{R}^L$.
Repeat
Calculate $u^l \in \partial h(x^l, z^l, w^l)$ via (17).
Solve the linear problem (16) to obtain $(x^{l+1}, z^{l+1}, w^{l+1})$.
$l \longleftarrow l + 1$
Until $\|(x^{l+1}, z^{l+1}, w^{l+1}) - (x^l, z^l, w^l)\| \le \epsilon$.

Theorem 2. *(Convergence properties of Algorithm DCA)*
(i) DCA generates the sequence (x^l, z^l, w^l) contained in $V(K)$ such that the sequence $\{f_\eta(x^l, z^l, w^l)$ is decreasing.
ii) The sequence (x^l, z^l, w^l) converges to $(x^*, z^*, w^*) \in V(K)$ after a finite number of iterations.
(iii) The point (x^*, z^*, w^*) is a critical point of Problem (15). Moreover if $x^*_i \ne \frac{1}{2}$ for all $i \in \{0, ..., NH\}$, then x^* is a local solution to (15).
(iv) For a number η sufficiently large, if at iteration r we have $x^r \in \{0, 1\}^{NH}$, then $x^l \in \{0, 1\}^{NH}$ for all $l \ge r$.

Proof. Immediate consequent of Theorem 1 in [1].

5   Numerical Result

The DCA is coded in C++. At each iteration, the solver CPLEX (version 11.2) is used to solve the linear problem. The efficiency of DCA is compared with the algorithm based on FTR algorithm given in [8]. All computational experiments are carried out on Core 2duo 3.0 GHz and 2.0 GB RAM.

We have tested the two algorithms on systems with an overall number of 100 equipments ($N = 100$). The parameters W_0 and W_1 are chosen as 1.0. The two horizons considered are respectively $H = 60$ days (2 months) and $H = 90$ days (3 months). The processor number is varied from 2% to 20% of the equipment number. The procedure used to assign the tasks to the processors enables balancing their load on the optimization time window. The results are presented in the figure 2 and the figure 3. The following notations have been used:

⋄ The average cost (or the mean cost) is calculated by $\frac{BestVal}{N}$, where $BestVal$ is the objective function value (total cost) and N is the number of equipments.

⋄ The average processing time or the mean time is computed by $\frac{T}{q}$, where T is the total time of processors utilization.

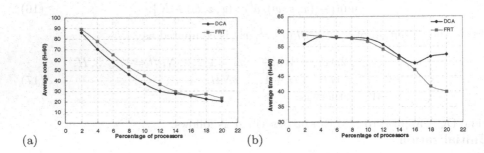

Fig. 2. Results with H=60

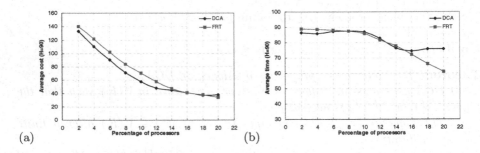

Fig. 3. Results with H=60

In the figure 2 (resp. the figure 3) we present the results obtained in the case $H = 60$ (resp. $H = 90$). The average cost is showed in the figures 2(a) and 3(a) while the average time is given in the figure 2(b) and 3(b).

We observe, from these results, that:

- The average cost decreases when the number of processors increases.

- In almost instances, the average costs provided by DCA are smaller than that of FTR Algorithm. Particularly, the differences are much important in the cases where $q = 4, 6, 8, 10, 12$. There is only one instance with $q = 20, H = 90$ where the average cost provided by DCA is greater than that of FTR Algorithm, but the difference is slight.

- The average processing time furnished by DCA is greater than FTR Algorithm in almost instances. Especially, when q=18, and q=20, DCA is much better. Note that the time of processors utilization is a criterion to evaluate the efficiency of processors utilization.

6   Conclusion

In this paper, we present the discrete formulation for minimizing the maintenance cost involving flow-time and tardiness penalty. A new and efficient approach based on DC Programming and DCA is proposed to solve the considered problem. The proposed DCA converges to a local solution after a finitely many iterations and it requires solving one linear program at each iteration. Moreover although the DCA is a continuous approach, it provides an integer solution. The computational results show that DCA overcomes FTR Algorithm, a recent efficient heuristic approach. In a future work we plan to combine DCA with global approaches. In addition to a more efficient tasks assignment procedure for balancing the load of the processors on the optimization time window, we will investigate the optimal number of processors that enables to minimize the maintenance cost.

References

1. Le Thi, H.A., Pham Dinh, T.: A Continuous approach for globally solving linearly constrained quadratic zero-one programming problem. Optimization 50(1-2), 93–120 (2001)
2. Le Thi, H.A., Pham Dinh, T.: The DC (difference of convex functions) Programming and DCA revisited with DC models of real world non convex optimization problems. Annals of Operations Research 133, 23–46 (2005)
3. Le Thi, H.A., Pham Dinh, T., Le Dung, M.: Exact penalty ind.c. programming. Vietnam Journal of Mathematics 27(2), 169–178 (1999)
4. Rockafellar, R.T.: Convex analysis, 1st edn. Princeton University Press, Princeton (1970)
5. Pham Dinh, T., Le Thi, H.A.: Convex analysis approach to d.c Programming: Theory, Algorithms and Applications. Acta Mathematica Vietnamica, dedicated to Professor Hoang Tuy on the occasion of his 70th birthday 22(1), 289–355 (1997)
6. Pham Dinh, T., Le Thi, H.A.: DC optimization algorihms for solving the trust region subproblem. SIAM J. Optimization 8, 476–505 (1997)
7. Le Thi, H.A., Pham Dinh, T.: A continuous approach for the concave cost supply problem via DC programming and DCA. Discrete Applied Mathematics 156, 325–338 (2008)
8. Adjallah, K.H., Adzakpa, K.P.: Minimizing maintenance cost involving flow-time and tardiness penalty with unequal release dates. Journal of Risk and Reliability, Part O 221(1), 57–66 (2007) (in press)
9. Adzakpa, K.P.: Maintenance of distributed systems: methods for real-time decision-making. PhD Thesis, University of Technology of Troyes, France (October 2004)
10. Burke, E.K., Smith, A.J.: Hybrid evolutionary techniques for the maintenance scheduling problem. IEEE 2000 Trans. on Power Systems 15(1), 122–128 (2000)
11. Derman, C., Lieberman, G.J., Ross, S.M.: On the optimal assignment of servers and a repairman. Journal of Applied Probability 17(2), 577–581 (1980)
12. Duron, C., Ould Louly, M.A., Proth, J.-M.: The one machine scheduling problem: Insertion of a job under the real-time constraint. European Journal of Operational Research 199, 695–701 (2009)
13. Gopalakrishnan, M., Mohan, S., He, Z.: A tabu search heuristic for preventive maintenance scheduling. Computers & Industrial Engineering 40(1-2), 149–160 (2001)
14. Lenstra, J., Kan, A.R., Brucker, P.: Complexity of machine scheduling problems. Annals of Discrete Mathematics 1, 343–362 (1977)

Cooperative Agents Based-Decentralized and Scalable Complex Task Allocation Approach Pro Massive Multi-Agents System

Zaki Brahmi[1], Mohamed Mohsen Gammoudi[2], and Malek Ghenima[3]

[1] Faculty of Sciences of Tunis, Tunisia
[2] High School of Statistics and Information Analysis of Tunis, Tunisia
[3] Higher School of Electronic Business of Manouba, Tunisia
Zaki.Brahmi@isigk.rnu.tn, mohamed.gammoudi@fst.rnu.tn,
Malek.Ghenima@escem.rnu.tn

Abstract. A major challenge in the field of Multi-Agent Systems is to enable autonomous agents to allocate tasks efficiently. In previous work, we have developed a decentralized and scalable method for complex task allocation for Massive Multi-Agent System (MMAS). The method was based on two steps: 1) hierarchical organization of agent groups using Formal Concepts Analysis approach (FCA) and 2) computing the optimal allocation. The second step distributes the tasks allocation process among all agent groups as follows:

   i. Each local allocator proposes a local allocation, then
  ii. The global allocator computes the global allocation by resolution of eventual conflict situations.

Nevertheless, a major boundary of the method used to compute the global allocation is its centralized aspect. Moreover, conflicts process is a greedy solution. In fact, if a conflict is detected steps i) and ii) are reiterated until a non conflict situation is attained. This paper extends our last approach by distributing the global allocation process among all agents. It provides a solution based on cooperation among agents. This solution prohibits generation of conflicts. It's based on the idea that each agent picks out its own sub-task.

Keywords: task allocation, Massive Multi-Agent, conflict, cooperation.

1 Introduction

A major challenge in the field of Multi-Agent Systems is to enable autonomous agents to allocate tasks efficiently. Tasks allocation is defined, in [1], as the ability of agents to self-organize in groups of agents in order to perform one or more tasks which are impossible to perform individually. In this work, we address the problem of task allocation in a cooperative environment. Common goal of agents is to maximize system's overall profit [2], [3]. This kind of allocation finds its applicability in many areas of the real world such as e-commerce [4], distributed delivery vehicles [5], grid computing [6], etc. These applications, generally, involve a large number of interacting agents. Thus, we

N.T. Nguyen, M.T. Le, and J. Świątek (Eds.): ACIIDS 2010, Part II, LNAI 5991, pp. 420–430, 2010.
© Springer-Verlag Berlin Heidelberg 2010

call this kind of MAS, Massive MAS (MMAS). A MMAS is a very large MAS, containing tens to hundreds of thousands of agents.

In the context of MMAS which is characterized by a large number of dynamic and heterogeneous agents, traditional approaches based on the negotiation between agents, or a single allocator agent, proved impracticality. Therefore, we think that a task allocation method for MMAS is judged efficient unless it's scalable, decentralized and dynamic. In [7], we have proposed a decentralized and scalable complex task allocation method for MMAS. This method is based on the idea of a group for each sub-task, agents which potentially, able to perform it, and then affect it to the appropriate agent which proposes the minimal cost.

For this finality, we have proposed in [8] a decentralized method which is based on two steps:

1. Hierarchical organization of agent groups using Formal Concepts Analysis approach (FCA),
2. Computing the optimal allocation

The second step, focus of this paper, proceeds as follows:

1. Each local allocator suggests a local allocation,
2. The global allocator computes the optimal global allocation by working out the union of all local allocations after resolution of eventual conflict situations. If a conflict is detected, the related allocation will be recalculated.

The global allocation is the union of the local allocations set proposed by each local allocator. The main difference of our method from others (e.g. [9]) is the fact that we don't have to generate all possible allocations. However, this method remains centralized at the level of computing the global optimal allocation. Moreover, the conflict process is a greedy algorithm.

The aim of this paper is to extend our previous approach by distributing the process of computing the optimal allocation among all agents based on the hypothesis: *non conflict will be generated in the task allocation processes*. Indeed, while being based on its Galois Sub-Hierarchy (GSH) and cooperation with other agents, each agent chooses the appropriate sub-task that ensures the global allocation optimality.

The rest of this paper is organized as follows: in the next section we present the formalisms on which we base our proposal. In Section 3, we present a brief reminder of our task allocation method for MMAS. Section 4 introduces our new distributed global allocation method. Finally, we conclude our work and we give some perspectives.

2 FCA Basic Concepts

We remind the mathematical foundations of Formal Concepts Analysis approach (FCA) as they are basic for this work. We give the following definitions, for more details, the interested reader is invited to see [10].

2.1 Formal Context

A formal context is a triplet *(O, A, I)* for which *O* is a set of objects (or entities), A is a set of attributes (or properties) and *I(P(O), P(A))* a binary relation between *O* and *A*. R associates an object to a property: *(o, a)* ∈ *I* when "*o* has the property *a*" or "the property *a* is applied to the object *o*".

In our task allocation problem, objects are agents and properties are sub-tasks. The incidence relation indicates the sub-tasks for which agents have the necessary capabilities to perform.

2.2 Correspondence of Galois

Given a formal context *R (O,A, I)*, we define two functions *f* and *g* making possible to express the correspondences between the subsets of objects *P(O)* and the subsets of attributes *P(A)* induced by relation *R*, as follows: *f* is the application which with any element *o* of *O* associates $f(o) = \{a \in A/ (o, a) \in I\}$, and g is the application which with any element *a* of *A* associates $g(a) = \{o \in O / (o, a) \in I\}$. These two applications constitute the Galois correspondence of the context *R*.

2.3 Concept

Given a context *R(O,A,I)* and *O1* ∈ *O* and *A1* ∈ *A*, the couple *C=(O1, A1)* is called a concept of *R* if and only if $f(O1) = A1$ and $g(A1) = O1$. *O1* is called the extent of the concept; *A1* is called the intent of the same concept.

For the proposed method, a formal concept is like an agents' group. It connects a set of agents (extent) to a set of sub-tasks (intent).

2.4 Galois Lattice (Lattice of Concepts)

The set *L* of all formal concepts, provided with order relation $\leq_l$: $(O2, A2) \leq_l (O1, A1)$, $\Leftrightarrow A1 \subseteq A2$ (or $O2 \subseteq O1$), has the mathematical structure of lattice and is called Galois lattices (or concepts) of the context *(O, A, I)*.

2.5 Object Concept (Resp. Attribute Concept)

An object (resp. an attribute) *x* is said to be introduced by a concept *C* if *x* is in the extent (resp. intent) of this concept and no ancestor (resp. descendant) of this concept contains *x* in it extent (resp. intent). *C* is called the object concept (resp. attribute concept) of *x*. Object (resp. attribute) introduced by a formal concept formed his reduced extent (resp. reduced intent). To obtain a reduced extent (resp. reduced intent) of a concept, we will remove from its extent (resp. intent) any formal object (resp. formal attribute) which is present in any extent of its ancestor (resp. descendant).

2.6 Galois Sub-Hierarchy

The Galois Sub-Hierarchy (GSH) was introduced by Godin [11] with aims to construct a hierarchy of classes [13], in order to reduce the number of concepts. A GSH deletes all concepts which are neither objects concept nor attributes concept. This means deleting concepts for which its reduced extension and intension are empty.

3   Complex Task Allocation Method

Before reminding our task allocation method, we present some definitions useful for its understanding and the formalization of the task allocation problem.

3.1   Problem Statement

The task allocation problem can be formulated as follow: Consider a set T of n tasks T = $\{T_1,.., T_n\}$, where each task T_i is composed of a set of sub-tasks ST_{ij}. Each task T_i has an execution cost $C(T_i)$. The costs of sub-tasks are not given. We suppose that the set of sub-task is equal to p. And given a set MAS of m agents MAS = $\{Ag_1,.. Ag_m\}$ where each agent Ag_i has a set of capacities to carry out one or several sub-tasks. The goal is to find the feasible and optimal allocation which forms a set of agents in order to perform the set of tasks. The following definitions will be exploited.

Definition 1: an allocation AL is feasible, if it satisfies the following conditions: Cn1) each agent is assigned to at least one sub-task. Cn2) each task is either fully allocated (one agent per sub-task) or unallocated. Cn3) for each task $T_i \in AL$, the total cost proposed by the agents to perform their component sub-tasks doesn't exceed the task's total cost $C(T_i)$.

Definition 2: the value of a feasible allocation AL is the sum of costs suggested by the agents in order to perform sub-tasks belonging to AL:

$$V(Al) = \sum_{ST_{ij} \in Al} C_{Ag_i}^{ST_{ij}} \tag{1}$$

Where, $C_{Ag_i}^{ST_{ij}}$ is the cost proposed by the agent Ag_i in order to execute the sub-task ST_{ij}.

Definition 3: An agent is responsible for a group G_i if it's the first in the list of agents which compose G_i.

Definition 4: An allocator agent is an agent responsible for its group. It's named global allocator if is a member in the group associated to the top concept of the Galois lattices in FCA approach. Otherwise, it is called local allocator.

Definition 5: An allocation $AL = \{<Ag_i, ST_{jk}> \mid Ag_i \in MAS$ and $ST_{jk} \in ST_j\}$ is a distinct set of assignments. The couple $<Ag_i, ST_{jk}>$ is the assignment of the agent Ag_i to the sub-task ST_{jk} where ST_{jk} belongs to the set of sub-tasks which composes the T_j task.

Definition 6: there is a conflict between two assignments Af_i and Af_j, if they don't check the Cn1 condition (see definition 1).

3.2   Method Groundwork

Our task allocation method [7], [8] is based on the definition of an allocation which is the assignments' set of each sub-task ST_{ij} to an agent $Ag_k \in$ MAS (see definition4) such as the cost proposed by Ag_k to execute ST_{ij} is the minimal one. Being based on that, our idea is to find for each sub-task an agents' group that can execute it, and then calculate the optimal allocation without calculating all possible allocations. Indeed, our method distributes the task allocation process by sharing the calculation between all responsible agents. We find the optimal allocation by computing the union of all local allocations after resolution of potential conflict situations. These allocations are based on a process of regrouping agents according to their capacities to execute a sub-task.

In another manner, the method is based on the following steps:

1. Hierarchical organization of agent groups using the FCA approach,
2. Each local allocator proposes a local allocation,
3. Global allocation computes the global allocation by resolution of potential conflict situations.

In our previous work [7], [8] it's the global allocator agent which resolves conflicts and computes the global allocation as described in the third step. The problem of conflict between two assignments comes from the fact that either condition Cn1 or condition Cn2 is not checked (see Definition 1 in section 3.1):

- Cn1: an agent A_i is assigned to at least one sub-task;
- Cn2: A sub-task is assigned to at least one agent.

The second condition is checked in the organization of agents step [8]. The definition of the order relation $\leq_l$ between two concepts $C1$ and $C2$ means that set of agents' group ($C1$.Ext) associated to concept $C1$ is included in the set of agents' group ($C2$.Ext) associated to concept $C2$. This implies that during the task allocation process, agent $A_i \in C1$.Ext and $A_i \in C2$.Ext, can be subject of conflict between the allocation proposed by the group related to $C1$ and the one proposed by the group related to $C2$.

To resolve a conflict, we have proposed in our last work a recursive algorithm that solves a conflict between a new allocation Al_i and another allocation Al_j already existing in the list of allocations that have been proposed. Indeed, after inserting a new allocation proposed by a local allocator, the global allocator checks if a conflict exists. If so, it chooses the allocation that will be modified. Then it will be inserted in the list of allocations called Allocation. Recursively, the process will be repeated until resolution of all conflicts. Thus, the principle of the algorithm implemented by the global allocator is as follows:

For each local allocation (Al_i) communicated by a local allocator we handle these four steps:

i. Check if there is a conflict between Al_i and other local allocation in the set Allocation of allocations.
ii. If there is a conflict, choose the allocation that will be modified,

iii.   Generate a new allocation Al_j and send it to the global allocator and go to 1)

iv.   Else insert Al_i in the allocation set Allocation

The complexity of this algorithm is of $O(\sum_{i=1}^{|ST|}(|G_i.Ext| * i + i^2))$ where $|G_i.Ext|$ represents the number of agents in the group G_i. Indeed, for each assignment a number i of conflict can be generated and for each conflict the assignment that will be modified implies generation of new assignment. Thus, the complexity of steps 1), 2) and 3) is in the order of $O(|G_i.Ext| *i+ i^2)$; where i represents the number of local allocations in the Allocation set for each iteration. In the worst case, the number of agents belongs to a given group is equal to ml. The allocations set cardinality is equal to the number of sub-tasks p. Thus, the complexity of resolving processes conflict for an assignment is evaluated to $O(p*m+ p^2)$.

In the next section we propose a new approach for global allocation based on the cooperation between agents without conflict generation. The goal is to maximize the time and minimize message complexity.

4 Cooperative Agents -Based-Global Allocation Process

4.1 Global Allocation Processes

In order to improve the conflict resolution processes time and decrease the number of messages communicated between the global agent and local agents, we propose in this work a method which guarantees that no conflict will be generated. This method is based on partial order relation $\leq$ and the cooperation between agents involved in the task allocation processes. Indeed, each agent A_i has as knowledge the set of groups wherein it's member. These groups compose the Galois Sub-Hierarchy of the agent A_i. Based on this knowledge, each agent can choose one sub-task by checking the two following conditions: i) No conflict will be generated, ii) The global allocation optimality is guaranteed.

In order to reach the first goal, an agent can choose only a sub-task that it's not chosen by other agents. For this finality, agents cooperate together in the choice process. Thus, each agent A_i carries out sort extension for each group $G_j \in SHG_i$. The sort is done on agents who compose a given group, according to their cost to execute the sub-task which is the intension of the given group. The assignment of an agent to a sub-task is based on the agent's position in the extension set $G_j.Ext$. Thus, we identify threes cases: The agent is on the head of one group, on the head of more than one group, not on the head of any group.

Case 1: if an agent A_i is on the head of one group, this means that it's the best agent which can execute the related sub-task. In this case, agent A_i withdraws itself from each group G_i where it belongs ($A_i \in G_i$) and marks the sub-task ST_{ij} as assigned. Then, it sends two messages to all agents belong to the group G_i related to concept C_i as $C_i.Int = \{ST_{ij}\}$.

Case 2: if an agent A_i is on the head of more than one group, it chooses the sub-task ST_{ij} which increases global allocation optimality. Then, A_i proceeds like first case. And preventing, accordingly, that the sub-task will be chosen by another's agent.

Case 3: the agent is not on the head of any group. In this case, it waits if it will be in the head of its group when others agents withdraw their selves and there's yet again not assigned sub-tasks. Thus, it will check the first or the second case.

This approach is valid only if each group G_i has one sub-task to execute; $|G_i.\text{Int}|=1$. For this finality, each agent A_i proceeds to generate its $GSHOInt_i$, for GSH One Intension organization. Definition 7 presents our $GSHOInt$ definition.

Definition 7: a Galois Sub-Hierarchy is called $GSHOInt$ if and only if each concept $C_i \in GSH$ has only one element on its intension: $|C_i.\text{Int}| = 1$.

The approach cannot be applicable unless if each group G_i has only one sub-task; $|G_i.\text{Int}| = 1$. For that, each agent A_i proceeds to modifying its organization GSH_i into $GSHOInt$. We define a GSH_i like $GSHOInt$ as follows:

Definition 8: A Galois Sub-Hierarchy is $GSHOInt$ if and only if each concept $C_i \in GSHOInt$ doesn't have more than one element on its intension; $|Ci.\text{Int}| = 1$.

To create its $GSHOInt$, each agent A_i proceeds to the partitioning of each group G_i (concept) $\in GSH_i$ into a number of $|G_i.\text{Int}|$ groups. Hence, for each $ST_{pj} \in G_i.\text{Int}$, a new group G_j is created. G_j has the same members as the group G_i and a just one sub-task: $G_j.\text{Ext} = G_i.\text{Ext}$ and $G_j.\text{Int} = ST_{pj}$. The groups then created have the same partial order characteristics such as their original groups: $\forall G_p \in GSH_i \leq (\text{resp.} \geq) G_i$ then $G_p \leq (\text{resp.} \geq) G_j \in GSHOInt_i$.

Accordingly, to guarantee the non conflict generation each agent A_i implements the following **NoConflict** algorithm:

NoConflict algorithm

```
1. Generate the GSHOInt_i
2. Sort each G_j.Ext ∈ GSHOInt_i
3. Compute G = {G_j ∈ SHGOInti | Ai is on the head of the
   G_j.Ext list and the sub-task ST ∈ G_j.Int is not
   marked}
   a. If |G| = 1, then
      • Choose the sub-task ST, and Mark(ST)
      • Send (Withdraw, ID) and Send(mark, ST, ID)
   b. If |G| > 1, then
      • Choose the sub-task while being based on a
        choice process and Mark(ST).
      • Send (Withdraw, ID) and Send(mark, ST, ID).
   c. If |G| = 0, then
      Wait for the reception of a message from the
      agent which occupies the head of its group.
      i. If the received message is Withdraw then
```

```
         //the sub-task is not market
         Go to 3)
     ii. If the received message is mark then
         //the sub-task is chosen and an agent is
         //withdrawn
         Delete the related group from its GSHOInt_i and
         Go to 3)
```

To choose a sub-task, each agent A_i uses the choice process which is founded on the following basis:

1. Select the sub-task $ST_{kj} = G_i.Int$ where the agent A_i is the unique member inside the group G_i.

2. Select the sub-task ST_{ki} that increases the global allocation optimality:

$$C_{A_i}^{ST_{kj}}/C(T_k) \qquad C_{A_i}^{ST_{pj}}/C(T_p)$$

We do favor the first selection to undertake the affectation of all the sub-tasks. In fact, if a group holds just one agent, that stands for its aloneness to execute the sub-task in query which arises from the definition of a concept that checks the Galois Correspondence (section 1.2). The unique group that can have just one member is the inferior group in the GSH. This member is the agent A_i (That holds the GSH_i) as $A_i \in \forall\ G_j.Ext \in GSH_i$. This justifies our hypothesis: *An agent privileges the first selection.*

Being A_i and A_j two agents, we have identified a particular case that occurs once verification of the following conditions:

✓ A_i and A_j belong to the same group Gp, such as Gp.Ext = { A_i, A_j }

✓ A_i and A_j have as indices in the list Gp.Ext 1 and 2 respectively.

Therefore, if both the two agents A_i and A_j make the decision to withdraw from the group G_p simultaneously without marking the sub-task in query, then the sub-task will never be affected. To illustrate this problem, Fig.1 presents the following $GSHOInt_2$ of the agent A_2:

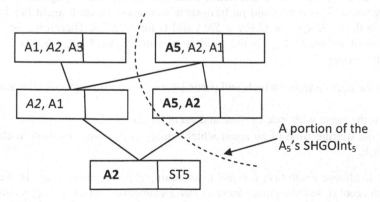

Fig. 1. agent A_2's GSHOInt

In this illustration, the agent A_2 couldn't choose except the sub-task ST_5 and the agent A_5 could choose either ST_2 or ST_3. The problem is fronted when A_5 decides to choose ST_2. In that case, the sub-task ST_3 would never be affected. The solution remains in the fact that the agent A_5 should wait until agent A_2 makes its decision. If A_2 decides to choose ST_5, then A_5 must obligatory according to the choice process that we have defined select the sub-task ST_3.

4.2 Complexity of the NoConflict Algorithm

The complexity o this algorithm depends of:

    i.   The complexity of the algorithm used to make the sort,
    ii. The sub-task's choice process,
    iii. The agent's time waiting to receive either a withdraw or mark message

For the sort algorithm, the best complexity is evaluated to $O(k.logk)$. Here, k is the number of agents that make the group G_j: $G_j.Ext$. In the worst case, the number of agents that are members of group G_j equals the number of agents implied in the global allocation process; m. This way, G_j will be the root group in the GSH.

Each agent A_i proceeds to sort all the groups that make its Sub-Hierarchy $GSHOInt_i$. In fact, in a $GSHOInt_i$, we have $|R(A_i)|$ groups. $|R(A_i)|$ is the number of sub-task that A_i can execute. In the worst case, a $SHGOInt$ has p groups and each group is composed by m agents. Thus, the complexity of the step i) is evaluated to $O(p*m*log(m))$.

The choice process is to search in the sub-tasks list the one appropriated to the agent. This means to search the minimal value in an array of $|R(A_i)|$ elements if the agent A_i is on the head of all groups that make its $GSHOInt_i$, which is the worst case. Thus, the complexity of this step is evaluated to $O(|R(A_i)|)$. It's to be noticed that an alone agent can perform the research in an array which size equals the number of the sub-tasks that it can execute. For the other agents which are not on the head of their groups, the number of groups to treat decreases according to the number of marked sub-tasks.

Each agent $A_{i,Ind}$ waits, in the worst case, a number of agents which is equal to $p-1$. Only the last agent that will be affected to a sub-task could wait $p-1$ agents. This is the case only when all agents could perform their sub-tasks; i.e each agent has just one group G in the GSH such as $G.Ext = SMA$ and $G.Int = \{ST_{ij}\}$. Therefore, the waiting complexity of an agent $A_{i,Ind}$ is the waiting sum of agents $A_{i,Ind-1}$, $A_{i,Ind-2}$ ..., $A_{i,1}$: $\sum_{j=1}^{Ind-1} W_j$; Where:

- Ind is the agent number which will participate in the optimal global allocation; $Ind \leq p$.
- $A_{i,Ind}$ is the agent with index $i \leq m$, and agent number Ind.
- W_j is the waiting time of an agent which equals to the time duration to choose a sub-task.

The time to choose a sub-task is equal to searching element in an array. In the worst case, each agent A_i searches in an array of $Ind-1$ elements. Thus, the $A_{i,Ind}$'s complexity choice process, in the worst case, is evaluated to $O(\sum_{j=1}^{Ind-1} p - j) = O(p \frac{p+1}{2})$

This way, the complexity of the algorithm is polynomial and inferior to the complexity of our previous approach. In the worst case, it's evaluated to:

$$O(p*(m \ log(m)+1+\frac{p+1}{2}))$$

Or, in the context of our work, the number of sub-tasks p is negligible compared to the number of agents; m, so, the complexity will be:

$$O(m \ log(m))$$

Message complexity, in the worst case, is evaluate to $O(m)$. Indeed, only agents that has choose a sub-task sends two messages (*withdraw* and *mark*) to all agents belong to these groups. In the worst case, the group's cardinality is equal to m. In this case only one group composes the agents' organization.

5 Conclusion

In our previous work we have developed a scalable and distributed task allocation method for MMAS in cooperative environments. An extension to this method is developed in this work through distributing the global allocation process. Each agent cooperates with other agents in order to calculate the allocation of one sub-task without generating any conflict. This novel method reduces the time and the number of messages communicated of our previously proposed task allocation method. As a future work, we intend to apply our approach on real cases such as the web services discovery.

References

1. Tosic, P.T., Agha, G.A.: Maximal Clique Based Distributed Group Formation For Task Allocation in Large-Scale Multi-Agent Systems. In: Ishida, T., Gasser, L., Nakashima, H. (eds.) MMAS 2005. LNCS (LNAI), vol. 3446, pp. 104–120. Springer, Heidelberg (2005)
2. Rahwan, T., Jennings, N.R.: An Algorithm for Distributing Coalitional Value Calculations among cooperating Agents. Artificial Intelligence Journal 171(8-9), 535–567 (2007)
3. Viswanathan, V., DesJardins, M.: A Model for Large-Scale Team Formation for a Disaster Rescue Problem. In: Proceedings of AAMAS 2005, LSMAS (2005)
4. Tsvetovat, M., Sycara, K.P., Chen, Y., Ying, J.: Customer coalitions in the electronic marketplace. In: Proceedings of the Fourth International Conference on Autonomous Agents, pp. 263–264 (2000)
5. Sims, M., Goldman, C.V., Lesser, V.R.: Self-Organization through Bottom-up Coalition Formation. In: AAMAS 2003, Melbourne, Australia, July 14-18 (2003)
6. Samee, U.K., Ishfaq, A.: Non-cooperative, Semi-cooperative, and Cooperative Games-based Grid Resource Allocation. In: Proceedings of IPDPS 2006, Rhodes Island, Greece (April 2006)
7. Brahmi, Z., Gammoudi, M.M.: Decentralized method for complex task allocation in massive MAS. In: Proc. of 1st Mediterranean Conference on Intelligent Systems and Automation (CISA 2008), vol. 1019, pp. 287–293 (2008)

8. Brahmi, Z., Gammoudi, M.M.: Hierarchical organization of agents based on Galois sub-hierarchy for complex tasks allocation in Massive MAS. In: Håkansson, A., Nguyen, N.T., Hartung, R.L., Howlett, R.J., Jain, L.C. (eds.) KES AMSTA 2009. LNCS (LNAI), vol. 5559, pp. 460–470. Springer, Heidelberg (2009)
9. Manisterski, E., David, E., Kraus, S., Jennungs, N.R.: Forming Efficient Agent Groups for Completing Complex Tasks. In: AAMAS 2006, Hakodate, Hokkaido-Japan (May 2006)
10. Ganter, B., Wille, R.: Formal Concept Analysis, Mathematical Foundation. Springer, Berlin (1999)
11. Godin, R., Mili, H.: Building and Maintaining Analysis-Level Class Hierarchies using Galois Lattices. In: Proc. of OOPSLA 1993, October 1993, vol. 28, pp. 394–410. ACM Press, New York (1993)
12. Montealegre Vazquez, L.E., Lopez, Y., Lopez, F.: An Agent-Based Model for Hierarchical Organizations. In: Proceedings of AAMAS (May 2006)
13. Gammoudi, M.M., Nafkha, I.: A formal method for inheritance graph hierarchy construction. Information Sciences 140(3-4), 295–317 (2002)

Mining Informative Rule Set for Prediction over a Sliding Window

Nguyen Dat Nhan[1], Nguyen Thanh Hung[2], and Le Hoai Bac[1]

[1] Faculty of Information Technology
University of Science, Ho Chi Minh City, Vietnam
[2] High School for the Gifted
Vietnam National University - Ho Chi Minh City, Vietnam
datnhan1z@yahoo.com, pthung@ptnk.edu.vn, lhbac@fit.hcmus.edu.vn

Abstract. We study the problem of mining informative (association) rule set for prediction over data streams. On dense datasets and low minimum support threshold, the generating of informative rule set does not use all mined frequent itemsets (FIs). Therefore, we will waste a portion of FIs if we run existing algorithms for finding FIs from data streams as the first stage to mine informative rule set. We propose an algorithm for mining informative rule set directly from data streams over a sliding window. Our experiments show that our algorithm not only attains high accurate results but also out performs the two-stage process, find FIs and then generate rules, of mining informative rule set.

Keywords: data streams, sliding window, informative rule set.

1 Introduction

Recently, data mining communities have focused on a new data model, in which data arrive in the form of continuous streams, called *data streams*, such as sensor data, web log and click-streams, telecom data, e-business.... Unlike mining static data, mining data streams poses many challenges. Data streams are unbounded and therefore we can not store the entire data in memory or even in secondary memory. New data continuously arrives, possibly at high speed, requires fast processing to keep up with high data arrival rate [4].

To the extent of our knowledge, all work on data streams only focus on mining FIs or frequent closed itemsets, which is just the first stage for mining association rules. Association rule set is known for its very large size and causes many difficulties when using it. The well-known effort to reduce the size of the association rule set is non-redundant association rule set [21] which uses frequent closed itemsets – a subset of FIs. Another effort is the *informative rule set* (IRS) [13] for prediction which only uses a portion of small-sized FIs. Hence, if we mine FIs from a data stream just for mining IRS will waste the portion of FIs which is not used for generating IRS.

In this paper, we combine the algorithm for mining IRS and an algorithm for mining FIs over data streams into an integrated algorithm for mining IRS over data

N.T. Nguyen, M.T. Le, and J. Świątek (Eds.): ACIIDS 2010, Part II, LNAI 5991, pp. 431–440, 2010.

streams. Our experiments show that the integrated algorithm have better performance compare to the generating of IRS from FIs mined by the algorithm for mining data streams and attains high accurate results.

The organization of the paper is as follows. We discuss the related work in Section 2. In Section 3 we give the background and introduction to mining data streams over a sliding window and informative rule set. We describe our algorithm in Section 4 and discuss our experimental result in Section 5. Section 6 concludes the paper.

2   Related Work

Most work on data streams mine FIs or frequent closed itemsets, and a few algorithms mine frequent maximal itemsets. These algorithms can be classified into three categories based on the model that they adopt: the *landmark window*, *sliding window* or *time-fading* model [5]. The landmark models mine entire data between a fixed time-stamp in the history and the present. The sliding window models mine a fixed number of recent transactions (or batches). In the time-fading window model, recent transactions are more important than previous ones. According to the number of transactions that are update each time, the algorithms are further categorized into *update-per-transaction* or *update-per-batch*.

The approaches in the landmark window model include *LossyCounting* [15], *DSM-FI* [12], *FDPM* [19], *estMax* [18] and *hMiner* [16]. any authors refer to mine FIs in the sliding window because it cost less memory and it is more reasonable to mine the most recent portion of data due to the older portion usually obsolete and may produce false effects to the knowledge discovered. The approaches in the sliding window include *Moment* [7,8], *CFI-Stream* [10], *IncMine* [6], *FPCFI-DS* [1] focus on mining frequent closed itemsets, and *MFI-TransSW* [11], *MineSW* [4], *Top-K Lossy Counting* [17] focus on mining FIs. Some approaches emphasize the more important of recent data by using the time-fading window model such as *estDec* [3], *FP-Streaming* [9]. Few algorithms such as *FPCFI-DS* [1], *Moment* [7,8] and *CFI-Stream* [10] are exact approaches, all other algorithms are support-approximate approaches. The reason is that exact methods require much more space and computation cost than approximate methods due to the combinatorial explosion. Algorithms in [4,6,9,12,15,19] start the process after each batch of transactions, but only [4,6] process all transactions in the batch at once (*update-per-batch*), the others process each transaction separately (*update-per-transaction*). All others algorithms process each transaction immediately as soon as it arrives (also called update-per-transaction). Cheng et al. [4] proves that update-per-batch is more efficient than update-per-transaction.

Association rule set and non-redundant rule set are generated from FIs and frequent closed itemsets, produce by the above algorithms. Unlike mining association rules, the process of mining IRS can be performed simultaneously with the progress of finding FIs and both will stop when it is sure that no more rules will be generated. In this paper, we incorporate IRS into *MineSW* (the state-of-the-art of update-per-batch algorithms) to achieve better algorithm for mining IRS on data streams.

3 Preliminaries

3.1 Introduction to Mining FIs over Data Stream

Let I be a set of *items*. A set $X = \{x_1, x_2, \ldots, x_n\} \subseteq I$ is called an *itemset* or *k-itemset* if X contains k items. A *transaction* is a couple (tid, T), where tid is the transaction identifier and T is an itemset over I. Transaction (tid, T) supports an itemset, X, if $T \supseteq X$. A *transaction data stream* (or data stream for short) is a sequence of incoming transactions. We denote the kth batch of transactions in the stream as B_k. A batch can be time-based or count-based. The batches are *time-based* if they contain all transactions in a fixed interval of time such as a minute, an hours, a day...; or the batches are *count-based* if they contains a fixed number of transaction in every batches. A *window* is a set of w successive batches, where w is the size of the window. A *sliding window* is a window that slides forward for every incoming batch. The *current (latest) window* $W = \langle B_{t-w+1}, B_{t-w+2}, \ldots, B_t \rangle$, where B_t is the *current (latest) batch*. The support of an itemset X in a window W or in a batch B, denote $s(X, W)$ or $s(X, B)$ is the number of transactions in W or B that support X. If the context is clear, we use $s(X)$ for short.

Given a minimum support threshold σ ($0 \leq \sigma \leq 1$) and a data stream, the problem of mining FIs over a sliding window, W, in the data stream is to find the set of all itemsets X over W that $s(X) \geq \sigma |W|$, where $|W|$ is the number of transactions in W.

To mine FIs over a data stream, it is necessary to keep not only FIs, but also the infrequent itemsets that are promising to become frequent later. Therefore, approximate algorithms use a relaxed minimum support threshold σ_r ($0 \leq \sigma_r \leq \sigma \leq 1$) to obtain extra set of itemsets that are potential to become frequent later [5]. With the use of relaxed minimum support threshold, the support of an itemset, X, that less than σ_r will be lost. Therefore, if X suddenly becomes frequent in the next window, the support we obtain for X is not accurate and we denote as $\tilde{s}(X)$.

Cheng et al. [4] propose an approximate update-per-batch algorithm, *MineSW*, for mining FIs over a sliding window. In the sliding window model, the first window must be initialized before mining FIs. *MineSW* use an existing non-streaming algorithm to mine all relaxed[1] FIs in each batch of the first window and then update their support if they already exist in the prefix tree, otherwise create new nodes for them. When a new relaxed FI appear or reappear in the tree, *MineSW* record the batch identifier (BID) for this itemset. The operations are similar for each new arrival batch in the sliding stage. After update information from the new batch, the prefix tree need to be traverse one to make sure that all itemsets still satisfy the relaxed minimum support for their life span. For example, an itemset X appear at BID $= k$, the latest batch is t ($k < t$), then X have a life span from k to t and is a relaxed FI only if $\tilde{s}(X) \geq \sigma_r(k, t)$, where $\sigma_r(k, t) = \sigma_r \times (|B_k| + |B_{k+1}| + \ldots + |B_t|)$. In the sliding stage, before updating the new arrival batch, the obsolete batch must be eliminated first. Relaxed FIs in the expiring batch will be mined again; and if their corresponding nodes exist in the prefix tree that have BID equal to the expiring BID, the expiring support will be eliminate. BID values

---

[1] *MineSW* uses a progressively increasing minimum support threshold function. In this paper we describe *MineSW* using normal relaxed minimum support threshold. Refer to [4] for more details.

are updated in the expiring process or the traversing process to make sure that no item-set life span is longer than the window size. For more details please consult [4].

3.2 Introduction to the Informative Rule Set

In the informative rule set (IRS) [13], the authors adopt the confidence priority model for prediction to exclude the more specific rules which have smaller confidence compare to their more general rule. Rule $X' \to y$ is more specific than rule $X \to y$ if $X' \supset X$ (proper subset). The IRS only includes 1-consequence rules; the rationale behind this is described in [13]. Assume that $X \to y$ is the most specific rule for y, then rules $X' \to y$ ($X' \supset X$) will not include in the IRS. If Z is the largest FI that produce rules (that means rule $Z \backslash z \to z$ is the most specific rule for some $z \in Z$ and have been excluded by its more general rule for other $z \in Z$) mining any $Z' \supset Z$ will waste time. For more details about the algorithm for mining IRS, please refer to [13].

4   Mining Informative Rule Set over a Sliding Window

In this section, we describe our proposed algorithm for mining informative rule set over a sliding window, called *MirsOswin*, which is an integrated algorithm forms by combining algorithm for mining IRS and *MineSW*. When mining IRS on a data stream, a rule exclude by the current window may appear on later window. Assume a k-itemsets Z is the largest FIs produce rules, we must maintain some superset Z' of Z in case of Z' will produce rules later in the stream. We propose to use a reserved margin, δ, to keep all Z' that $Z' \supset Z$ and $|Z'| \leq |Z| + \delta$.

The processing of first w-1 batches in *MirsOswin* is similar to *MineSW* because it does not involve mining IRS. But when the wth batch arrive and complete the first window and the process must be changed to mine IRS simultaneously with the update of itemsets mined from the new batch. We use *dEclat* [20] for mining relaxed FIs in each batch, but because of the nature of mining IRS, we transform *dEclat* from depth-first approach to breadth-first approach to allow the integration of IRS into *MineSW*.

Algorithm *MirsOswin*
Input: A data stream, σ, σ_r, δ, w
         and γ (minimum confidence, but can be omit [14])
Output: An approximate IRS for each sliding window.
Initialize the first window

1. Initialize an empty prefix tree T.
2. For each of the first w-1 batches, B_i ($1 \leq i < w$),
3.   Mine all FIs, for each mined FIs, X, check if there is a node n for X exists in T.
4.   If exists, update $n.\tilde{s}(X) = n.\tilde{s}(X) + \tilde{s}(X, B_i)$
5.   If $n.\tilde{s}(X) < \sigma_r(n.BID, i)$
6.   Remove X and stop mining supersets of X.
7.   Else, create a new node n for X with $n.BID = i$ and $n.\tilde{s}(X) = \tilde{s}(X, B_i)$
8.   For each node n in the tree T
9.   If $n.\tilde{s}(X) < \sigma_r(n.BID, i)$, then remove n from T.

10.UpdateAndMineIRS(T,B_w,σ,σ_r,δ,w,γ) //Finalize, mine IRS from the first window

Slide the window

11.Repeat for each slide

12.   RemoveExpiredBatch(T,B_{t-w+1},σ,σ_r,δ,w)

13. $t = t + 1$

14.   UpdateAndMineIRS(T,B_t,σ,σ_r,δ,w,γ)

RemoveExpiredBatch(T, B_e, σ, σ_r, δ, w)

Input: The current prefix tree T, the expired batch B_e, σ, σ_r, δ, w

Output: The prefix tree T with information from B_e removed.

1. Mine all FIs from B_e, for each mined frequent itemset, X

2.   If exists a node n in T for X and $n.BID = e$

3. $n.\tilde{s}(X) = n.\tilde{s}(X) - \tilde{s}(X, B_e)$

4.   If $n.\tilde{s}(X) = 0$, remove n from T and stop mining supersets of X.

5.   Else, set $n.BID = e+1$.

6.   Else, stop mining supersets of X

UpdateAndMineIRS(T, B_t, σ, σ_r, δ, w, γ)

Input: The current prefix tree T, the new arrival batch B_t, σ, σ_r, δ, w, γ

Output: The prefix tree T with information from B_t updated and IRS.

1. $R = \{\}$ // The empty IRS

2. $l = 1$ and $T_l = \{\}$ // Contain nodes for l-itemsets mined from B_t.

3. Mine relaxed frequent 1-itemsets from B_t

4.   For each mined relaxed frequent 1-itemset, X, check

5.   If exists a node n for X in T

6. $n.\tilde{s}(X) = n.\tilde{s}(X) + \tilde{s}(X, B_t)$

7.   If $t - n.BID + 1 > w$ then update $n.BID = t - w +1$

8.   If $n.\tilde{s}(X) < \sigma_r(n.BID, t)$ then

9.   Remove n from T and repeat step 4 for the next X

10.   Else, create a new node n for X with $n.BID = t$ and $n.\tilde{s}(X) = \tilde{s}(X, B_t)$

11. $n.margin = -1$

12. $T_l = T_l \cup \{n\}$

13.Repeat if T_l is not empty

14.   (T_{l+1},R_{l+1}) = MineFI_IRS(T_l, σ, σ_r, δ, w, γ)

15. $R = R \cup R_{l+1}$

16. $l = l+1$

17.   For each node n of l-itemset $X \in T$ and $X \notin T_l$

18.   If $t - n.BID + 1 > w$ then $n.BID = t - w +1$

19.   If $n.\tilde{s}(X) < \sigma_r(n.BID, t)$ then

20.   Remove n from T and repeat step 17 for the next node

21.   Else

22. $R = R \cup$ MineIRS(n, T, γ)

23.Return R

The first 12 lines of UpdateAndMineIRS mine all relaxed frequent 1-itemsets in B_t and place into T_l as candidates for mining larger relaxed FIs. All larger relaxed FIs and informative rules appear in B_t will be mined by MineFI_IRS. Lines 17 to 22 traverse the prefix tree, T, to process the itemsets do not appear in B_t but appear in T before. Work in breath-first model, MineFI_IRS was called to mine all $(l+1)$-itemsets from l-itemsets before mining larger itemsets. The process continues mining larger itemsets until no more relaxed FIs produced. MineFI_IRS (for mining each layer of nodes) and MineIRS (for mining IRS from a FI) just do the job specifies in [13] and preserve a margin of δ itemsets beyond the largest produce-rule itemsets.

MineFI_IRS$(T, T_l, \sigma, \sigma_r, \delta, w, \gamma)$
Input: $T, T_l, \sigma, \sigma_r, \delta, w, \gamma$
Output: T_{l+1}, the IRS R_l mined from T_l.

1. $T_{l+1} = \{\}$
2. $R_{l+1} = \{\}$
3. Mine relaxed frequent $(l+1)$-itemsets from T_l, for each mined itemset, X
4.   If exits a node n for X in T
5. $n.\tilde{s}(X) = n.\tilde{s}(X) + \tilde{s}(X, B_t)$
6.   If $t - n.BID + 1 > w$ then $n.BID = t - w + 1$
7.   If $n.\tilde{s}(X) < \sigma_r(n.BID, t)$ then
8.   Remove n from T and repeat line 2 for the next X
9.   Else, create a new node n for X with $n.BID = t$ and $n.\tilde{s}(X) = \tilde{s}(X, B_t)$
10.   If $n.\tilde{s}(X) < \sigma \times |W|$ then set $n.margin = 1$
11.   Else $R_{l+1} = R_{l+1} \cup$ MineIRS(n, T, γ)
12. $T_{l+1} = T_{l+1} \cup \{n\}$
13. Return (T_{l+1}, R_{l+1})

MineIRS(n, T, γ)
Input: A node n, T
Output: Rules produce by $n.X$

1. n.Z = n.X
2. $n.margin = \infty$
3. For each $X \in P_l(n.X)^2$
4.   If not exists a node $m \in T$ that $m.X = X$ then remove n and return NULL.
5.   Set $n.margin = min(n.margin, m.margin)$
6.   If m is restricted then mark n restricted and $n.Z = n.Z \cap m.Z$
7.   Else $n.Z = n.Z \cap (m.Z \cup n.X \backslash m.X)$
8. If $n.margin > 0$ then $n.margin = n.margin + 1$
9. If $n.margin > \delta$ then remove n and return NULL.
10. If n is not restricted
11.   If $\exists m \in T_{[l-1]}^3$ that $m.X \subset n.X$ and $n.\tilde{s}(X) = m.\tilde{s}(X)$ then

---

[2] $P_l(A) = \{A' | A' \subset A, |A'| = l\}$, where A is a $(l+1)$-itemset.
[3] $T_{[l-1]}$ is a set of all nodes for $(l-1)$-itemsets in the prefix tree.

12.   Mark n restricted [**].

13.For each $z \in n.Z$

14.   If $\exists m \in T_{[l-1]}$ that $m.X\backslash z \subset n.X\backslash z$ and $\tilde{s}((m.X \setminus z) \cup \neg z) = \tilde{s}((n.X \setminus z) \cup \neg z)$

15. $n.Z = n.Z\backslash z$

16.If n is restricted and $n.Z$ is empty, set $n.margin = 1$.

17.Return qualified rules from n to R_n.

(**) Note: In [13], after mark n restricted (line 12), the authors also remove the item $n.X\backslash m.X$ from $n.Z$. But the usage of Lemma 6 in [13] is misplaced. The correct usage of this Lemma is the intersection in line 6.

5   Experimental Evaluation

In this section, we perform experiments for evaluating the recall and precision of IRS mine by *MirsOswin*, and compare the performance to the generation of IRS from FIs mined by *MineSW*. Both algorithm are implemented in C++, built by MS Visual Studio 2005 and perform on a PC with Intel Core 2 Duo 2.53 GHz CPU, 3.8GB of RAM under Window XP Professional. We run tests on two synthetic datasets T10P4 and T15P6 generated by IBM synthetic data generator, where 10 and 15 (4 and 6) are the average transaction size (the maximal FI), each dataset consist of 3M transactions and 10K unique items. Each dataset are simulated as a data stream with a sliding window consists of 20 batches, each batch consist of 50K transactions and we have 41 consecutive slides over the entire dataset. In all experiments, we set $\sigma_r = 0.1\sigma$ as in [15] and [4]. Reference [4] suggests that *MineSW* can be used with $\sigma_r = 0.5\sigma$ or larger, but state that $\sigma_r = 0.1\sigma$ should use in application that require high accurate. Mining IRS requires high accurate result because a small change in support can include or exclude many rules.

First, we evaluate the recall and the precision of IRS mine by *MirsOswin* by comparing with IRS (complete IRS) generate from complete FIs set mined from each window. Fig.1 shows the average recall for each window when $\delta = 0$, 2 and 3, there is no difference between the results of $\delta = 1$ and $\delta = 2$. The average recall per window is at least 99% for both datasets when δ is set to 3. Hence, all other experiments will be perform with $\delta = 3$. The precision is always 100%, that means all informative rule mined by *MirsOswin* appear in the complete IRS.

Second, we evaluate the average running time per slide. Fig.2 shows that when test on dataset T10P4 *MirsOswin* only perform better than *MineSW*+IRS (total time use for mining FIs by *MineSW* and time use for generating IRS from FIs directly from RAM) at the lowest support value. On denser dataset, T15P6, *MirsOswin* out perform *MineSW*+IRS in a wider range. The tests show that integrated algorithm performs better on dense dataset and low support thresholds. The reason is obvious because at low support threshold or on dense dataset, the portion of FIs pruned by IRS becomes larger and hence *MirsOswin* process much less itemsets than *MineSW*. This reason reveals in our third test on memory consumption.

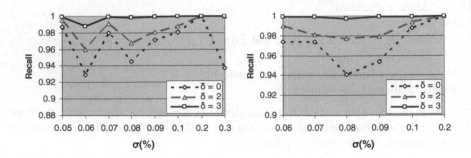

Fig. 1. Recall with different reserved margins for T10P4 (left) and T15P6 (right)

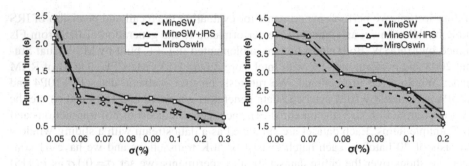

Fig. 2. Average processing time per slide for T10P4 (left) and T15P6 (right)

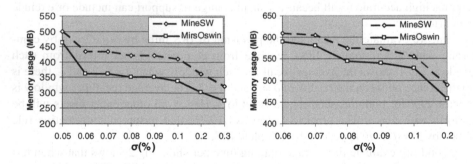

Fig. 3. Memory consumption for T10P4 (left) and T15P6 (right)

Fig.3 shows that *MirsOswin* consumes less memory than *MineSW*. Note that, *MirsOswin* use more memory for each node in the prefix tree to mine both FIs and IRS simultaneously, while *MineSW* only mine FIs. But this memory usage for sliding stage only. In the initializing stage, *MineSW* consumes less memory than *MirsOswin* because *MineSW* use the progressively increasing minimum support threshold function while *MirsOswin* does not. We can introduce the progressively increasing function into the initializing stage of *MirsOswin* to reduce some memory consumption.

6   Conclusion

In this paper, we study the problem of mining IRS set over data streams and propose an algorithm for mining IRS over a sliding window, *MirsOswin*. Our experiments prove that *MirsOswin* attain high accuracy and out performs the two-stage process of generating IRS from FIs mined by a streaming algorithm on dense datasets and low support thresholds.

Acknowledgement

We specially thank James Cheng for providing us the code of *MineSW*.

References

1. An, F., Du, J., Yan, Y., Liu, D., Huang, K.: An Efficient Algorithm for Mining Closed Frequent Itemsets in Data Streams. In: Proceedings of the 2008 IEEE 8th International Conference on Computer and Information Technology Workshops (2008)
2. Calders, T., Dexters, N., Goethals, B.: Mining Frequent Itemsets in a Stream. In: Seventh IEEE International Conference on Data Mining, Omaha, NE, pp. 83–92 (2007)
3. Chang, J.H., Lee, W.S.: Finding recent frequent itemsets adaptively over online data streams. In: Proceedings of the ninth ACM SIGKDD international conference on Knowledge discovery and data mining, Washington, D.C., pp. 487–492 (2002)
4. Cheng, J., Ke, Y., Ng, W.: Maintaining Frequent Itemsets over High-Speed Data Streams. In: Ng, W.-K., Kitsuregawa, M., Li, J., Chang, K. (eds.) PAKDD 2006. LNCS (LNAI), vol. 3918, pp. 462–467. Springer, Heidelberg (2006)
5. Cheng, J., Ke, Y., Ng, W.: A survey on algorithms for mining frequent itemsets over data streams. Knowledge and Information Systems 16(1), 1–27 (2008)
6. Cheng, J., Ke, Y., Ng, W.: Maintaining frequent closed itemsets over a sliding window. Journal of Intelligent Information Systems 31(3), 191–215 (2008)
7. Chi, Y., Wang, H., Yu, P.S., Muntz, R.R.: Moment: maintaining closed frequent itemsets over a stream sliding window. In: Fourth IEEE International Conference on Data Mining, Brighton, UK, November 1-4, pp. 59–66 (2004)
8. Chi, Y., Wang, H., Yu, P.S., Muntz, R.R.: Catch the moment: maintaining closed frequent itemsets over a data stream sliding window. In: Knowledge and Information Systems, October 2006, vol. 10, pp. 265–294. Springer, Heidelberg (2006)
9. Giannella, C., Han, J., Pei, J., Yan, X., Yu, P.S.: Mining Frequent Patterns in Data Streams at Multiple Time Granularities. In: Data mining: next generation challenges and the future directions, pp. 191–212. MIT/AAAI Press (2004)
10. Jiang, N., Gruenwald, L.: CFI-Stream: mining closed frequent itemsets in data streams. In: Proceedings of the 12th ACM SIGKDD international conference on Knowledge discovery and data mining, Philadelphia, PA, USA, pp. 592–597 (2006)
11. Li, H.F., Ho, C.C., Shan, M.K., Lee, S.Y.: Efficient Maintenance and Mining of Frequent Itemsets over Online Data Streams with a Sliding Window. In: IEEE International Conference on Systems, Man and Cybernetics, 2006, Taipei, pp. 2672–2677 (2006)
12. Li, H.F., Lee, S.Y., Shan, M.K.: DSM-FI: an efficient algorithm for mining frequent itemsets in data streams. In: Knowledge and Information Systems, October 2008, vol. 17, pp. 79–97. Springer, New York (2007)

13. Li, J., Shen, H., Topor, R.: Mining the informative rule set for prediction. Journal of Intelligent Information Systems 22(2), 155–174 (2004)
14. Li, J., Zhang, Y.: Direct Interesting Rule Generation. In: Proceedings of the Third IEEE International Conference on Data Mining, p. 155 (2003)
15. Manku, G.S., Motwani, R.: Approximate frequency counts over data streams. In: Proceedings of the 28th international conference on Very Large Data Bases, 2002, Hongkong, China, pp. 346–357 (2002)
16. Wang, E.T., Chen, A.L.: A novel hash-based approach for mining frequent itemsets over data streams requiring less memory space. Data Mining and Knowledge Discovery 19(1), 132–172 (2009)
17. Wong, R.C.W., Fu, A.W.C.: Mining top-K frequent itemsets from data streams. Data Mining and Knowledge Discovery 13(2), 193–217 (2006)
18. Woo, H.J., Lee, W.S.: estMax: Tracing Maximal Frequent Itemsets over Online Data Streams. In: Proceedings of the 2007 Seventh IEEE International Conference on Data Mining, vol. 21(10), pp. 1418–1431 (2007)
19. Yu, J.X., Chong, Z., Lu, H., Zhou, A.: False positive or false negative: mining frequent itemsets from high speed transactional data streams. In: Proceedings of the Thirtieth international conference on Very large data bases, Toronto, Canada, vol. 30, pp. 204–215 (2004)
20. Zaki, M.J., Gouda, K.: Fast vertical mining using diffsets. In: Proceedings of the ninth ACM SIGKDD international conference on Knowledge discovery and data mining, Washington DC, pp. 326–335 (2003)
21. Zaki, M.J.: Mining Non-Redundant Association Rules. In: Data Mining and Knowledge Discovery, November 2004, vol. 9(3). Springer, Netherlands (2004)

Using Text Classification Method in Relevance Feedback

Zilong Chen[1] and Yang Lu[2]

[1] State Key Lab. of Software Development Environment, BeiHang University
HaiDian District, Beijing, P.R.China
chenzl@nlsde.buaa.edu.cn
[2] School of Software and Microelectronics, Peking University
HaiDian District, Beijing, P.R.China
sheep@pub.ss.pku.edu.cn

Abstract. In modern Information Retrieval, traditional relevance feedback techniques, which utilize the terms in the relevant documents to enrich the user's initial query, is an effective method to improve retrieval performance. In this paper, we re-examine this method and show that it does not hold in reality – many expansion terms identified in traditional approaches are indeed unrelated to the query and harmful to the retrieval. We then propose a Text Classification Based method for relevance feedback. The classifier trained on the feedback documents can classify the rest of the documents. Thus, in the result list, the relevant documents will be in front of the non-relevant documents. This new approach avoids modifying the query via text classification algorithm in the relevance feedback, and it is a new direction for the relevance feedback techniques. Our Experiments on TREC dataset demonstrate that retrieval effectiveness can be much improved when text classification is used.

Keywords: relevance feedback, text classification, vector space model, SVM.

1 Introduction

In modern Information Retrieval (IR), search is formalized as a problem of document ranking based on degree of matching between query terms and document terms. It is possible for the search to differentiate correctly between the relevant and the non-relevant documents for the query if user can identify all of the terms that describe the subject of interest. But, most users find it difficult to formulate queries which are well designed for retrieval purposes. In fact, a user's original query statement will typically consist of just a few terms germane to the topic. It is usually too short to describe the information need accurately. Many important terms can be absent from the query, leading to a poor coverage of the relevant documents. Spink et al. [2] observe that users have to reformulate their search queries 40% to 52% of the time in order to find what they want.

This difficulty suggests that the first query formulation should be treated as an initial attempt to retrieve relevant information. Following that, the documents initially retrieved could be examined for relevance by user, and then new improved query

N.T. Nguyen, M.T. Le, and J. Świątek (Eds.): ACIIDS 2010, Part II, LNAI 5991, pp. 441–449, 2010.

formulations could be constructed in the hope of retrieving additional useful documents. Such query reformulation involves two basic steps, expanding the original query with new terms and reweighting the terms in the expanded query.

Traditional relevance feedback technique uses these approaches to improve the initial query formulation based on feedback information from the user. The basic idea is to reformulate the query such that it gets closer to the relevant documents. In vector space model, Rocchio method and some variants are widely used. The main idea of these approaches are that add the terms which appear in the relevant documents to the initial query and remove the ones which appear in the non-relevant documents at the same time. However, the usefulness of the terms which appear in the relevant documents is different. Good expansion terms will improve the effectiveness, bad expansion terms can hurt it and neutral expansion terms have little impact on the retrieval effectiveness. Ideally, we would like to use only good expansion terms to expand queries. But in reality, many expansion terms identified in traditional approaches are indeed unrelated to the query and harmful to the retrieval.

From another perspective, refer to the early work of Salton, we can view the IR problem as one of clustering. We think of the documents as a collection C of objects and think of the user query as a (vague) specification of a set A of objects. In this scenario, the IR problem can be reduced to the problem of determining which documents are in the set A and which ones are not (i.e., the IR problem can be viewed as a clustering problem). So shall we use the text classification algorithm in IR process?

Normally, it is hard to directly use the text classification algorithm in IR because the query is usually too short to describe the relevant documents accurately. It is usually too vague as a specification of relevant documents. After the documents initially retrieved have been examined for relevance, we can obtain more features to distinguish between relevant documents and non-relevant ones, so we can utilize text classification algorithms in relevance feedback to avoid the problem of query modification. The feedback process could be repeated, thus, we can use each feedback documents to modify the classifier until the user is satisfied with the results.

In this paper, we propose a relevance feedback algorithm based on the text classification algorithm. In particular, we present a two-stage approach for this algorithm. In the first stage, we train a classifier on the feedback documents. In the second stage, we utilize the classifier trained in the first stage to classify the rest documents and then provide a ranked list of the relevant documents.

The rest of the paper is organized as follows. In Section 2, we present a brief overview of related work about the relevance feedback. In Section 3, we describe the details of the text classification algorithm used in the relevance feedback process. The evaluation results are shown in Section 4. At the end, Section 5 concludes with a summary and makes suggestions for future work.

2   Related Work

The application of relevance feedback for the Vector Space Model considers that the term-weight vectors of the documents identified as relevant (to a given query) have similarities among themselves (i.e., relevant documents resemble each other). The basic idea is to reformulate the query in order to get closer to the term-weight vector space of the relevant documents.

Definition. For a given query q,

D_r: set of relevant documents, as identified by the user, among the retrieved documents;

D_n: set of non-relevant documents among the retrieved documents;

C_r: set of relevant documents among all documents in the collection;

$|D_r|, |D_n|, |C_r|$: number of documents in the sets D_r, D_n, and C_r, respectively;

α, β, γ: tuning constants.

Consider first the unrealistic situation in which the complete set C_r of relevant documents to a given query q is known in advance. In such a situation, it can be demonstrated that the best query vector for distinguishing the relevant documents from the non-relevant documents is given by,

$$\overline{q_{opt}} = \frac{1}{|C_r|}\sum_{\forall \overline{d_j} \in C_r} \overline{d_j} - \frac{1}{N-|C_r|}\sum_{\forall \overline{d_j} \notin C_r} \overline{d_j} \tag{1}$$

The problem of this formulation is that the relevant documents which compose the set C_r are not known a priori. The natural way to avoid this problem is to formulate an initial query and to incrementally change the initial query vector.

A classic way to calculate the modified query is described in a paper by Rocchio in 1971 [8]. The main idea of the way is that add the terms which appear in the relevant documents to the initial query, at the same time remove the terms which appear in the non-relevant documents from the initial query.

$$\text{Standard\_Rochio: } \overline{q_m} = \alpha \overline{q} + \frac{\beta}{|D_r|}\sum_{\forall \overline{d_j} \in D_r} \overline{d_j} - \frac{\gamma}{|D_n|}\sum_{\forall \overline{d_j} \in D_n} \overline{d_j} \tag{2}$$

Usually, the information contained in the relevant documents is more important than the information provided by the non-relevant documents [4]. This suggests making the constant γ smaller than the constant β. Experiments demonstrate that the performance is best when β=0.75 and γ=0.25.

Now, almost all the relevance feedback techniques are based on the expressions (2) and its variations. Iwayama [7] considers that the similarity of relevant documents is higher than non-relevant documents in the initial retrieval results, so we can firstly use text clustering algorithm before feedback. We firstly cluster around the top N (usually N sets 20-30) documents in the initial results and sum the similarity of each category. Then we feedback top K documents from the category, which displays highest similarity sum. He also suggests bringing query to each document before cluster as an improvement. Amo [1] considers that treating the top N documents of the initial retrieval results as same will lose some useful information. Some rank index of the initial results will be not used. They design some smooth functions to weight the document, the weight of document decreases when position index of the document increases and the sum weight of all the documents is 1. Therefore, the documents will be more important in the feedback when its position is in the front.

All of those methods have the same problem that is adding some non-relevance terms appearing in the relevant documents to the new query, while they use the terms in the relevant documents to enrich the user's initial query. As a result, we must choose good terms to expand query. There have been some work on this, paper [5] propose a method to integrate a term classification process for predicting the usefulness of expansion terms. Tan [9] even proposes a term-feedback technique, and it is based on the terms clustering to form some subtopic clusters which will be examined by users.

3   Text Classification Based Method for Relevance Feedback

In this section, we will introduce how to take advantage of the text classification algorithm in the relevance feedback process to improve the retrieval performance in details.

3.1   Problem Formulation

We formulate the problem which we should solve in a similar way as presented in [12]. Given a query Q and a document collection C, a retrieval system returns a ranked list of documents L. Li denotes the i-th ranked document in the ranked list. User should examine the top f documents for relevance in each feedback process. The goal is to study how to use these feedback documents: $N = \{L_1, ..., L_f\}$, to rerank the rest documents in the original ranked list: $U = \{L_{f+1}, L_{f+2}, ...\}$. In this paper, we set f = 20.

3.2   Procedure of Text Classification Algorithm Applied in Relevance Feedback

We train a classifier on the feedback documents collection N, classify the rest documents U and give a ranked list of the relevant documents which classified by the classifier.

The performance of the classifier trained on the feedback documents, depends on the distribution of these documents in all documents. If feedback documents include few documents which belong to the ideal boundary between the relevant documents and non-relevant ones, the performance of the classifier is poor. The non-relevant document may be classified as relevant document and ranked in the front of result list. However, after user examines the top 20 document in the next cycle, these mistakes can be corrected. After several repeat, the performance of the classifier will be satisfied.

We name this text classification feedback algorithm as SRFA(Support vector machine based Relevance Feedback Algorithm). The flow of SRFA is given as follow. The procedure is illustrated in Figure 1.

Step 1. Given query Q, calculate the cosine value between document vectors and query vector, and the user is presented with a ranked list of the retrieval documents: L.

Step 2. User examines the top 20 documents in L. Get the feedback documents N.

Step 3. Add feedback documents N to the samples space and Train the classifier on it.

Step 4. Classify the rest documents U by the classifier trained in the Step 3.

Step 5. Get the relevant documents, which classified by the classifier, and rank them as L in the next cycle.

Step 6. Repeat step 2—5 until the user is satisfied with the results.

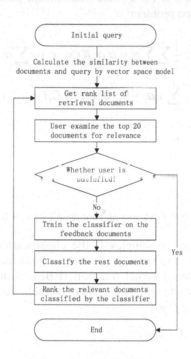

Fig. 1. Procedure of SRFA

In the Step 5, we should design an algorithm to rank the relevant documents because we cannot rank the documents based on the classifier. In this paper, we propose to use the following method to rank the documents: Rank documents based on the original cosine value between the document vector and the query vector.

To measure the performance of our algorithms, we use two standard retrieval measures: (1) Mean Average Precision (MAP), which is calculated as the average of the precision after each relevant document is retrieved, reflects the overall retrieval accuracy. (2) Precision at 20 documents (P@20), which is not well average and only gives us the precision for the first 20 documents. It reflects the utility perceived by a user who may only read up to the top 20 documents on the first page. This evaluation scheme is also applied in [11].

In above algorithm, we utilize SVM(support vector machine) as classification algorithm, because SVM is a technique motivated by statistical learning theory and has been successful applied to numerous classification tasks. The key point is to separate two classes by a decision surface with maximum margin. The extensive discussion of

SVMs can be found in [10]. In this paper, we should consider SVM in the binary classification setting as follow.

1. Given training data $T = \{(x_1, y_1), ..., (x_l, y_l)\}$, where $x_i \in X = R^n$, $y_i \in Y = \{1, -1\}$, $i = 1, ..., l$;

2. Choose kernel function $K(x, x')$ and punishment parameter C, construct and solve the follow optimization problem

$$\min \frac{1}{2}\sum_{i=1}^{l}\sum_{j=1}^{l} y_i y_j \alpha_i \alpha_j K(x_i, x_j) - \sum_{j=1}^{l} \alpha_j$$

$$s.t. \sum_{i=1}^{l} y_i \alpha_i = 0, \quad 0 \le \alpha_i \le C, i = 1, ..., l$$

Get the optimum solution $\alpha^* = (\alpha_1^*, ..., \alpha_l^*)^T$;

3. Select a positive component α_j^* smaller than C of α^*, and calculate $b^* = y_j - \sum_{i=1}^{l} y_i \alpha_i^* K(x_i, x_j)$;

4. Get the decision function $f(x) = \text{sgn}(\sum_{i=1}^{l} y_i \alpha_i^* K(x_i, x) + b^*)$;

We can see that, if given the samples T, the classification decision function f(x) only depends on the kernel function $K(x, x')$ and the punishment parameter C. In this paper, we use the radial-based kernel function (RBF). Some research results have demonstrated its effectiveness[3, 6]. This kernel function is defined as $K(x_i, x_j) = \exp(- \| x_i - x_j \| / 2\gamma^2)$.

In this paper, we set $\gamma = 0.1$, $C = \infty$.

4   Experiments and Results

4.1   Experimental Dataset

To evaluate our SVM based method for relevance feedback described in the previous section, we experimented with a TREC dataset. It is from the ROBUST track of TREC 2004 which includes all documents on TREC disks 4&5. It contains about 556000 news articles and 311410 documents marked whether relevant with each query or not. On average, each document includes 467 terms. We use all the 249 queries as our base query set.

4.2   Comparison of Different Relevance Feedback Algorithm

In experiment, 3 kinds of algorithms are used on the same TREC data set. Only top 20 documents were examined for relevance. We use P@20 and MAP to measure the performance of different algorithm. Repeat the feedback process in 10 times. The experimental results are shown as below.

Original: Only use vector space model and do not use any relevance feedback techniques.

Rochio: Same as Original, except using Standard_Rochio as relevance feedback.

SRFA: Same as Original, except using SRFA as relevance feedback.

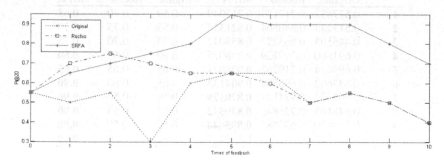

Fig. 2. P@20 of Original, Rochio and SRFA

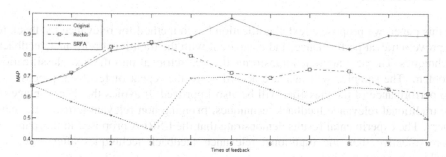

Fig. 3. MAP of Original, Rochio and SRFA

We can discover that the distribution of P@20 and MAP is irregular because the Original method does not use any feedback information techniques. Thus, the performance of the Original method is not satisfactory, while the relevant documents are not at the top of retrieval results.

Rochio method is effective at the beginning. However, after repeating feedback process 2-3 times, the P@20 and MAP values are decreased obviously because it has brought some non-relevant terms in the new query. The two values are lower than the Original algorithm after repeating 10 times.

The improvement is not obvious at the beginning by using SRFA because of only few samples. After repeating feedback process 2-3 times, the classification results are satisfied and the values of P@20 and MAP become higher than Rochio method obviously. Especially in the 5th feedback process, the values of P@20 and MAP over the maximum 0.9 show that the top 20 documents are almost relevant. It means the high retrieval performance. It demonstrates that the disadvantages of traditional feedback techniques mentioned in previous section.

More experimental data is shown in the following table.

Table 1. MAP and P@20 of Original, Rochio and SRFA

Times	MAP			P@20		
	Original	Rochio	SRFA	Original	Rochio	SRFA
1	0.577632	0.717179	0.712049	0.50	0.70	0.65
2	0.513233	0.840804	0.821355	0.55	0.75	0.70
3	0.445503	0.863277	0.855042	0.30	0.70	0.75
4	0.691860	0.799139	0.888025	0.60	0.65	**0.80**
5	0.699494	0.719262	**0.978021**	0.65	0.65	**0.95**
6	0.637602	0.695509	**0.904133**	0.65	0.60	**0.90**
7	0.566303	0.734144	**0.870379**	0.50	0.50	**0.90**
8	0.648165	0.725842	**0.830542**	0.55	0.55	**0.90**
9	0.639410	0.639683	**0.895644**	0.50	0.50	**0.80**
10	0.497076	0.618253	**0.893677**	0.40	0.40	**0.70**

5 Conclusion

In this paper, we propose a text classification based method for relevance feedback to improve retrieval performance, and compare it with the traditional relevance feedback techniques. The new approach transforms the text retrieval into the text classification problem. The samples will increase according to the repeat of feedback process, so the performance of the classifier will be also improved. It avoids the disadvantage of the traditional relevance feedback techniques, bringing non-relevant terms in the new query. The experimental results demonstrate that the I-SRFA proposed in this paper is more effective than the traditional relevance feedback techniques after repeating feedback process 2-3 times.

There are several interesting research directions that may further improve the effectiveness of text classification based method for relevance feedback. First one is that, research on the distribution of the relevant documents and non-relevant ones can help us to find more effective kernel function in SVM classification. Second one is that, the disadvantage of vector space model is that index terms are assumed to be mutually independent. Thus, the consideration of term dependencies may help us to improve the ranking performance in the relevant documents classified by the classifier.

Acknowledgment. We acknowledge support from State Key Lab. of Software Development Environment, BeiHang University's programs. These programs ID are 2005CB321901, 2005CB321903. We also appreciate valuable suggestions from anonymous reviewers.

References

1. Amo, P., Ferreras, F.L., Cruz, F., et al.: Smoothing functions for automatic relevance feedback in information retrieval. In: Proc of the 11th International Workshop on Database and Expert Systems Applications, pp. 115–119 (2000)

2. Spink, A., Jansen, B.J., Wolfram, D., Saracevic, T.: From e-sex to e-commerce: Web search changes. IEEE Computer 35(3), 107–109 (2002)
3. Bishop, C.: Patten recognition and machine learning. Springer, Berlin (2006)
4. Salton, G., McGill, M.J.: Introduction to Modern Information Retrieval. McGraw-Hill, New York (1983)
5. Cao, G., Nie, J.-Y., Gao, J., Robertson, S.: Selecting good expansion terms for pseudo-relevance feedback. In: Proc of ACM SIGIR Conference on Research and Development in Information Retrieval (2008)
6. Hsu, C.W., Chang, C.C., Lin, C.J.: A practical guide to support vector classification (2003-08-10/2004-11-10),
   http://www.csie.ntu.edu.tw/~cjlin/papers/guide/guide.pdf
7. Iwayama, M.: Relevance feedback with a small number of relevance judgements. In: Proc of ACM SIGIR Conference on Research and Development in Information Retrieval, pp. 10–16 (2000)
8. Rocchio, J.J.: Relevance feedback in information retrieval//Salton G. In: The Smart Retrieval System: Experiments in Automatic Document Processing, pp. 313–323. Prentice Hall, New Jersey (1971)
9. Tan, B., Velivellia, Fang, H., et al.: Term feedback for information retrieval with language models. In: Proc of ACM SIGIR Conference on Research and Development in Information Retrieval, pp. 263–270 (2007)
10. Vapnik, V.: Statistical Learning Theory. John Wiley & Sons Inc., New York (1998)
11. Shen, X., Zhai, C.: Active feedback in ad hoc information retrieval. In: Proc. of ACM SIGIR conference on Research and development in information retrieval, March 2005, pp. 55–66 (2005)
12. Wang, X., Fang, H., Zhai, C.: Improve retrieval accuracy for difficult queries using negative feedback. In: CIKM, pp. 991–994 (2007)

Fast and Near Optimal Parity Assignment in Palette Images with Enhanced CPT Scheme

Nguyen Hai Thanh[1] and Phan Trung Huy[2]

[1] Ministry of Education and Training
nhthanh@moet.gov.vn
[2] Hanoi University of Technology
huyfr2002@yahoo.com

Abstract. In the area of palette image steganography, by Parity Assignment approach, we introduce two fast and near Optimal Parity Assignment (nOPA) methods. The first is called rho-method which does not take any reorder procedures on color space of palettes. The second is called for-loop method which is the fastest and simplest one. The experimental results on 1000 palette images randomly selected show that this algorithm runs faster than the algorithms in OPA approach, while the costs (of changing colors) of two methods are equal. Comparison of the rho-, for-loop and EzStego methods with the OPA method is presented which shows the advantages of the two nOPA methods over the third. Among these the rho- method is the best, and the costs of the rho- and OPA method are equal. To prevent steganalysis, base on our nOPA methods, we propose a practical scheme by an enhanced CPT scheme for palette images.

Keywords: palette images, steganography, rho-method, nOPA method, enhanced CPT scheme.

1 Introduction

In the area of steganography, for hiding secret data in palette images, by *Parity Assignment (PA)* approachs due to [4], for some chosen distance d on colors (see [2,3]), ones try to partition the color palette P (in given order) of a given image G (such as in 8bpp bmp file format) into two disjoint parts, P_1 and P_2, by a parity function Val: $P \rightarrow E$ from P onto the set $E=\{0,1\}$, so that all colors in P_1 are assigned with the value 1, and in the rest P_2, with the value 0. That means, $P_1 = Val^{-1}(1)$ and $P_2 = Val^{-1}(0)$. Together with the partition ones need to find a function *PNext*: $P \rightarrow P$ which satisties $PNext(P_1) \subseteq P_2$ and $PNext(P_2) \subseteq P_1$ and minimizes the color distances $d(PNext(c), c)$ for all $c \in P$. Once this claim is satisfied, it implies immediately the minimality of *the cost of the method - which is seen as the sum* $\sum_{c \in P} d(PNext(c),c)$. We obtain a *PNext* function which is *an optimal parity assignment (OPA)* [4] for this palette image and make use of this to hide some bit data stream S. In the step of hiding secret data S into G, consecutively each bit b of S is "hidden" in a pixel having color x as follows (the pixels can be arranged in G by some linear order): x is kept intact if $b=0$, otherwise x is changed to $PNext(x)$. In the extracting step, to obtain S, each secret bit b can be

N.T. Nguyen, M.T. Le, and J. Świątek (Eds.): ACIIDS 2010, Part II, LNAI 5991, pp. 450–459, 2010.

extracted from the color x of the pixel by taking $b=Val(x)$. The optimality leads to the high quality of stego-image G' obtained from G after changing colors. For example, the method in [4] provides an interesting algorithm to get an OPA for palette images in a manner similar to the Kruskal algorithm in finding a minimal cover forest of the complete graph G_P of the color space P. However, in the algorithm due to [4], one step which consumes a large amount of time is the step to reorder the set of edges in ascending direction. This step takes on average time complexity of $O(N^2 LogN^2)$ with $N=n.(n-1)/2$ edges, where n is the color number of P, if ones use quick sort procedure as a good selection. The *EzStego method* [5,6] provides another example of PA ap-proach. Indeed, by virtue of the major step in the algorithm of this method, on the selected hamintonian cycle H of G_P (P has 256 vertices), on a chosen direction, H is ordered and this leads to the fact that P is reordered to a new palette P', and a *PNext* function on colors in P' (i.e. in P also) is natural obtained by this new order, that is $P'=\{c_0,c_1,..,c_{n-1}\}$ where c_k and c_{k+1} are two neigbour vertices on H, $k=0,1,..,n-1$ mod n. This implies that a parity function Val is defined on P' and also on H. One color $c=c_k$ by this order has the value $Val(c)=k$ mod 2 which is the *least signal bit (LSB)* of c. Hence the next color $c'=c_{(k+1) \bmod n}$ of the color c is one of two next vertices of c on H, $c'=PNext(c)$, $Val(c')\neq Val(c)$. Then the LSB method is applied on each pixel having color x in the data area of the image G, to hide one secret bit b by changing x to color $x'=PNext(x)$ for this pixel, if the least signal bit of color x differs from b. Otherwise x is kept intact. It is often the case that H is not minimal, the *PNext* function is not OPA and the cost of EzStego method is larger than the optimal. If we consider the whole process of sorting again P' to P, the run-time for the full algorithm is a bit larger. In this paper, we propose two methods by *near OPA approach (i.e. methods for which the ratio of theirs costs to the cost of OPA method are not larger than $1+\varepsilon$, for some selected error ε)*. The first is called *rho-method* and the second is called *for-loop method* by simplicity of its content: only to use for-loops to get the result. The rho-method is a very near OPA one in theoretical aspect but it seems to be the best in practice. One major advantage of our methods is that they do not take any reorder procedure on the set of edges in the color palette space. The experimental results (see the last part 4) on 1000 palette images randomly selected show this fact. In our meth-ods, the algorithms run faster than the algorithms mentioned in [4,5,6] while the costs of the rho- and OPA methods in [4] are equal. To compare with the OPA one, the average ratio of the costs of others methods to the cost of OPA method in 1000 palette images are taken and showed in the table 1 in the part 4. For the rho-method the ratio is 1.0000, for the for-method is 1.2087 and for the EzStego method it is larger, it equals 1.9278. This situation can be explained in the figure 1 below, when we process some palette in which some clusters of colors are rather far from each others

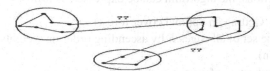

Fig. 1. The cost of EzStego method is much larger than the OPA's cost in cases the palette P is partitioned into some clusters which are far from each others

For this reason, the rho-method seems to provide the best results among nOPA and OPA methods mentioned above, in practical aspects, and some error ε choosen between 0.2 and 0.93 can be considered as a reasonable option. The rho-method expresses advantages whenever we need to process many palette images in a short time. In case ones need to use some schemes like CPT [1] to prevent steganalysis, specially to histogram-based attacks (see for examples some interesting analysis in [7,8], if the alpha ratio of the number of changed pixels to the number of total pixels of a given palette image is lower than 0.1, it is very difficult to guess if the image contains hidden data), combining with our methods the schemes can provide a higher quality of stego palette images with shorter run-time of algorithms. Since by applying these schemes to a concrete palette, ones can gain the small alpha ratio while amount of total hidden bits for real problems is large enough. For these purposes, the part 2 of the paper is devoted for representation of two methods. In the part 3, combining with our nOPA methods, an enhanced CPT scheme (ECPT for short) for palette images is introduced. In most cases in each block of pixels our ECPT scheme gives one hidden bit more than the original CPT scheme [1]. In combining with OPA or nOPA methods, to prevent steganalysis, this scheme gives much higher quality of stego palette images than those are processed by only OPA or nOPA methods, since only a small number of colors in stego images are changed randomly.

2 OPA and nOPA Methods

2.1 Common Step

Given a palette image G with the color palette P of $n \geq 2$ colors on which an order is defined. We can write $P = \{c_1, c_2, .., c_n\}$ by this order.

Step 0) Choose some distance function d on the set of all colors (for instance see [2,3]): the more x is "similar" to y, the smaller $d(x,y)$ is. P is considered as the set of vertices of the weight complete undirected graph G_P combined with P. Each pair (c_i, c_j) is an edge of G_P having the weight $d(c_i, c_j)$. We define at first the optimal *Next* function (for all mentioned methods): *Next*: $P \rightarrow P$ by the condition $d(c, Next(c))$ $= Min_{c \neq x \in P} d(c, x)$.

Let us recall briefly the essence of OPA method due to [4] in the weight complete undirected graph G_P combined with the palette P of a palette image G.

2.2 OPA Method

In terms of the minimal cover forest problem (optimal cover forest) and of the notions *PNext* and *Val* functions, the algorithm can be expressed as follows.

Algorithm 1. (Getting an optimal cover forest)
Step 1) Sort the set of all edges E by ascending order on weights.
Step 2) (Inition)
+ Build a subset C of vertices of P with two vertices $c_i \neq c_j$ so that the edge (c_i, c_j) has the smallest weight.

+ Build *PNext* and *Val* on C by: $PNext(c_i)=c_j$, $PNext(c_j)=c_i$, set randomly $Val(c_i)=0$ or 1 and $Val(c_j)=1$ XOR $Val(c_i)$.

+ Build a forest F_P by choosing randomly c_i or c_j as the root of the first new tree in F_P.

Step 3) (loop)

```
while C ≠ P do
    a)  Try to get one smallest edge (c_i,c_j) so that at least
        one of two c_i or c_j does not belong to C;
    b)  Three cases happen
        +  c_i, c_j∉C: put  C(new)  =  C ∪{c_i,c_j};PNext(c_i)=c_j,
           PNext(c_j)=c_i. Choose randomly c_i or c_j as the root
           of the new tree T={c_i, c_j}; set randomly Val(c_i)= 0
           or 1; Val(c_j)=1 XOR Val(c_i).
        +  c_i∈C, c_j∉C: put  C(new)  =  C ∪ {c_j}; PNext(c_j)=c_i;
              Val(c_j)=1 XOR Val(c_i).
           {PNext(c_i), Val(c_i) are defined in some previous
           loops already}
        +  c_j∈C, c_i∉C: C(new)  =  C ∪ {c_i}; PNext(c_i)=c_j;
           Val(c_i) =1 XOR Val(c_j).
    Endwhile;
```

Step 4) Return C, FP, $PNext$;

Remark 1. In fact, the loops in step (3) above are processed in a similar way as in the steps in the Kruskal algorithm. By the OPA method we can obtain the minimal cover forest F_P of the graph G_P with following property:

a) Each tree in F_P has at least two vertices.
b) The $Val(c)$ for each root c in the forest can be randomly set by 0 or 1. On each tree, for each non-roof vertex x, $Val(x)=1$ XOR $Val(PNext(x))$.
c) $d(c,PNext(c)) = $ Min $_{c≠x∈P}d(c,x)$, that means $PNext=Next$, therefore it is minimal.

2.3 Rho-Method

We call *a rho tree* a sequence $T=(x_1,x_2,..,x_k)$ of vertices of G_P which defines a path satisfying: either it contains a unique cycle defined by the conditions: $x_i≠x_j$ $\forall i≠j < k$ and $x_k=x_t$ for some $1≤t≤k$, or it does not contain any cycle (i.e. $x_i≠x_j$ $\forall i≠j$). *The rho-tree T is said be good* if it does not contain any cycle or the cycle $x_t,x_{t+1},..x_k$ having even number of vertices, that is $k-t$ is even. A good rho-tree is called *optimal* if it satisfies $d(x,y)=$Min $_{x≠c∈P}d(x,c)$ for any edge(x,y) along the path of this rho-tree.

A *rho-forest* can be defined inductively as follows
- The smallest rho-forest is the empty set.
- A rho-tree is a rho-forest.
- Disjoint union of two rho-forests is also a rho-forest.
- *Hooking* of a rho-tree $T = x_1,x_2,..x_k$ to a rho-forest F can be understood as the case whenever x_k belong to F and the rests, $x_1,x_2,..,x_{k-1}$ do not. In that case we obtain an expanded rho-forest $F'= F$ union T.

- If F contains all vertices of G_P we say F is a *cover rho- forest*, further more if F satisties $d(x,y)=$Min$_{x \neq c \in P} d(x,c)$ for any edge (x,y) of F, we call F the *minimal cover rho- forest* or *optimal cover rho- forest*. The *cost of a rho-forest F* is the sum of all weights of the edges in F.
- One cover rho- forest or cover forest F *is called near optimal* (*with small selected error ε*) if the ratio of its cost to the cost of one minimal cover forest is not larger than $1+ε$.

It is easily seen (see Proposition 2 in the part 2) that if an optimal cover rho- forest F exists, the cost of F is exactly the cost of minimal cover forest in OPA method.

Without taking any reorder procedure on the set of edges of G_P, we can get a near optimal rho-forest F for the graph G_P by following algorithm.

Algorithm 2. (finding near optimal cover rho- forest)

Step 1) (Inition) Given the palette P. Set the rho-forest F as the empty set. Put firstly the set of rest vertices C as P.

Step 2) (Define the *Next* function)

```
For i =1 to n do
  Begin
      +) Compute all d(xᵢ,xⱼ), i≠j, xᵢ,xⱼ∈P;
      +) Find an y =xₜ, i≠t, so that
        d(xᵢ,xₜ) = Min ₁≤i≠j≤n d(xᵢ,xⱼ), set Next(xᵢ)=y;
  End;
```

Step 3) Try to get a first near optimal rho – tree T and build the *Pnext* , *Val* functions on P inductively by

```
  a) Take randomly an x∈C; Set y₁=x; Set Val(y₁)=0 or 1
randomly
      Mark y₁ as both the root and the roof of T. Set imax=1;
  b) (loop), suppose yᵢₘₐₓ is defined and belongs to T already)
      While T is still updated do
        Begin
          b1) Find x=Next(yᵢₘₐₓ),
          b2) If x ∉ T, set yᵢₘₐₓ₊₁ =x and add yᵢₘₐₓ₊₁ into T.
              update C = C-{x}, imax =imax+1;
              Set PNext(yᵢₘₐₓ)=Next(yᵢₘₐₓ) =yᵢₘₐₓ₊₁ ;
              Val(x)=1 XOR Val(yᵢₘₐₓ);
          b3) If x = Next(yᵢₘₐₓ)= yₜ for some yₜ∈T,t<imax, that
              is y returns back first time to meet one vertex
              yₜ in T.
              + Mark yᵢₘₐₓ as the roof of T;
              + Two cases happen:
          b31) T is a good rho-tree, assign T as the rho-
              forest F; Set PNext(yᵢₘₐₓ)= yₜ;
          b32) Otherwise, Set PNext(yᵢₘₐₓ)= yᵢₘₐₓ₋ ₁;
              + Stop updating T;
        End; {While- updating T}
```

Step 4) (loop-taking completion of F, for all x in F , $Val(x)$, $Pnext(x)$ are defined, taking the same substeps as in Step 2 for the vertices in C, try to find a new good rho-tree T, or hook T into F to expand F, and loop until $C =\varnothing$)

```
While C ≠ Ø do
  Begin
    a) Choose ramdomly x∈C; Set imax=imax+1;yimax=x;
       Mark yimax as the root of a new rho-tree T;
       Set Val(yimax)=0 or 1 randomly.
    b)(loop,suppose yimax is defined  and belongs to F)
       While T is still updated do
         Begin
           b1) find x=Next(yimax),
           b2) if (x ∉ C) and (T hooks into F at x)
               + Set Pnext(yimax) = x; Mark yimax as the roof of
                 T; {x not in T}
               + If Val(yimax) = Val(x) then
                 Val(yt) = Val(yt) XOR 1; {inverting Val(yt),
                 ∀yt ∈ T, Else keep Val(yt) intact, ∀yt ∈ T}
               + ExitWhile;{stop adding this T}
             b3)  if x∈C and  x∉T :
               + set yimax+1 =x and add yimax+1 into T.
               | Set PNext(yimax)=Next(yimax) -yimax+1 ;
                 Val(yimax+1)=1 XOR Val(yimax);
               + update C=C-{x}; imax =imax+1;
           b4)If  x= Next(yimax) = yt for some yt ∈ T, t<imax,
               {x returns back to meet yt ∈ T}
               + mark yimax as the roof of T;
               + Two cases happen:
             b41)imax-t is even:T is good, assign T as a member
                 of F and set PNext(yimax)= yt;
             b42) imax-t is odd: Set PNext(yimax)= yimax- 1;
               (Modified T becomes good with Pnext).
               + Stop updating this T;
         End;{While- updating this T}
       End;{While C ≠ Ø }
Step 5) Return F, Pnext, Val;
```

By the algorithm, if the cases (b32) and (b43) do not happen, without any difficulty we deduce by inductition that $Pnext(y) = Next(y)$ for all y in the rho-forest F as the result of the algorithm. Hence we obtain

Proposition 1. *If the cases (b32) and (b42) do not happen in the algorithm 2 for any palette image G, the results of this algorithm give an optimal OPA.*

Let us remark that in the results of algorithm 1 if F is a forest tree and the pallete P is reordered (whenever one vertex x is added to the set C). By making reference to this order on P in the algorithm 2 to choose roots of rho-trees T at first steps 3(a), 4(a), it is verified without any difficulty that F can be recovered as one result of the algorithm 2. Hence we deduce

Proposition 2. *Give any palette images G. The set of optimal forests obtained as results of algorithm 1 is a subset of the set of rho-forests obtained as the results of the algorithm 2.*

Let us remark that the steps 3, 4 in algorithm 2 take a time complexity of $O(n)$, where n is the number of vertices of the graph.

The step 2 in the algorithm 2 can be considered as a common step of all algorithms 1,2 and of the next algorithms. This step takes roughly a time complexity about $O(n^2)$ for all algorithms. Therefore it is easily verified

Theorem 3. *The algorithm 2 take a time complexity of $O(n^2+n)$. The steps 2,3 in this algorithm take a time complexity of $O(n)$ to obtain a near cover rho-forest F, where n is the number of vertices of graph.*

Remark 2. In practice, our experimental results (processing 1000 palette images and showed in Table 1 in the part 4) present the fact that all cover rho-forests obtained by the algorithm 2 are optimal cover forests, they provide exactly OPAs on these images.

2.4 For-Loop Method

a) Reorder P by some way; suppose $P=\{x_1,x_2..x_n\}$

b) (Try to build a new function *Pnext* near the optimal *Next* function obtained in the step 2, algorithm 2).

```
For i =1 to n do
  Begin
        +) Compute all d(xᵢ,xⱼ) i≠j;
        +) Find an y =xₜ, i≠t, i+t is odd, so that
           d(xᵢ,xₜ) = Min 1≤i≠j, ≤n, j+i is odd d(xᵢ,xⱼ);
        +) Set Pnext(xᵢ)=y;
        +) Set Val(xᵢ)= i mod 2;
  End;
```

c) In hiding step, for each pixel having a color x in the data area of the stego image G, *Pnext* can be used to hide a secret bit b by changing x to color $x'=Pnext(x)$ at this pixel if $Val(x)\neq b$. Otherwise keep x intact.

d) In the extracting step to obtain hidden bits from G, one can take a hidden bit $b=Val(x)$ from the color x of chosen pixel in the data area of G.

e) This algorithm runs fast, *the cost of this algorithm* can be defined on the *Pnext* function as *the sum $\sum_{x\in P} Pnext(x)$.*

f) Our experimental results in the part 4 showes that, in processing 1000 palette images, the average ratio of the cost of for-loop method to the cost of OPA method is 1.2087. Hence this method provides an error 0.2087 in practice, this is smaller than an error 0.9278 gained by EzStego method. Obviously the for-loop method is the simplest.

Proposition 4. *The algorithm 3 take time complexity of $O(n(n-1)/2)$, where the basic step is the computation the distance of two colors.*

3 ECPT Scheme for Palette Images

In this part, as application of nOPA methods, to prevent steganalysis, we introduce an *enhanced CPT scheme (ECPT for short)* which is a modification of the *CPT scheme* for binary images [1] to palette images, as a practical scheme. This scheme can provide stego palette images with high quality, since only a small number of colors need to be changed in a stego images, and they are chosen randomly, while the number of hidden bits is large enough for real problems.

+ Given a palette image G whose data area is partitioned into blocks of pixels F_t which are considered as matrices of the same size $m \times n$, $1 \leq t \leq p$ for some integer p. Combined with these blocks of the image are two matrices K, W of the same size $m \times n$: K is a binary key matrix K whose elements are randomly chosen, W is a weight matrix whose elements W_{ij} are integers chosen randomly with the restriction (3.1) below.

+ For this image suppose that some fast nOPA method is selected, such as the rho-method. This means that some *Pnext, Val* functions are selected.

+ Consider a concrete block $F = (F_{ij})_{m \times n}$ (for the sake of simplicity we ignore the sub-index of F). Each entry F_{ij} is assigned with a color of a pixel. Hence, with each entry F_{ij} presenting a value $x \in P$, $Val(F_{ij})$ and $Pnext(F_{ij})$ stand for $Val(x)$ and $Pnext(x)$... In our ECPT scheme the block F can be used to embed $r = \lfloor \log_2(m.n) \rfloor + 1$ hidden bits, while by the original CPT scheme [1], in F ones can hide $\lfloor \log_2(m.n+1) \rfloor$ hidden bits. That means in most cases, using ECPT scheme in each block ones can hide one bit more than using the original CPT. For simplicity, F_{ij} stands for both- the value and entry, depending on concrete situations. *Changing (or inverting) F_{ij} in F* means that F_{ij} is changed to $F_{ij}' = Pnext(F_{ij})$. The operation $\oplus$ is the bitwise exclusive-OR (XOR) on two equal-size binary matrices. The operation $\otimes$ is the pairwise multiplication on equal-size integer matrices.

+ We consider such a block F as a matrix of colors in the palette P of size $m \times n$.

+ Let u be some integer, $0 < u \leq mn$. Set $t = 2u + 1$.

3.1 Parameters for ECPT Scheme

Hiding secret bits: ECPT input consists of following.

   a) An integer b, $0 \leq b \leq t-1$, *b is considered as secret data which need to be embed-ded into F in this scheme.* Also we can present b as an element in the additive group $\mathbb{Z}_t$ of integers modulo t.

   b) The matrix $F = (F_{ij})_{m \times n}$ of colors in the palette P that is going to hide b.

   c) The binary key matrix K of the same size $m \times n$: $K = (K_{ij})_{m \times n}$.

   d) The integer weight matrix W of the same size $m \times n$: $W = (W_{ij})_{m \times n}$ satisfying

$$\{ W_{ij}, 1 \leq i \leq m, 1 \leq j \leq n \} = \{ 1, 2, \ldots, u \}. \tag{3.1}$$

ECPT output: the new matrix $F = (F_{ij})_{m \times n}$ of colors in P that has been embedded b.

Algorithm 3. (embedding *b* in *F*)

```
Step 1) Compute T = Val(F) ⊕ K;

Step 2) Compute S = SUM[T ⊗ W] mod t, 0 ≤ S ≤ t-1, where
```

$$\text{SUM}[T \otimes W] = \sum_{1 \le i \le m} \sum_{1 \le j \le n} T_{ij} \times W_{ij} \bmod t. \tag{3.2}$$

Step 3) (Try to invert at most two entries in F so that

$$S = \text{SUM}[T \otimes W] \bmod t = b)$$
+ Set $\alpha = b - S \bmod t$.
+ We need to find F_{ij} such that S is added α after inverting F_{ij}, that is $b = S(new) \bmod t$.
For this aim, define the set

$$
\begin{aligned}
S_\alpha &= \{F_{ij} \mid T_{ij} = 0, Val(W_{ij}) = \alpha \quad \bmod t\} \\
\text{or} &= \{F_{ij} \mid T_{ij} = 1, Val(W_{ij}) = t-\alpha \bmod t\} \\
\text{or} &= S_\beta \text{ if } \beta = \alpha \bmod t \text{ for any integers } \alpha, \beta
\end{aligned} \tag{3.3}
$$

+ If $\alpha = 0$ (i.e. $S = b$), keep F intact and exit, otherwise go to the next step 4.

Step 4) ($\alpha \ne 0$) two cases happen:
+) $S_\alpha \ne \varnothing$: randomly invert an arbitrary $F_{ij} \in S_\alpha$ and exit.
+) $S_\alpha = \varnothing$: go to the next step 5.

Step 5) Find the smallest natural number $h, h > 1$ satisfying $S_{h\alpha} \ne \varnothing$ and $S_{\alpha-h\alpha} \ne \varnothing$ (the existence of h due to Lemma 6 below), randomly invert an $F_{ij} \in S_\alpha$ and another $F_{ij} \in S_{\alpha-h\alpha}$ to increase S a total $h\alpha + (\alpha - h\alpha) \bmod t$, that is
$$S (new) = S + h\alpha + (\alpha - h\alpha) \bmod t = b \bmod t.$$
Step 6) End.

Extracting hidden bits from F: This stage is simple, since we can extract the hidden data b by applying the equation $b = \sum_{1 \le i \le m} \sum_{1 \le j \le n} T_{ij} \times W_{ij} \bmod t$ with the restriction $0 \le b \le t-1$.

The correctness of this ECPT scheme is deduced directly from the lemma below. At first, from the definition equation (3.3) of the set S_α, obviously we have

Lemma 5. *If $S_\alpha = \varnothing$ and $\alpha \ne 0 \bmod t$, then $S_{t-\alpha} \ne \varnothing$.*

Lemma 6. *If $S_\alpha = \varnothing$ and $\alpha \ne 0 \bmod t$, then there exists a natural number h, $h > 1$ such that $S_{h\alpha} \ne \varnothing$. Moreover, if h is the smallest natural number h, $h > 1$ satisfying $S_{h\alpha} \ne \varnothing$, then $S_{\alpha-h\alpha} \ne \varnothing$.*

Proof. We can view α as an element of the additive group $\mathbb{Z}_t$. Suppose that k is the order of α, in the cyclic subgroup $\langle \alpha \rangle$ generated by α. By definition, $k\alpha = 0 \bmod t$ and $(k-1)\alpha \ne 0 \bmod t$. Consider now the sequence of the sets S_α, $S_{2\alpha}$, $S_{3\alpha}$... ,$S_{k\alpha}$. Using $k\alpha = t = 0 \bmod t$ we deduce $S_{k\alpha} = S_t$ and $(k-1)\alpha = t-\alpha \bmod t$. This implies from definition that $S_{(k-1)\alpha} = S_{t-\alpha}$, therefore $S_{t-\alpha}$ belongs to this sequence. By Lemma 5, $S_{t-\alpha} \ne \varnothing$, this shows that $h = k-1$ is the number we need: $S_{h\alpha} \ne \varnothing$, and therefore $h > 1$. Now, suppose h is the smallest natural number h, $h > 1$ satisfying $S_{h\alpha} \ne \varnothing$, then $S_{(h-1)\alpha} = \varnothing$. By lemma 5 it implies $S_{t-(h-1)\alpha} \ne \varnothing$. Since $\alpha - h\alpha = t-(h-1)\alpha \bmod t$, we obtain $S_{\alpha-h\alpha} = S_{t-(h-1)\alpha} \ne \varnothing$.

Example. In a palette image 8bpp G having N blocks of size 2×4, with 8N data bytes, to hide 4N bits or about $N/2$ bytes, at most the colors of 2N pixels to be changed.

4 Experimental Results

In this part, we present some results of the nOPA methods (the rho-, for-loop and EzStego methods) in comparison with the optimal OPA method [4]. In the table 1 below we show the average ratio costs of rho-, for-loop and EzStego methods to the cost of OPA method by processing 1000 palette images (in 8bpp bmp file format) choosen randomly. The experimental results show that in practice, the rho-method is the best one, its cost equals to the cost of OPA method, but it provides a faster running time as mentioned in the part 2 above.

Table 1. Avg. ratio of the costs of nOPA methods in 1000 palette images 8bpp

Rho-Method	For-loop method	EzStego method
1.0000	1.2087	1.9278

References

1. Chen, Y., Pan, H., Tseng, Y.: A secret of data hiding scheme for two-color images. In: IEEE symposium on computers and communications (2000)
2. Colour metric, http://www.compuphase.com/index_en.htm
3. Alman, D.H.: Industrial color difference evaluation. Color Research and Application 18, 137–139 (1993)
4. Fridrich, J., Du, R.: Secure Steganographic Methods for Palette Images. In: Pfitzmann, A. (ed.) IH 1999. LNCS, vol. 1768, pp. 47–60. Springer, Heidelberg (2000)
5. Machado, R.: EZStego[EB/OL], http://www.stego.com
6. Por, L.Y., Lai, W.K., Alireza, Z., Ang, T.F., Su, M.T., Delina, B.: StegCure: A comprehensive Steganographic Tool using Enhanced LSB Scheme. Wseas Transactions on Computers 7(8) (August 2008), ISSN: 1109-2750 1309
7. Zhang, X., Wang, S.: Vulnerability of pixel-value differencing steganography to histogram analysis and modification for enhanced security. Pattern Recognition Letters 25, 331–339 (2004)
8. Zhang, X., Wang, S.: Analysis of Parity Assignment Steganography in palette Images. In: Khosla, R., Howlett, R.J., Jain, L.C. (eds.) KES 2005. LNCS (LNAI), vol. 3683, pp. 1025–1031. Springer, Heidelberg (2005)

Solving QoS Routing Problems by DCA

Ta Anh Son[1,3], Le Thi Hoai An[2], Djamel Khadraoui[3], and Pham Dinh Tao[1]

[1] Laboratory of Modelling, Optimization & Operations Research,
National Institute for Applied Sciences - Rouen
BP 08, Place Emile Blondel F 76131 France
[2] Laboratory of Theoretical and Applied Computer Science (LITA)
Paul Verlaine - Metz University, Ile du Saulcy, 57045, Metz, France
[3] CRP Henri Tudor, 29 avenue John F. Kennedy, 1855 Kirchberg, Luxembourg

Abstract. The Quality of Service (QoS) routing emphasizes to find paths from source to destination satisfying the QoS requirements. In this paper, we consider the five problems in QoS Routing called the MCP (Multi-Constrained Path) and the MCOP (Multi-Constrained Optimal Path). They are all NP-hard problems. We first formulate them as Binary Integer Linear Programs (BILP) and then investigate a new solution method based on DC (Difference of Convex functions) programming and DCA (DC Algorithms). Preliminary numerical results show that the proposed algorithm is promising: it is more robust than CPLEX, the best solver for BILP.

Keywords: Multi-Constrained Path, Multi-Constrained Optimal Path, QoS routing, DC programming, DC Algorithm, Exact Penalty, Binary Integer Linear Programming (BILP).

1 Introduction

In recent years, transmissions of multimedia content over the communication networks have many challenges. In order that these transmissions might work properly, QoS measures like bandwidth, delay, jitter and packet loss, etc., need to be controlled. These issues give rise to the problem of routing in the network where the requirements of QoS are satisfied.

In general, the routing includes two entities: the routing protocol and the routing algorithm. The routing protocol has the task of capturing the state of the network and its available network resources, and of disseminating this information throughout the network. The routing algorithm uses this information to compute an optimal path from a source to a destination.

The QoS Routing Problem consists of finding an optimal path from a source to a destination subject to QoS constraints (e.g., time delay, cost or packets loss constraints). It is well-known that this problem is NP- hard [16]. Many approaches for solving the QoS routing problems have been proposed (see e.g. [5], [10], [11], [8], etc.). Almost of the existing approaches for solving MCP and MCOP are based on classical methods for solving the shortest path problem such as Dijkstra and Ford-Bellman algorithms (e.g., [10],[6],[16]). The second approaches are

N.T. Nguyen, M.T. Le, and J. Świątek (Eds.): ACIIDS 2010, Part II, LNAI 5991, pp. 460–470, 2010.
© Springer-Verlag Berlin Heidelberg 2010

based on network flow algorithms (e.g., [12]). In addition, there are several poly-nomial ϵ-approximate solution methods for these problems (e.g.,[7], [15], [9]). However, as farther as we know, none of them is considered as a BILP problem.

In this paper we provide a new approach for QoS based routing problems. We formulate the five QoS routing problems in the form of BILP problems. The routing algorithm consists in solving the BILP based on DC programming and DCA.

DC programming and DCA introduced by Pham Dinh Tao in 1984 and ex-tensively developed by Le Thi Hoai An and Pham Dinh Tao since 1994 (see [3], [13], [14] and references therein) is an efficient approach for nonconvex continuous optimization. They address a general DC program of the form

$$\alpha = \inf\{\{(x) := g(x) - h(x) \; : \; x \in \mathbb{R}^n\} \quad (P_{dc})$$

where g and h are lower semi-continuous proper convex functions on $\mathbb{R}^n$. Such a function f is called DC function, and $g - h$, DC decomposition of f while the convex functions g and h are DC components of f. It should be noted that a constrained DC program whose feasible set C is convex can always be trans-formed into a unconstrained DC program by adding the indicator function χ_C of C ($\chi_C(x) = 0$ if $x \in C, +\infty$ otherwise) to the first DC component g. Based on the DC duality and the local optimality conditions, the idea of DCA is simple: each iteration k of DCA approximates the concave part $-h$ by its affine ma-jorization (that corresponds to taking $y_k \in \partial h(x^k)$) and minimizes the resulting convex function

$$\min\{g(x) - h(x^k) - \langle x - x^k, y^k \rangle : x \in \mathbb{R}^n\} \quad (P_k)$$

to obtain x^{k+1}. The construction of DCA involves the convex DC components g and h but not the DC function f itself. Moreover, a DC function f has infinitely many DC decompositions $g - h$ which have a crucial impact on the qualities (speed of convergence, robustness, efficiency, globality of computed solutions,...) of DCA.

The DCA has been successfully applied to real world non convex programs in various fields of applied sciences (e.g.,[3], [13], [14]). It is one of the rare efficient algorithms for non smooth non convex programming which allows solving large-scale DC programs.

In [2] a continuous approach based on DCA has been developed for the well-known linearly constrained quadratic zero-one programming

$$\alpha = \min\left\{f(x) := \frac{1}{2}x^T Cx + c^T x : Ax \leq b, x \in \{0,1\}^n\right\}, \quad (01QP)$$

In this paper we show how to adapt this approach for solving the QoS based routing problems formulated as BILP. The proposed DCA for BILP enjoys sev-eral advantages: it converges to a local (integer) solution after a finitely many iterations, and requires only the solution of a few number of linear programs. Moreover, it is worth to mention that, although the DCA is a continuous ap-proach working on a continuous domain, it provides an integer solution. This is unusual in continuous approaches and is original property of the proposed DCA. The efficiency of DCA is compared with CPLEX, the best solver for BILP. The computational results on several test problems show that DCA is more robust than CPLEX.

The paper is organized as follows. In Section 2 we introduce the five QoS routing problems and formulate them in the form of BILP. The solution method based on DC programming and DCA for solving these problems is developed in Section 3 while the numerical results are reported in Section 4. Finally, conclusions are presented in Section 5.

2   Problem Statement and Mathematical Formulation

In this paper, the network-state information is assumed to be temporarily static, and this information is known at each node. We perform the second task in QoS routing, namely computing paths with given multiple QoS constraints.

Let $G = (N, E)$ be the graph which presents a network topology, with the set of nodes N and the set of arcs E. The number of nodes, the number of arcs, and the number of QoS measures are denoted, respectively by n, m and p. Each arc is characterized by a p-dimensional arc weight vector. Each vector composes p nonnegative QoS weight components which are $(w_i(u, v), i = 1, \cdots, p, (u, v) \in E)$.

The QoS measures of a path can either be additive, multiplicative, or min/max. The path weight of additive measures (e.g., delay, jitter or cost) equals to the sum of the QoS weights of the arcs on the path. The multiplicative measures (e.g., packet loss) can be transformed into additive measures by using logarithm function. The path weight of min/max measures (e.g., min/max bandwidth) presents the minimum/maximum of the QoS weights defining the path. Constraints on min/max QoS measures can easily be settle by omitting all arcs (or disconnected nodes), which do not satisfy the requested QoS constraints. In practical, constraints on additive QoS measures are more difficult. In this paper, QoS measures are assumed to be additive.

Definition 1. ((MCP) [4]) Consider a network $G = (N, E)$. Each arc $(u, v) \in E$ is specified by p additive QoS weights $w_i(u, v) \geq 0, i = 1, \cdots, p$. Given p constraints L_i, the problem consists of finding a path P from a source node s to a destination node t such that:

$$\sum_{(u,v) \in P} w_i(u, v) \leq L_i, \ \forall i = 1, \cdots, p.$$

A path P that satisfies the QoS constraints is a feasible path. The MCOP problem consists of finding the optimal path. If the QoS constraints are eliminated from the MCP and the MCOP, the resulting problems can be solved by Dijkstra or Ford-Bellman algorithms. However, the MCP and the MCOP are NP-hard problems.

By assumption, the state of the network (the routing table at each node, and the QoS constraints) is considered as known. Moreover, the measurement of a path is defined by the number of nodes belonging to it.

In this section, we introduce the formulation of five problems in QoS routing. Problem 1 consists of finding a feasible path. Problem 2 deals with the finding of an optimal path. Problem 3 and Problem 4 aim to find, respectively, k-arc disjoint paths and the maximal number of arc disjoint paths such that if the QoS constraints are satisfied. Problem 5 consists of finding k feasible arc disjoint paths $P_1, \cdots, P_k$, each of them satisfying the QoS constraints.

Problem 1. *Finding a feasible path P.*

Define the binary variable y_{ij} as:

$$y_{ij} = \begin{cases} 1 \text{ if } (i,j) \in P \\ 0 \text{ otherwise.} \end{cases}$$

Then a path from s to t must satisfy the following constraints:

$$\sum_{(s,v) \in E} y_{sv} - \sum_{(u,s) \in E} y_{us} = 1,$$
$$\sum_{(u,v) \in E} y_{uv} - \sum_{(v,\ell) \in E} y_{v\ell} = 0, \ \forall v \in N \backslash \{s,t\},$$
$$\sum_{(t,v) \in E} y_{tv} - \sum_{(u,t) \in E} y_{ut} = -1,$$

The QoS constraints can be formulated as:

$$\sum_{(u,v) \in E} w_i(u,v) y_{uv} \leq L_i, \ i = 1, \cdots, p. \tag{1}$$

If the value of $w_i(u,v)$ is fixed to each arc $(u,v) \in E$ (e.g., the cost for sending one message or one unit data on the arc (u,v), or the time delay on the arc (u,v)), then the QoS constraints are linear constraints. When $w_i(u,v)$ is also a variable, this problem becomes more complex.

Let us define a_{ij} as follows:

$$a_{ij} = \begin{cases} 1 \text{ if } (i,j) \in E, \\ -1 \text{ if } (j,i) \in E, \\ 0 \text{ otherwise.} \end{cases}$$

Furthermore, let $A = (a_{ij})$ (resp. $W = (w_i(u,v))$) be a $m \times n$ (resp. $p \times n$) matrix, let $b = (1, 0, \cdots, 0, -1)^T \in \mathbb{R}^m$, $y = (y_{ij})_{(i,j) \in E}^T \in \mathbb{R}^n$, and $L = (L_1, \cdots, L_p)^T \in \mathbb{R}^p$.

Then, the QoS constraints (1) become $Wy \leq L$. Finally, Problem 1 can be formulated as

(P1) finding $y \in \{0,1\}^n$ such that $Ay = b$ and $Wy \leq L$.

Problem 2. *Finding an optimal path P satisfying the QoS constraints.*

Clearly, by adding the objective function into (P1) we obtain the mathematical formulation of this problem:

$$\text{(P2)} \quad \min \left\{ \sum_{(i,j) \in E} y_{ij} : Ay = b \ \text{ and } \ Wy \le L, y \in \{0,1\}^n \right\}.$$

Problem 3: *Finding k feasible arc disjoint paths which satisfy the QoS constraints.*

Let us define the binary variable y_{ij} as:

$$y_{ij} = \begin{cases} 1 \text{ if } (i,j) \in E \text{ belongs to the set of } k \text{ arc disjoint paths,} \\ 0 \text{ otherwise.} \end{cases}$$

Let $b = (k, 0, \cdots, 0, -k)^T \in \mathbb{R}^n$. Problem 3 can be formulated as:

$$\text{(P3) finding } y \in \{0,1\}^n \text{ such that } Ay = b \text{ and } Wy \le L,$$

with the matrixes A, W, L being defined above.

Problem 4. *Finding the maximal number of arc disjoint paths from s to t which satisfy the QoS constraints.*

Let $d = (d_{(i,j) \in E})^T \in \mathbb{R}^n$, where

$$d_{ij} = \begin{cases} 1 \text{ if } (i,j) \in E \text{ is outgoing arc from } s, \\ -1 \text{ if } (i,j) \in E \text{ is ingoing arc to } s, \\ 0 \text{ otherwise.} \end{cases}$$

Problem 4 can be then formulated as

$$\text{(P4)} \quad \max\langle d, y \rangle$$

subject to:

$$\sum_{(s,v) \in E} y_{sv} - \sum_{(u,s) \in E} y_{us} = -\left[\sum_{(t,v) \in E} y_{tv} - \sum_{(u,t) \in E} y_{ut} \right],$$

$$\sum_{(u,v) \in E} y_{uv} - \sum_{(v,\ell) \in E} y_{v\ell} = 0, \ \forall v \in N \backslash \{s,t\},$$

$$Wy \le L \text{ and } y \in \{0,1\}^n.$$

In wireless networks, the reliability (defined as the success packet delivery) is a key factor. In the case where an arc changes and fails, the network is unreliable. The arising problem is to find an alternative path when the initial one is damaged.

Problem 5. *Finding k feasible arc disjoint paths $P_1, \cdots, P_k$, each path satisfies the QoS constraints.*

Define the binary variable $y_{ij\ell}$ as:

$$y_{ij\ell} = \begin{cases} 1 \text{ if } (i,j) \in P_\ell, \ \ell = 1, \cdots, k, \\ 0 \text{ otherwise.} \end{cases}$$

Then, similarly to Problem 1, we have the following constraints:

$$\sum_{(s,v)\in E} y_{sv\ell} - \sum_{(u,s)\in E} y_{us\ell} = 1, \qquad (i)$$

$$\sum_{(u,v)\in E} y_{uv\ell} - \sum_{(v,q)\in E} y_{vq\ell} = 0, \ \forall v \in N\backslash\{s,t\}, \qquad (ii)$$

$$\sum_{(t,v)\in E} y_{tv\ell} - \sum_{(u,t)\in E} y_{ut\ell} = -1. \qquad (iii)$$

On the other hand, the next constraint ensures that each arc $(i,j) \in E$ belongs to at most one path P_ℓ, $(\ell = 1, \cdots, k)$:

$$\sum_{\ell=1}^{k} y_{ij\ell} \leq 1, \ \text{for each } (i,j) \in E. \qquad (iv)$$

The QoS constraints for each path P_ℓ, $(\ell = 1, \cdots, k)$ are expressed as

$$\sum_{(u,v)\subset P_\ell} w_i(u,v) \leq L_i, \ i = 1, \cdots, p. \qquad (v)$$

Problem 5 can be formulated as:

(P5) finding $y_{ij\ell} \in \{0,1\}((i,j)\in E, \ \ell = 1, \cdots, k)$ such that $(i)-(v)$ are satisfied.

We observe that (P1), (P3) and (P5) are MCP problems while (P2) and (P4) are of the form MCOP. They are all formulated as BILP. In the sequence, we investigate a new approach based on DCA for solving Problem 2. The other problems can be solved in a similar way.

3   Solving QoS Routing Problems by DCA

In this section, we consider the MCOP Problem (P2). By using an exact penalty result, we can reformulate the MCOP in the form of a concave minimization program. The exact penalty technique aims at transforming the original problem (P2) into a more tractable equivalent DC program. Let $K := \{y \in \mathbb{R}^n : Ay \leq b, \ Wy \leq L, \ y \in [0,1]^n\}$. The feasible set of (P2) is then $S = \{y : y \in K, y \in \{0,1\}^n\}$.

Let us consider the function $p : \mathbb{R}^n \to \mathbb{R}$ defined by:

$$p(y) = \sum_{i=1}^{n} y_i(1 - y_i).$$

It is clear that $p(y)$ is concave and finite on K, $p(y) \geq 0 \ \forall y \in K$ and that:

$$\{y : y \in S\} = \{y : y \in K, p(y) \leq 0\}.$$

Hence problem (P2) can be rewritten as:

$$\min\left\{\sum_{i=1}^{n} y_i : y \in K, p(y) \leq 0\right\}.$$

The following theorem can then be formulated.

Theorem 1. *Let K be a nonempty bounded polyhedral convex set, f be a finite concave function on K and p be a finite nonnegative concave function on K.*

Then there exists $\eta_0 \geq 0$ such that for $\eta > \eta_0$ the following problems have the same optimal value and the same solution set:

$$(P_\eta) \qquad \alpha(\eta) = \min\{f(y) + \eta p(y) : y \in K\},$$
$$(P) \qquad \alpha = \min\{f(y) : y \in K, p(y) \leq 0\}.$$

Furthermore

- If the vertex set of K, denoted by $V(K)$, is contained in $x \in K : p(y) \leq 0$, then $\eta_0 = 0$.
- If $p(y) > 0$ for some y in $V(K)$, then $\eta_0 = \min\left\{\frac{f(y)-\alpha(0)}{S_0} : y \in K, p(y) \leq 0\right\}$, where $S_0 = \min\left\{p(y) : y \in V(K), p(y) > 0\right\} > 0$.

Proof. The proof for the general case can be found in [1].

From Theorem 1 we get, for a sufficiently large number η $(\eta > \eta_0)$, the equivalent concave minimization problem (P2):

$$\min\{f_\eta(y) := \sum_{i=1}^{n} y_i + \eta p(y) : y \in K\},$$

which is a DC program of the form:

$$\min\{g(y) - h(y) : y \in \mathbb{R}^n\}, \tag{2}$$

where: $g(y) = \chi_K(y); h(y) = -f_\eta(y) = -\sum_{i=1}^{n} y_i - \eta p(y)$.

We have successfully transformed an optimization problem with integer variables into its equivalent form with continuous variables. Notice that (2) is a polyhedral DC program where g is a polyhedral convex function (i.e., the pointwise supremum of a finite collection of affine functions).

DCA applied to the DC program (2) consists of computing, at each iteration k, the two sequences $\{y^k\}$ and $\{z^k\}$ such that $z^k \in \partial h(y^k)$ and y^{k+1} solves the next linear program of the form (P_k) (see Introduction)

$$\min\left\{g(y) - \langle y - y^k, z^k \rangle : y \in \mathbb{R}^n\right\} \Leftrightarrow \min\{-\langle y, z^k \rangle : y \in K\}. \tag{3}$$

From the definition of h, a sub-gradient $z^k \in \partial h(y^k)$ can be computed as follows:

$$z^k = \nabla h(y^k) = 2\eta y^k - (\eta + 1)e, \tag{4}$$

where $e = (1, \cdots, 1)^T \in \mathbb{R}^n$.

The DCA scheme applied to (2) can be summarized as follows:

Algorithm 1
Initialization:
  Choose a initial point y^0, set $k = 0$;
  Let ϵ_1, ϵ_2 be sufficiently small positive numbers;
Repeat
  Compute z^k via (4);
  Solve the linear program (3) to obtain y^{k+1};
  $k \leftarrow k + 1$;
Until either $\|y^{k+1} - y^k\| \leq \epsilon_1(\|y^k\| + 1)$ or $|f_\eta(y^{k+1}) - f_\eta(y^k)| \leq \epsilon_2(|f_\eta(y^k)| + 1)$.

Theorem 2. *(Convergence properties of Algorithm DCA)*

- DCA generates the sequence $\{y^k\}$ contained in $V(K)$ such that the sequence $\{f_\eta(y^k)\}$ is decreasing.
- The sequence $\{y^k\}$ converges to $y^* \in V(K)$ after a finite number of iterations.
- The point y^* is a critical point of Problem (2). Moreover if $y_i^* \neq \frac{1}{2}$ for all $i \in \{0,\ldots,n\}$, then y^* is a local solution to (2).
- For a number η sufficiently large, if at iteration r we have $y^r \in \{0,1\}^n$, then $y^k \in \{0,1\}^n$ for all $k \geq r$.

Proof. Immediate consequences of the DCA applied to concave quadratic zero-one programming whose proof can be found in [2].

4   Numerical Simulation

The algorithm has been coded in C++ and implemented on a Intel Core 2 CPU 0.53 Ghz, RAM 2GD. The directed graph $G = (V, E)$ is randomly generated with M-vertices and N-links. The number of QoS constraints is P.

The datasets are randomly generated in a similar way as the data used in [6]. We consider two sets of data. In the first dataset, the $w(i,j)$ $((i,j) \in E)$ are random values in $[0,10]$ and L_i are random values in $[50,59]$. In the second dataset, we consider three QoS constraints ($P = 3$): the cost, the time delay, and the packet loss. Denote $w_1(u,v), w_2(u,v), w_3(u,v)$ (resp. L_1, L_2, L_3) are, respectively, the values of cost, time delay and packet loss on arc (u,v) (resp. the bound of cost, the bound of time delay and the bound of packet loss). We generate five sub-data sets. Each sub-data set is a random value in a range as presented below:

i) sub-data 1: (M,N) = (100,200), $w_1(u,v) \in [0,200]$, $w_2(u,v) \in [0,50ms]$, $w_3(u,v) \in [0\%,5\%]$, $L_1 \in [600, 660]$, $L_2 \in [150, 165ms]$ and $L_3 \in [5\%,10\%]$.
ii) sub-data 2: the sub-data 1 with the following modifications: (M,N)= (200,400), $w_3(u,v) \in [0\%,2\%]$, $L_3 \in [1\%,5\%]$.
iii) sub-data 3: the sub-data 2 with the following modifications (M,N) = (500,1000), $L_1 \in [2000,2050]$, $L_2 \in [500, 515ms]$.
iv) sub-data 4: the sub-data 3 with the following modifications (M,N)= (1000,2000), $L_3 \in [5\%,10\%]$.
v) sub-data 5: the sub-data 4 with the following modifications (M,N)= (5000,7000), $L_1 \in [7000,7050]$, $L_2 \in [3000, 3050ms]$, $L_3 \in [0\%,15\%]$.

Table 1 and Table 2 present, respectively, the numerical results for the first dataset and the second dataset. In these tables, OBJ-DCA, OBJ-CP9.0, ITE-DCA and DATA stand for, respectively, the objective value obtained by DCA, the one given by CPLEX 9.0, the number of iterations of DCA and the name of the generated data (For example, for the "DATA1", it consists of the generated data for a problem, where $M = 100, N = 200, P = 10$, the values of $w_i(u,v)$ and L_i are given in a file "data1.txt".).

Table 1. Comparative results between DCA and Cplex 9.0

	DATA	OBJ-DCA	OBJ-CP9.0	ITE-DCA		DATA	OBJ-DCA	OBJ-CP9.0	ITE-DCA
	DATA1	7	7	2	M=500	DATA22	3	3	4
M=100	DATA2	9	9	3	N=7000	DATA23	3	2	4
N=200	DATA3	5	5	3	P=500	DATA24	2	2	2
P=10	DATA4	5	4	2	M=500	DATA25	2	2	3
	DATA5	6	6	3	N=10000	DATA26	3	3	3
	DATA6	5	5	2	P=500	DATA27	3	3	3
	DATA7	7	6	3	M=1000	DATA28	3	3	4
M=200	DATA8	6	6	2	N=10000	DATA29	3	3	3
N=500	DATA9	2	2	2	P=100	DATA30	3	3	3
P=20	DATA10	xxx	xxx	3	M=2000	DATA31	5	5	2
	DATA11	8	8	3	N=10000	DATA32	5	5	4
	DATA12	7	7	2	P=100	DATA33	4	4	3
	DATA13	5	5	2	M=2000	DATA34	2	2	4
	DATA14	6	6	3	N=12000	DATA35	4	4	3
	DATA15	2	2	3	P=100	DATA36	5	5	2
M=1000	DATA16	3	3	3	M=5000	DATA37	4	4	3
N=5000	DATA17	5	4	2	N=2000	DATA38	5	5	4
P=20	DATA18	5	5	3	P=100	DATA39	6	6	3
	DATA19	6	5	3	M=7000	DATA40	6	xxx	3
	DATA20	4	4	4	N=25000	DATA41	2	xxx	4
	DATA21	7	6	3	P=100	DATA42	7	7	3

Table 2. Comparative results between DCA and Cplex 9.0

	DATA	OBJ-DCA	OBJ-CP9.0	ITE-DCA		DATA	OBJ-DCA	OBJ-CP9.0	ITE-DCA
	DATA43	5	5	3	M=1000	DATA56	7	6	3
M=100	DATA44	5	5	4	N=2000	DATA57	8	8	4
N=200	DATA45	4	4	3		DATA58	9	9	4
	DATA46	4	4	2		DATA59	9	9	3
	DATA47	4	4	3		DATA60	xxx	xxx	3
	DATA48	4	4	3		DATA61	xxx	xxx	4
M=200	DATA49	3	3	3		DATA62	12	12	3
N=400	DATA50	5	5	4	M=5000	DATA63	16	12	4
	DATA51	6	6	3	N=7000	DATA64	16	15	4
	DATA52	3	3	4		DATA65	14	14	3
M=500	DATA53	5	5	3		DATA66	10	10	4
N=1000	DATA54	6	6	3		DATA67	14	14	4
	DATA55	7	7	3		DATA68	19	19	4

From the numerical results, we observe that:

- DCA always provides an integer solution and it converges after a few number of iterations.
- In the most cases, the objective values given by DCA and CPLEX are the same: 33/41 problems in the first dataset and 21/24 problems in the second

dataset (note that the feasible set of "DATA10" in the first dataset and of "DATA60" and "DATA61" of the second dataset are empty). For the remaining cases the difference is small (one).
- DCA works on all test problems while CPLEX 9.0 fails in some cases.

5   Conclusion and Future Work

In this paper, we introduced the BILP formulation for five QoS routing problems. An efficient approach based on DC programming and DCA is proposed for solving Problem 2. The computational results obtained show that this approach is efficient and original as it can give integer solutions while working in a continuous domain. The problems 1, 3, 4, 5 can be similarly solved. In a future work we plan to combine DCA and Branch-and-Bound or Branch-and-Cut Algorithm for globally solving these problems.

References

1. Hoai An, L.T., Pham Dinh, T., Muu, L.D.: Exact penalty in d.c. programming. Vietnam Journal of Mathematics 27(2), 169–178 (1999)
2. Hoai An, L.T., Pham Dinh, T.: A Continuous approach for globally solving linearly constrained quadratic zero-one programming problem. Optimization 50(1-2), 93–120 (2001)
3. Hoai An, L.T., Pham Dinh, T.: The DC (difference of convex functions) Programming and DCA revisited with DC models of real world non convex optimization problems. Annals of Operations Research 133, 23–46 (2005)
4. Kuipers, F.A., Van Mieghem, P.F.A.: Conditions That Impact the Complexity of QoS Routing. IEEE/ACM Transactions on Volume Networking 13(4), 717–730 (2005)
5. Kuipers, F.A., Korkmaz, T., Krunz, M., Van Mieghem, P.: Overview of constraint-based path selection algorithms for QoS routing. IEEE Commun. Mag. 40(12), 50–55 (2002)
6. Chen, S., Nahrstedt, C.: On Finding Multi-constrained Paths. International Journal of Computational Geometry and Applications (1998)
7. Hassin, R.: Approximation Schemes for the Restricted Shortest Path Problem. Mathematics of Operations Research 17(1), 36–42 (1992)
8. Liu, G., Ramakrushnam, K.G.: A*Prune: an algorithm for finding K shortest paths subject to multiple constraints. In: IEEE INFOCOM 2001, April 2, 2001, pp. 743–749 (2001)
9. Lorenz, D.H., Orda, A., Raz, D., Shavitt, Y.: Efficient QoS Partition and Routing of Unicast and Multicast. In: Proceedings IEEE/IFIP IWQoS, Pittsburgh, PA (June 2000)
10. Mieghem, P.V., Kuipers, F.A.: On the complexity of QoS routing. Computer Communications 26(4), 376–387 (2003)
11. Mieghem, P.V., Kuipers, F.A.: Concepts of exact QoS routing algorithms. IEEE/ACM Trans. on Networking 12(5), 851–864 (2004)
12. Orda, A., Sprintson, A.: Efficient algorithm for computing disjoint QoS paths. In: IEEE INFOCOM 2004, March 1, 2004, pp. 727–738 (2004)

13. Pham Dinh, T., Hoai An, L.T.: Convex analysis approach to DC programming: Theory, Algorithms and Applications. Acta Mathematica Vietnamica, dedicated to Professor Hoang Tuy on the occassion of his 70th birthday 22(1), 289–357 (1997)
14. Pham Dinh, T., Hoai An, L.T.: DC optimization algorithms for solving the trust region subproblem. SIAM J. Optimization 8, 476–505 (1998)
15. Warburton, Q.: Approximation of Pareto Optima in Multiple Objective Shortest Path Problems. Operations Research 35, 70–79 (1987)
16. Yuan, X., Liu, X.: Heuristic algorithms for multi-constrained quality of service routing. In: Proceedings of INFOCOM 2001, vol. 2, pp. 844–853 (2001)

Perceptual Watermarking Using a Multi-scale JNC Model

Phi-Bang Nguyen, Marie Luong, and Azeddine Beghdadi

L2TI Laboratory, University Paris 13, Galilée Institute, 99 Ave. J. B. Clement
93430 Villetaneuse, France
{nguyen,luong,beghdadi}@univ-paris13.fr

Abstract. In this paper, we propose a novel watermarking method based on a multi-scale JNC (Just Noticeable Contrast) map. The main idea consists in using this JNC map to determine the optimal strength for embedding the watermark providing an invisible and robust watermarking scheme. The image is first decomposed into a multi-scale representation using the scale space decomposition. A multi-scale band limited contrast is then computed. The watermark is inserted into these contrast images. An experimental evaluation is carried out to demonstrate the performance of the proposed technique in terms of transparency and robustness to a wide variety of attacks.

Keywords: Watermarking, Contrast, Human Visual System, JNC, Scale Space.

1 Introduction

In the last decade, digital watermarking has involved an increasing number of smart applications including the digital right managements and digital media business. Watermarking technology is the efficient technology mostly used for copyright protection, media tracing and identification. In some applications e.g. the mobile e-commerce with wireless media streaming and download through mobile terminal, the watermarking technology has a critical role to play. It can be used in a biometrics-based authentication system to provide user identity information for data identification purposes. The digital watermarking consists in hiding imperceptibly information called watermark, in a host digital data. The watermark can only be detected and extracted using a secret key. The main but conflicting requirements that a watermarking system need to fulfill are the imperceptibility (the watermark must not deteriorate the visual quality of the host data) and the robustness (the watermark cannot be removed by either unintentional manipulation or malevolent attack). To solve the trade-off problem between these two criteria, the best solution is to take into account the human visual system (HVS) characteristics into the design of the watermarking system, as transparency is generally evaluated by human observers. The main objectives of a perceptual watermarking scheme mainly consist in taking advantage of the HVS characteristics for detecting perceptually relevant zones to insert the watermark or for determining a maximum energy that the watermark can reach without affecting its transparency.

N.T. Nguyen, M.T. Le, and J. Świątek (Eds.): ACIIDS 2010, Part II, LNAI 5991, pp. 471–480, 2010.

Generally, major perceptual watermarking approaches that exploit the human vision characteristics proposed to use some implicit properties of the HVS to determine an adaptive strength for embedding the watermark. By using the fact that the HVS is less sensitive in high activity regions (texture, edge) than in flat regions, the watermark strength can be simply adapted and weighted by the local contrast of host pixel as proposed in [1] by Do *et al.*, or by the local variance of image regions using the so-called NVF (Noise Visibility Function) [2]. In [3], the watermark is simply embedded in the blue channel of a RGB color image to which the human eye is known to be less sensitive. However, most of these implicit methods require an additional global factor to control the watermark strength. The drawback of such an approach is that this global factor is empirically determined and hence depends on image characteristics. Another strategy consists to use the HVS's properties in an explicit manner to determine the visual detection threshold, generally called JND (Just Noticeable Difference) for embedding watermark signal. This detection threshold can be computed by taking advantage of some characteristics of the HVS, such as contrast sensitivity [4], luminance masking [5], [6], contrast masking [7]. In [8], an isotropic local contrast is used in the contrast masking model of Legge and Foley [7] to determine the contrast threshold for each pixel. The JND is then computed as the product of this contrast threshold with the low pass filtered image. Wolfgang *et al.* [9] also proposed a watermarking scheme using the JND model developed by Watson for DCT coefficients quantization [10].

In this paper, a novel robust and imperceptible watermarking scheme is proposed, using a new JNC model which is based on the JND model developed in [11]. This JNC is computed for each pixel at each difference of Gaussian (DoG) scale by incorporating the most relevant HVS properties such as contrast sensitivity, luminance adaptation and contrast masking. However, in contrast to [11], we propose a multi-scale contrast, which is based on the contrast defined by Peli [12] but adapted for scale space context. The watermark is weighted by its JNC value and then embedded into different scales of this contrast scale space By this way, the method achieves a good trade-off between robustness and transparency.

The paper is organized as follows: in section 2, we present the multi-scale JNC model. Section 3 introduces the watermarking method. The experimental results are reported in section 4 and finally, we give the concluding remarks in section 5.

2   The Proposed Multi-scale JNC Model

In this section, a multi-scale JNC model is proposed for the perceptual weighting of the watermark. It takes into account three main characteristics of the HVS: contrast sensitivity, luminance masking and contrast masking. This JNC model is similar to the JND (Just Noticeable Difference) model developed for the Laplacian pyramid [11] but adapted for the scale space context.

2.1   Scale Space Decomposition

The scale space decomposition has been shown to be a powerful tool for analyzing image into multiscale representation. It has been proved that the Gaussian is the unique kernel to generate a linear scale space [13]. This uniqueness is due to the requirement that new structures must not be created from a fine scale to any coarser scale. In order to reduce complexity, we performed the scale space decomposition for only one octave. Following descriptions in [14], the Gaussian scale space at level s is defined as:

$$G_s(x, y) = g_{\sigma_s} * I(x, y) \tag{1}$$

where g_{σ_s} is an isotropic Gaussian kernel with the standard deviation σ_s computed as: $\sigma_s = \sigma_0.2^{s/S}$, σ_0 is the base scale level and $s = 0 \rightarrow S$, S is the maximum level (here $S=5$). As a result, the Gaussian scale space represents the same information of the image at different levels of scale in which two nearby levels are separated by a multiplicative factor k ($k = 2^{1/S}$). The DoG scale space $DoG_s(x,y)$ is then computed by taking the difference of two adjacent Gaussian scales as in [16]:

$$DoG_s(x, y) = G_{s+1}(x, y) - G_s(x, y) \approx (k-1)\sigma_0^2 \Delta^2 G \tag{2}$$

In the frequency point of view, the Gaussian scale space can be considered as a successive sequence of lowpass filtered images while DoG images are a set of quasi bandpass images with peak frequencies $f_0/\sigma_0,..., f_0/\sigma_S$ where f_0 is the spatial peak frequency of the original image. Here, we fixed $f_0=32$ cpd (cycles/degree) by deriving from the viewing distance, the image size and the screen's dot-pitch.

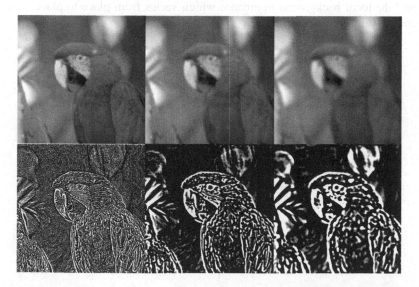

Fig. 1. Gaussian Scale Space at levels 1, 3, 5 and DoG Scale Space at levels 0, 2, 4 of the "Parrots" image

However, we noted that the DoG decomposition does not provide an inverse transformation which is needed for a reconstruction of watermarked image. Therefore, we replaced the first Gaussian scale level by the original image before computing the

DoG scale space. This modification allows a perfect image reconstruction as shown through equation (12) in Section 3. As illustration, the Gaussian Scale Space and the DoG Scale Space are shown in Fig. 1.

2.2 The Proposed Multi-scale Band Limited Contrast

In this section, we proposed a new contrast model, which is based on the contrast definition of Peli [12] but adapted for the multi-scale representation. It is an adaptive version of the local band limited contrast defined below:

$$C(x, y) = \frac{BP(x, y)}{LP(x, y)} \tag{3}$$

where $BP(x,y)$ is the bandpass-filtered version of the image and $LP(x,y)$ is the low pass filtered version of the image containing all energy below the pass band $BP(x,y)$ (cf. figure 1).

In the context of scale space decomposition, it is calculated as follows:

$$C_s(x, y) = \frac{DoG_s(x, y)}{G_{s+1}(x, y)} \tag{4}$$

The use of this contrast's definition is more appropriate for the subband processing below because it provides a local contrast that depends on the local energy at that band and the local background luminance which varies from place to place in image and band to band in the spectrum.

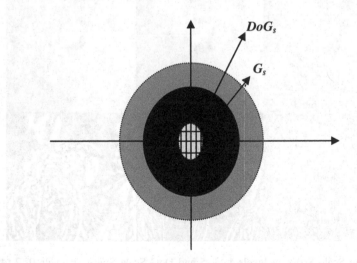

Fig. 2. Multi-bandpass illustration of the scale space decomposition

2.3 Incorprating of Contrast Sensitivity and Luminance Adaptation

The CSF describes the variation of HVS's sensitivity as a function of spatial frequency (in cpd) and therefore has to be applied in frequency domain. To adapt this phenomenon for spatial domain, we refer to the approach in [11]. It is known that the early visual stages of HVS work as a multi-channel analyzer and the CSF measured by psychophysical experiments is believed to be the envelope of these channels response. For each channel, the contrast threshold at a given frequency f could be expressed as contrast threshold at the peak frequency $CT(f_s^{peak})$ weighted by the contribution of this channel:

$$CT_s(f) = CT_s(f_s^{peak}).\alpha_s(f) \quad \text{where} \quad \alpha_s(f) = \frac{DoG_s(f)}{\sum_{s=0}^{S-1} DoG_s(f)} \tag{5}$$

where $\alpha_s(f)$ is the weighting function and $DoG_s(f)$ is the DoG response of the sth channel. In spatial domain, the local contrast threshold is redefined as follows:

$$CT_s(x, y) = CT(f_s^{peak}).\alpha_s(x, y) \tag{6}$$

where $DoG_s(x, y)$ is the sth channel's DoG response at the pixel (x,y), $CT(f_s^{peak})$ is the contrast threshold at the peak frequency of the channel, computed as its inverse CSF.

In this paper, we used the Barten's CSF [8] thanks to its flexibility and relative simplicity. The Barten's CSF is described as follows:

$$CSF(f) = a.f.\exp(-b.f).\sqrt{1 + c.\exp(b.f)} \tag{7}$$

$$a = \frac{540(1 + 0.7/L)^{-0.2}}{1 + \frac{12}{w(1 + f/3)}} \quad b = 0.3(1 + 100/L)^{0.15} \quad c = 0.06$$

where w is the angular picture size in degrees and L is the global luminance in cd/m^2. Previous works treat the contrast sensitivity and the luminance adaptation separately by taking the CSF at a fixed level of illumination followed by a model of luminance masking. However, as can be inferred from equation (5), the CSF exhibits a non-separable dependence between spatial frequency and luminance level. Therefore, we incorporated these two mechanisms into one stage by defining the global luminance L as follows:

$$L(x, y) = L_0 + L_I(x, y) \tag{8}$$

where L_0 is the ambient luminance and L_I is the local luminance computed, for each pixel, from the corresponding Gaussian value in the $(s+1)^{th}$ level $G_{s+1}(x, y)$ followed by a grayscale to luminance transformation as follows:

$$L_I(x, y) = \max(L_{max}(\frac{G_{s+1}(x, y)}{255})^\gamma, L_{min}) \tag{9}$$

where L_{max} and L_{min} are respectively the maximum and minimum luminance of the display, whereas γ is the exponent factor used in the gamma correction of the display.

The computed global luminance, $L(x,y)$, is then placed in equation (4) and (5) to account for luminance adaptation.

2.4 Contrast Masking

Contrast masking refers to the phenomenon whereby the visibility of a signal is reduced by the presence of another. Here, we propose a contrast masking model inspired from the Legge-Foley's model [7] but we use the new contrast definition for this model:

$$JNC_s(x,y) = \begin{cases} CT_s(x,y) & if \; |C_s(x,y)| \le CT_s(x,y) \\ CT_s(x,y).\left(\dfrac{|C_s(x,y)|}{CT_s(x,y)} \right)^\varepsilon & otherwise \end{cases} \qquad (10)$$

where $DoG_s(x,y)$ is the sth channel DoG response at the pixel (x,y), ε is a factor describing the degree of masking, $0.6 \le \varepsilon \le 1$ [7]. As illustration, the contrast images and the JNC maps for three scale levels of the "Parrots" image are shown in Fig. 3.

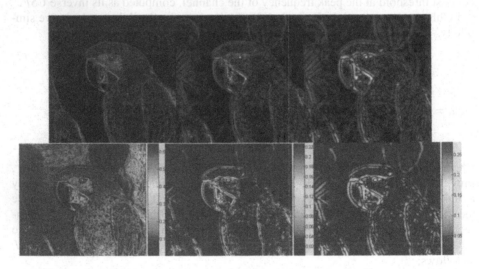

Fig. 3. Multi-scale contrast images (top) and its corresponding JNC maps (bottom) of the "Parrots" image at scales 0, 2, 4

3 The Proposed Watermarking Method

3.1 Embedding Scheme

At each level, a different bipolar pseudo-random sequence $W^i \in \{-1,1\}$ with zero mean and variance one is inserted to the contrast image using the following rule:

$$C_{sw}(x,y) = \begin{cases} C_s(x,y) + JNC_s(x,y).W(x,y) \\ C_s(x,y) \qquad otherwise \end{cases} \tag{11}$$

If $Cs(x,y) > JNCs(x,y)$, where $JNC_s(x,y)$ is the just-noticeable-difference threshold at the position (x,y) at level s computed from the previous section.

The watermarked image is then reconstructed by:

$$I_w = \sum_{s=0}^{S-1} C_{sw}.G_{s+1} + G^S, \qquad S = 5 \tag{12}$$

3.2 Detection Scheme

This correlation value is then compared to a decision threshold Th to decide whether the watermark is present or not. The decision is made according to the hypothesis test below:

H_0: the image is not watermarked with W
H_1: the image is watermarked with W.

Be noted that $Cor_s > Th$ for just one level is enough to determine the presence of the watermark in the test image. According to Neyman-Pearson criterion, this threshold is computed by fixing the false alarm. Up to now, optimum detector is only developed for AWGN channel (for which the work is modeled by some simple distribution law in case of no attack or Gaussian distributed attack). Because of this difficulty, we consider only the random watermark false alarm. In this case, the detector output follows a Gaussian distribution law. Hence, the false alarm is computed as:

$$P_{fa} = P\left(Cor > Th \mid H_0\right) = \frac{1}{2} erfc\left(\frac{Th}{\sqrt{2}\sigma_{H_0}}\right) \tag{13}$$

The decision threshold is $Th = 3.9\sqrt{2}\sigma_{H_0}$ for a false alarm of 10^{-8}. σ_{H_0} is directly estimated on the non-watermarked image using 10000 different watermark sequences.

4 Experimental Results

As shown in Fig. 4, the original and the watermarked images are perceptually undistinguishable. Further experiments are also carried out on a variety of natural images to validate the performance of our method in terms of robustness and imperceptibility. Due to the limited space and in order to facilitate the comparison, we only report the results for a set of 8 images.

Fig. 4. Original (left) and watermarked image (right)

4.1 Transparency Evaluation

In order to evaluate the quality of watermarked images, the DSIS (Double Stimulus Impair Scale) subjective test is performed on 15 observers (researchers in image processing domain) with the test condition according to [15]. The grading scale consists of 5 levels as 5: imperceptible, 4: perceptible but not annoying, 3: slightly annoying, 2: annoying, 1: very annoying. Besides, objective assessment is also performed using three quality metrics, the SSIM (Structural Similarity Index Measure) [16], the so-called Watson metric [17] which measures the Total Perceptual Error (TPE) between the original and the watermarked image and a wavelet-based metric (WAV) [18]. The results in Table 1 show that the proposed algorithm provides a good imperceptibility. However, objective measures do not correlate with the subjective measures for some images. This is mainly due to the fact that textured images ("Baboon") tend to facilitate visual masking even for a high distortion level measured by these metrics.

Table 1. Imperceptiblity Evaluation. Be noted that the TPE shows the distortion level while the PSNRwav and the SSIM shows the fidelity between the original and the watermarked image.

Image	Lena	Barbara	Peppers	Parrots	Baboon	Man	Couple	Clown
DSIS	4.2	4.2	4.87	4.6	5	4.87	4.73	4.67
SSIM	0.988	0.982	0.986	0.992	0.976	0.976	0.983	0.982
TPE	0.035	0.067	0.041	0.024	0.096	0.086	0.057	0.052
WAV	23.21	17.95	23.43	20.7	16.13	19.95	20.85	22.21

4.2 Robustness Evaluation

The robustness of our algorithm is tested via some common attacks from Stirmark [19] and Checkmark [17] benchmarks. A comparison with our previous method [1] and the Digimarc product [21] (which is available as plug-in in Photoshop CS3) is also performed. In [1], the watermark is embedded in "invisible" regions and the watermark strength is chosen according to the Moon-Spencer's contrast [20]. The results are shown in Table 2 in which the parameters represent the break down limit of the methods (i.e the strongest attacks that the watermark still survives). We can see that, the new method outperformed our previous method and the Digimarc's, especially for severe attacks like Jpeg compression until 3%, random cropping until 0.5%, etc.

Table 2. Robustness Evaluation

Attack Type	Our method	Method [1]	Digimarc
Random Cropping	0.5%	13%	34%
Jpeg compression	QF=3%	QF=15%	QF=19%
Jpeg 2000 compression	0.08 bpp	0.9 bpp	0.8 bpp
Gaussian Noise	$\sigma = 70\%$	$\sigma = 10\%$	$\sigma = 4\%$
Wienner filtering	Ok	Ok	Ok
Median filtering	Filter Size=5x5	3x3	Failed
Sharpening	Ok	Failed	Failed
Blurring	Ok	Failed	Failed
Bit plan reduction	Ok	Failed	Failed
Histogram Equalization	Ok	Ok	Ok
Rescale (50%)	Ok	Ok	Ok

5 Conclusion and Perspectives

The paper presents a novel contrast based image watermarking operating in the scale space. Such watermarking method presents additional advantages over published watermarking schemes in terms of robustness and transparency. The employed multi-scale JNC model is simple and efficient by exploiting principal characteristics of the HVS. Firstly, the contrast sensitivity and the luminance adaptation are incorporated via the Barten's CSF. Secondly, a new contrast masking model is adapted from the model of Legge and Foley for the scale space context. The experimental results show that the proposed method has a good performance in terms of robustness and imperceptibility. As perspective, an extension of the method to video watermarking will be considered.

Acknowledgement. This work is financially supported by the Regional Council of Ile-de-France in the framework of the HD3D-IIO project of the Cap Digital competitiveness cluster.

References

1. Do, Q.B., Beghdadi, A., Luong, M., Nguyen, P.B.: A Perceptual Pyramidal Watermarking Technique. In: Proc. of IEEE ICME, Hannover, Germany, pp. 281–284 (2008)
2. Voloshynovskiy, S., Herrigel, A., Baumgärtner, N., Pun, T.: A stochastic approach to content adaptive digital image watermarking. In: Pfitzmann, A. (ed.) IH 1999. LNCS, vol. 1768, pp. 211–236. Springer, Heidelberg (2000)
3. Kutter, M., Jordan, F., Bossen, F.: Digital Signature of Color Images using Amplitude Modulation. In: Proc. of SPIE EI 1997, San Jose, California, pp. 518–526 (1997)

4. Barten, P.G.J.: Evaluation of Subjective Image Quality with the Square-Root Integral Method. Journal of the Optical Society of America A: Optics, Image Science, and Vision 7(10), 2024–2031 (1990)
5. Larson, G.W., Rushmeier, H., Piatko, C.: A Visibility Matching Tone Reproduction Operator for High Dynamic Range Scenes. IEEE Transactions on Visualization and Computer Graphics, 291–306 (1997)
6. Chou, C.H., Li, Y.C.: A perceptually Tuned Subband Image Coder Based on the Measure of Just-Noticeable-Distortion Profile. IEEE Transaction on Circuits and Systems for Video Technology 5 6, 467–476 (1995)
7. Legge, G.E., Foley, J.M.: Contrast Masking in Human Vision. Journal of the Optical Society of America 70, 1458–1471 (1980)
8. Geisler, W.S., Perry, J.S.: A Real Time Foveated Multiresolution System for Low-Bandwidth Video Communication. In: Proc. of SPIE, vol. 3299, pp. 294–305 (1998)
9. Wolfgang, R.B., Podilchuk, C.I., Delp, E.J.: Perceptual Watermarks for Digital Images and Video. Proc. of the IEEE 87(7), 1108–1126 (1999)
10. Watson, A.B.: DCT quantization matrices visually optimized for individual images. In: Proc. of SPIE Int. Conf. Human Version, Visual Processing, and Digital Display - IV, pp. 202–216 (1993)
11. Nguyen, P.B., Beghdadi, A., Luong, M.: Perceptual watermarking using pyramidal JND maps. In: Proc. of 10th IEEE International Symposium on Multimedia, Berkeley, CA, USA, pp. 418–423 (2008)
12. Peli, E.: Contrast in complex images. Journal of Opt. Soc. Am. A 7(10), 2032–2040 (1990)
13. Lindeberg, T.: Scale-Space Theory in Computer Vision. Kluwer Academic Publishers, Dordrecht (1994)
14. Lowe, D.G.: Distinctive image features from scale-invariant keypoints. International Journal of Computer Vision 60, 91–110 (2004)
15. ITU-R BT.500-10 Recommendation, Methodology for the Subjective Assessment of the Quality of Television Pictures, Draft Revision
16. Wang, Z., Bovik, A.C., Sheikh, H.R., Simoncelli, E.P.: Image Quality Assessment: from Error Visibility to Structural Similarity. IEEE Transactions on Image Processing 13(4), 600–612 (2004)
17. Checkmark Benchmark,
    http://watermarking.unige.ch/Checkmark/index.html
18. Beghdadi, A., Pesquet-Popescu, B.: A New Image Distortion Measure Based on Wavelet Decomposition. In: Proc. of 7th IEEE ISSPA, Paris, France, vol. 2, pp. 485–488 (2003)
19. Stirmark Benchmark,
    http://www.petitcolas.net/fabien/watermarking/stirmark
20. Iordache, R., Beghdadi, A., Viaris de Lesegno, P.: Pyramidal perceptual filtering using Moon and Spencer's contrast. In: Proc. of International Conference on Image Processing, vol. 3, pp. 146–149 (2001)
21. Digimarc's Watermarking Technologies,
    https://www.digimarc.com/solutions/dwm.asp

Author Index